S0-AKJ-305

# DYNAMICS OF
# MOLECULAR LIQUIDS

# DYNAMICS OF MOLECULAR LIQUIDS

**WALTER G. ROTHSCHILD**
**Ford Motor Company**

**A Wiley-Interscience Publication**

**JOHN WILEY & SONS**

**New York   Chichester   Brisbane   Toronto   Singapore**

CHEMISTRY,
7120 - 3060

Copyright © 1984 by John Wiley & Sons, Inc.

All rights reserved. Published simultaneously in Canada.

Reproduction or translation of any part of this work
beyond that permitted by Section 107 or 108 of the
1976 United States Copyright Act without the permission
of the copyright owner is unlawful. Requests for
permission or further information should be addressed to
the Permissions Department, John Wiley & Sons, Inc.

*Library of Congress Cataloging in Publication Data:*

Rothschild, Walter G., 1924–
    Dynamics of molecular liquids.

    "A Wiley-Interscience publication."
    Bibliography: p.
    Includes index.
    1. Liquids.   2. Molecular dynamics.   3. Molecular
rotation.   4. Vibrational spectra.   I. Title.

QC145.2.R67   1983            530.4'2      83-10203
ISBN 0-471-73971-5

Printed in the United States of America

10   9   8   7   6   5   4   3   2   1

QC
145
.2
R67
1984
CHEM

*"Toutes les théories sont des fictions qui se vérifient en se réfutant."*

O. Mannoni, *Un commencement qui n'en finit pas*,
Le Seuil, Paris 1979.

# PREFACE

Infrared and Raman spectroscopists have long been fascinated by the difference between gaseous and liquid-phase spectra and what it tells about molecular motion in the condensed phase. Before 1964 useful quantitative theories were nonexistent. Only qualitative conclusions about molecular motion could be drawn from frequency and bandwidth data of some simple liquids because the spectra looked similar to those of the compressed gases.

This situation changed with the publication of R. G. Gordon's papers on the fundamental connection between spectral parameters of a liquid and rotational motion of its molecules. The lasting merit of these publications lies in their clear, elegant, concise, yet thorough presentation of the significance of the "Heisenberg picture" in its capability of obtaining quantitative dynamic information from spectral data. Since the procedure requires only simple numerical Fourier transformation, Gordon's work gave a tremendous uplift to studies of molecular dynamics. In subsequent years the approach was not only extended to include atomic motion (vibrational relaxation) but also spawned a still growing number of significant theoretical developments in the dynamics of liquids that occur on the microscopic (atomic, molecular) level.

It was fortunate that these theories arose in a period when laser-excited Raman scattering had become as familiar as infrared spectroscopy, when data-handling computers had become common, and when people were freer to work for the sake of pure knowledge and scientific adventure.

In this book I give an account of major approaches, results, and developments of the dynamics of molecular liquids as they relate to Heisenberg "fluctuation spectroscopy." I base the discussion on subjectively chosen but representative experimental and theoretical literature examples that deal with condensed-phase motion on the molecular level. I emphasize the motional phenomena—not the spectroscopic aspects—and discuss the trends, the relations, and the scope and growth of this scientific discipline. Subjects reach from simple theoretical models to promising and not yet widely accepted concepts.

The content of the book is organized into three chapters. An introductory theoretical chapter (Chapter 1) serves as a guide to the reader. The expert, anxious to brush up and to extend, or the graduate student is urged to read it first.

The second chapter discusses rotational motion. The subjects are classified under random-collisional (A) and nonrandom-collisional events (B). The theoretical treatment is enlarged in the sections as demanded by the data.

The third chapter deals with principal aspects of vibrational relaxation, including a section on nonrandom-collision effects. The first part of the chapter contains vibrational relaxation of two-level systems, and the second part contains vibrational relaxation of multilevel systems. In this chapter the discussion stresses the kinship of vibrational and rotational relaxation. The complexities of the subjects culminate with this chapter.

The matter discussed requires familiarity with basic quantum mechanics. Any spectroscopic background is helpful. The demands on the reader's attention and dedication increase as the book progresses. I think that the mathematics are easier than the physical ideas. These take some getting accustomed to since they are rarely taught in graduate school.

I have tried never to leave the reader stranded by the frustrating "as can be shown." Important relations and significant theoretical concepts are developed, and only obvious as well as tedious computations are suppressed. The omitted algebra can be checked with pencil and paper. Extreme mathematical rigor is not attempted. The reader will find enlightenment on the fine points in 11 appendices and the references to the literature of original papers and monographs. Fluctuation spectroscopy furnishes not only a unique method for the exploration of dynamic aspects of condensed phases but also presents an introduction to the theories of molecular statistics and of irreversible phenomena. In connection with readily obtainable spectral data, the methods and theories discussed in this book teach the application of function transforms, projection operators, time-dependent perturbation theory, commutator algebra, cumulants, Wigner matrices, master equations, and more.

Owing to lack of space, I have not included several deserving theories of rotational and vibrational relaxation, I have restricted the discussion to spectroscopic data from infrared, far-infrared, Raman, and some Rayleigh work. I have only touched upon dynamics of atomic liquids, molecular dynamics calculations (computer simulations), and spectroscopic measurement techniques. At any rate, the discussions in this book prepare the reader for appreciating the literature that is not dealt with here or that will appear in the future.

Serious writing of this book began during 1976–1977 when I was guest professor at the Laboratoire de Spectroscopie Infrarouge, Université de Bordeaux I, and at the 1. Physikalisches Institut (now Institut für Experimentalphysik), Universität Wien. I am expressing here my deep-felt gratitude to my hosts, Prof. J. Lascombe of Bordeaux, Prof. F. Kohler (now Dortmund), Prof. P. Weinzierl, and Dr. H. A. Posch of Wien, for the opportunity and the challenge in continuing this enterprise.

It has given me great pleasure to make acquaintances and friends among active members of this scientific discipline. I acknowledge the many benefits received from mutual visits, meetings, seminars, stimulating discussions, and correspondence.

My interest in condensed-phase spectroscopy began in 1964 with work at the Scientific Research Laboratory of the Ford Motor Company in Dearborn, Michigan. These were the golden years of science under the guidance of an enlightened management of this progressive industrial concern. I shall always appreciate having experienced this exciting time.

Walter G. Rothschild

*Detroit, Michigan*
*September 1983*

# CONTENTS

## APPENDICES

CHAPTER 1

# FUNDAMENTALS OF VIBRATIONAL-ROTATIONAL FLUCTUATION SPECTROSCOPY

## 1. GENERAL THEORY

### 1.1. Spectra, Dynamic Processes, and Fourier Transforms

A spectrum displays intensity peaks at certain frequencies. These frequencies are characteristic of the particular energy levels involved in the observed transitions. Figure 1.1 shows, as typical example, the mid-infrared spectrum of liquid deuterochloroform ($CDCl_3$). The spectrum was scanned rapidly for purposes of compound identification and analysis. Under similar spectroscopic conditions, a researcher might explore solvent-induced frequency shifts, perform temperature-dependent molecular conformation analyses, determine the fundamental frequencies for normal coordinate calculations, and so forth. At any rate, in such studies the investigator is interested in the peak frequencies and intensities of the bands since these quantities are used to solve the particular eigenvalue problem.

Figure 1.2$a$ and $b$ show a rescan of the spectral regions 2400–2100 and 1160–1060 cm$^{-1}$ but at a slower speed of the grating drive. This lets us see clearly that the bands have an inherent intensity profile or contour; in other words, different bands possess different frequency distributions $I(\omega)$ and bandwidths. (See Note 1 at end of chapter.)

Whereas peak frequencies of vibrational bands of liquid-phase molecules depend on *static* parameters (force constants, atomic masses, bond distances and angles, electric charges), band profiles and bandwidths depend on *dynamic* parameters from atomic and molecular motions. To describe dynamic processes,

1

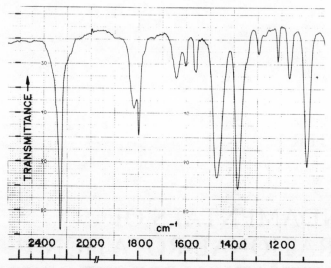

**Figure 1.1.**   Infrared survey spectrum of liquid chloroform-$d$ ($CDCl_3$) at 30°C. Note change of abscissa scale at 2000 cm$^{-1}$.

it is natural to choose time $t$ as independent variable. We remember that a Fourier transform,

$$F(p) = \int_{-\infty}^{\infty} dq \exp(ipq) f(q),$$

decomposes a function of independent variable $p$ into an integral sum over a continuous range of an associated variable $q$. Identifying $p$ with $\omega$ and $q$ with $t$, we can transform the experimental spectral profile $I(\omega)$ from the frequency into the time domain by

$$\int_{-\infty}^{\infty} d\omega\, I(\omega) \exp(-i\omega t) = f(t). \tag{1.1a}$$

Four principal questions arise:

1. What is the magnitude of characteristic times that contribute to the integrand of Eq. 1.1$a$?
2. What is the physical meaning of $f(t)$ in terms of quantum-mechanical operators?
3. How can we use $f(t)$ to explore quantitatively dynamics of the condensed phase?
4. What precautions must be taken to avoid uncertainties from multiple oscillators (overlapping bands)?

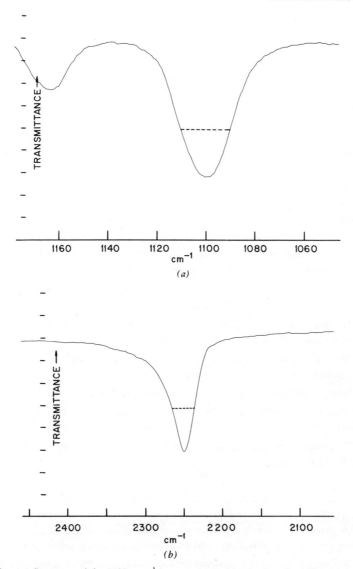

**Figure 1.2.**   (*a*) Spectrum of the 1100-cm$^{-1}$ contour (a combination band) of Fig. 1.1, scanned at decreased grating drive speed. The halfwidth of the contour amounts to 20 cm$^{-1}$. (*b*) Spectrum of the 2250-cm$^{-1}$ contour (C-D stretching fundamental) of Fig. 1.1, scanned at decreased grating drive speed. The halfwidth of the contour is about 30 cm$^{-1}$.

We write the absolute frequency $\omega$ as sum and difference,

$$\omega = \omega_0 \pm \omega'.$$

$\omega_0$ is the frequency of the peak of the band. $\omega'$ is the frequency increment to higher $(+)$ and lower$(-)\omega$, counted from $\omega_0$ of the band contour $I(\omega_0 \pm \omega')$. We split off the modulation factor $\exp(i\omega_0 t)$. It is uninteresting since $\omega_0$ is constant. Therefore,

$$\int_{-\infty}^{\infty} d\omega \, I(\omega) \exp\{-i(\omega - \omega_0)t\} = f(t).$$

The largest contributions to the integrand come from $0 \leq \omega' t \leq 2\pi$. Clearly, the smallest allowable $\omega'$, $\omega'_{min}$, is the spectral resolution. Commonly, it is about 3 cm$^{-1}$ or $0.6 \times 10^{12}$ radian sec$^{-1}$. We identify $\omega'_{max}$ with the halfwidth since $I(\omega')$ drops rapidly for $\omega'$ larger than the halfwidth. Usually halfwidths are in the range 5–100 cm$^{-1}$ or $1 \times 10^{12}$–$2 \times 10^{13}$ radian sec$^{-1}$ (see Fig. 1.1). Hence typical $\omega'_{max}$ and $\omega'_{min}$ correspond to time periods between fractions of a picosecond and several picoseconds. This range covers the following dynamical phenomena:

1. Average duration of an intra- or intermolecular vibrational energy migration step ($\sim 1 \times 10^{-12}$ sec).
2. Periods between near-instantaneous rotational jumps in strongly anisotropic media [$(1-5) \times 10^{-12}$ sec].
3. Average duration of free rotation steps between collisions in light, near-spherical molecules ($\sim 0.2 \times 10^{-12}$ sec).
4. Intervals between random changes of the frequency of an oscillator in a molecule [$(0.1-0.5) \times 10^{-12}$ sec].
5. Phonon lifetimes of excited vibrational levels ($> 1 \times 10^{-12}$ sec).
6. Lifetimes of configurations of small molecular clusters [$(2-5) \times 10^{-12}$ sec].
7. Times between hard collisions and durations of sticky collisions [$(0.1-0.5) \times 10^{-12}$ sec].

We conclude from these examples that $f(t)$ is appreciable during time intervals that reflect the scope of a wide variety of dynamical phenomena in molecular liquids. A priori, a many-body interaction potential determines these dynamical processes. Such a potential, which depends on time, angular momentum, and spacial coordinates, is a complicated quantity. To write it down and use it is impossible. We must design approximations and see how well they work. Approximations retain a physical picture, such as outlined in the foregoing examples 1–7, whereas rigorous mathematical formulations lack the aspect of familiar "touchable" objects.[1]

In the following sections of this chapter we answer the remaining questions posed. We discuss how to extract quantitative dynamical information from spectral data with the help of established theory. We demonstrate how to identify individual dynamical processes by selecting the proper spectral technique and the importance of choosing appropriate molecular systems and avoiding overlapping contours.

Since the experimental frequency sampling interval is finite, Eq. 1.1$a$ reads

$$f(t) \approx \sum_{\Delta\omega} I(\omega)\exp(-i\omega t)\,\Delta\omega. \qquad (1.1b)$$

Nevertheless, we use integrals throughout.

## 1.2.  Statistical Theory of Band Shape Analysis

The frame of our approach is the statistical-mechanical theory of irreversible processes. In particular, we consider concepts known as "linear response theory" and "fluctuation-dissipation theorem".[2] These concepts, well known in nuclear magnetic relaxation, had not attracted the attention of infrared and Raman spectroscopists until the publication of Gordon's and Shimizu's papers.[3]

There are many analogies between nuclear magnetic resonance relaxation and infrared-Raman relaxation. Since many readers have some knowledge of nuclear magnetic relaxation but lack familiarity with infrared-Raman relaxation, it is helpful to recall the following:

1.  In a nuclear magnetic resonance experiment a constant magnetic field removes the degeneracy of a few spins. This causes an excess of the population of the upper spin state with respect to the macroscopic equilibrium value (absence of a magnetic field).

2.  In a liquid medium, the excited spins return to equilibrium by coupling to randomly fluctuating magnetic fields. Such magnetic fields persist owing to static inhomogeneities over molecular distances. Dynamic variations of the magnetic fields are caused by the fluctuations of molecular orientations and positions.

3.  Linear response theory and the fluctuation-dissipation theorem predict that the induced magnetization responds linearly to the fluctuating magnetic fields and that the excited spins dissipate their excess energy irreversibly to the kinetic molecular motions.

The depopulation of an excited vibrational level of an oscillator is the analogous relaxation process in infrared and Raman spectroscopy. Here, the oscillator couples to the periodic radiation field of the incident infrared or laser light. As is well known from time-dependent perturbation theory, such interaction causes transitions between vibrational energy levels. This process permits us to study molecular motion in liquids because the spectrometer acts as an "instigator" as well as "examiner" of the relaxation process. The spectrometer is the instigator because it prepares, either by absorption or by scattering of radiant energy, a nonequilibrium distribution of vibrational energy levels in a few molecules. The spectrometer is the examiner because it observes the dissipation of the excess energy of the excited oscillators to the molecular motions.

For instance, let us consider a two-level scheme of ground and excited oscillators ("the system") in a solvent ("the medium"). The medium exerts random, fluctuating forces on the oscillator by the continually changing orientational and translational coordinates of its molecules.

1. We assume all system oscillators (phonons) initially in their equilibrium state. At this point, we know nothing about the dynamics of the medium or the fluctuating forces between system and medium.

2. We now subject the system to the radiation field of an infrared spectrometer. This causes a certain number of oscillators to make instantaneous transitions to the excited level. At this point, the upper oscillator level contains a phonon population in excess of the equilibrium value that corresponds to the temperature of the medium. This excess energy is equivalent to the radiant energy absorbed. We can say that the spectrometer has generated or isolated a certain number of fluctuations of oscillator energies out of the initial equilibrium state.

3. Subsequently, the forces between the system and the medium induce the return of the excess upper-level oscillator population to the ground level. The duration of this reequilibration process, the "lifetime" $\tau$ of the upper system level, leads to a bandwidth increment $\Delta\omega$. (We can estimate $\Delta\omega$ by the energy-time uncertainty relation[4a] $\tau\Delta E \sim \hbar$ or $\tau\Delta\omega \sim 1$.) Consequently, the radiation-induced transitions between ground and upper system levels show a frequency distribution, a band profile, around the unperturbed oscillator frequency. From the evaluation of this oscillator band profile we predict the time-dependence of the oscillator-medium interactions. In turn, from the oscillator-medium interactions we draw conclusions about the dynamical effects in the medium. We can say that the spectrometer has examined the molecular motions in the medium through the changes in the distribution of the oscillator frequency. The oscillator band profile serves as "probe" of the molecular motion of the medium.

In the following sections we develop these ideas quantitatively.

### 1.3.   Fluctuation-Dissipation Theorem

The macroscopic polarization $P$ of a nonconducting medium due to a traversing electromagnetic field $E$ is

$$P = \chi''E, \tag{1.2}$$

where $\chi''$ is the imaginary part of the complex electric susceptibility $\chi(\omega) = \chi'(\omega) + i\chi''(\omega)$. Usually, an absorption measurement is expressed by a frequency-dependent macroscopic absorption coefficient $\sigma(\omega)$ or extinction coefficient $\kappa(\omega)$. They are related to $\chi''$ by

$$2n(\omega)\kappa(\omega) = \chi''(\omega) = \epsilon''(\omega),$$

$$\sigma(\omega) = 2\omega\kappa(\omega) = \frac{\chi''(\omega)\omega}{n(\omega)}, \tag{1.3}$$

where $1 + \chi(\omega) = \underset{\sim}{\epsilon}(\omega)$. $\underset{\sim}{\epsilon}(\omega)$ is the complex dielectric constant $\underset{\sim}{\epsilon}(\omega) = \epsilon'(\omega) + i\epsilon''(\omega)$. $n(\tilde{\omega})$, the refractive index, is the real part of the complex refractive index $\underset{\sim}{n}(\omega) = n(\omega) + i\kappa(\omega)$.

Equation 1.2 shows that the polarization of the medium depends linearly on the external field. The response $P$ of the system to $E$ is determined by $\chi''$ or $\epsilon''$. Hence such relations are our starting point.

We define by $\Delta B(t)$ the response of a quantum-mechanical system variable $B(t)$ to an external radiation field of amplitude $E(t)$. $B(t)$ and $E(t)$ are coupled by the perturbation Hamiltonian $H_e(t) = -AE(t)$. Operator $A$, which may be time-dependent, is a system operator.

We introduce the "response function" or "aftereffect function" $\Phi_{BA}(t)$. Its subscripts remind us that it depends solely on system variables. The ensemble-averaged response of $B(t)$ to $E(t)$, $\langle \Delta B(t) \rangle$, is written (see Eq. 1.2)[2]

$$\langle \Delta B(t) \rangle = \int_{-\infty}^{t} dt' \, \Phi_{BA}(t - t') E(t'). \tag{1.4}$$

The aftereffect function $\Phi_{BA}(t)$ describes the response of the system variable $B(t)$ to a delta-pulse of the external field $E(t)$ (see Note 2).

Equation 1.4 characterizes a linear and causal response of a microscopic system variable $B$ to field $E$. The field is turned on in the "infinite" past—when the system is in equilibrium:

$$\langle \Delta B(t \rightarrow -\infty) \rangle = 0.$$

The observed response $\langle \Delta B(t) \rangle$ at time $t$ to past disturbances by $E$ at $t'$ ($t' \leq t$) is given by the product of the aftereffect function $\Phi_{BA}(t)$ and the input $E(t)$, averaged over all input impulses from the infinite past until $t$. $\langle \Delta B(t) \rangle$ is the time convolution of $\Phi$ and $E$.

To connect the microscopic $\langle \Delta B(t) \rangle$ to the macroscopic $\underset{\sim}{\chi}(\omega)$, we write

$$\langle \Delta B(t) \rangle = \underset{\sim}{\chi}_{BA}(\omega) E(\omega, t), \tag{1.5}$$

where

$$\underset{\sim}{\chi}_{BA}(\omega) = \chi'_{BA}(\omega) + i\chi''_{BA}(\omega)$$

$$= \int_{0}^{\infty} dt \, \Phi_{BA}(t)\exp(i\omega t)$$

is the Fourier-Laplace transform of $\Phi_{BA}(t)$ (see Note 3).

We now express $\chi''_{BA}(\omega)$ through $\Phi_{BA}(t)$, in terms of the quantum-mechanical system variables $B$ and $A$:

1.   We remember that the average value of the time development of any quantal variable $B(t)$ is obtained by performing the trace operation

$$\langle B(t) \rangle = \mathrm{Tr}\,\rho(t) B(0), \tag{1.6}$$

$$\rho(t) = \frac{\exp(-H/kT)}{\sum \exp(-H/kT)}. \tag{1.7}$$

$\rho(t)$ is the density matrix.[4b] $H$ is the total Hamiltonian. $H$ is the sum of Hamiltonian $H_0$ of the unperturbed system and perturbation Hamiltonian $H_e(t)$,

$$H(t) = H_0 + H_e(t). \tag{1.8}$$

2.   We obtain $\rho(t)$ by integrating the Liouville equation[5]

$$\frac{d\rho(t)}{dt} = -\left(\frac{i}{\hbar}\right)[H(t), \rho(t)] \qquad [\,,\,] = \text{commutator} \tag{1.9}$$

up to first-order perturbation (linear response).

3.   We assume that the system is in equilibrium before the time-dependent perturbation is turned on. Therefore,

$$\langle B(t \to -\infty) \rangle = 0,$$

$$\rho(-\infty) \equiv \rho_0 = \frac{\exp(-\beta H_0)}{\mathrm{Tr}(-\beta H_0)}, \tag{1.10}$$

with $\beta = (kT)^{-1}$. Equation 1.10 tells us that the average of the fluctuations of the system variables is zero in the absence of the external perturbation.[2c] We set ($\hbar = 1$)

$$\bar{\rho}(t) \equiv \exp(iH_0 t)\rho(t)\exp(-iH_0 t) \tag{1.11}$$

since it facilitates the solution of Eq. 1.9.[6] We solve Eq. 1.11 for $\rho(t)$,

$$\rho(t) = \exp(-iH_0 t)\bar{\rho}(t)\exp(iH_0 t) \tag{1.12}$$

and differentiate. This gives

$$\left(\frac{d}{dt}\right)\rho(t) \equiv \dot{\rho}(t) = -iH_0\exp(-iH_0 t)\bar{\rho}(t)\exp(iH_0 t)$$

$$+ \exp(-iH_0 t)\dot{\bar{\rho}}(t)\exp(iH_0 t)$$

$$+ \exp(-iH_0 t)\bar{\rho}(t)iH_0\exp(iH_0 t) \tag{1.13}$$

$$= \exp(-iH_0 t)\dot{\bar{\rho}}(t)\exp(iH_0 t)$$

$$- i[H_0, \exp(-iH_0 t)\bar{\rho}(t)\exp(iH_0 t)].$$

4. We insert Eq. 1.8 and Eq. 1.12 into the commutator of the Liouville equation (Eq. 1.9), set the result equal to Eq. 1.13, and solve for $(d/dt)\bar{\rho}(t)$ by multiplying from the left by $\exp(iH_0 t)$ and from the right by $\exp(-iH_0 t)$. We get

$$\dot{\bar{\rho}}(t) = \exp(iH_0 t)\{-i[H_0 + H_e(t), \exp(-iH_0 t)\bar{\rho}(t)\exp(iH_0 t)]$$

$$+ i[H_0, \exp(-iH_0 t)\bar{\rho}(t)\exp(iH_0 t)]\}\exp(-iH_0 t)$$

$$= i[\exp(iH_0 t)A\exp(-iH_0 t), \bar{\rho}(t)]E(t),$$

remembering $H_e(t) = -AE(t)$. Integration from $t = -\infty$ to $t$ yields

$$\bar{\rho}(t) = \rho_0 + i\int_{-\infty}^{t} dt'[\exp(iH_0 t')A\exp(-iH_0 t'), \bar{\rho}(t')]E(t'). \quad (1.14)$$

5. To obtain $\rho(t)$, we insert Eq. 1.11 into both sides of Eq. 1.14 and multiply by $\exp(-iH_0 t)$ from the left and $\exp(iH_0 t)$ from the right. The result is

$$\rho(t) - \rho_0 = i\int_{-\infty}^{t} dt'[\exp\{-iH_0(t - t')\}A\exp\{iH_0(t - t')\}, \rho(t')]E(t').$$

We are only concerned with linear response. Since $AE(t)$ is a first-order term, we can substitute $\rho(t)$ by $\rho_0$. [The density matrix $\rho_0$ contains only the equilibrium Hamiltonian $H_0$ ("canonical distribution").] Therefore we write Eq. 1.6

$$\langle \Delta B(t) \rangle = i\int_{-\infty}^{t} dt' \mathrm{Tr} \exp\{-iH_0(t - t')\}[A, \rho_0]\exp\{iH_0(t - t')\}E(t')B$$

$$= i\int_{-\infty}^{t} dt' \mathrm{Tr} \exp\{iH_0(t - t')\}B\exp\{-iH_0(t - t')\}[A, \rho_0]E(t').$$

$$(1.15)$$

(We have made use of the invariance of a trace operation to cyclic permutation, have set $B(0) = B$, and recalled that $[H_0, \rho_0] = 0$.) Since

$$\exp\{iH_0(t - t')\}B\exp\{-iH_0(t - t')\} = B(t - t'),$$

we write the average linear response $\langle \Delta B(t) \rangle$ of system variable $B(t)$

$$\langle \Delta B(t) \rangle = i\int_{-\infty}^{t} dt' \mathrm{Tr} B(t - t')[A, \rho_0]E(t')$$

$$= \int_{-\infty}^{t} dt' \Phi_{BA}(t - t')E(t'). \quad (1.16)$$

(See Eq. 1.4.) Therefore (Note 4)

$$\Phi_{BA}(t) = i\operatorname{Tr}[A, \rho_0]B(t)$$

$$= i\operatorname{Tr}\rho_0[B(t), A]$$

$$\equiv i\langle[B(t), A]\rangle. \qquad (1.17)$$

We conclude: The aftereffect or response function of the dynamical system variable $B(t)$ is given by the ensemble average of the commutator of $B(t)$ and the system-dependent part $A$ of the first-order external perturbation Hamiltonian $H_e(t) = -AE(t)$.

$\Phi_{BA}(t)$ has the following important properties that we will use explicitly or implicitly:[2]

1. $\Phi_{BA}(t)$ is real since $\langle\Delta B(t)\rangle$ is an observable.
2. $\Phi_{BA}(t)$ is a function of time difference only ("translational time invariance") as it does not matter when the external perturbation is imposed on the equilibrium system. Hence for a shift of $t$ by $+t'$,

$$i\langle[B(t), A]\rangle = i\langle B(t''), A(t')]\rangle \qquad (1.18a)$$

with $t'' = t' + t$.

3. Variables $B$ and $A$ are single-particle operators.
4. The system eventually approaches equilibrium:

$$\lim_{t \to \infty} \Phi_{BA}(t) = 0.$$

5. The time reversal relation of $\Phi_{BA}(t)$ is

$$\Phi_{BA}(-t) = -\Phi_{AB}(t). \qquad (1.18b)$$

This follows from the commutation rule and property 2:

$$i\langle[B(-t), A]\rangle = -i\langle[A, B(-t)]\rangle = -i\langle[A(t), B]\rangle.$$

Now we are ready to compute the absorption coefficient. The rate of loss of radiant energy through absorption in the system is given by the negative rate of change of $H_e(t) = -AE(t)$. We average over one cycle of the radiation field $E(t) = (1/2)[E\exp(i\omega t) + E^*\exp(-i\omega t)]$[2c]

$$-\frac{\omega}{2\pi}\int_0^{2\pi/\omega} dt\, \frac{\partial}{\partial t}H_e(t) \equiv Q(\omega)$$

$$= \frac{\omega}{2\pi}\int_0^{2\pi/\omega} dt\langle A(t)\rangle\frac{\partial}{\partial t}E(t). \qquad (1.19)$$

(The asterisk denotes complex conjugate.) We then introduce $\langle A(t)\rangle$ through

Eq. 1.5 as follows:

$$\langle A(t)\rangle = \underset{\sim}{\chi}_{AA}(\omega)E(\omega,t)$$

$$= \left(\tfrac{1}{2}\right)\{\underset{\sim}{\chi}_{AA}(\omega)E\exp(i\omega t)+\underset{\sim}{\chi}^{*}_{AA}(\omega)E^{*}\exp(-i\omega t)\}.$$

We insert this expression into Eq. 1.19, differentiate $E(t)$, and integrate between 0 and $2\pi/\omega$. We obtain the power dissipation $Q(\omega)$,

$$Q(\omega)=\frac{i\omega}{8\pi}\int_{0}^{2\pi/\omega}dt\,\omega\{\underset{\sim}{\chi}_{AA}[E^{2}\exp(2i\omega t)-EE^{*}]$$

$$+\underset{\sim}{\chi}^{*}_{AA}[E^{*}E-E^{*2}\exp(-2i\omega t)]\}$$

$$=\frac{i\omega}{4}\{\underset{\sim}{\chi}^{*}_{AA}-\underset{\sim}{\chi}_{AA}\}|E|^{2}=\frac{\omega}{2}\chi''_{AA}|E|^{2} \qquad (1.20)$$

or

$$\frac{Q(\omega)}{|E|^{2}}=\frac{\omega}{2}\,\mathrm{Im}\,\underset{\sim}{\chi}_{AA}=\frac{\omega}{2}\,\mathrm{Im}\int_{0}^{\infty}dt\exp(i\omega t)\Phi_{AA}(t).$$

(Im = imaginary.) The last equality in this relation follows from Eq. 1.5.
To verify $\chi''$, we use

$$2i\chi''=(\underset{\sim}{\chi}-\underset{\sim}{\chi}^{*}).$$

With

$$\underset{\sim}{\chi}_{AA}(\omega)=\int_{0}^{\infty}dt\exp(i\omega t)\Phi_{AA}(t)$$

and

$$\underset{\sim}{\chi}^{*}_{AA}(\omega)=\int_{0}^{\infty}dt\exp(-i\omega t)\Phi_{AA}(t)$$

$$=-\int_{-\infty}^{0}dt\exp(i\omega t)\Phi_{AA}(t)$$

(see Eq. 1.18b and replace $t$ by $-t$), we get

$$2\chi''(\omega)=\left(\frac{1}{i}\right)\int_{-\infty}^{0}dt\exp(i\omega t)\Phi_{AA}(t)+\left(\frac{1}{i}\right)\int_{0}^{\infty}dt\exp(i\omega t)\Phi_{AA}(t)$$

$$=\int_{-\infty}^{\infty}dt\exp(i\omega t)\langle[A(t),A(0)]\rangle. \qquad (1.21)$$

Therefore, according to Eq. 1.3, we write

$$\chi''(\omega) = \sigma(\omega)n(\omega)\omega^{-1} = \left(\frac{1}{2}\right)\int_{-\infty}^{\infty} dt \exp(i\omega t)\langle[A(t), A(0)]\rangle. \quad (1.22)$$

This is a form of the fluctuation-dissipation theorem.[2]

To gain insight into the meaning of Eq. 1.22, we imagine a small "source system" which is capable of providing a periodic force and delivering energy, and a large "dissipative system" which is capable of absorbing the energy from the source system.[7] We consider the dissipative system as an equilibrium "heat bath" (simply "bath") or "lattice" (to borrow the term from nuclear magnetic relaxation) of infinite heat capacity. We assume that dissipative and source systems are first isolated and separated and that the source system receives some internal energy. The isolated source system then performs periodic motions since the internal energy of the isolated source system is, by definition, unperturbed by physical processes (such as collisions). The source system possesses dynamic coherence: Its rotational motion is purely kinetic (no hindering torques), its vibrational motion is purely harmonic (no interaction).

We now allow source and dissipative systems to interact. The initial coherence of the periodic motion of the source system gradually and irreversibly vanishes by merging into the random noise of the thermal equilibrium fluctuations of the dissipative system. The dissipative system, through its random motions, causes the coherence loss of the source system.

We observe this behavior through band analysis. In many examples we are able to identify an initial time interval where the coherence of the system is determined by unhindered motion, and a later time (correlation time) after which the motion is smeared out within the random fluctuations of the lattice.

It is important to realize that we do not observe the time dependence of system variable $A(t)$. We observe manifestations and characteristics of $A(t)$ through time-dependent aftereffect functions $\langle[A(t), A(0)]\rangle$. The average motions in the equilibrium system are the same at any instant. Their sum effect amounts to zero; each transition is balanced by its reverse transition. The spectrometer, however, singles out the individual event by coupling the radiation field to a system variable ("fluctuation spectroscopy"). Out of the entire system the spectrometer picks the subsystem that is absorbing or scattering radiation; the remainder of the system acts as perturbing environment owing to its fluctuating intermolecular forces. Nevertheless, we need not concern ourselves with the specifics of the interaction between radiation field and system. Only system-dependent operators of this process are of interest. (As we demonstrate in the next section, a relevant operator is the transition moment tensor.)

In the following section we discuss how to simplify Eq. 1.22 by writing the commutator as a "one-sided" average $\langle A(t)A(0)\rangle$. Such quantities are called "correlation functions." They are the basis for our discussions of dynamic phenomena of the condensed phase. On one hand, we obtain the correlation functions by numerical Fourier inversion of experimental band contours; on the other hand, we predict the correlation functions by theoretical modeling.

## 1.4.  Correlation Functions: Some General Aspects

We develop the commutator of Eq. 1.22 in terms of a complete set of eigenfunctions of the total system Hamiltonian $H$. First, we split $H$ into a zero-order part $H_0$—the unperturbed (equilibrium) system—and into a small time-dependent perturbation term $H_1(t)$,

$$H = H_0 + H_1(t). \tag{1.23}$$

$H_1(t)$, a random operator of the dissipative system, causes the irreversible return of the prepared or source system to equilibrium. At present we do not specify $H_1(t)$ further.

Second, we identify system operator $A$ as the total electric displacement of the system under the action of external field $E(t)$.[8] In other words, we assign $A(t)$ to the active system dipole moment operator $\mu(t)$. Developing the commutator of Eq. 1.17 in a complete set of eigenstates $|n\rangle$ of $H$, we get

$$\langle[\mu(t), \mu(0)]\rangle = \text{Tr}\rho_0[\mu(t), \mu(0)]$$

$$= \sum_{nm} \{\rho_n(n|\exp(iHt)\mu(0)\exp(-iHt)|m)(m|\mu(0)|n)$$

$$- \rho_m(m|\mu(0)|n)(n|\exp(iHt)\mu(0)\exp(-iHt)|m)\}$$

$$= \sum_{nm} (\rho_n - \rho_m)(n|\mu(0)|m)(m|\mu(0)|n)\exp(i\omega_{nm}t),$$

with

$$H|n) = E_n|n),$$

$$\omega_{nm} = (E_n - E_m),$$

$$\hbar = 1.$$

Since

$$\rho_m = \rho_n\exp(-\beta\omega_{mn}); \qquad \beta = \frac{1}{kT},$$

we obtain

$$\langle [\mu(t), \mu(0)] \rangle$$

$$= \sum_{nm} \rho_n [1 - \exp(-\beta\omega_{mn})](n|\mu(0)|m)(m|\mu(0)|n)\exp(i\omega_{nm}t).$$

Inserting this result into Eq. 1.22 and remembering the identity

$$\left(\frac{1}{2\pi}\right)\int_{-\infty}^{\infty} dt \exp(iXt) = \delta(X), \tag{1.24}$$

where $\delta(X)$ is the delta function of argument $X$, we get (omitting uninteresting constant factors)

$$\frac{\sigma(\omega)n(\omega)}{\omega} = \sum_{nm} [1 - \exp(-\beta\omega_{mn})]\rho_n(n|\mu(0)|m)(m|\mu(0)|n)\,\delta(\omega + \omega_{nm})$$

$$= \sum_{nm} [1 - \exp(-\beta\omega)]\rho_n(n|\mu(0)|m)(m|\mu(0)|n)\,\delta(\omega + \omega_{nm}).$$

The second equality arises from the property of the delta function $\delta(\omega + \omega_{nm})$. Dividing by the exponential factor, we obtain

$$\frac{\sigma(\omega)n(\omega)}{\omega[1 - \exp(-\beta\omega)]} \equiv \mathscr{I}(\omega)$$

$$= \sum_{nm} \rho_n(n|\mu(0)|m)(m|\mu(0)|n)\,\delta(\omega + \omega_{nm}). \tag{1.25}$$

$\mathscr{I}(\omega)$ is called "spectral (infrared) density." The spectral density is more significant than the intensity since it represents a transition probability (as we discuss later). The spectral (infrared) density contains the refractive index $n(\omega)$, a slowly varying function over the frequency range of weak transitions. Therefore, we compute the absorption coefficient from the incident $T_0$ and transmitted $T$ intensity by

$$\sigma(\omega) \propto \ln\left\{\frac{T_0(\omega)}{T(\omega)}\right\}. \tag{1.26}$$

(For intense bands the frequency dependence of the reflected intensity is appreciable and Eq. 1.26 breaks down.[9])

Only a few more steps are needed to transform Eq. 1.25 into a useful expression for a one-sided average of $\langle [\mu(t), \mu(0)] \rangle$. We merely retrace the steps that led to Eq. 1.25 from Eq. 1.22—but in reversed order. We then rewrite Eq. 1.25 successively as

$$\mathcal{I}(\omega) = \int_{-\infty}^{\infty} dt \sum_{nm} \rho_n (n|\mu(0)|m)(m|\mu(0)|n) \exp\{i(\omega + \omega_{nm})t\}$$

$$= \int_{-\infty}^{\infty} dt \exp(i\omega t) \sum_{nm} \rho_n (n|\exp(iHt)\mu(0)\exp(-iHt)|m)(m|\mu(0)|n)$$

$$= \int_{-\infty}^{\infty} dt \exp(i\omega t) \mathrm{Tr}\, \rho_0 \{\mu(t)\mu(0)\}$$

$$= \int_{-\infty}^{\infty} dt \exp(i\omega t) \langle \mu(t)\mu(0) \rangle. \tag{1.27}$$

We invert the result and obtain

$$\langle \mu(t)\mu(0) \rangle = \int_{-\infty}^{\infty} d\omega \exp(-i\omega t) \mathcal{I}(\omega),$$

$$\langle \mu(0)\mu(t) \rangle = \int_{-\infty}^{\infty} d\omega \exp(i\omega t) \mathcal{I}(\omega). \tag{1.28}$$

In words: *The spectral density of a band contour and the correlation function of the relevant transition moment operator are mutual Fourier transforms.*

What is the physical picture behind $\langle \mu(0)\mu(t) \rangle$? We see that $\langle \mu(0)\mu(t) \rangle$ is an average scalar product of the observed transition moment operator taken at two different times. $\langle \mu(0)\mu(t) \rangle$ correlates the value of $\mu(t')$ at an arbitrary instant $t = t'$ with the value of $\mu(t'')$ at arbitrary $t = t'' - t'$ later, averaged over the whole system. It is not difficult to see how such products give us information on molecular and atomic motions: $\mu(t)$ is fixed in the molecule or within a molecular group of the molecule, or—as a composite quantity—within a cluster of a few molecules. Variations in molecular positions and orientations as well as variations of vibrational amplitudes and phase relations determine the time dependence of $\langle \mu(0)\mu(t) \rangle$. Since we must average over the system (we only observe averages), $\langle \mu(0)\mu(t) \rangle$ is a statistical quantity.

Other useful formulations of the correlation function of $\mu(t)$, their general properties, and several important theorems are discussed in Appendix A.

We always assume that the solution of the zero-order problem under Hamiltonian $H_0$—the equilibrium system—is known. With

$$H_0 |n) = E_n^0 |n),$$

$$H_0 |m) = E_m^0 |m),$$

$$\omega_{mn}^0 = \frac{(E_m^0 - E_n^0)}{\hbar},$$

we reduce Eqs. 1.28 to

$$\mathscr{I}(\omega) = \int_{-\infty}^{\infty} dt \exp\{i(\omega - \omega_{mn}^0)t\}\langle \mu^I(t)\mu^I(0)\rangle. \qquad (1.29)$$

Here,

$$\mu^I(t) = \exp\left(\frac{iH_0 t}{\hbar}\right)\mu(t)\exp\left(-\frac{iH_0 t}{\hbar}\right), \qquad (1.30)$$

$$H_1^I(t) = \exp\left(\frac{iH_0 t}{\hbar}\right)H_1(t)\exp\left(-\frac{iH_0 t}{\hbar}\right), \qquad (1.31)$$

is a so-called interaction representation (or picture) of $\mu(t)$: The time development of operator $\mu$ depends only on $H_0$.[5] (See also Eq. 1.11; we discuss and use the interaction representation in more detail later on.) At any rate, Eqs. 1.29 and 1.30 require that we take the exponential in the Fourier transforms with respect to the unperturbed frequency $\omega_{mn}^0$, the "band center."

To sum up, we have written useful forms of function $f(t)$ and band profile $I(\omega)$ [see Eq. 1.1a] in terms of quantum-mechanical system-averaged operator products and macroscopic spectral quantities. In the following section we specify the correlation functions derivable from purely rotational, purely vibrational, and vibration-rotation spectra of infrared, Raman, and Rayleigh data.

## 1.5.  Rotational and Vibrational Relaxation: Statistical Independence

Equation 1.29 and its associated relations (Appendix A) look deceptively simple. Their numerical evaluation is also simple: An integration procedure of summing rectangular areas in an inner $\omega$-loop and an outer $t$-loop is sufficient (see Eq. 1.1b).

On the other hand, it is difficult to attach the correct physical meaning to the result. After all, nothing keeps us from Fourier-inverting any experimental band profile. Whether it makes sense is another matter.

Two principal conditions to arrive at meaningful results are as follows:

1.  Different types of relaxation processes are not statistically related: Cross-correlation functions between different relaxation processes are zero over the time intervals of interest. It is possible to relax this restriction in particular instances. But a description of cross-correlation effects succeeds only through a considerable amount of theoretical sophistication that is not always matched by the quality of the spectral data.

2.  We observe the *entire* band contour of *one* specific oscillator: Band contours of different vibrational modes must not overlap. It is even necessary to reduce band broadening caused by isotope-induced oscillator multiplicities. [Vibrational modes that involve the Cl atoms in natural chloroform (Cl-35 and

Cl-37, abundance ratio $3:1$) are a good example.[10]] This "single oscillator" condition is one of the most difficult to meet experimental obstacles to a meaningful analysis. In favorable cases overlapping band contours can be disentangled by numerical-theoretical methods.[11] Simple curve-fitting procedures are unsatisfactory.

Of course, we can reduce our expectations by evaluating the band*width*. A bandwidth is equivalent to a correlation *time* (see Section 1.1). Considering correlation times entails loss of information since we no longer observe the time development of a correlation but its average. The advantage of infrared and Raman fluctuation spectroscopy over nuclear magnetic or dielectric relaxation, namely, the observation and evaluation of a susceptibility over a continuous *range* of frequencies, is thereby surrendered.

However, our immediate concern is to define and characterize vibrational and rotational correlation functions. We first specify in Eq. 1.28 or 1.29 the relevant transition moment operator $\mu(t)$. We develop $\mu(t)$ into a series in terms of the atomic displacement coordinates $Q^{\nu}$ (normal coordinate) of vibrational mode $\nu$,

$$\mu(t) = \mu_0(t) + \left[\frac{\partial\mu(t)}{\partial Q^{\nu}(t)}\right]_{Q^{\nu}=0} Q^{\nu}(t) + \cdots \qquad (1.32a)$$

For a vibrational-rotational transition the relevant term is

$$\left[\frac{\partial\mu(t)}{\partial Q^{\nu}(t)}\right]_{Q^{\nu}=0} Q^{\nu}(t) \equiv \mathbf{m}^{\nu}(t). \qquad (1.32b)$$

We rewrite Eq. 1.29 in terms of a complete set of uncoupled *vibrational-rotational* eigenstates $|vr\rangle = |v\rangle|r\rangle$), labeling the molecular-fixed axis direction of $\mathbf{m}$ by subscript $i$. Since $\rho_0 \equiv \rho_v^0\rho_r^0$, we get

$$\mathcal{I}_i^{\nu}(\omega) = \int_{-\infty}^{\infty} dt \operatorname{Tr}\rho_{vr}^0 (vr|m_i^{\nu}(t)|v'r')(v'r'|m_i^{\nu}(0)|vr)$$

$$\times \exp\left[i(\omega - \omega_{vr}^0)t\right]$$

$$= \int_{-\infty}^{\infty} dt \operatorname{Tr}\rho_v^0 \operatorname{Tr}\rho_r^0 (v|m^{\nu}(t)|v')(r|\hat{m}_i^{\nu}(t)|r')$$

$$\times (v'|m^{\nu}(0)|v)(r'|\hat{m}_i^{\nu}(0)|r)\exp\left[i(\omega - \omega_{vv'}^0 - \omega_{rr'}^0)t\right]. \qquad (1.33)$$

In the last equality of Eq. 1.33 we have written vector $\mathbf{m}_i^{\nu}$ as product of its unit vector $\hat{\mathbf{m}}_i^{\nu}$ and its modulus $m^{\nu}$,

$$\mathbf{m}_i^{\nu}(t) = \hat{\mathbf{m}}_i^{\nu}(t)m^{\nu}(t)$$

$$m^{\nu}(t) = |\mathbf{m}_i^{\nu}(t)|$$

$$\hat{\mathbf{m}}_i^{\nu}(t) = \mathbf{m}_i^{\nu}(t)/m^{\nu}(t). \qquad (1.34)$$

Clearly $\hat{\mathbf{m}}_i$ and $m$ denote the angle-dependent and amplitude-dependent factor, respectively, of the infrared transition moment vector. Therefore we assign $\hat{\mathbf{m}}_i(t)$ to the purely orientational coordinate and $m(t)$ to the purely vibrational coordinate of the fluctuating vibrational-rotational transition moment. Performing the trace operations in Eq. 1.33 gives

$$\mathscr{I}_i^{\nu}(\omega) = \int_{-\infty}^{\infty} dt \exp\{-i(\omega - \omega_0)t\}\langle \hat{\mathbf{m}}_i^{\nu}(0)\hat{\mathbf{m}}_i^{\nu}(t)\rangle \langle m^{\nu}(0)m^{\nu}(t)\rangle, \quad (1.35)$$

where $\omega_{\nu r}^0 \equiv \omega_0$. Equation 1.35 expresses the assumed independence of vibrational and rotational relaxation. The average of the product is equal to the product of the averages,

$$\langle m^{\nu}(0)\hat{\mathbf{m}}_i^{\nu}(0)m^{\nu}(t)\hat{\mathbf{m}}_i^{\nu}(t)\rangle = \langle m^{\nu}(0)m^{\nu}(t)\rangle \langle \hat{\mathbf{m}}_i^{\nu}(0)\hat{\mathbf{m}}_i^{\nu}(t)\rangle. \quad (1.36)$$

Or: Ensemble averaging is performed separately over vibrational and rotational coordinates.

For a vibration-rotation Raman contour we replace the first-rank dipole moment tensor $\boldsymbol{\mu}$ by the second-rank polarizability tensor $\boldsymbol{\alpha}$,

$$\boldsymbol{\alpha} = \boldsymbol{\alpha}_0 + \left(\frac{\partial \boldsymbol{\alpha}}{\partial Q^{\nu}}\right)_{Q^{\nu}=0} Q^{\nu} + \cdots \quad (1.37)$$

We write again the vibrational-rotational correlation function as a product of a scalar (amplitude) and a tensorial (angle-dependent) factor,

$$\langle \alpha^{\nu}(0)\alpha^{\nu}(t)\rangle_{\text{vib}}\langle \boldsymbol{\alpha}^{\nu}(0)\boldsymbol{\alpha}^{\nu}(t)\rangle_{\text{rot}}. \quad (1.38)$$

We assume that the purely vibrational correlation functions for the infrared and Raman tensors are identical (except unimportant constant factors) for one and the same mode,

$$\langle \alpha^{\nu}(0)\alpha^{\nu}(t)\rangle = \langle m^{\nu}(0)m^{\nu}(t)\rangle. \quad (1.39)$$

[The purely angle-dependent correlation function of the Raman tensor $\boldsymbol{\alpha}(t)$ is more complicated than the purely angle-dependent correlation function of the infrared tensor $\boldsymbol{\mu}(t)$. We discuss this in detail later.]

We summarize assumptions and results as follows:

1.  The vibrational-rotational transition moment operator is fixed within the molecular coordinate system. A normal coordinate is assigned to the observed vibrational-rotational band. [We neglect transition moments in molecular clusters (collision-induced transition moments).[12]]

2.  The transition moment operator is written as product of an angle-dependent and a magnitude-dependent factor.

3. The time development of the angle-dependent factor, which depends on the rank of the transition moment tensor and its molecule-fixed direction, is assigned to a coordinate of the *molecular reorientational* fluctuations ("rotational relaxation").

4. The time development of the magnitude-dependent factor, which is independent of the rank of the transition moment tensor, is assigned to the coordinate of the fluctuations of the *atomic displacement amplitudes* of a normal mode ("vibrational relaxation").

The molecular axis (or axes) direction of the orientational factor of the transition moment tensor should be known. Otherwise evaluation of the orientational motion of the molecule in terms of its rotational correlation function is not informative. We discuss this in the next section.

## 2.  ROTATIONAL RELAXATION: FIRST-ORDER TENSORS

### 2.1.  Orientational Correlation Functions from Infrared Vibration-Rotation Spectra

We discussed in Section 1.5 that the dynamics of the orientational motion can be obtained through the correlation function $\langle \hat{m}_i(0)\hat{m}_i(t) \rangle$. We defined $\hat{m}_i(t)$ as tensor, of unit length, along the direction $i$ of the vibrational-rotational transition moment of the molecule. The direction of axis $i$ is known, in relation to the three (orthogonal) principal molecular axes of inertia, in molecules of tetragonal or higher-symmetry point groups. [The direction of the transition moment for lower-symmetry point groups may be established through normal coordinate calculations.]

Depending on the point group symmetry of the molecule, we can select as many as three independent molecule-fixed directions $(i, j, k)$ among the allowed transition moments. In tetrahedral molecules there is only one infrared-active vibration-rotation species. For axial symmetry (symmetric tops) there are two active species, allowing us to select two orthogonal molecular axis directions. For tetragonal (or lower) symmetry we can pick out the maximum of three different axes directions.

In particular, we picture the *orientational* motion of axis direction $i$ of $\hat{m}$ to arise from *rotation* of the molecule around the two other axes $(j, k)$. Since the spacial direction of axis $i$ is left invariant when the molecule rotates around $i$ itself, we deal with only two of the three principal moments of inertia in our experimental and theoretical analyses. For this reason we call the angular fluctuations of the chosen axis of $\hat{m}$ "reorientational" or "orientational" fluctuations (rather than "rotational" fluctuations).

We refer the time development of the orientational fluctuations of the specified axis $i$ to an external, laboratory-fixed, coordinate frame. Function

$\langle \hat{m}(t')\hat{m}(t'') \rangle \equiv G(t''-t')$ or its related forms (see Appendix A) therefore describe the ensemble-averaged direction of molecule axis $i$ at an arbitrary time $t''$ with respect to its direction at time $t''-t'$ earlier. For a first-rank tensor transition moment (infrared), $G(t''-t')$ characterizes the average cosine through which the observed molecular axis of $\hat{m}(t)$ has reoriented during $t''-t'$ owing to molecular rotation around the other two orthogonal axes.[3b, 13] [$G(t''-t')$ is more complicated for a second-rank tensor transition moment (Raman) of a vibration-rotation mode.]

Figure 1.3 gives a graphic example. The rotating body—a dipolar symmetric top such as $CH_3I$, $CHCl_3$, $CH_3CN$—is represented by its inertial ellipsoid, by the direction $\hat{\mu}$ of the permanent molecular dipole moment, and by the direction $\hat{m}$ of the vibrational-rotational transition moment. Figure 1.3a shows a "parallel" band: The allowed vibrational symmetry species is $A_1$ (totally symmetric). $\hat{m}(t)$ lies along the threefold molecular symmetry axis ( = $c$ in this example). We assume that at $t = t_0$ the inertial axis system $a, b, c$ is coincident

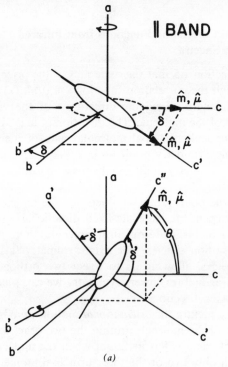

(a)

**Figure 1.3.** Inertial axis system of a dipolar symmetric top ($CH_3I$, $CClH_3$, $CH_3CN$). The $C_3$ symmetry axis, which contains the direction of the permanent dipole moment $\mu$ and the smallest moment of inertia, is drawn along axis $c$. ( a ) *Parallel band*: The permanent dipole moment and the vibrational transition moment reorient together. ( b ) *Perpendicular band*: The permanent dipole moment stays invariant during the first reorientation step of the vibrational transition moment, polarized along one of the two equivalent perpendicular axes $a, b$.

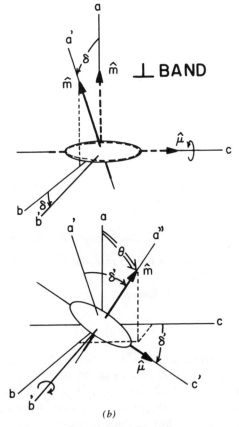

(b)

**Figure 1.3.**  (*Continued*)

with the space-fixed reference frame $X, Y, Z$ (not shown). At $t = t_1 > t_0$ axis $c$ has reoriented through angle $\delta$ whereas the molecule has rotated around axis $a$ by the same angle. At $t = t_2 > t_1$ axis $c$ has further reoriented, through angle $\delta'$, whereas the molecule has rotated around axis $b'$ by $\delta'$. As a result, axis $c$ has reoriented by angle $\theta$, the molecule has rotated around two of its inertial axes, and $\hat{m}$ now points along space direction $c''$. During both orientational steps the molecular permanent dipole moment $\mu$ has been tilted out of the spacial orientations it had at $t = t_0$ and $t = t_1$.

In Fig. 1.3$b$ we describe the analogous steps for a "perpendicular" vibrational-rotational mode (doubly degenerate, species $E$) of the symmetric top. The transition moment lies along any of the two (equivalent) perpendicular inertial axes. The first reorientational step of $\hat{m}$ by angle $\delta$ involves molecular rotation around axis $c$ (in this example). We note that the space-fixed direction of the molecular permanent dipole moment stays invariant. The subsequent reorientational step of $\hat{m}$ involves rotation of the molecule around axis $b'$ through angle $\delta'$. This step is equivalent to any of the reorientational steps for the parallel band (shown in Fig. 1.3$a$). Consequently the permanent dipole moment is tilted during the motion that begins at $t = t_1$ and ends at $t = t_2$.

The example demonstrates the usefulness of evaluating the rotational components of vibrational-rotational spectra since we can characterize the anisotropy of the rotational motion and how it is influenced by polarity of the liquid. Furthermore, we can construct linear combinations of the basic (independent) orientational correlation functions.[14] In our example of a symmetric top, we write combinations of $G_{A_1}(t)$ and $G_E(t)$. Thereby we mix different, preselected molecule-fixed orientations of $\hat{\mathbf{m}}$. The result is a correlation function of an average molecular reorientation axis. Such a correlation function offers interesting aspects and revealing insights into orientational dynamics since the generated axis direction moves with the molecule *and* within the molecule. We defer further elaboration until we discuss a concrete example.

## 2.2. Orientational Correlation Functions from Far-Infrared Spectra

The transition moment operator for an infrared purely rotational transition is equivalent to the molecular permanent dipole moment operator $\boldsymbol{\mu}_0$. Hence, the purely orientational correlation function is given by (see Eq. 1.32$a$; we omit the subscript 0)

$$G(t) = \langle \boldsymbol{\mu}(0) \cdot \boldsymbol{\mu}(t) \rangle. \tag{1.40}$$

This orientational correlation function should be identical with $G(t) = \langle \hat{\mathbf{m}}(0) \cdot \hat{\mathbf{m}}(t) \rangle$ (Eq. 1.35) if we consider a parallel-polarized band of a symmetric top (for instance species $A_1$, $A'$ of point group $C_{3v}, C_{3h}$) or of a linear molecule (species $\Sigma_u^+$ of point groups $D_{\infty h}$ and $C_{n\infty}$). (Vibrational species of lower-symmetry molecules qualify but do not interest us here.) In other words, whenever $\boldsymbol{\mu}$ and $\mathbf{m}$ describe in the same molecule the same molecule-fixed axis, we should expect the same $G(t)$ regardless of whether it is from the infrared vibration-rotation or pure rotation contour.

However, $\langle \boldsymbol{\mu}(0) \cdot \boldsymbol{\mu}(t) \rangle$ restricts us to one axis reorientation mode. We can no longer select up to three independent molecular axis directions as we may have done under $\langle \hat{\mathbf{m}}(0) \cdot \hat{\mathbf{m}}(t) \rangle$.

Second, the Fourier transform of a purely rotational spectral density $\mathscr{I}(\omega)$ (Eq. 1.29) is taken with respect to $\omega_0 = 0$. Note that pure rotational infrared spectra are in the far-infrared frequency range.

Third, the observed orientational correlation function $\langle \boldsymbol{\mu}(0) \cdot \boldsymbol{\mu}(t) \rangle$ might contain cross correlations between dipoles on different molecules since dipolar (or other) forces between adjacent molecules can result in few-molecule clusters possessing a composite dipole moment $\boldsymbol{\mu}(t) = \Sigma_i \boldsymbol{\mu}^{(i)}(t)$. The dipole moment $\boldsymbol{\mu}^A$ at arbitrary time $t''$ on molecule $A$ feels the value of dipole moment $\boldsymbol{\mu}^B$ at time $t'$ on molecule $B$. We write

$$G(t) = \left\langle \boldsymbol{\mu}^A(0) \cdot \boldsymbol{\mu}^A(t) \right\rangle + \sum_{A \neq B} \left\langle \boldsymbol{\mu}^A(0) \cdot \boldsymbol{\mu}^B(t) \right\rangle. \tag{1.41}$$

The summation terms over unequal indices account for pairwise cross correlation.

The autocorrelation term in Eq. 1.41, $\langle \mu^A(0) \cdot \mu^A(t) \rangle$, is equivalent to $\langle \hat{m}(0) \cdot \hat{m}(t) \rangle$ if both refer to the same molecular axis. Experimentally, cross correlation is minimized by dilution in a nonpolar medium.

Why did we not consider cross correlation in Section 1.5 in $\langle \mathbf{m}(0) \cdot \mathbf{m}(t) \rangle$? Because any *orientational* cross correlation $\langle \hat{m}^A(0) \cdot \hat{m}^B(t) \rangle$ in the infrared (or Raman) vibrational-rotational correlation function (Eqs. 1.36 and 1.38)

$$\langle \mathbf{m}^A(0) \cdot \mathbf{m}^B(t) \rangle \approx \langle m^A(0) m^B(t) \rangle \langle \hat{m}^A(0) \cdot \hat{m}^B(t) \rangle$$

is obliterated by the precipitous drop of the *vibrational* cross-correlation factor[14]

$$\langle m^A(0) m^B(t) \rangle \sim 0 \quad \text{for all } t.$$

On the other hand, in ordered solids $\langle m^A(0) m^B(t) \rangle \neq 0$. Therefore orientational cross correlation $\langle \hat{m}^A(0) \cdot \hat{m}^B(t) \rangle$ is conceivable in locally highly structured liquids or in disordered media where local, crystallinelike order persists for sufficiently long times.

## 3.  ROTATIONAL RELAXATION: SECOND-ORDER TENSORS

### 3.1.  Raman and Rayleigh Scattering: General Aspects

The scattering of light by the atoms of a molecule is a coupled radiative process. First, an incident optical photon is absorbed by an atom of the molecule through dipole interaction with the electronic level. Subsequently a photon is emitted spontaneously by a second dipole interaction.[15] If the excited atom returns to its electronic-vibrational ground state, the wavelengths of emitted and incident photons are identical. The scattering process is called "elastic": it represents *Rayleigh* scattering. If the emitted light has undergone a frequency shift from its incident frequency, the process is called "inelastic": it leads to *Raman* scattering. For red-shifted (down-frequency) scattered light, the energy balance creates a phonon. This is the *Stokes* Raman process. For blue-shifted scattered light, the energy balance leads to the destruction of a phonon. This is the *anti-Stokes* Raman process. Evidently, an anti-Stokes process can only originate from an excited vibrational level (usually the first on account of the Boltzmann factor). Because of the relatively weak intensity of anti-Stokes lines above 800 cm$^{-1}$, we are only concerned with the Stokes process.

To modify the theory of a resonant absorption process (see Section 1.3) for a scattering phenomenon, we recall, first, that the incident photon induces an electronic dipole moment; second, that we observe a spontaneously emitted (equals scattered) radiation flux in the optical frequency region. (The wavelength of the blue $Ar^+$ laser is 0.488 microns. In contrast, the fundamental range of infrared/far-infrared radiation stretches from about 3 to 1000 microns.) Therefore, we substitute the system dipole moment operator $\mu_g$ by the system polarizability tensor $\alpha_{gg'}$ in terms of space-fixed or molecule-fixed directions ($g, g' = X, Y, Z$; $i, j = x, y, z$). The first subscript characterizes the direction of the inducing polarization by the external field, the second denotes the direction of the resulting polarization in the medium. For instance (summation over equal indices is implied),

$$P_g = \alpha_{gg'} E_{g'} \propto \chi''_{gg'} E_{g'}.$$

(Compare with Eq. 1.2; note the linear response.)

Second, we substitute the absorption coefficient by the spectral density of spontaneously emitted light. As is well known, this quantity is characterized by the Einstein $A$-coefficient, which itself is proportional to the Einstein $B$-coefficient for stimulated emission or absorption—in other words, to $\chi''(\omega)$. The proportionality constant of the respective intensities is equal to the fourth power of the emitted frequency $(\omega_L - \omega_0)^4$, if $\omega_L$ designates the laser angular frequency. (Remember, we are only interested in Stokes processes.) Since the frequency of the laser light is a factor of 10 higher than that of the vibrational band center, we can omit the slowly varying $(\omega_L - \omega_0)^4$.

An overriding advantage to prefer the (experimentally and theoretically) complicated Raman scattering process over infrared absorption measurements is that a polarizability tensor can be decomposed into a sum of two terms,

$$\begin{pmatrix} \alpha_{ii} & \alpha_{ij} & \alpha_{ik} \\ \alpha_{ji} & \alpha_{jj} & \alpha_{jk} \\ \alpha_{ki} & \alpha_{kj} & \alpha_{kk} \end{pmatrix} = \begin{pmatrix} \frac{1}{3} & 0 & 0 \\ 0 & \frac{1}{3} & 0 \\ 0 & 0 & \frac{1}{3} \end{pmatrix} \bar{\alpha}$$

$$+ \begin{pmatrix} \alpha_{ii} - \frac{1}{3}\bar{\alpha} & \alpha_{ij} & \alpha_{ik} \\ \alpha_{ji} & \alpha_{jj} - \frac{1}{3}\bar{\alpha} & \alpha_{jk} \\ \alpha_{ki} & \alpha_{kj} & \alpha_{kk} - \frac{1}{3}\bar{\alpha} \end{pmatrix}$$

$$\equiv \tfrac{1}{3}\bar{\alpha}\mathbf{I} + \beta_{ij}, \tag{1.42}$$

where $\mathbf{I}$ is the unit tensor and $\bar{\alpha} = \mathrm{Tr}(\alpha_{ii} + \alpha_{jj} + \alpha_{kk})$. The trace of the diagonal tensor elements $\alpha_{ii}$ is a coordinate-independent, scalar quantity. The diagonal tensor $\bar{\alpha}\mathbf{I}$ is therefore independent of the orientation of the molecule with respect to a fixed reference frame. Consequently we can split a scattering correlation function into an angle-independent (from $\bar{\alpha}\mathbf{I}$) and into an angle-dependent term (from $\beta_{ij}$). Note from Eq. 1.42 that tensor $\beta_{ij}$ has zero trace.

Experimentally we accomplish the decomposition by observing the scattered light at parallel and normal polarization geometries relative to the polarization of the incident radiation field. From the normal geometry we obtain the angle-dependent, from the parallel geometry we obtain the angle-independent intensity component.

This, then, is the great a priori advantage of performing a Raman scattering experiment. In the inelastic process it leads to a separation of purely vibrational and purely orientational effects. In the elastic process it offers the possibility of studying translational as well as angle-dependent motions. (We are interested in the former.) We outline in the next section the mathematical formulation of the polarization experiment.

### 3.2.  Raman-Rayleigh Scattering: The VV-VH Experiment

According to our discussion in the previous section, we can immediately write, in general terms, the scattered intensity of a Raman-Rayleigh band. We first replace dipole moment $\mu$ (see Eq. 1.32) by the polarizability $\alpha$ (Eq. 1.37). Second, we remember that the two indices of the polarizability (Eq. 1.42) designate the direction of the inducing and the resulting polarization (or induced dipole moment). The spontaneous Stokes emission cross section (per unit solid angle and frequency interval) is then

$$I_{ij}(\omega) \propto (\omega_L - \omega_0)^4 \int_{-\infty}^{\infty} dt \exp(i\omega t) \left\langle \alpha_{ij}^A(t)\alpha_{ij}^B(0) \right\rangle, \qquad (1.43)$$

where the superscripts $A, B$ account for the possibility of pairwise cross correlation of the dynamic variable from adjacent molecules $A$ and $B$. Making use of Eqs. 1.38, 1.39, and 1.41 and attaching superscripts 0 to denote the permanent polarizability term (Eq. 1.37), we split up the general second-order tensor scattering correlation function as

$$\left\langle \alpha_{ij}^A(t)\alpha_{ij}^B(0) \right\rangle = \left\langle \alpha_{ij}^{0A}(t)\alpha_{ij}^{0B}(0) \right\rangle + \left\langle \alpha_{ij}^\nu(t)\alpha_{ij}^\nu(0) \right\rangle \left\langle \alpha^\nu(t)\alpha^\nu(0) \right\rangle$$

$$= \left\langle \alpha_{ij}^{0A}(t)\alpha_{ij}^{0A}(0) \right\rangle + \sum_{A \neq B} \left\langle \alpha_{ij}^{0A}(t)\alpha_{ij}^{0B}(0) \right\rangle$$

$$+ \left\langle \alpha_{ij}^\nu(t)\alpha_{ij}^\nu(0) \right\rangle \left\langle m^\nu(t)m^\nu(0) \right\rangle. \qquad (1.44)$$

The first two terms of the final relation are the auto- and cross-correlation terms of the elastic (Rayleigh) scattering contribution. The third term denotes the Raman vibration-rotation autocorrelation function of vibrational mode $\nu$. (Uninteresting constant factors are omitted.)

We first discuss the Raman vibration-rotation correlation function for transition between vibrational levels $n$ and $m$. Using Eq. 1.42, we rewrite Eq.

1.44 as

$$\int_{-\infty}^{\infty} d\omega \exp\{i(\omega_{mn}^0 - \omega)t\} I_{ij}^{\nu}(\omega)$$

$$= G_{ij}^{\nu}(t)$$

$$= \langle [\beta^{\nu}(t) + \tfrac{1}{3}\bar{\alpha}][\beta^{\nu}(0) + \tfrac{1}{3}\bar{\alpha}]\rangle \langle m^{\nu}(t)m^{\nu}(0)\rangle$$

$$= \langle \beta^{\nu}(t)\beta^{\nu}(0) + \tfrac{1}{9}\bar{\alpha}^2 + \tfrac{1}{3}\bar{\alpha}[\beta^{\nu}(t) + \beta^{\nu}(0)]\rangle \langle m^{\nu}(t)m^{\nu}(0)\rangle$$

$$= \langle \beta^{\nu}(t)\beta^{\nu}(0) + \tfrac{1}{9}\bar{\alpha}^2 \rangle \langle m^{\nu}(t)m^{\nu}(0)\rangle. \tag{1.45}$$

[We remember that $\langle A(t)A(0)\rangle = \mathrm{Tr}\,\rho_0\{A(t)A(0)\}$ (see Eq. 1.6 and 1.17) and $\mathrm{Tr}\,\rho_0\beta_{ij}(t) = 0$ since $\beta$ is a random variable.]

Second, we demonstrate how to disentangle Eq. 1.45 experimentally into a vibrational and orientational factor.[16, 17] We define a laboratory-fixed coordinate system $X, Y, Z$ with the molecule at its origin $0, 0, 0$. The incident laser radiation of amplitude $E_0$ and polarization vector $\mathbf{k}_0$ is $Z$-polarized and propagates in the $X$-direction toward $0, 0, 0$. We observe the scattered light of polarization vector $\mathbf{k}_s$ at a "90-degree geometry," for instance, by looking along the $Y$-axis (Fig. 1.4).

The molecule-fixed ellipsoid of the polarizability tensor is in an arbitrary position determined by polar and azimuthal angles $\theta = \arccos(Z, z)$ and $\varphi = \arccos(X, x)$ with respect to $X, Y, Z$. We assume axial symmetry for $\alpha$ (it will turn out that we do not gain useful information of a lower-symmetry molecular system) and that we have diagonalized the polarizability tensor to its molecule-fixed axes $x, y, z$ ($\alpha_{ij} = \alpha_{ji}$). This reduces the five independent components of $\alpha$ to two. The components of the optically induced dipole moment $\mu$ (or polarization $\mathbf{P}$) are $\mu_i = \alpha_i^{\nu} E_i$ (see the previous section). Hence (see Fig. 1.4)

$$\mu_z = \alpha_z^{\nu} E_0 \cos\theta$$

$$\mu_x = 0$$

$$\mu_y = \alpha_y^{\nu} E_0 \cos\left(\theta + \frac{\pi}{2}\right) = -\alpha_y^{\nu} E_0 \sin\theta.$$

Projection of $\mu_y$ and $\mu_z$ onto the laboratory frame $X, Y, Z$, gives the following:

**1.**

$$\mu_Z = \mu_z \cos\theta + \mu_y \cos\left(\theta + \frac{\pi}{2}\right) = \alpha_z^{\nu} E_0 \cos^2\theta + \alpha_y^{\nu} E_0 \sin^2\theta,$$

which equals

$$\left(\tfrac{1}{3}\right)\mathrm{Tr}\,\alpha^{\nu} E_0 + \left(\alpha_z^{\nu} - \alpha_x^{\nu}\right)E_0\left(\cos^2\theta - \tfrac{1}{3}\right)$$

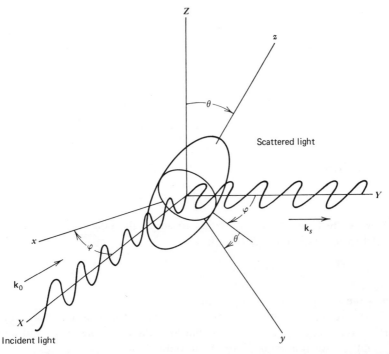

**Figure 1.4.** Molecule-fixed polarizability ellipsoid $x, y, z$ relative to the space-fixed $X, Y, Z$ coordinate system of incident $(\mathbf{k}_0)$ and scattered $(\mathbf{k}_s)$ radiation. The polarizer can be rotated around axis $Y$. The figure shows $\mathbf{k}_0$ polarized in the $Z$-direction, $\mathbf{k}_s$ polarized in the $X$-direction. [From H. D. Dardy, The Catholic University of America, Washington, D.C., Doctoral Dissertation 1972, University Microfilms, Ann Arbor, MI, No. 72–21, 565.]

with $(\alpha_x^\nu = \alpha_y^\nu)$

$$\text{Tr}\,\alpha^\nu = \alpha_x^\nu + \alpha_y^\nu + \alpha_z^\nu = 2\alpha_x^\nu + \alpha_z^\nu.$$

**2.**

$$\mu_X = \mu_z \cos\left(\frac{\pi}{2} - \theta\right)\cos\left(\frac{\pi}{2} - \varphi\right) + \mu_y \cos\theta\cos\left(\frac{\pi}{2} - \varphi\right)$$

$$= \left(\alpha_z^\nu - \alpha_x^\nu\right)E_0\cos\theta\sin\theta\sin\varphi.$$

**3.**

$$\mu_Y = \left(\alpha_z^\nu - \alpha_x^\nu\right)E_0\cos\theta\sin\theta\cos\varphi.$$

Of course, the induced dipole moments are functions of $\omega$.

Let us first turn the polarizer into a position that admits $\mathbf{k}_s$ to propagate only in the $Z$-direction $(\mathbf{k}_s \| \mathbf{k}_0)$ and observe the total intensity scattered into space direction $Y$. Since (in a liquid) the molecular $x, y, z$ system assumes all possible orientations with respect to the laboratory $X, Y, Z$ frame with equal probability, we average the scattered intensity isotropically over $\theta, \varphi$ $(E_0 = 1)$ as follows:

$$\langle I_Z^\nu \rangle_{\theta, \varphi} = \left( \frac{1}{4\pi} \right) \int_0^{2\pi} \int_0^\pi d\theta \, d\varphi \sin\theta \left\{ \frac{1}{3} \operatorname{Tr} \alpha^\nu + (\alpha_z^\nu - \alpha_x^\nu) \left( \cos^2\theta - \frac{1}{3} \right) \right\}^2 .$$

This is the $Z$-$Z$ polarized component. Integration yields

$$\langle I_Z^\nu \rangle = \left( \tfrac{4}{45} \right) (\alpha_z^\nu - \alpha_x^\nu)^2 + \tfrac{1}{9} (\operatorname{Tr} \alpha^\nu)^2 \equiv \langle I_{VV}^\nu \rangle.$$

Now we turn the polarizer by $90°$ to admit the polarization of $\mathbf{k}_s$ that is parallel to space direction $X$ $(\mathbf{k}_s \perp \mathbf{k}_0)$:

$$\langle I_X^\nu \rangle = \left( \tfrac{1}{15} \right) (\alpha_z^\nu - \alpha_x^\nu)^2 \equiv \langle I_{VH}^\nu \rangle.$$

This is the $Z$-$X$ polarized component.

Voilà. We merely subtract from the $Z$-$Z$ polarized band contour $\langle I_{VV}^\nu(\omega) \rangle$ (vertical-vertical, or $\|$) an amount equal to $\frac{4}{3}$ the $Z$-$X$ polarized band contour $\langle I_{VH}^\nu(\omega) \rangle$ (vertical-horizontal, or $\perp$) to obtain

$$\langle I_{VV}^\nu(\omega) \rangle - \tfrac{4}{3} \langle I_{VH}^\nu(\omega) \rangle = \left( \tfrac{1}{9} \right) (\operatorname{Tr} \alpha^\nu)^2$$

$$\equiv \langle I_{iso}^\nu(\omega) \rangle, \tag{1.46}$$

a scattered intensity component which is obviously independent of the molecular orientation ("isotropic").

The VV and VH correlation functions are (see Eq. 1.45)

$$\int_{-\infty}^\infty d\omega \{ I_{VV}^\nu(\omega) - \tfrac{4}{3} I_{VH}^\nu(\omega) \} \exp\{ i(\omega_{mn}^0 - \omega)t \} = \langle m^\nu(t) m^\nu(0) \rangle$$

$$\equiv G_{iso}^\nu(t), \tag{1.47}$$

$$\int_{-\infty}^\infty d\omega \, I_{VH}^\nu(\omega) \exp\{ i(\omega_{mn}^0 - \omega)t \} = \langle \beta_{ij}^\nu(t) \beta_{ij}^\nu(0) \rangle \langle m^\nu(t) m^\nu(0) \rangle$$

$$\equiv G_{VH}^\nu(t), \tag{1.48}$$

where we recall that $\beta_{ij}$ is the traceless, anisotropic part of the polarizability tensor (with respect to normal coordinate $Q$). (We have left out the uninteresting constant factors.) Finally,

$$G_{aniso}^\nu(t) = \langle \beta_{ij}^\nu(t) \beta_{ij}^\nu(0) \rangle = \frac{G_{VH}^\nu(t)}{G_{iso}^\nu(t)} \tag{1.49}$$

and

$$I_{VH}^v(\omega) = \int_{-\infty}^{\infty} d\omega'' \, I_{aniso}^v(\omega'') I_{iso}^v(\omega - \omega'').$$    (1.50)

This is the outcome of the VV–VH experiment. It yields the pure vibrational $G_{iso}^v(t)$ and pure orientational $G_{aniso}^v(t)$ correlation functions.[18] Such a feat is impossible for infrared spectra since a first-rank tensor is not separable into a sum of an isotropic and an angular part.

A correct result of this separation procedure relies on the validity of the basic assumptions enumerated several times in this and previous sections— particularly on the validity of Eq. 1.38. Furthermore, the vibration-rotation contour must evidently belong to a tensor with a nonvanishing trace (see Eq. 1.45). This eliminates (unfortunately) all nontotally symmetric vibrational species under the molecular point group of the particular system.[19] For example, the procedure is inapplicable to doubly degenerate vibrational modes of symmetric top molecules. These modes are depolarized (no Z-Z component).

The experimental requirements for a clean separation—which we do not discuss in detail—are severe. If the Raman bandwidths of $I_{VV}$ and $I_{VH}$ are comparable, the method may be inaccurate since this may indicate that statistical interference between orientational and vibrational correlation functions is significant owing to the near-equal weight of both relaxation processes. Evidently, we should prefer systems where one or the other relaxation process predominates—depending on our interest in orientational or in vibrational relaxation phenomena. Very strongly polarized band contours are conveniently evaluated for their pure vibrational relaxation processes. The $C \equiv N$ stretching fundamental of acetonitrile, $CH_3CN$, is a good example.[20] Weakly polarized band profiles lend themselves better to a determination of rotational relaxation processes, particularly if $I_{aniso}$ is broader than $I_{iso}$. The symmetric C–Cl deformation fundamental of chloroform, $CHCl_3$, is a good example here.[21]

There are systems where the simple analysis leads to nonsensical results since strong coupling between orientational and vibrational coordinates causes Eq. 1.48 to fail. The Raman band profile of the O–D stretching mode of solute HDO in $H_2O$ is an example.[22] The experimental bandwidths (at half-peak height) follow the inequalities $\Delta\nu_{VV} > \Delta\nu_{VH}$ and $\Delta\nu_{iso} > \Delta\nu_{VH}$. Application of the VV–VH experiment therefore leads to the result that the convolution of two band contours (see Eq. 1.50) predicts an overall band profile that is narrowed. This indicates that the individual contributory relaxation processes in this system are not statistically independent but are strongly correlated.

To make this clearer, we select the Raman autocorrelation from Eq. 1.44, suppressing sub- and superscripts and uninteresting constants. Then, according to Eq. 1.42,

$$G_{ij}^v(\text{vib-rot}) = \langle \alpha_{ij}^v(t) \alpha_{ij}^v(0) \rangle = \langle [\bar{\alpha}(t) + \beta(t)][\bar{\alpha}(0) + \beta(0)] \rangle.$$

In terms of coupled vibration-rotation states $|vr) \neq |v)|r)$, this gives

$$G(\text{vib-rot}) = G_{vr}(t)$$

$$= \text{Tr}\,\rho_{vr}\{\bar{\alpha}(t)\bar{\alpha}(0) + \beta(t)\beta(0) + \beta(t)\bar{\alpha}(0) + \bar{\alpha}(t)\beta(0)\}$$

$$= \text{Tr}\,\rho_{vr}\{\bar{\alpha}(t)\bar{\alpha}(0) + \beta(t)\beta(0) + 2\beta(t)\bar{\alpha}(0)\}.$$

[The last equality follows from Eq. A.8 of Appendix A and from the time-shift invariance of (classical) correlation functions.] Since we now assume intense statistical correlation between vibrational and rotational relaxation, it follows that $\rho_{vr} \neq \rho_v \rho_r$. Consequently, the VV-VH experiment can no longer give a clean separation into the vibrational-rotational and pure vibrational correlation functions (see Eq. 1.47 and 1.48)

$$\text{Tr}\,\rho_r[\beta(t)\beta(0)]\text{Tr}\,\rho_v[\bar{\alpha}(t)\bar{\alpha}(0)]$$

and

$$\text{Tr}\,\rho_v[\bar{\alpha}(t)\bar{\alpha}(0)].$$

First, there is the interference term

$$\text{Tr}\,\rho_{vr}[\beta(t)\bar{\alpha}(0)].$$

Second, $\langle\beta(t)\beta(0)\rangle$ and $\langle\bar{\alpha}(t)\bar{\alpha}(0)\rangle$ are averaged over coupled vibration-rotation states. In other words, the respective Fourier transforms of $I_{VV}^v - (\frac{4}{3})I_{VH}^v$ and of $I_{VH}^v$ represent two vibrational-orientational Raman correlation functions with different admixtures of *both* angle-independent ("translatory") and angle-dependent ("orientational") components.

With respect to the Rayleigh process, we deal only with the VH intensity in the range of several $\text{cm}^{-1}$ off the laser line to about $300\ \text{cm}^{-1}$ ("depolarized Rayleigh" or "depolarized Raman wing" spectrum). We are concerned with Rayleigh relaxation data that relate to results from Raman and infrared relaxation. Hence we omit Brillouin scattering[23] and shear wave scattering.[24] From Eq. 1.44 we then have

$$\int_{-\infty}^{\infty} d\omega \exp\{i(\omega_L - \omega)t\}I_{VH}^0(\omega) = \left\langle \alpha_{ij}^{0A}(t)\alpha_{ij}^{0A}(0) \right\rangle + \sum_{A \neq B} \left\langle \alpha_{ij}^{0A}(t)\alpha_{ij}^{0B}(0) \right\rangle,$$

$$(1.51)$$

where $\alpha_{ij}^0(t)$ is the permanent polarizability (see Eq. 1.37) and $\omega_L$ the laser frequency. Cross correlation between the dynamic variables on adjacent molecules $A$ and $B$ is important here for the same reasons as in the far infrared (see Eq. 1.41).

At this point in our discussions it is useful to include a section on general transition probabilities, detailed balance, spectral density, microscopic reversibility, and the golden rule.

### 3.3.  Transition Rates and Probabilities, Spectral Density, Microscopic Reversibility, and the Golden Rule

The principles of fluctuation spectroscopy can also be described in terms of transition rates between system states. The transitions are induced by random forces or potentials in the system.

The formulation can be developed to any desired order of time-dependent perturbation. It therefore includes the more complicated radiative processes of Raman and Rayleigh scattering (spontaneous scattering) and picosecond pulsing (induced scattering).

We first consider the probability per unit time for an active (or "probe") molecule to make a transition between two vibrational-rotational-translational levels $s, s'$ under the influence of a periodic perturbation $H_1(t)$. According to well-known quantum-mechanical principles, this probability is given by the squared modulus $|a_{ss'}(t)|^2$ of the coefficients of the state vector $s(t)$,

$$s(t) = \sum_{s'} a_{s'}(t)|s'),\qquad(1.52)$$

where $|s)$ is a set of orthonormal eigenvectors of the total system Hamiltonian $H = H_0 + H_1(t)$.[25] Transforming $s(t)$ and $H_1(t)$ into the interaction representation (Eqs. 1.30 and 1.31)

$$s^I(t) = \exp\left(\frac{iH_0 t}{\hbar}\right)s(t),$$

$$H_1^I(t) = \exp\left(\frac{iH_0 t}{\hbar}\right)H_1(t)\exp\left(-\frac{iH_0 t}{\hbar}\right),$$

replaces in Schrödinger's equation the total Hamiltonian $H$ by the perturbation Hamiltonian $H_1(t)$,[26]

$$i\hbar\left(\frac{d}{dt}\right)s^I(t) = H_1^I(t)s^I(t).\qquad(1.53)$$

We insert Eq. 1.52 into Eq. 1.53, multiply from the left by $|s)$, and sum over all $|s')$. Employing the known solutions of the zero-order problem,

$$H_0|s) = E_s^0|s),\qquad H_0|s') = E_{s'}^0|s'),$$

we obtain

$$ i\hbar \sum_{s'} \left( \frac{d}{dt} \right) a_{s'}^{I}(t)(s|s') = \sum_{s'} (s| H_1^I(t)|s') a_{s'}^I(t), $$

$$ i\hbar \left( \frac{d}{dt} \right) a_s^I(t) = \sum_{s'} (s| H_1(t)|s') \exp\left[ \frac{i(E_s^0 - E_{s'}^0)t}{\hbar} \right] a_{s'}^I(t). \quad (1.54) $$

Now we develop $a_s^I(t)$ into a perturbation series of zero, first, second,..., order,

$$ a_s = a_s^{(0)} + \lambda a_s^{(1)} + \lambda^2 a_s^{(2)} + \cdots \quad (1.55) $$

This transforms Eq. 1.54 into the recursion relation

$$ i\hbar \left( \frac{d}{dt} \right) a_s^{I,(n+1)}(t) = \sum_{s'} (s| H_1(t)|s') \exp\left( i\omega_{ss'}^0 t \right) a_{s'}^{I,(n)}(t), \quad (1.56) $$

since $H_1(t) = \lambda H_1(t)$ (linear response). We assume that the unperturbed system is in one definite initial state $a_{s'}(t=0) = 1$. For $H_1(t)$ we take

$$ H_1(t) = -(\tfrac{1}{2})\{A^\dagger \exp(i\omega t) + A \exp(-i\omega t)\}, \quad (1.57) $$

where $A^\dagger$ designates the adjoint of $A$ (Note 4). Inserting Eq. 1.57 into Eq. 1.56, setting $a_{s'}^{I,(0)}(t) = 1$, and integrating, we obtain

$$ a_{ss'}^{I,(1)}(t) = -(2i\hbar)^{-1} \int_0^t dt' \{(s| A^\dagger|s') \exp[ i( \omega_{ss'}^0 + \omega)t'] $$

$$ + (s| A|s') \exp[ i( \omega_{ss'}^0 - \omega)t']\} $$

$$ = (2\hbar)^{-1} \left[ \frac{(s| A^\dagger|s')}{(\omega_{ss'}^0 + \omega)} \right] \{\exp[ i( \omega_{ss'}^0 + \omega)t] - 1\} $$

$$ + (2\hbar)^{-1} \left[ \frac{(s| A|s')}{(\omega_{ss'}^0 - \omega)} \right] \{\exp[ i( \omega_{ss'}^0 - \omega)t] - 1\}. \quad (1.58) $$

We have attached the additional subscript $s'$ to coefficient $a_s$ since we do not yet wish to sum over all states $s'$. Keeping only the near-resonant term of Eq. 1.58, we obtain

$$ a_{ss'}^{I,(1)}(t) a_{ss'}^{I,(1)*}(t) = \left| a_{ss'}^{(1)}(t) \right|^2 $$

$$ = (4\hbar^2)^{-1} |(s| A|s')|^2 2\{1 - \cos[ (\omega_{ss'}^0 - \omega)t]\}(\omega_{ss'}^0 - \omega)^{-2}. $$

$$ (1.59) $$

Dividing the trigonometric factor in Eq. 1.59 by time $t$, we note that its long-time limit,

$$\lim_{t \to \infty} \left( \{1 - \cos[(\omega_{ss'}^0 - \omega)t]\} \left[ (\omega_{ss'}^0 - \omega)^2 t \right]^{-1} \right) = \pi \delta(\omega_{ss'}^0 - \omega), \quad (1.60)$$

represents one of the definitions of the delta function.[27] Therefore the probability per unit time that the periodic perturbation $H_1(t)$ induces a transition $s' \to s$ is

$$\frac{|a_{ss'}^{(1)}(t)|^2}{t} \equiv w_{s' \to s}(\omega)$$

$$= \left( \frac{\pi}{2\hbar^2} \right) |(s|A|s')|^2 \delta(\omega_{ss'}^0 - \omega)$$

$$= \left( \frac{\pi}{2\hbar} \right) |(s'|A^\dagger|s)|^2 \delta(E_s^0 - E_{s'}^0 - \hbar\omega). \quad (1.61)$$

This is the golden rule. [We have inserted the proper constant factors according to the definition of $H_1(t)$.] Note that Eq. 1.61 is developed to first-order perturbation and that it is valid for times after initial "transients" have decayed (Eq. 1.60).

The reverse quantity $w_{s \to s'}(-\omega)$ is

$$w_{s \to s'}(-\omega) = w_{s' \to s}(\omega), \quad (1.62)$$

because $|(s|A|s')|^2 = |(s'|A^\dagger|s)|^2$ and $\delta(\omega_{ss'}^0 - \omega) = \delta(\omega_{s's}^0 + \omega)$. This is an example of the principle of microscopic reversibility.

We now write the probability per unit time that the active molecule has transferred a quantum $\hbar\omega$ to the lattice regardless of initial and final states. It is

$$W(\omega) = \left( \frac{\pi}{2\hbar} \right) \sum_{ss'} \rho_{s'} |(s'|A^\dagger|s)|^2 \delta(E_s^0 - E_{s'}^0 - \hbar\omega)$$

with

$$\rho_s = \rho_{s'} \exp(-\beta\omega_{ss'}^0), \qquad \hbar\omega_s^0 > \hbar\omega_{s'}^0. \quad (1.63)$$

The inverse probability $W(-\omega)$ (the active molecule has taken a quantum $\hbar\omega$ from the lattice) is (see Eq. 1.61)

$$W(-\omega) = \left( \frac{\pi}{2\hbar^2} \right) \sum_{ss'} \rho_s |(s|A|s')|^2 \delta(\omega_{ss'}^0 - \omega)$$

$$= \left( \frac{\pi}{2\hbar^2} \right) \sum_{ss'} \rho_{s'} \exp(-\beta\omega) |(s|A|s')|^2 \delta(\omega_{ss'}^0 - \omega). \quad (1.64)$$

Consequently,

$$W(-\omega) = \exp\left(-\frac{\hbar\omega}{kT}\right)W(\omega). \tag{1.65}$$

Equation 1.65 is an example of the principle of detailed balance (see also Appendix A): In equilibrium the flux of upward transitions must balance the flux of downward transitions.[2c]

Now compare Eq. 1.63 with Eq. 1.27 (recalling the Fourier development of the delta function—see Eq. 1.24): Except for uninteresting constants Eq. 1.63 and 1.27 are identical. Hence the spectral density $\mathscr{I}(\omega)$ [or $\mathscr{I}(-\omega)$] equals the probability $W(\omega)$ [or $W(-\omega)$] per unit time that the active molecule has transferred (or taken) a quantum $\hbar\omega$ to (or from) the lattice regardless of initial and final states.

We can recast relation 1.63 into a more useful form by applying the methodology described in Section 1.4. It gives ($\hbar = 1$)

$$W(\omega) = \left(\frac{1}{4}\right)\int_{-\infty}^{\infty} dt \exp[i(\omega_{ss'} - \omega)t] \sum_{ss'} \rho_{s'}(s'|A^{\dagger}(t)|s)(s|A(0)|s').$$

$$\tag{1.66a}$$

We see that (1) the sum in Eq. 1.66a represents the correlation function of a stochastic perturbation Hamiltonian $A(t)$ under coupled molecule-lattice energy states $|s\rangle$, (2) the integral represents the Fourier transform of this correlation function taken with respect to the resonance frequency $\omega_{ss'}$,

$$W(\omega) = \left(\frac{1}{4}\right)\int_{-\infty}^{\infty} dt \exp[i(\omega_{ss'} - \omega)t]\langle A^{\dagger}(t)A(0)\rangle. \tag{1.66b}$$

The time-dependent perturbation operator of the system $A(t)$ is prescribed by the nature of the molecular environment and its dynamical behavior. In other words, Eq. 1.66 is related to Eq. 1.29 in physical content and meaning. Equation 1.66 is useful for checking a theoretical model of a relaxation process against experimental spectral data.

It is instructive to compare Eq. 1.66 with the corresponding formulation for nuclear magnetic relaxation (for instance, with Eq. 23 on page 273 of Ref. 6). Whereas the spectral band profile measurements lead to a transition probability (equals spectral density) which is frequency dependent over a relatively wide and continuous range, the nuclear magnetic relaxation data (performed at selected frequencies) do not yield the correlation function but only its time integral—the correlation time

$$\tau = \int_{0}^{\infty} dt\langle A^{\dagger}(t)A(0)\rangle. \tag{1.67}$$

This is easy to see. Since $\tau$ is real, we form (see Eqs. A.18 and A.9$b$ in Appendix A and assume that $A(t)$ is Hermitian)

$$\tau = \int_0^\infty dt \, \langle A(t)A(0) \rangle = \left(\frac{1}{2}\right) \mathrm{Re} \int_{-\infty}^\infty dt \, \langle A(t)A(0) \rangle$$

$$= \left(\frac{1}{2}\right) \int_{-\infty}^\infty \int_{-\infty}^\infty d\omega \, dt \, \mathscr{I}(\omega) \exp(-i\omega t)[1 + \exp(-\beta\omega)]$$

$$= \left(\frac{1}{2}\right) \int_{-\infty}^\infty dt \exp(-i\omega t) \int_{-\infty}^\infty d\omega \, \mathscr{I}(\omega)[1 + \exp(-\beta\omega)]$$

$$= \pi \int_{-\infty}^\infty d\omega \, \mathscr{I}(\omega)[1 + \exp(-\beta\omega)] \, \delta\omega$$

$$= 2\pi \mathscr{I}(\omega = 0)$$

$$= 2\pi \mathscr{I}(\omega_{ss'}). \tag{1.68}$$

We state the calculation in words as follows: The correlation time is proportional to the spectral density at the band center.

With the help of the golden rule we can establish the transition rate expression for a Raman or Rayleigh scattering process as easily as we did for the infrared absorption process. We recall that in the scattering process an electronic dipole moment is first induced by interaction of an incident photon $\hbar\omega_{in}$ with the electron of the molecular scatterer which is in initial state $I$ (coupled electronic-vibration-rotation-translational) of energy $E_I$. This interaction leads to a manifold of "virtual," nonresonant states $V$ of the system of energy $E_V$. (By "nonresonant" is meant that $\hbar\omega_{in}$ is far from any resonance frequency $E_i - E_I$ of the scatterer.) These virtual states then decay, under emission of a photon (the scattered photon) of energy $\hbar\omega_{sc}$, into their final coupled state $F$ of energy $E_F$ of the scatterer. The total process also takes place in the reversed order: State $I$ is raised to state $V$ by emitting the scattered photon $\hbar\omega_{sc}$, followed by step $V \rightarrow F$ under absorption of the incident photon $\hbar\omega_{in}$. These two possibilities are related to the products of photon creation $a^\dagger$ and annihilation operators $a$, $a_{sc}^\dagger a_{in}$ and $a_{in} a_{sc}^\dagger$, respectively (see p. 283 of Ref. 15).

The energy balance of the process is

$$E_F + \hbar\omega_{sc} = E_I + \hbar\omega_{in}.$$

If $\hbar\omega_{sc} < \hbar\omega_{in}$, the system has taken up energy $E_F - E_I$ during the scattering process (Stokes process). If $\hbar\omega_{sc} > \hbar\omega_{in}$, the system has given off the energy $E_F - E_I$ (anti-Stokes).

The transition matrix element for the total scattering process is (second-order perturbation theory[28,29])

$$(F|I) = C \sum_V \left\{ \frac{(F|\boldsymbol{\varepsilon}^{(\text{in})} \cdot \mathbf{p}|V)(V|\boldsymbol{\varepsilon}^{(\text{sc})} \cdot \mathbf{p}|I)}{E_I + \hbar\omega_{\text{in}} - E_V} \right.$$

$$\left. + \frac{(F|\boldsymbol{\varepsilon}^{(\text{sc})} \cdot \mathbf{p}|V)(V|\boldsymbol{\varepsilon}^{(\text{in})} \cdot \mathbf{p}|I)}{E_I - \hbar\omega_{\text{sc}} - E_V} + \delta_{FI}(\boldsymbol{\varepsilon}^{(\text{in})} \cdot \boldsymbol{\varepsilon}^{(\text{sc})}) \right\} \qquad (1.69)$$

with $E_I \neq E_V$. Here $C$ is a constant, $\boldsymbol{\varepsilon}$ is a unit vector along the polarization direction of the light, and $\mathbf{p}$ is the total electronic momentum. The energy terms in the denominator, $(E_I + \hbar\omega_{\text{in}}) - E_V$ and $(E_I - \hbar\omega_{\text{sc}}) - E_V$, correspond to the energy difference between initial and virtual states for process $a_{\text{sc}}^\dagger a_{\text{in}}$ and $a_{\text{in}} a_{\text{sc}}^\dagger$.

We decompose the initial, virtual, and final states into product wave functions of nuclear and electronic coordinates (adiabatic or Born-Oppenheimer approximation). We are interested in transitions where initial and final electronic states of the scatterer are the electronic ground state 0 since then the difference between incident and scattered photon energy shows up as phonon. This subjects the scattering process to the condition

$$E_{\text{min}} \gg \hbar\omega_{\text{in}},$$

where $E_{\text{min}}$ is the electronic excitation energy of the lowest optical absorption transition $E_i - E_0$.

We further assume

$$\omega_{\text{in}} \gg |\omega_{\text{in}} - \omega_{\text{sc}}| = \omega_{vv'} \qquad (v, v' = \text{vibrational}).$$

This condition is readily satisfied by the $Ar^+$ laser light source and the spectral region of vibrational fundamentals.

We now rewrite Eq. 1.69 in the form of a rate expression (see Eq. 1.61). For the process corresponding to product $a_{\text{sc}}^\dagger a_{\text{in}}$ we obtain[30]

$$w_{I \rightarrow F} = \sum_V \left| \frac{(\psi_F|H_1|\psi_V)(\psi_V|H_1|\psi_I)}{E_I - E_V} \right|^2 \delta(E_F - E_I), \qquad (1.70)$$

where $\psi_V$ is the virtual (intermediate) state, $\psi_I$ and $\psi_F$ are the initial and final states (electronic-vibration-rotation-radiation). $H_1(t)$ is the perturbation Hamiltonian.

Using the adiabatic approximation, we set

$$\psi_I = |\psi_{\text{adiabatic}})_I |\psi_{\text{rad}})_I$$

$$= \varphi_\gamma(\mathbf{r}, \mathbf{R}) \chi_{\gamma n}(\mathbf{R}) |n_{k_1} \ldots, n_{k_2} \ldots).$$

The subscript $I$ denotes an initial product state which is composed of the electronic eigenstate $\varphi_\gamma(\mathbf{r}, \mathbf{R})$ as function of electronic $\mathbf{r}$ and nuclear $\mathbf{R}$ coordinates, of the nuclear state $\chi_{\gamma n}(\mathbf{R})$, and of the *rad*iation field with incident $n_{k_1}$ and scattered $n_{k_2}$ photons. Greek and italic subscripts characterize electronic and nuclear levels, respectively.

For final state $\psi_F$ we accordingly find

$$\psi_F = \varphi_\gamma(\mathbf{r}, \mathbf{R}) \chi_{\gamma m}(\mathbf{R}) | n_{k_1} - 1 \ldots, n_{k_2} - 1 \ldots).$$

For the virtual or intermediate state we get

$$\psi_V = \varphi_\delta(\mathbf{r}, \mathbf{R}) \chi_{\delta l}(\mathbf{R}) | n_{k_1} - 1 \ldots, n_{k_2} \ldots).$$

Now we use the simplifying scattering conditions introduced previously to set the energy terms in the denominator of Eq. 1.70, recalling that the separation between electronic states (in a nonconductor) exceeds the separation of vibrational-rotational levels. We write the initial energy

$$E_I = \hbar\omega_{in} + \Phi^0 + E_n, \qquad (1.71a)$$

where $\Phi^0$, $E_n$ denote the electronic ground state and a nuclear energy level.

The energy of the intermediate state is

$$E_V = \Phi_\delta^0 + E_{\delta l}. \qquad (1.71b)$$

(Note that the incident photon, $\hbar\omega_{in}$, is absorbed.)

The final energy level is

$$E_F = \hbar\omega_{sc} + \Phi^0 + E_m. \qquad (1.71c)$$

The energy difference $E_I - E_V$ is therefore

$$E_I - E_V = \hbar\omega_{in} + \Phi^0 + E_n - \Phi_\delta^0 - E_{\delta l}$$

$$\approx \hbar\omega_{in} + \Phi^0 - \Phi_\delta^0$$

$$= \hbar\omega_{in} + \hbar\omega_{0\delta}, \qquad (1.71d)$$

where $\hbar\omega_{0\delta}$ is the (purely) electronic excitation energy. The delta function of Eq. 1.70 equals

$$\delta(\hbar\omega_{sc} - \hbar\omega_{in} + E_m - E_n), \qquad (1.72)$$

where we have again assumed that the system returns to its original ground electronic state after the scattering event (see Eqs. 1.71$a, c$). Therefore $\hbar(\omega_{sc} - \omega_{in})$ is equal to the vibrational energy difference.

For the perturbation operator, we use

$$H_e = -\left(\frac{e}{m_e}\right)\mathbf{p}\cdot\mathbf{A},$$

where $\mathbf{p}$ is the electronic momentum operator and $\mathbf{A}$ is the vector potential of the radiation field (see Note 5). We express $\mathbf{p}$ by

$$\mathbf{p} = m_e\left(\frac{d}{dt}\right)\mathbf{r},$$

$$\left(\frac{e}{m_e}\right)\mathbf{p} = \left(\frac{d}{dt}\right)e\mathbf{r} = \left(\frac{d}{dt}\right)\boldsymbol{\mu},$$

where $\boldsymbol{\mu}$ is the electronic dipole moment, $e$ the electronic charge, $m_e$ the electronic mass, and $\mathbf{r}$ the displacement vector. We write the equation of motion for $\boldsymbol{\mu}$ (see, for instance, Eq. 1.31),

$$\left(\frac{d}{dt}\right)\boldsymbol{\mu} = \left(\frac{e}{m_e}\right)\mathbf{p} = \left(\frac{i}{\hbar}\right)[H,\boldsymbol{\mu}], \tag{1.73}$$

in terms of the total Hamiltonian (Heisenberg picture),[26] insert this relation into Eq. 1.69, and operate with $H$ on the product electronic-vibrational-rotational system eigenstates. This gives (writing only the numerator of the term and setting $\hbar = 1$)

$$\sum_{\delta l}\boldsymbol{\varepsilon}^{(sc)}\big(\chi_{\gamma m}|(\varphi_\gamma|i[H,\boldsymbol{\mu}]|\varphi_\delta)|\chi_{\delta l}\big)$$

$$\times \boldsymbol{\varepsilon}^{(in)}\big(\chi_{\delta l}|(\varphi_\delta|i[H,\boldsymbol{\mu}]|\varphi_\gamma)|\chi_{\gamma n}\big)$$

$$= -\sum_{\delta l}(E_{\gamma m} - E_{\delta l})(E_{\delta l} - E_{\gamma n})\big(\chi_{\gamma m}|(\varphi_\gamma|\boldsymbol{\varepsilon}^{(sc)}\cdot\boldsymbol{\mu}|\varphi_\delta)|\chi_{\delta l}\big)$$

$$\times \big(\chi_{\delta l}|(\varphi_\delta|\boldsymbol{\varepsilon}^{(in)}\cdot\boldsymbol{\mu}|\varphi_\gamma)|\chi_{\gamma n}\big), \tag{1.74}$$

where $H|\chi_{\alpha a}) = E_{\alpha a}|\chi_{\alpha a})$, and so on.

We define a dipole (transition) moment operator

$$\mathcal{M}^{\gamma\delta}_{(\lambda)} = \boldsymbol{\varepsilon}^{(\lambda)}\cdot\mathcal{M}^{\gamma\delta}(\mathbf{R})$$

$$= \big\langle\big(\varphi_\gamma|\boldsymbol{\varepsilon}^{(\lambda)}\cdot\boldsymbol{\mu}(\mathbf{r},\mathbf{R})|\varphi_\delta\big)\big\rangle_r. \tag{1.75}$$

$\mathcal{M}$ represents an electronic dipole moment that depends on the nuclear coordinate $R$ and is to be averaged over electronic coordinate $r$. Then Eq. 1.74

reads after summing over $|\chi_{\delta l}\rangle\langle\chi_{\delta l}|$,

$$-\left(\omega_{\gamma m,\,\delta l}\right)\left(\omega_{\delta l,\,\gamma n}\right)\left(\chi_{\gamma m}|\,\boldsymbol{\varepsilon}^{(sc)}\cdot\boldsymbol{\mathscr{M}}^{\gamma\delta}(\mathbf{R})|\chi_{\delta l}\right)$$

$$\times\left(\chi_{\delta l}|\,\boldsymbol{\varepsilon}^{(in)}\cdot\boldsymbol{\mathscr{M}}^{\delta\gamma}(\mathbf{R})|\chi_{\gamma_n}\right).$$

Finally we set $\gamma = 0$ since we consider the ground electronic state only. The complete rate expression (Eq. 1.70) is then

$$\left|\sum_\delta \omega_{0\delta}^2 (\hbar\omega_{in} - \hbar\omega_{\delta 0})^{-1}\left(\chi_{0m}|\,\boldsymbol{\varepsilon}^{(sc)}\cdot\boldsymbol{\mathscr{M}}^{0\delta}(\mathbf{R})\boldsymbol{\varepsilon}^{(in)}\cdot\boldsymbol{\mathscr{M}}^{\delta 0}(\mathbf{R})|\chi_{0n}\right)\right|^2 \rho(E_{f_v})$$

$$= \left|\left(\chi_{0m}|\,\boldsymbol{\varepsilon}^{(sc)}\cdot\boldsymbol{\alpha}(\mathbf{R})\cdot\boldsymbol{\varepsilon}^{(in)}|\chi_{0n}\right)\right|^2 \rho(E_{f_v}), \tag{1.76}$$

with the polarizability $\boldsymbol{\alpha}$

$$\boldsymbol{\alpha}(\mathbf{R}) \equiv \sum_\delta \frac{\omega_{0\delta}^2}{\hbar\omega_{in} - \hbar\omega_{\delta 0}}\boldsymbol{\mathscr{M}}^{0\delta}(\mathbf{R})\boldsymbol{\mathscr{M}}^{\delta 0}(\mathbf{R}).$$

[We have substituted the delta function (Eq. 1.72) by an energy-conserving density of final vibrational states $\rho(E_{f_v})$.]

The foregoing description is not exhaustive or rigorous. For instance, we have not quantized the radiation field (see Ref. 30, pp. 282–287). Nevertheless, this outline shows the complexity of the optical scattering phenomenon that is of particular interest to our discussion.

Since the radiation field (or the purely harmonic factor) need no longer be explicitly considered (see Eq. 1.57), from now on we define any perturbation operator as a pure system operator.

### 3.4. Orientational Correlation Functions from Raman Vibration-Rotation Spectra

**3.4.1. Raman Scattering Cross Section.** We write the expression for the Raman scattering cross section (which corresponds to Eq. 1.63 for the absorption regime) as[31]

$$\lambda^4\left(\frac{d\sigma}{d\Omega}\right) = \sum_{jf}\rho_j\left|\left(j|\,\boldsymbol{\varepsilon}^{(in)}\cdot\boldsymbol{\beta}^v\cdot\boldsymbol{\varepsilon}^{(sc)}|f\right)\right|^2 \delta(\omega_{sc} - \omega_{in} + \omega_v + \omega_f - \omega_j). \tag{1.77}$$

$d\sigma/d\Omega$ is the differential cross section for light scattering per molecule per solid angle $d\Omega$ for a transition between vibrational levels $v = v$ and $v = v'$ ($v'$ is usually $= v + 1$) of energy difference $\hbar\omega_v$. $2\pi\lambda$ ($\lambda = \lambda/2\pi$) is the wavelength of the scattered light. $\beta^v$ is the off-diagonal (anisotropic) component of the derivative of the polarizability tensor $\alpha$ (Eq. 1.76) with respect to the normal coordinate of the vibrational mode (see Eqs. 1.37 and 1.42). $\omega_{in}$ and $\omega_{sc}$ are the (angular) frequencies of the incident and scattered photons with polarization direction $\varepsilon^{(in)}$ and $\varepsilon^{(sc)}$, respectively. $\hbar\omega_j$ and $\hbar\omega_f$ are the initial and final energies of the rotation-translational states of the system. The delta function expresses the energy conservation during the scattering event,

$$\omega_{in} + \omega_j = \omega_{sc} + \omega_v + \omega_f.$$

Comparing Eq. 1.77 with Eq. 1.61 shows that we have developed Eq. 1.77 one step further since we have split the system vector $|s\rangle$ into (statistically uncorrelated) products $|s\rangle = |v\rangle|j\rangle$ of vibrational $v$ and rotational-translational $j$ states (sometimes called "internal" and "external" states). The formulation of Eq. 1.77 therefore corresponds to Eq. 1.49: We assume that the vibrational correlation function $G_{iso}(t)$ is either known or near-constant during the evolution of the rotational relaxation process.

We assume (Eq. 1.10) that the system is in equilibrium before a scattering event,

$$\rho_n = \exp\left(-\frac{\hbar\omega_n}{kT}\right) \Big/ \sum_n \exp\left(-\frac{\hbar\omega_n}{kT}\right).$$

The total Hamiltonian is $H(t) = H_0 + H_1(t)$. $H_0 \equiv H_v$ operates solely on the internal levels $|v\rangle$. Therefore

$$H|v\rangle)|j\rangle = E_{vj}|v\rangle)|j\rangle; \qquad H_v|v\rangle = E_v|v\rangle.$$

We now follow the procedure applied for Eqs. 1.23–1.25 and get

$$\lambda^4\left(\frac{d\sigma}{d\Omega}\right) = \left(\frac{1}{2\pi}\right)\sum_{jf}\rho_j\left(j|\varepsilon^{(in)}\cdot\beta^v\cdot\varepsilon^{(sc)}|f\right)\left(f|\varepsilon^{(in)}\cdot\beta^v\cdot\varepsilon^{(sc)}|j\right)$$

$$\times \int_{-\infty}^{\infty} dt\exp\left(-i\omega_j t\right)\exp\left(i\omega_f t\right)\exp\left[i(\omega_{sc} - \omega_{in} + \omega_v)t\right]$$

$$= \left(\frac{1}{2\pi}\right)\sum_{jf}\rho_j\left(j|\varepsilon^{(in)}\cdot\beta^v(0)\cdot\varepsilon^{(sc)}|f\right)$$

$$\times \left(f|\exp(iHt)\varepsilon^{(in)}\cdot\beta^v(0)\cdot\varepsilon^{(sc)}\exp(-iHt)|j\right)$$

$$\times \int_{-\infty}^{\infty} dt\exp\left[i(\omega_{sc} - \omega_{in} + \omega_v)t\right], \qquad (1.78)$$

with

$$\beta^v(t) = \mathrm{Tr}\,\rho_v(v|\exp(iH_v t)\beta(t)\exp(-iH_v t)|v),$$

$$\omega_v = E_{v'} - E_v; \qquad \hbar = 1.$$

(Note that $\beta^v$ is in the interaction picture under $H_v$.)

Notice the analogy between Eq. 1.78 and the relation for an absorption process (Eqs. 1.27 and 1.29) if we set

$$\omega_{\mathrm{sc}} - \omega_{\mathrm{in}} + \omega_v = -\omega'$$

where $\omega'$ is the frequency displacement from the center of the vibrational-rotational Raman contour. Hence,

$$\lambda^4\left(\frac{d\sigma}{d\Omega}\right) \equiv \mathcal{I}(\omega')$$

$$= \left(\frac{1}{2\pi}\right)\int_{-\infty}^{\infty} dt \exp(-i\omega' t)\left\langle\left[\varepsilon^{(\mathrm{in})}\cdot\beta^v(0)\cdot\varepsilon^{(\mathrm{sc})}\right]\left[\varepsilon^{(\mathrm{in})}\cdot\beta^v(t)\cdot\varepsilon^{(\mathrm{sc})}\right]\right\rangle$$

$$= \int_{-\infty}^{\infty} dt \exp(i\omega' t)\langle\beta^\nu(t)\cdot\beta^\nu(0)\rangle. \tag{1.79}$$

In the last relation we omitted uninteresting constants, presumed the proper averaging over polarization direction $\varepsilon^{(\mathrm{in})}$ and $\varepsilon^{(\mathrm{sc})}$ (as discussed in Section 3.2), and substituted superscript $v$ by $\nu$ to indicate that we consider mainly fundamental transitions. Furthermore we replaced $t$ by $-t$ and applied the time-shift invariance of the correlation function (see Appendix A). (At any rate, such a fine point matters little since we assume classical Raman rotational correlation functions.)

### 3.4.2. Raman Orientational Correlation Functions and Molecular Symmetry.

In this section we derive and describe several frequently encountered purely orientational (classical) Raman correlation functions $\langle\hat{\beta}(0)\cdot\hat{\beta}(t)\rangle$. The symbol indicates normalization to unity for $t = 0$.

Depending on the symmetry species of a vibrational mode we expect different forms of $\langle\hat{\beta}(0)\cdot\hat{\beta}(t)\rangle$ since the symmetry species identifies the axis direction of the vibrational-rotational transition moment in the molecule. We have discussed this aspect in Section 2.1 for infrared vibrational-rotational contours. The situation for the Raman is conceptually identical although it is more complex since we deal with a second-order tensor. The result depends on a combination of several coordinate directions.

The literature describes two useful methods for establishing the Raman orientational correlation functions. (We are only concerned with autocorrelation functions.) One method employs Cartesian, molecule-fixed axes directions to identify the orientational correlation function.[17b] It has the advantage of describing the detailed form of $\langle \hat{\beta}(0) \cdot \hat{\beta}(t) \rangle$ for any desired Raman tensor. However, for several frequently encountered molecular symmetries the correlation functions mix axes directions in a way that renders useful experimental exploitation and theoretical predictions difficult.

In the other approach[18b] the Cartesian tensor elements are written as combinations of spherical coordinates. This naturally leads to orientational correlation functions with the full Wigner rotation matrices as basis set. Their expansion coefficients are related to linear combinations of the transformed Raman tensor elements. This approach therefore describes the reorientational motion of the tensor function through the rotation of the molecular frame with respect to the laboratory-fixed reference coordinate system. This permits comparison of experimental and theoretically modeled orientational correlation functions on the basis of the molecular point group symmetry.

We first discuss the approach in terms of Cartesian coordinates since it is easy to understand.

As illustration we use the totally symmetric (polarized) modes of species $A_1, A', A, A_{1g}$ of a symmetric top of point group $C_{3v}$, $C_{3h}$ and $D_{3h}$, $C_3$, and $D_{6h}$, respectively. The form of the Raman tensor $\alpha$ is[32]

$$\begin{pmatrix} a & & \\ & a & \\ & & b \end{pmatrix},$$

where $a, b, c$ stand for the Cartesian tensor elements $\alpha_{xx}, \alpha_{yy}, \alpha_{zz}$ in the molecular frame. (Only nonvanishing elements are shown.)

The irreducible representation of this tensor for the totally symmetric species enumerated under the indicated molecular point groups is

$$\alpha_{xx} + \alpha_{yy}, \alpha_{zz}.$$

The linear combination $\alpha_{xx} + \alpha_{yy}$ as well as the term $\alpha_{zz}$ are symmetric with respect to all symmetry operations under these groups.

We recall that the transformation properties of a Raman tensor element $\alpha_{pq}$ are identical to those of the product $p \cdot q$ of the translation coordinate components $p$ and $q$.[33] Thus it is easy to write the correlation function $\langle \beta(0) \cdot \beta(t) \rangle$ of the symmetry species of the vibrational mode of interest. We merely express the polarizability tensor elements as products of unit vectors along the transition *dipole* moments (see also Eq. 1.76), using the symbol **m**. We carry the molecular axes directions $p, q = x, y, z$ as superscripts and the two indices of the Raman tensor elements as subscripts $i, j = 1, 2, 3$.

The anisotropic term of this tensor is evidently (see Eq. 1.42)

$$\left(\tfrac{1}{3}\right)\begin{pmatrix} a - b & & \\ & a - b & \\ & & 2(b - a) \end{pmatrix}.$$

Hence the element $\beta_{ij}$ is, according to our prescription, given by

$$\beta_{ij} = \left(\tfrac{1}{3}\right)(a-b)m_i^x m_j^x + \left(\tfrac{1}{3}\right)(a-b)m_i^y m_j^y$$
$$+ \left(\tfrac{2}{3}\right)(b-a)m_i^z m_j^z,$$

where $z$ is assigned to be along the threefold symmetry axis of the symmetric top. Adding and subtracting $\left(\tfrac{1}{3}\right)(a-b)m_i^z m_j^z$ and using the orthogonality condition

$$m_i^x m_j^x + m_i^y m_j^y + m_i^z m_j^z = \delta_{ij},$$

leads to

$$\beta_{ij} = (b-a)\left[m_i^z m_j^z - \tfrac{1}{3}\delta_{ij}\right].$$

Now we form $\beta_{ij}(0)\beta_{ij}(t)$ and sum over all $i, j = 1, 2, 3$:

$$\langle \boldsymbol{\beta}(0)\cdot\boldsymbol{\beta}(t)\rangle = (b-a)^2\left\langle \sum_{i,j=1}^{3} \left[m_i^z(0)m_j^z(0) - \frac{1}{3}\delta_{ij}\right] \right.$$

$$\left. \times \left[m_i^z(t)m_j^z(t) - \frac{1}{3}\delta_{ij}\right]\right\rangle$$

$$= (b-a)^2\left\langle \sum_{i,j=1}^{3}\left[-\frac{1}{3}\delta_{ij}m_i^z(0)m_j^z(0) - \frac{1}{3}\delta_{ij}m_i^z(t)m_j^z(t)\right.\right.$$

$$\left.\left. + \frac{1}{9}\delta_{ij}\delta_{ij} + m_i^z(0)m_j^z(0)m_i^z(t)m_j^z(t)\right]\right\rangle$$

$$= (b-a)^2\left\langle -\frac{1}{3} - \frac{1}{3} + \frac{3}{9} + \sum_{i,j=1}^{3} m_i^z(0)m_j^z(0)m_i^z(t)m_j^z(t)\right\rangle, \quad (1.80)$$

since

$$\sum_i m_i^q(0)m_i^q(0) = 1.$$

We assume classical Raman orientational correlation functions; hence $m_i^z(0)$ and $m_j^z(t)$ commute (see Eq. A.11 of Appendix A). Therefore after normalization we obtain

$$\langle \hat{\boldsymbol{\beta}}(0)\cdot\hat{\boldsymbol{\beta}}(t)\rangle \equiv \hat{G}(t)$$

$$= \left(\frac{1}{2}\right)\left\langle 3\sum_{i,j=1}^{3}\left[m_i^z(0)m_i^z(t)m_j^z(0)m_j^z(t)\right] - 1\right\rangle$$

$$= \left(\frac{1}{2}\right)\left\langle 3[\hat{\mathbf{m}}^z(0)\cdot\hat{\mathbf{m}}^z(t)]^2 - 1\right\rangle. \quad (1.81a)$$

We recast this result into its well-known form

$$\hat{G}(t) = \left\langle P_2\left[\hat{\mathbf{m}}^z(0)\cdot\hat{\mathbf{m}}^z(t)\right]\right\rangle, \qquad (1.81b)$$

where $P_2$ = Legendre polynomial of order 2.

Incidentally, the orientational correlation functions of the symmetry species $\Sigma^+$ and $\Sigma_g^+$ of point groups $C_{\infty v}$ and $D_{\infty h}$ (linear molecules) likewise belong to Eqs. 1.81. Of course, direction $z$ then falls together with the molecular axis.

The correlation functions for lower-symmetry groups, such as rhomboidal molecules, are considerably more complicated. The Raman tensor for the totally symmetric species $A_1$ and $A_g$ of $C_{2v}$ and $D_{2h}$ is

$$\begin{pmatrix} a & & \\ & b & \\ & & c \end{pmatrix} \qquad xx, \, yy, \, zz.$$

Following the procedure outlined previously for the symmetric and linear tops —with the sole difference that now $m_i^x m_j^x \neq m_i^y m_j^y$—we find

$$\langle \hat{\boldsymbol{\beta}}(0)\cdot\hat{\boldsymbol{\beta}}(t)\rangle = \hat{G}(t)$$

$$= C\Big\langle (b-a)^2 P_2\left[\hat{\mathbf{m}}^y(0)\cdot\hat{\mathbf{m}}^y(t)\right] + (c-a)^2 P_2\left[\hat{\mathbf{m}}^z(0)\cdot\hat{\mathbf{m}}^z(t)\right]$$

$$+ (b-a)(c-a)\{P_2\left[\hat{\mathbf{m}}^y(0)\cdot\hat{\mathbf{m}}^z(t)\right] + P_2\left[\hat{\mathbf{m}}^z(0)\cdot\hat{\mathbf{m}}^y(t)\right]\}\Big\rangle,$$

$$(1.82)$$

with $C = (a^2 + b^2 + c^2 - ac - ab - bc)^{-1}$.

Clearly, it is difficult to see how such an expression could be fruitfully evaluated. We note here the (rarely mentioned) disadvantage of orientational Raman correlation functions: They fail to be useful for lower-symmetry molecules. After all, point group $C_{2v}$ contains several rather interesting molecular systems, for instance, methylene chloride, $CH_2Cl_2$, which poses no difficulties in the infrared.[34]

The situation becomes worse for doubly degenerate vibrational Raman modes of symmetric top molecules of, for instance, point group $C_{3v}$. This group comprises an important set of molecules of structure $YZX_3$, chloroform, acetonitrile, methyl iodide, their fluorinated analogues, their trimethylated analogues, and so on. The tensor (note it is traceless since the mode is depolarized) is

$$\begin{pmatrix} a & & \\ & -a & b \\ & b & \end{pmatrix} \quad \text{and} \quad \begin{pmatrix} & -a & -b \\ -a & & \\ -b & & \end{pmatrix} \qquad \begin{array}{c} (xx - yy, \, xy) \\ (xz, \, yz). \end{array}$$

We only give the final, rather complicated and pretty hopeless result (it requires knowledge of the absolute values of the tensor elements—as does Eq. 1.82):

$$\hat{G}(t) = \left(\frac{1}{a^2 + b^2}\right)\Big\langle a^2\left[\hat{m}^x(0)\cdot\hat{m}^x(t)\right]^2$$

$$+ b^2\{\left[\hat{m}^x(0)\cdot\hat{m}^x(t)\right]\left[\hat{m}^z(0)\cdot\hat{m}^z(t)\right] + \left[\hat{m}^x(0)\cdot\hat{m}^z(t)\right]^2\}$$

$$+ 2ab\left[\hat{m}^x(0)\cdot\hat{m}^y(t)\right]\left[\hat{m}^x(0)\cdot\hat{m}^z(t)\right]\Big\rangle. \tag{1.83}$$

Again, we have put molecular direction $z$ along the $C_3$ axis.

There is a notable exception to the cumbersome forms of $\hat{G}(t)$ for $E$ modes of symmetric top molecules, namely, species $E'$ of point group $D_{3h}$ and $E_{2g}$ of group $D_{6h}$. (For instance, cyclopropane belongs to the former.) The tensor is

$$\begin{pmatrix} a & & \\ & -a & \\ & & \end{pmatrix} \quad \text{and} \quad \begin{pmatrix} & a & \\ a & & \\ & & \end{pmatrix} \quad xx - yy,\ xy,$$

and transforms as

$$\beta_{ij} = a\left(m_i^x m_j^x - m_i^y m_j^y\right)$$

and

$$\beta_{ij} = a\left(m_i^x m_j^y + m_i^y m_j^x\right).$$

Instead calculating the total $\hat{G}(t)$ from the (normalized) sum of the two $\Sigma_{ij}[\beta_{ij}(0)\beta_{ij}(t)]$, we merely set $b = 0$ in Eq. 1.83 and obtain the simple result

$$\hat{G}(t) = \Big\langle\left[\hat{m}^x(0)\cdot\hat{m}^x(t)\right]^2\Big\rangle.$$

Since $x$ and $y$ are equivalent by symmetry, this yields

$$\hat{G}(t) = \Big\langle\left[\hat{m}^p(0)\cdot\hat{m}^p(t)\right]^2\Big\rangle, \qquad p = \begin{cases} \text{any direction} \\ \text{in the degenerate} \\ x\text{-}y \text{ plane.} \end{cases} \tag{1.84}$$

Incidentally, many unknown correlation functions may be written down immediately by comparing the transformation properties of their Raman tensors[19] with those belonging to already derived correlation functions of the same degree of degeneracy. If the respective tensors have the same transformation properties, the Cartesian correlation functions are closely related. For instance, the tensors for the doubly degenerate $E'$ mode of point group $C_{3h}$ are on first sight rather complicated[17b]:

$$\begin{pmatrix} a & b & \\ b & -a & \\ & & \end{pmatrix}, \quad \begin{pmatrix} b & a & \\ a & -b & \\ & & \end{pmatrix}.$$

However, they transform as $xx - yy$, $xy$ and, upon normalization, yield the correlation function

$$\hat{G}(t) = \left\langle \left[ \hat{\mathbf{m}}^p(0) \cdot \hat{\mathbf{m}}^p(t) \right]^2 \right\rangle \qquad (1.85)$$

(see Eq. 1.84). (Clearly, degenerate correlation functions may be composed from sums of known *nondegenerate* correlation functions that furnish the required tensor transformation properties of the degenerate mode.)

In addition, we list here the Raman rotational correlation functions for a doubly degenerate and a triply degenerate vibrational mode of a spherical top molecule (species $E$ and $F_2$ of point group $T_d$). The derivation of the correlation function for species $E$ has been described in detail.[35] It is

$$\hat{G}(t) = \left\langle P_2 \left[ \hat{\mathbf{m}}^q(0) \cdot \hat{\mathbf{m}}^q(t) \right] \right\rangle, \qquad \left( 2\alpha_{zz} - \alpha_{xx} - \alpha_{yy}, \alpha_{xx} - \alpha_{yy} \right),$$

$$(1.86a)$$

where $q$ is any of the three equivalent, twofold symmetry axes of the spherical molecule. Equation 1.86a equals $\hat{G}(t)$ for the polarized (nondegenerate) species of a symmetric top molecule (Eq. 1.81b) but refers to a different molecule axis direction.

The Raman orientational correlation function for the triply degenerate vibrational mode of species $F_2$ and its tensor is[17b]

$$\hat{G}(t) = \left\langle \left[ \hat{\mathbf{m}}^q(0) \cdot \hat{\mathbf{m}}^q(t) \right]^2 + \left[ \hat{\mathbf{m}}^q(0) \cdot \hat{\mathbf{m}}^p(t) \right]^2 \right\rangle, \qquad \left( \alpha_{xy}, \alpha_{xz}, \alpha_{yz} \right),$$

$$(1.86b)$$

where $q$ is again any of the three equivalent twofold symmetry axes and $p$ is any of the others.

Reference 17b gives an extensive list of Raman orientational correlation functions of additional molecular point group systems. A glance at the form of these functions convinces that their application is restricted. They are useful only to describe the short initial time period of the rotational relaxation (the various cross-axis terms are relatively small). In some instances the combination of the $G(t)$ from all the allowed vibrational species of the molecule "cancels" the cross-axis terms.

The transformation of the Cartesian correlation functions $\langle \boldsymbol{\beta}(0) \cdot \boldsymbol{\beta}(t) \rangle$ to their spherical combinations is now readily performed.[18b] First, we express the Cartesian molecule-fixed $\beta_{ij}$ in spherical notation,[36]

$$\beta_0 = \left( \tfrac{1}{6} \right)^{1/2} \left[ 2zz - xx - yy \right]$$

$$\beta_{\pm 1} = \mp \left( \tfrac{1}{2} \right) \left[ xz + zx - i(yz + zy) \right]$$

$$\beta_{\pm 2} = \mp \left( \tfrac{1}{2} \right) \left[ xx - yy \pm i(xy + yx) \right]. \qquad (1.87)$$

(We recall that the subscript of $\beta$ represents the allowed quantum numbers $m$

of the eigenvalues of the $z$-component of the angular momentum.) Next, we orient the molecular frame through the Euler angle $\Omega(\alpha, \beta, \gamma)$ so that it coincides with the laboratory reference frame. In other words, we transform from the original spherical basis to a rotated spherical basis. As is well known,[37] a unitary transformation to perform this operation is given by the Wigner rotation matrices $D_{ij}^{(2)}(\Omega)$. We write, omitting the order index,

$$\beta_n(\Omega) = \sum_{m=-2}^{2} D_{nm}(\Omega)\beta_m(MF),$$    (1.88)

where $\beta_m(MF)$ are the spherical tensor elements in their molecular frame (see Eq. 1.87).

Now we pick an appropriate Raman scattering geometry to give us an orientational correlation function by Fourier transformation of the measured spectral density: We choose the geometry of the VH experiment discussed in Section 3.2. In our nomenclature, this particular experiment selects the element $\beta_{zx}$. Hence from Eq. 1.87 we find

$$\beta_{zx}(\Omega) = \beta_{-1}(\Omega) - \beta_{+1}(\Omega).$$

(We could have chosen any other scattering geometry as long as the polarizer is turned to select a depolarized component of the scattered radiation. Of course, it makes no difference for a depolarized contour.) Equation 1.88 is then transformed into

$$\beta_{zx}(\Omega) = \sum_m \left[ D_{-1m}(\Omega) - D_{1m}(\Omega) \right] \beta_m(MF).$$

The correlation function is, consequently,

$$\langle \beta_{zx}^*(\Omega')\beta_{zx}(\Omega) \rangle = \sum_{mm'} \left\langle \left[ D_{-1m'}^*(\Omega') - D_{1m'}^*(\Omega') \right] \right.$$
$$\left. \times \left[ D_{-1m}(\Omega) - D_{1m}(\Omega) \right] \beta_{m'}^*(MF)\beta_m(MF) \right\rangle,$$    (1.89)

where the starred quantities denote complex conjugates and $\Omega$, $\Omega'$ is shorthand for $\Omega(t)$, $\Omega(t')$.

The isotropy of the liquid implies that the original angle between the chosen axis directions of laboratory- and molecule-fixed coordinate systems is immaterial; the correlation functions depend only on angle differences, not on absolute angles.[38] Designating the angle difference by $\Delta\Omega$, we write $\Omega' = \Omega + \Delta\Omega$. According to the addition theorem[39] we find

$$D_{nm'}^*(\Omega') = \sum_{n'} D_{nn'}^*(\Omega)D_{n'm'}^*(\Delta\Omega).$$    (1.90)

To simplify the evaluation of the ensemble average of the correlation function over all orientations we separate the time dependence of the angular set $\Omega(t)$ into a purely time-dependent factor and into expansion coefficients that depend only on the magnitude of the angular quantities. We define the (isotropic) conditional probability density $P(\Delta\Omega, t)$ that the single molecule has oriented through an angle between $\Delta\Omega$ and $\Delta\Omega + d(\Delta\Omega)$ during time $t$ when its original orientation was $\Omega$ at time $t = 0$. We set

$$P(\Delta\Omega, t) = \sum_{lk} D_{kk}^{(l)}(\Delta\Omega) f_k^{(l)}(t), \qquad (1.91)$$

which furnishes a convenient expansion of $P(\Delta\Omega, t)$ in terms of the Wigner rotation matrices of angle $\Delta\Omega$ and the time-dependent factor $f_k^{(l)}(t)$.

The correlation function Eq. 1.89 is constructed from $P(\Delta\Omega, t)$ by way of the general relation[40]

$$\left\langle \beta_{zx}^*[\Omega(t')] \beta_{zx}[\Omega(t)] \right\rangle$$

$$= \frac{1}{8\pi^2} \int\int d\Omega\, d(\Delta\Omega)\, P(\Delta\Omega, \tau) \beta_{zx}^*(\Omega') \beta_{zx}(\Omega), \qquad (\tau = t - t')$$

$$(1.92)$$

where

$$\frac{1}{8\pi^2} = \left\{ \int_0^\pi d\alpha \sin\alpha \int_0^{2\pi} d\beta \int_0^{2\pi} d\gamma \right\}^{-1}$$

is the a priori probability that the molecular frame has the initial orientation $\Omega$ at $\tau = 0$. Expressing in Eq. 1.92 the Cartesian components by their spherical expansion (Eq. 1.89), substituting the conditional probability $P(\Delta\Omega, t)$ by its expansion (Eq. 1.91), and replacing the Wigner matrix $D_{nm}$ of angle $\Omega'$ by expansion in terms of angles $\Omega$ and $\Delta\Omega$ (Eq. 1.90), gives

$$\left\langle \beta_{zx}^*[\Omega(t')] \beta_{zx}[\Omega(t)] \right\rangle = \frac{1}{8\pi^2} \int\int d\Omega\, d(\Delta\Omega) \sum_{lk} D_{kk}^{(l)}(\Delta\Omega) f_k^{(l)}(\tau)$$

$$\times \sum_{mm'n'} \{ D_{-1n'}^*(\Omega) D_{n'm'}^*(\Delta\Omega) - D_{1n'}^*(\Omega) D_{n'm'}^*(\Delta\Omega) \}$$

$$\times \{ D_{-1m}(\Omega) - D_{1m}(\Omega) \} \beta_{m'}^*(MF) \beta_m(MF). \qquad (1.93)$$

Multiplying out we obtain products of the form (we now set $t' = 0$, $\tau = t$)

$$\sum_{\substack{mm' \\ lkn'}} f_k^{(l)}(t) \beta_{m'}^*(MF) \beta_m(MF) \int d\Omega\, D_{1n'}^*(\Omega) D_{1m}(\Omega)$$

$$\times \int d(\Delta\Omega)\, D_{n'm'}^*(\Delta\Omega) D_{kk}^{(l)}(\Delta\Omega).$$

These, according to the orthogonality properties of the Wigner matrices,[41] yield terms proportional to (recall that $l = 2$)

$$\sum_{\substack{mm' \\ lkn'}} \beta^*_{m'}(MF)\beta_m(MF)f_k^{(l)}(t)\delta_{l2}\delta_{m'k}\delta_{kn'}\delta_{11}\delta_{n'm}.$$

Hence

$$G(t) = \langle \boldsymbol{\beta}(0)\cdot\boldsymbol{\beta}(t)\rangle = C\sum_{m=-2}^{2}|\beta_m(MF)|^2 f_m^{(2)}(t), \qquad (1.94)$$

where $C$ is a constant.

Equation 1.94 shows that an orientational Raman correlation function of a vibration-rotation contour is a linear combination of up to five independent time-dependent functions $f_m^{(2)}(t)$. Their expansion coefficients are the squares of the spherical Raman tensors in the molecular coordinate system. Of course, without further assumptions about the nature of the orientational motion we know nothing about the $f(t)$ except that they must be well behaved [$f(0) = 1$, $f(t \rightarrow \infty) = 0$].

From the transformation properties of the $\beta_m(MF)$ under the point group of the molecule we can predict how many of the $m$-indexed terms in Eq. 1.94 contribute to $G(t)$. We need only substitute the element $\beta_{ij}$ in the Cartesian coordinate expressions of $\beta_0, \beta_{\pm 1}, \beta_{\pm 2}$ (Eq. 1.87) by the appropriate elements $a, b, \dots$ of their Raman tensors. The nonvanishing $f_m(t)$ for the vibrational symmetry species discussed here are listed in Table I. [Reference 18b gives an extensive collection of $f(t)$ for additional molecular point groups and vibrational symmetry species. Few of these are useful.] As an example we express Eq. 1.83—which represents $\langle \hat{\boldsymbol{\beta}}(0)\cdot\hat{\boldsymbol{\beta}}(t)\rangle$ of an $E$-mode of a $C_{3v}$ molecule in Cartesian coordinates—in its spherical form. Replacing $f_m(t)$ by $G^{(2, m)}(t)$ we find

$$\hat{G}(t) = \frac{\displaystyle\sum_{m=1,2}|\beta_m(MF)|^2 G^{(2, m)}(t)}{\displaystyle\sum_{m=1,2}|\beta_m(MF)|^2}, \qquad (1.95)$$

with

$$|\beta^E_{\pm 1}|^2_{C_{3v}} = (\tfrac{1}{4})|\alpha_{xz} + \alpha_{zx} \pm i(\alpha_{yz} + \alpha_{zy})|^2$$

$$= (\tfrac{1}{4})\{[-2b]^2 + (0+0)^2\} = b^2,$$

$$|\beta^E_{\pm 2}|^2_{C_{3v}} = (\tfrac{1}{4})|\alpha_{xx} - \alpha_{yy} \pm i(\alpha_{xy} + \alpha_{yx})|^2$$

$$= (\tfrac{1}{4})\{[a-(-a)]^2 + (0+0)^2\} = a^2.$$

**Table I.  Assignments of the Nonvanishing Orientational Raman Time Correlation Functions $f_m(t)$ to Vibrational Symmetry Species of Vibration-Rotation Raman Contours of Selected Molecular Point Groups**[a]

| Molecular Point Group | Vibrational Symmetry Species | Spherical Molecule-Fixed Anisotropic Tensor | Orientational Raman Correlation Function |
|---|---|---|---|
| $C_{3v}$ | $A_1$ | | |
| $C_{3h}, D_{3h}$ | $A'$ | | |
| $C_3$ | $A$ | $\beta_0$ | $f_0(t)$ |
| $D_{6h}$ | $A_{1g}$ | | |
| $C_{\infty v}, D_{\infty h}$ | $\Sigma^+, \Sigma_g^+$ | | |
| $C_{2v}$ | $A_1$ | | $f_0(t), f_{+2}(t),$ |
| $D_{2h}$ | $A_g$ | $\beta_0, \beta_{+2}, \beta_{-2}$ | $f_{-2}(t)$ |
| $C_{3v}$ | $E$ | $\beta_{\pm 2}, \beta_{\pm 1}$ | $f_{\pm 2}(t), f_{\pm 1}(t)$ |
| $D_{3h}, D_{6h}$ | $E', E_{2g}$ | | |
| $C_{3h}$ | $E'$ | $\beta_{\pm 2}$ | $f_{\pm 2}(t)$ |
| $T, T_h$ | $E, E_g$ | $\beta$ | $f(t)$ |

[a] From Ref. 18b.

We have not yet assigned the $G^{(2, m)}(t)$ to a model of orientational motion. Equation 1.95 merely shows that the experimental evaluation of the band contour of a doubly degenerate vibrational-rotational Raman mode of a symmetric top molecule, such as acetonitrile or chloroform, yields a rotational correlation function that mixes indices $m$ (also denoted by $k$) = 1, 2. We discuss later whether there are systems that permit an approximate resolution of the two component correlation functions.

### 3.5. Orientational Correlation Functions and Anisotropic Pair Correlation Factors from Depolarized Rayleigh Wing Scattering Spectra

To obtain an anisotropic Rayleigh correlation function we replace in Eq. 1.79 the derivative of the anisotropic polarizability $\beta^v$ of vibrational mode $v$ by the permanent anisotropic polarizability $\beta^0$ and set $\omega'$ equal to the frequency displacement from the laser line. Factor $\lambda^4$ is then equal to the fourth power of the wavelength divided by $2\pi$ of the laser radiation. The resulting second-order tensor pure orientational correlation function is the (Raman) analogue of the first-order tensor (far-infrared) correlation function of Eq. 1.41. Experimentally, we perform a VH scattering experiment over a frequency range from the laser line to about 300 cm$^{-1}$.

We write the depolarized Rayleigh spectral density

$$\mathscr{I}_{VH}(\omega') = \int_{-\infty}^{\infty} dt \exp(-i\omega't) \Big\{ \big\langle \beta_{zx}^{0A}[\Omega(0)]\beta_{zx}^{0A}[\Omega(t)] \big\rangle$$

$$+ \sum_{A \ne B} \big\langle \beta_{zx}^{0A}[\Omega(0)]\beta_{zx}^{0B}[\Omega(t)] \big\rangle \Big\}, \qquad (1.96)$$

where superscripts $A$ and $B$ denote different molecules. Since the permanent polarizability of an isolated spherical molecule vanishes, only the cross-correlation term (also called pair correlation, or distinct term),

$$\sum_{A \ne B} \big\langle \boldsymbol{\beta}^{0A}(0) \cdot \boldsymbol{\beta}^{0B}(t) \big\rangle,$$

survives in such molecular systems. In fact, interest in pair correlations constitutes the main motive for discussing depolarized Rayleigh scattering of condensed-phased systems in this book.

Taking the zero-time average of the distinct Rayleigh correlation function $\langle \sum_{A \ne B} \boldsymbol{\beta}^{0A}(0) \cdot \boldsymbol{\beta}^{0B}(t) \rangle$ yields the so-called static orientational pair correlation factor or generalized Kirkwood g-factor,[42,43]

$$g_2 = 1 + \sum_{i \ne 1}^{N-1} \big\langle D_{00}^{(2)}(\Omega^1) D_{00}^{(2)*}(\Omega^i) \big\rangle \big\langle D_{00}^{(2)}(\Omega^1) D_{00}^{(2)*}(\Omega^1) \big\rangle^{-1}. \qquad (1.97)$$

Here $D(\Omega^1)$ and $D(\Omega^i)$ are Wigner rotation matrices of the orientations of molecules $A \equiv 1$ and $B \equiv i$ in terms of the time-independent Eulerian angles $\alpha_i, \beta_i, \gamma_i$. Summation is over all contributing molecules; $g_2$ furnishes information on static distributions of orientations among $N$ molecules. We discuss this in Chapter 2.

Another useful correlation factor obtainable from depolarized Rayleigh spectra is the so-called dynamic orientational pair correlation between $N$ molecules,[42]

$$J_2 = 1 + \sum_{i \ne 1}^{N-1} \left[ \int_0^{\infty} dt \big\langle \dot{D}_{00}^{(2)}(\Omega^1) \dot{D}_{00}^{(2)*}(\Omega^i, t) \big\rangle^+ \right.$$

$$\left. \times \left\{ \int_0^{\infty} dt \big\langle \dot{D}_{00}^{(2)}(\Omega^1) \dot{D}_{00}^{(2)*}(\Omega^1, t) \big\rangle^+ \right\}^{-1} \right]. \qquad (1.98)$$

$J_2$ contains correlations between angular velocities (indicated by the dots = time derivatives). The time evolution of the correlations in Eq. 1.98 represents a so-called memory function (indicated by the superscript plus). We discuss this in Chapter 2 at the appropriate time.

The form of the correlation functions (Eq. 1.96) is determined by the point group symmetry of the molecular system of interest. For symmetric top molecules (we have no opportunity to deal with Rayleigh spectra of lower-symmetry molecules) $\beta_{zx}^0$ transforms as $zz - xx$. This is the totally symmetric species (see Eq. 1.87 and note that $xx = yy$). Consequently, for a symmetric top the autocorrelation term for the permanent polarizability (Eq. 1.96) is equivalent to the orientational autocorrelation of the vibrationally induced polarizability along the molecular $C_3$-axis (Eq. 1.81).

This concludes the sections on the introductory theoretical discussion of orientational relaxation phenomena. In Chapter 2 we evaluate experimental literature material, elaborating and extending these theories as required.

## 4. VIBRATIONAL RELAXATION PROCESSES AND THEIR CORRELATION FUNCTIONS

### 4.1. General Aspects

Historically, work in infrared and Raman fluctuation spectroscopy of condensed systems was first directed toward exploration of vibrational relaxation.[44] But this subject soon lost favor and was treated with benign neglect during the meteoric rise of interest in orientational motion.

Now the tide has turned to a preference of investigations of vibrational relaxation. This is not surprising since vibrational relaxation processes are more numerous and varied than rotational relaxation effects. A priori, each vibrational mode is associated with a unique vibrational correlation function. There are, at least, $3N - 6$ of such fundamental modes in a $N$-atom molecule—not to forget overtones, summation bands, difference bands, and hot bands. Furthermore, the values of the force constants of vibrational modes cover a wide range, leading to stretching and deformation motions of diverse nature in each molecule. As is well known, certain vibrational motions can be considered "group motions," such as the carbon-hydrogen stretches. Since their force constant is strong and the oscillatory motion is practically isolated in one bond, uniformity of vibrational relaxation behavior in otherwise different molecules may be expected here. On the other hand, there are vibrational modes that, because of bond polarity (for instance, the motions involving the $C=O$ bond), depend strongly on the environment.

Consequently, the perturbation Hamiltonian $H_1(t)$ in a vibrational relaxation process is a multiparameter quantity. To predict a vibrational correlation function

$$\phi(t) = \langle m(0)m(t) \rangle \tag{1.99}$$

and to compare the result with the experiment, we might proceed by first defining $H_1(t)$, then solve the equation of motion for $m(t)$,

$$\left(\frac{d}{dt}\right)m^I(t) = \frac{i}{\hbar}\left[H_0, m^I(t)\right],$$

and finally take the trace

$$\langle m(t)m(0)\rangle = \mathrm{Tr}\,\rho_0\left[m(t)m(0)\right]. \qquad (1.100)$$

[Note that $m(t)$ and $H_1(t)$ are in the interaction representation—see Eqs. 1.30, 1.31, and 1.53.[26]]

Let us introduce the different vibrational relaxation processes.

1. *Vibrational Energy Relaxation (Vibrational Dissipation).* The phonons of the excited vibrational level of an intramolecular vibrational mode return to ground level by dissipating the energy difference into lattice phonons of average energy $kT$. The process is analogous to a $T_1$-process in nuclear magnetic relaxation.

2. *Resonance Vibrational Energy Transfer.* The phonon population loss of an upper oscillator level is the subsequent phonon population gain in the same oscillator level but on an adjacent molecule. In other words, the downward transition in one oscillator causes the corresponding upward transition in its twin. The resonance vibrational transfer process is analogous to the dipole-dipole broadening by like spins through a "flip-flop" motion.

The energy relaxation (or dissipation) and the resonance energy transfer process lead to a true lifetime of a vibrational level. We obtain this lifetime through (see Section 1.2)

$$\tau \approx \frac{1}{\Delta\omega}, \qquad (1.101)$$

where $\Delta\omega$ is the incremental bandwidth due to the process in question. On the other hand, the nature of the molecular motions that induce the vibrational lifetimes due to energy dissipation and resonance transfer processes is different. Consider first vibrational energy dissipation to the lattice. Note that the usual energy of the vibrational quantum is $600$–$3000$ cm$^{-1}$ (for fundamentals). This compares to an average of approximately $200$ cm$^{-1}$ for the random heat motions of the molecules. Hence the probability of finding a rotational-translational lattice fluctuation of about $10\,kT$ is $5 \times 10^{-5}$. Consequently lifetimes of excited vibrational levels that lose their excess energy by way of dissipation to the heat motions of the bath are relatively long. The process requires the fastest components of the lattice motion. Vibrational energy dissipation therefore probes the effectiveness and importance of strong, inelastic collisional events. Raising the temperature increases the rate of this process.

Consider next vibrational resonance energy transfer. This process conserves vibrational energy. It does not require the intervention of the *motions* of the lattice. In fact, if molecular motion takes place at a faster rate than the resonance energy transfer step, the rate of the resonance effect is diminished. This motional effect, which is the vibrational analogue of "motional narrowing" in nuclear magnetic relaxation, is discussed extensively in Chapter 3.

A priori, we surmise that manifestations of resonance vibrational energy transfer pervade vibrational relaxation since the molecules in a liquid nearly touch. However, is it usually found that the contribution of this process to band broadening is minor.

The resonance phenomenon gives us information on the frequency and importance of low- to average-energy collisions in the medium and on vibrational energy migration.

3. *Vibrational Dephasing.* This ubiquitous phenomenon is a pure phase relaxation process: The lifetime of the excited level has no meaning and is considered infinitely long (a delta function in the vibrational energy). The use of the word "dephasing" is unfortunate since there is a phase loss (loss of the initial coherence of the autocorrelation of the vibrational amplitude) during any type of vibrational relaxation mechanism. As we just discussed, in processes 1 (dissipation) and 2 (resonance transfer) the phase loss is caused by the environment-induced annihilation of the excited phonon and the subsequent "cut-off" of the instantaneous phase of the oscillating amplitude motion. In the vibrational dephasing process, however, the observed band profile and its associated correlation function arise from a quasi-continuous distribution of vibrational energy levels about a mean. The distribution is induced by the effects of locally different environment-oscillator perturbations which shift the oscillator energy levels. We therefore deal with a vibrational band broadening, hence a vibrational relaxation process, which is caused by a certain probability distribution of instantaneous vibrational frequency shifts $\omega_1(t)$,

$$\hat{\phi}(t) = \left\langle \exp i \int_0^t dt\, \omega_1(t) \right\rangle, \qquad (1.102)$$

with $\langle \omega_1(t) \rangle = 0$. Indeed, we can consider $\omega_1(t)$ as a fluctuating time-dependent phase angle of zero average which modulates the unperturbed, sharp (long-lived) oscillator frequency $\omega_0$ ("carrier wave").

Molecular motion is not a prerequisite for vibrational dephasing; it occurs in a rigid as well as mobile molecular environment. Motional narrowing plays, however, an important role. It permits us to elucidate the time scale of intervening interference phenomena ("collisions") that scramble the molecular environment; the dephasing process probes local static and dynamic properties.

It is usually difficult to write a meaningful and useful expression for the intermolecular, coordinate- and time-dependent, interaction Hamiltonian $H_1(t)$ which couples the oscillator to the molecular environment. It is often simpler to define or guess a workable expression for the autocorrelation function of $H_1(t)$ (see also Eq. 1.66b). We use this approach in Chapter 3.

The dephasing process is the vibrational analogue of a $T_2$-process in nuclear magnetic relaxation.

4. *Intramolecular Vibrational Relaxation.*   This process, in which vibrational energy moves from the originally excited mode to a different mode in the same molecule, can be expected to prevail for anharmonic modes in large molecules. Intramolecular vibrational relaxation has not received the attention it merits. Evidently, the process is important in reaction kinetics and dissociation phenomena in the condensed phase, particularly if higher excited modes are generated and probed.

So far, we have assumed a two-level system of one lower and one upper vibrational level. Usually, vibrational relaxation involves several vibrational levels since (particularly in larger molecules) a vibrational level may be populated (depopulated) by transitions from (to) upper and lower-lying states, the energy balance being removed or furnished by the lattice. In this situation a "master equation" or "rate balancing equation" is constructed which gives the rate of change of the occupation (or phase) of the chosen vibrational state in terms of its pairwise interactions with other vibrational levels. (This is treated in Chapter 3.) Such a scheme can be written to include the simultaneous occurrence of several different vibrational relaxation processes. It is therefore necessary to disentangle such situation by experiment. For instance, to undo the band broadening due to a vibrational resonance energy transfer process, we repeat the experiment by diluting the active oscillator by its isotopically substituted homologue. If this succeeds in causing a significant frequency shift of the band profile (a C-H stretch of a molecule diluted in its deuterated species), the interoscillator coupling between the active molecule and its neighbors-in-resonance is removed.

Finally, let us reemphasize the relationship of relaxation in vibrational and nuclear magnetic resonance spectra (NMR) by comparing the change in the respective quantum numbers ($v$ = vibrational, $s$ = spin) during the various processes.

1.   Vibrational energy relaxation, $\Delta v = 1$ (fundamental); $T_1 - \text{NMR}$, $\Delta s = 1$.
2.   Resonance vibrational energy transfer, $\Delta v_1 = 1(1 \rightarrow 0)$ and $\Delta v_2 = -1(0 \rightarrow 1)$; NMR dipole-dipole "flip-flop", $\Delta s_1 = 1$, $\Delta s_2 = -1$. For both processes, we find $\sum_i \Delta v_i = \sum_i \Delta s_i = 0$ for all molecules $i$.
3.   Vibrational dephasing, $\Delta v_i = 0$ for each $i$; $T_2 - \text{NMR}$, $\Delta s_i = 0$ for each $i$.

## 4.2.   Separation of Vibrational and Rotational Relaxation Phenomena

The separation of the purely vibrational and purely rotational contributions of a Raman spectral density of a polarized vibration-rotation band through a VV-VH polarization experiment has been described in Section 3.2. The accurate determination of the purely orientational and purely vibrational correlation functions from infrared or depolarized Raman profiles is yet another difficult experimental problem since the VV-VH experiment is no longer useful.

To obtain at least an approximate idea of the vibrational and orientational contributions, it was suggested[45] to freeze rotational motion by cooling the system to its solid and to assign the remaining bandwidth to vibrational relaxation at *any* temperature. The method was attacked as useless since vibrational relaxation processes are temperature dependent. Nevertheless, we are of the opinion that this simple technique has merit if performed carefully. (We demonstrate later that the temperature variations in some situations are mild over a wide range.)

It is safer, however, to choose the molecular system that discriminates rotational or vibrational relaxation. Frequently, this can be predicted by simple and obvious considerations. For instance, recall that a molecule of $N$ atoms possesses (at most) three rotational degrees of freedom and (at least) $3N - 6$ vibrational degrees of freedom. Hence in larger molecules there are more avenues of vibrational than rotational relaxation. In addition, larger molecules reorient more slowly than do smaller molecules because of their mass (moments of inertia) and shape (steric hindrance).

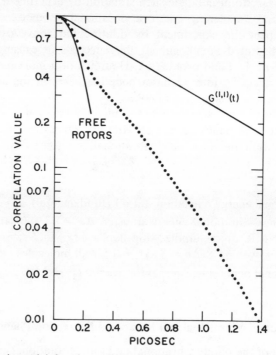

**Figure 1.5.** Experimental infrared correlation function (dots) of the doubly degenerate ($E$) 884-cm$^{-1}$ fundamental of liquid CH$_3$I at 30°C. The parabola shows the fastest possible effective initial decay of the combined rotational motions around the symmetry axis and around one of the two equivalent perpendicular axes. (Function $G^{(1,1)}(t)$ is discussed in Section 2.6 of Chapter 3.)

This situation is demonstrated by the correlation data in Figs. 1.5 and 1.6. (The figures contain additional information needed later.) Figure 1.5 displays the infrared correlation function obtained by Fourier inversion of the depolarized 884-cm$^{-1}$ fundamental (species $E$) of $CH_3I$ (methyl iodide), a symmetric top ($C_{3v}$).[46] Figure 1.6 shows the infrared and Raman correlation functions obtained by Fourier inversion of the respective in-plane fundamental at about 1030 cm$^{-1}$ of quinoline, $C_9H_7N$.[47] Quinoline is planar ($C_s$); it consists of two fused benzene rings with one CH group replaced by N.

We know the position of the vibrational-rotational transition moment vector $\mathbf{m}(t)$ within the molecular inertial coordinate system of $CH_3I$: $\mathbf{m}(t)$ lies along any one of the two equivalent perpendicular inertial axes (orthogonal to the C—I bond). We recall that the infrared orientational correlation function $\hat{G}_E(t) = \langle \hat{\mathbf{m}}(0) \cdot \hat{\mathbf{m}}(t) \rangle$ characterizes the molecular rotational motion around the $C_3$-axis and around one of the two perpendicular axes (see Fig. 1.3$b$). The component rotation around the $C_3$-axis should be fast. First, the direction of the permanent dipole moment of $CH_3I$ stays invariant during this motion; angle-dependent dipole-dipole forces are therefore ineffective in exerting a

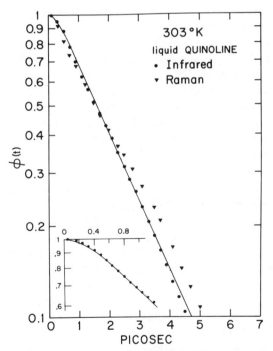

**Figure 1.6.** Correlation functions from the infrared and the Raman contours, respectively, of an in-plane deformation fundamental of liquid quinoline, $C_9H_7N$. The contour is peaked at 1033 cm$^{-1}$, its symmetry species (point group $C_s$) is $A'$. [W. G. Rothschild, *J. Chem. Phys.* **65**, 455 (1976).]

torque (friction). Second, the moment of inertia for this rotational component is small since it involves the three light H atoms and their small distance from the $C_3$-axis. Consequently, the infrared correlation function $\langle \mathbf{m}(0)\cdot\mathbf{m}(t)\rangle$ in Fig. 1.5 should contain a considerable contribution of rotational relaxation since motion involves fast "spinning" around the $C_3$-axis of the molecule.

On the other hand, the infrared and Raman correlation functions of the 1030-cm$^{-1}$ mode of quinoline arise from vibrational relaxation (dephasing) since we know—from $^{14}$N quadrupolar NMR relaxation measurements[48]—that liquid quinoline has an orientational correlation time of about $120 \times 10^{-12}$ sec and we see—from Fig. 1.6—that the infrared and Raman correlations have decayed long before. Note that both infrared and Raman correlation functions are near identical, in accordance with our assumption (Eq. 1.39).

### 4.3. Stimulated Vibrational Relaxation: Picosecond Pulsing

In the last 10 years a new experimental and theoretical approach to vibrational relaxation has joined established methods: "Picosecond Pulsing" by stimulated Raman scattering technique.[49] This involves a nonlinear interaction between the system variables and the strong incident laser radiation field. Such field is capable of pumping into an upper vibrational level a phonon population and a vibrational amplitude coherence which are in slight excess of their equilibrium values. In other words, the system is prepared in a nonequilibrium state that possesses, loosely speaking, relatively more extensive regions of coherent subsystems than those generated by linear interaction with a weak radiation field ("spontaneous relaxation," see Section 1.3).

Our basic assumptions about linear response, however, remain valid. The actual relaxation processes (return to equilibrium of the excited system) take place through a spontaneous, ordinary Raman relaxation process (anti-Stokes) whose time evolution is interrogated by a weak probe pulse. In other words, the pumping effect produces a slight phonon overpopulation and excess vibrational amplitude coherence by a stimulated, nonlinear (in the electric field strength) Stokes process; relaxation is observed through a spontaneous, linear (in the electric field strength) anti-Stokes process.

Experimentally, the weak probe pulse is injected, after the pump pulse has passed, with a delay appropriate to the time evolution of the relaxation process. We therefore observe, with increasing times (probe pulse delay), the decay of the anti-Stokes signal intensity originating from the upper level as function of its momentary phonon population or amplitude phase coherence (depending on the experimental conditions). We read off a true phonon lifetime or true vibrational dephasing time.

The method is useful since it permits us to examine concurrent vibrational relaxation processes and to circumvent interference by rotational relaxation. For instance, by stimulated Raman scattering we can obtain the lifetime of an upper vibrational level (process 1, Section 4.1) and the lifetime for the decay of the phase coherence of the oscillator amplitude by vibrational resonance energy transfer (process 2) or by pure dephasing (process 3), even if these mechanisms occur simultaneously or concurrently with significant rotational relaxation. (We recall the experimental difficulties for separating vibrational from rotational relaxation—see Sections 3.2 and 4.2.)

In the following discussion we outline (classically) the amplification process that puts excess phonons and phase coherence into a vibrational level.[50] We designate by subscript $M$ a molecular property and by subscript $L$ a property of the laser field. Subscript $S$ designates "Stokes." We write the equation of motion of an induced, electric dipole oscillator (for instance, the atoms and electrons of the C-H group) of polarizability $\alpha$ (Eq. 1.76) in terms of normal coordinate $Q$, eigenfrequency $\omega_M$, and reduced mass $\mu_M$. Since the oscillator is driven by an intense electric external field $E(t)$, we assume that the Lagrangian contains—in addition to the harmonic potential term $U_h = \mu_M \omega_M^2 Q^2$—an anharmonic potential term $U_{an}$ which arises from interaction of the dipole moment $p(t)$ induced by the field $E(t)$ (see beginning of Section 3.1),

$$p(t) = \alpha\{Q(t)\}E(t),$$

with the field $E(t)$. Writing for $U_{an}$ the well-known relation

$$U_{an} = -\int dE\,\alpha E(t) = -(\tfrac{1}{2})\alpha E^2,$$

we obtain for the Lagrangian equation of motion (see Note 6)

$$\left(\frac{d^2}{dt^2}\right)Q(t) + \Gamma_M\left(\frac{d}{dt}\right)Q(t) + \omega_M^2 Q(t) = (2\mu_M)^{-1}\left[\alpha_{(1)}\{Q\}E\right]E,$$

$$(1.103)$$

where $\alpha_{(1)} = (\partial\alpha/\partial Q)|_{Q_0}$ (see Eq. 1.37), $\mu_M\omega_M^2 Q^2$ is the kinetic restoring energy, and $\Gamma_M\mu_M(d/dt)Q(t)$ the frictional force with damping constant $\Gamma_M$.

Equation 1.103 is solved by an iterative procedure (the details are not of interest) for $Q(t)$ or the polarization $P(t) = V^{-1}\Sigma_j p_j(t)$, where $V$ is a volume element. The solution leads to a general expression in terms of time-dependent susceptibilities $\chi^{(n)}$ of increasing order $n$ of iteration

$$P_i(t) = \int_{-\infty}^{\infty} dt_1 \chi_{ij}^{(1)}(t_1) E_j(t - t_1)$$

$$+ \int_{-\infty}^{\infty} dt_1 \int_{-\infty}^{\infty} dt_2 \chi_{ijk}^{(2)}(t_1, t_2) E_j(t - t_1) E_k(t - t_2)$$

$$+ \int_{-\infty}^{\infty} dt_1 \int_{-\infty}^{\infty} dt_2 \int_{-\infty}^{\infty} dt_3 \chi_{ijkl}^{(3)}(t_1, t_2, t_3) E_j(t - t_1) E_k(t - t_2) E_l(t - t_3)$$

$$+ \cdots$$

$$(1.104)$$

where

$$\chi^{(1)}(t_1) \propto Q(t_1),$$

$$\chi^{(2)}(t_1, t_2) \propto \int_{-\infty}^{\infty} d\tau\, Q(\tau) Q(t_1 - \tau) Q(t_2 - \tau), \text{ etc.} \tag{1.105}$$

The solution of Eq. 1.103, carried up to third order of iteration, reads

$$P(t) = C \exp\left(-\frac{\Gamma_M t}{2}\right) \cos\left\{\left(\omega_M^2 - \frac{1}{4}\Gamma_M^2\right)^{1/2} t + \varphi\right\} \alpha_{(1)} E(t)$$

$$+ (2\mu_M)^{-1} \alpha_{(1)} E(t) \int_{-\infty}^{\infty} d\tau\, \mathscr{A}_M(\tau) \alpha_{(1)} E^2(t - \tau)$$

$$\equiv P^{(1)}(t) + P^{(3)}(t) \tag{1.106}$$

with

$$\mathscr{A}_M(t) = \left(\omega_M^2 - \frac{\Gamma_M^2}{4}\right)^{-1/2} \exp\left(-\frac{\Gamma_M t}{2}\right) \sin\left[\left(\omega_M^2 - \frac{\Gamma_M^2}{4}\right)^{1/2} t\right].$$

($C$ and $\varphi$ are constants of integration.)

The linear term $P^{(1)}$ of Eq. 1.106 represents the spontaneous (ordinary) Raman effect: Set $\Gamma_M \ll \omega_M$ and $E(t) = E(\omega_L)\cos(\omega_L t)$, and find

$$P(t) = C\alpha E(\omega_L)\cos(\omega_M t + \varphi)\cos(\omega_L t)$$

$$= \left(\tfrac{1}{2}\right) C\alpha E(\omega_L)\{\cos[(\omega_L - \omega_M)t - \varphi] + \cos[(\omega_L + \omega_M)t + \varphi]\},$$

where $\omega_L - \omega_M$ and $\omega_L + \omega_M$ are the Stokes and anti-Stokes frequencies. (Recall that we give a classical derivation.)

The third-order iterative term $P^{(3)}$ describes the polarization of the medium due to the lowest-allowed nonlinear interaction between the laser field and the system. (Nonlinear, even effects vanish in unstructured liquids.)[50]

We now show how $P^{(3)}$ leads to an amplification of a Stokes signal. [Remember that a (weak) initial Stokes wave is present on account of the term $P^{(1)}$.] First, we transform $P^{(3)}$ into the frequency domain. We obtain two contributions:

$$P^{(3)}(\omega_S) = \chi^{(3)}(\omega_S; \omega_S, \omega_L, -\omega_L)|E(\omega_L)|^2 E(\omega_S),$$

$$P^{(3)}(\omega_L) = \chi^{(3)}(\omega_L; \omega_L, \omega_S, -\omega_S)|E(\omega_S)|^2 E(\omega_L). \tag{1.107}$$

The notation $\omega_i; \omega_j, \omega_k, \omega_l$ means $\omega_i = \omega_j + \omega_k + \omega_l$. The quantity $\chi^{(3)}(\omega)$ denotes a third-order iterative susceptibility (it is a fourth-rank tensor—see Eq. 1.104) with the property

$$\chi^{(3)}(\omega_S; \omega_S, \omega_L, -\omega_L) = \chi^{(3)*}(\omega_L; \omega_L, \omega_S, -\omega_S)$$

$$= \frac{\alpha_{(1)}}{4\mu_M} \tilde{\mathscr{A}}_M^*(\omega_L - \omega_S), \qquad (1.108)$$

where $\tilde{\mathscr{A}}_M(\omega)$ is the Laplace-Fourier transform of $\mathscr{A}_M(t)$ (Eq. 1.106),[51]

$$\tilde{\mathscr{A}}_M(\omega) = \{-\omega^2 + i\Gamma_M \omega + \omega_M^2\}^{-1}. \qquad (1.109)$$

A few explanatory remarks about the frequency arguments of $\chi^{(3)}$ in Eqs. 1.107 and 1.108 are helpful here. We first remember that we deal with discrete frequencies. In fact, $P^{(3)}(\omega) = P(\omega_j, \omega_k, \omega_l)$ means that $\chi^{(3)}$ is defined in $(\omega_j, \omega_k, \omega_l)$ space. Hence, the amplitude of $P^{(3)}$ can be caused by three, two, or one discrete field strength amplitudes,

$$P^{(3)}(\omega) \propto \chi^{(3)}(\omega; \omega_j, \omega_k, \omega_l)E(\omega_j)E(\omega_k)E(\omega_l), \qquad (1.110)$$

with (see above) $\omega = \omega_j + \omega_k + \omega_l$.

Two limiting cases are not of interest to us but merit mentioning. The first involves three different $\omega$; it is important in photon mixing. The second involves only one frequency, $\omega'$, which leads to the two contributions

$$P^{(3)}(\omega') \propto \chi^{(3)}(\omega_j \to \omega', \omega_k \to \omega', \omega_l \to -\omega') \times E(\omega')E(\omega')E^*(\omega'),$$

$$P^{(3)}(3\omega') \propto \chi^{(3)}(\omega_j \to \omega', \omega_k \to \omega', \omega_l \to \omega') \times E(\omega')E(\omega')E(\omega').$$

These are significant in "self-focusing" and "third-harmonic generation," respectively.

The interesting combination for the stimulated Stokes Raman process combines two frequencies (laser and Stokes line). A priori, this leads to the three combinations

$$|2|\omega'| + |\omega''||, |2|\omega'| - |\omega''||, \quad \text{and } |\omega'|.$$

Hence, we may write three $P^{(3)}$, namely $P^{(3)}(\omega = 2\omega' + \omega'')$, $P^{(3)}(\omega = 2\omega' - \omega'')$, and

$$P^{(3)}(\omega) = P^{(3)}(\omega = \omega' + \omega'' - \omega'')$$

$$\propto \chi^{(3)}(\omega_j \to \omega', \omega_k \to \omega'', \omega_l \to -\omega'')$$

$$\times E(\omega')E(\omega'')E^*(\omega''). \qquad (1.111)$$

The relations remain unchanged upon permutations among the $\omega_j, \omega_k, \omega_l$ and the $\omega', \omega''$, respectively.

Only the last relations of the set, where $\omega'$ is either identified with $\omega_L$ or with $\omega_S$, is of importance. To see this, reflect on a situation where the frequency difference between laser and Stokes lines is in near resonance with a particular molecular vibrational eigenfrequency of the irradiated system,

$$|\omega_L - \omega_S - \omega_M| \sim 0.$$

As Eqs. 1.108 and 1.109 show, the susceptibility then becomes *purely imaginary*:

$$\chi^{(3)}(\omega_S; \omega_S, \omega_L, -\omega_L) = \chi^{(3)*}(\omega_L; \omega_L, \omega_S, -\omega_S)$$

$$= -\chi^{(3)}(\omega_L; \omega_L, \omega_S, -\omega_S)$$

$$= i|\chi_R|, \quad \text{with } \omega_L = \omega_M + \omega_S, \qquad (1.112)$$

where the subscript $R$ denotes resonance.

The consequence of this relation is seen by writing the change of amplitude of both laser and Stokes waves as they travel, say along the $z$-direction, through the medium. Assuming that $\chi^{(3)}$ is polarized along direction $x$, Maxwell's equations[50] yield the corresponding change of the electrical amplitude $E_x(\omega_n, z)$ as

$$\left(\frac{d}{dt}\right)\bar{E}_x(\omega_S, z) = -iC\left[\frac{\omega_S^2}{k(\omega_S)}\right]\chi_x^{(3)}(\omega_S)|\bar{E}_x(\omega_L, z)|^2\bar{E}_x(\omega_S, z)$$

$$\left(\frac{d}{dt}\right)\bar{E}_x(\omega_L, z) = -iC\left[\frac{\omega_L^2}{k(\omega_L)}\right]\chi_x^{(3)}(\omega_L)|\bar{E}_x(\omega_S, z)|^2\bar{E}_x(\omega_L, z),$$

where $C$ is a constant and $k(\omega_n)$ the frequency-dependent wave vector component along $z$. The bar denotes that $E_x$ varies slowly with wavelength. We note that these relations show no phase dependence. Insertion of Eq. 1.112 and integration yields the pair

$$\bar{E}_x(\omega_S, z) = \bar{E}_x(\omega_S, 0)\exp\left\{C\left[\frac{\omega_S^2}{k(\omega_S)}\right]|\chi_R|\int_0^z dz'|\bar{E}_x(\omega_L, z')|^2\right\}$$

$$\bar{E}_x(\omega_L, z) = \bar{E}_x(\omega_L, 0)\exp\left\{-C\left[\frac{\omega_L^2}{k(\omega_L)}\right]|\chi_R|\int_0^z dz'|\bar{E}_x(\omega_S, z')|^2\right\}.$$

$$(1.113)$$

In other words, during the interaction of two waves that traverse a Raman-active medium, the low-frequency wave is (exponentially) amplified, the high-frequency wave is weakened. For each destroyed laser photon a Stokes photon is created. Since $\omega_L \gg \omega_S$, the energy difference $\hbar(\omega_L - \omega_S) \sim \hbar\omega_M$ creates a phonon in the liquid. This, in brief, is the principle of amplifying a Stokes wave in the medium. Together with the method of interrogating the decay of the resulting upper level phonon population and phase coherence excess by a weak probe pulse, as described at the beginning of this section, it comprises the stimulated Raman effect observed through anti-Stokes scattering. At this point in our discussion this should suffice for giving a general idea about the process. We go into the necessary detail when we deal with specific examples and experiments.

## 5. THEORETICAL MODELING OF ROTATIONAL AND VIBRATIONAL RELAXATION

Theoretical modeling of infrared, far-infrared, Raman, and Rayleigh relaxation has common as well as different traits. (It is dealt with extensively in Chapters 2 and 3.)

Among the theoretical approaches common to both phenomena are the "Langevin" and the "memory function" methods. The classical Langevin equation[2b,c] is a differential equation for the velocity $v$ of a particle under the influence of the surrounding medium (Brownian motion). The effect of the medium is divided into a "dynamical friction" $-\Gamma v$ and a "fluctuating part." The dynamical friction is a property that depends on particle and medium coordinates (medium viscosity; mass, size, and velocity of the particle). The fluctuating part is an instantaneous velocity-independent, statistically defined force (its autocorrelation decays much faster than that of the particle velocity) that causes random changes in the particle velocity.

To solve the Langevin equation, a probability distribution for the occurrence of the dynamical quantity $X$ of interest (molecular angular velocity for rotational, oscillator amplitude for vibrational relaxation) is written. The random force is defined in the appropriate coordinates (for instance, an intermolecular torque for rotational, an intermolecular force for vibrational relaxation).

The memory function approach extends the Langevin method by introducing a frequency-dependent friction ("damping") coefficient $\Gamma(\omega)$ and non-instantaneous random forces.[2b] This replaces $-\Gamma X$ by the convolution $-\int_0^t dt' \, \Gamma(t-t') X(t')$, thereby introducing summation over all "events" occurring prior to time $t$ (see also Eq. 1.4).

A priori, the direct approach of theoretically modeling a correlation function or the density matrix (see also Eq. 1.6) is to solve the equation of motion in its various representations (Heisenberg, Schrödinger, or interaction picture). This method is convenient, however, only for vibrational relaxation; it is too difficult to devise the angle- and angular momentum-dependent (and workable) interaction Hamiltonian required for orientational motion. At any rate, once the formal solution of the equation of motion for the vibrational phase, amplitude, phonon occupation operator, or their respective density matrix, is established, various perturbation methods are available: for instance, the "cumulant expansion" and the transformation to an integro-differential equation under simplifying, problem-related assumptions.

Rotational relaxation is conveniently modeled by a chain of stochastic orientational and collisional events and their probability distributions. The explicit form of the event probability functions conforms to the quality of the experimental data. Thereby, we can study orientational phenomena of a wide range of complexity, with the added advantage of having established a detailed physical picture of the orientational motion on the molecular level.

## 6. NOTES

1.  The symbol $\omega$ stands for the angular frequency in units of radian per second; $\omega = 2\pi\nu$, where $\nu$ is the linear frequency in hertz (cycles per second). The unit $cm^{-1}$ denotes the number of wavelengths per centimeter. It is proportional to an energy. For instance, 3000 $cm^{-1}$ corresponds to $\omega = 3000 \times 2\pi c$ ($c$ = velocity of light), approximately $5.7 \times 10^{14}$ radian $sec^{-1}$, or to $\nu = 3000c$, approximately $9 \times 10^{13}$ $sec^{-1}$.

2.  An electric resistance-capacitance (RC) circuit is an example of a system with a simple aftereffect function. Imposing on the circuit a delta (short) input voltage pulse yields an exponentially decreasing output voltage (aftereffect function).

3.  Denoting the Laplace transform $\mathscr{L}$ by a superposed tilde sign and the Laplace variable by $s$, we write ("convolution theorem") Eq. 1.4

$$\mathscr{L}\left\{ \int_0^t dt' \Phi_{BA}(t-t')E(t') \right\} = \tilde{\Phi}_{BA}(s)\tilde{E}(s),$$

where

$$\tilde{\Phi}_{BA}(s) = \int_0^\infty dt\, \Phi_{BA}(t)\exp(-st).$$

Substituting $i\omega$ for $-s$ and applying Eq. 1.2 yields Eq. 1.5.

4.  We remember from the definition of the scalar product $(A|B)$ that $(A|B) = \Sigma(a_i^\dagger|b_j)$, where $^\dagger$ denotes the Hermitian conjugate (adjoint), that

$$\Phi_{BA}(t) = i\left\langle \left[ B(t)^\dagger, A \right] \right\rangle.$$

However, we will deal mainly with operators representing physically observable quantities. Then, as is well known, $A = A^{\dagger}$. We only show the dagger sign when the non-Hermitian property is important to the discussion.

5. The development of matrix elements $(F|I)$ from the electron-radiation interaction Hamiltonian,

$$H_e = -\frac{e}{m_e}\mathbf{p}\cdot\mathbf{A} + \frac{e^2}{2m_e^2}|\mathbf{A}|^2,$$

where $\mathbf{A}$ is the usual vector potential and $m_e, e$ the electronic mass and charge, is straightforward. The interested reader is referred to Ref. 29, pp. 142–143, 190–192. We also note that the contribution of the term in $A^2$ of the relation is negligible—see Ref. 15, pp. 282–284. Incidentally, in this reference the convention $E_I = 0$ is used.

Note that the symbol $\mathbf{A}$ is used, throughout the literature, for the electromagnetic vector potential. Hence it should not be confused with the system variable $A$ introduced in Section 1.3.

6. The notation $\alpha\langle Q(t)\rangle$ and $\alpha_{(1)}\langle Q\rangle$ shall imply that polarizabilities $\alpha, \alpha_{(1)}$ are functions of $Q(t)$ and $Q$, respectively.

The Lagrangian is the difference of kinetic and potential energies. For the Lagrangian equation of motion, see L. Pauling, *Introduction to Quantum Mechanics*, McGraw-Hill, New York, 1935.

# CHAPTER 2

# ROTATIONAL RELAXATION
# AND ORIENTATIONAL MOTION

## A. Random Collision Processes

### 1. GENERAL PRINCIPLES: FREE ROTORS, DEPHASING, AND COLLISIONS

In this chapter we discuss experimental orientational correlation functions obtained from vibration-rotation and pure rotation spectra. It is our aim to compare the experimental functions with theoretical orientational correlation functions based on current models of rotational relaxation.

Rotational relaxation differs from vibrational relaxation by two important points. First, rotational transition energies are small compared to vibrational excitation energies encountered in the mid- or near-infrared and the Raman. This requires ensemble averaging over a Boltzmann distribution of rotational energies. Second, it is useful to include the rotational kinetic energy in the description of rotational relaxation. This requires that theoretical models of rotational relaxation are constructed in terms of orientational (positional) and kinetic (angular momentum) coordinates.

Ensemble averaging over rotational energies in an (classical) ensemble of freely rotating molecules leads to the loss of the initial coherence of the orientational autocorrelation function. Although the axis direction $i$ of each free rotor reorients at constant frequency $\omega_i$ and thus projects amplitude $\cos \omega_i t$ onto a laboratory-fixed axis direction, the amplitude autocorrelation of the classical (Maxwellian) ensemble of free rotors is "smeared out" since the amplitude is averaged over all $\omega_i$ ("dephasing"). To show an example, we take a polar linear molecule whose axis $i$ reorients around the space-fixed direction of its angular momentum vector $\mathbf{J}_i$ with constant angular velocity $\omega_i$, cutting angle $\omega_i t \equiv \eta_i(t)$ during time $t$. Since free rotors do not interact, $\mathbf{J}_i$—which is normal to the molecular axis—is a constant of motion. The direction of $\mathbf{J}$ (with respect to a space-fixed reference system) is arbitrary. All $\mathbf{J}_i$ are isotropically distributed on the surface of a unit sphere, and the average over all directions of $\mathbf{J}$ yields a constant factor (which cancels upon normalization of the correlation function). Hence the autocorrelation function is simply (Eq. 1.41, Chapter 1)

$$\langle \eta_i(0)\eta_i(t)\rangle = \langle \mu_i(0)\mu_i(t)\rangle = \langle \cos(\omega_i t)\rangle = \left\langle \cos\left(\frac{J_i t}{I}\right)\right\rangle \qquad (2.1)$$

where $I$ is the moment of inertia.

We average over all magnitudes of $\mathbf{J}$ with the classical (normalized) Maxwell distribution function (the continuous equilibrium distribution of rotational energies)

$$P(J)\,dJ = (IkT)^{-1}\exp\left(-\frac{J^2}{2kIT}\right)J\,dJ$$

or

$$P(\omega)\,d\omega = \left(\frac{kT}{I}\right)^{-1}\exp\left(-\frac{I\omega^2}{2kT}\right)\omega\,d\omega, \qquad (2.2)$$

where

$$\int d\omega\,\omega^2 P(\omega) = \frac{2kT}{I} \equiv \langle \omega^2\rangle,$$

$kT/I =$ mean-squared angular classical velocity. This gives[52]

$$\langle\cos(\omega_i t)\rangle = \frac{I}{kT}\int_0^\infty \cos(\omega_i t)\omega_i \exp\left(-\frac{I\omega_i^2}{2kT}\right)d\omega_i. \qquad (2.3)$$

The limiting time evolution of the orientational correlation functions of the ensemble of free rotors is readily established.

**1.** For short $t$, Eq. 2.3 is approximated by

$$\langle\cos(\omega t)\rangle = \frac{I}{kT}\int_0^\infty d\omega\left\{1 - \frac{\omega^2 t^2}{2} + \cdots\right\}\omega\exp\left(-\frac{I\omega^2}{2kT}\right)$$

$$= 1 - \left(\frac{kT}{I}\right)t^2 \pm \cdots$$

$$= 1 - \frac{\langle\omega^2\rangle t^2}{2} \pm \cdots \qquad (2.4)$$

2. For long $t$, $\langle \cos(\omega_i t) \rangle \sim 0$ since appreciable contributions to the integral in Eq. 2.3 require that $\omega \to 0$ for $t \to \infty$.

3. For intermediate times, Eq. 2.3 gives negative correlation values. Physically, this means that at the time point of maximum negative correlation most of the molecules have turned through an angle greater than $\pi/2$.[52]

It is noteworthy that the autocorrelation of quantum-mechanical free rotors (free rotors described by a distribution function of discrete values of $|\mathbf{J}|$) periodically regains its original coherence. However, this is of no practical interest to us since we deal with classical rotors.

It is easy to understand that the phase loss of the autocorrelation of rotors in liquids decays eventually at a slower rate than that of free rotors since intermolecular torques exert a friction on the orientational motion. In other words, hindered rotors lose the memory of their original orientation at a slower rate than free rotors. Clearly, for long times

$$\langle \cos(\omega t) \rangle_{t \to \infty} = 0 \qquad (2.5)$$

since, finally, the molecules no longer remember their initial orientational position (regain of equilibrium).

The concept of an ensemble of free rotors is useful since it permits us to construct orientational theoretical correlation functions for real liquids. For instance, we may model the dynamics in terms of alternating events of a free rotation step around an axis direction and a perturbation step that arrests the molecule for a certain period at a certain orientation. The simplest concept is to consider the perturbing event as instantaneous binary collision. In other words, the period of motional arrest is "infinitely short." This is indeed a simplified picture since the magnitude of intermolecular torques exerted during an instantaneous time interval is correspondingly "infinitely large." (Later in the text we present a theoretical model that admits collisional encounters of arbitrary duration.) However, let us not worry about this now and proceed to propose how instantaneous collisional events can affect free orientational motion.

It is ingenious to let such a "collision" act on the rotational angular momentum instead of the orientational ( = positional) coordinates of the molecule. This eliminates the tremendous difficulty of having to model the dynamics of a realistic collision process in a condensed phase; we need only to consider the *effects* of a collision on the angular momentum since a change in its magnitude and orientation causes a corresponding change in the spacial orientation of the molecule. Clearly, it is easier to assume the likely outcome of a collision (based on some preconceived idea) than to model it.

Furthermore, the dynamic processes that determine the correlation functions of the orientational positions of the molecule frequently are on a time scale different from the events that determine the correlation of its rotational angular momentum. This leads to considerable simplifications in the formulation of the theory and has the added advantage of giving access to data that are not comprehensive or accurate enough to merit full evaluation.

For instance, to describe the angular rotational motion between collisions we may use a variable of the angular velocity (Eq. 2.1),[3b]

$$\eta(t) = \int_0^t dt' \, \omega(t'), \tag{2.6}$$

where $\eta(t)$ is the angle through which the molecule has rotated with constant $\mathbf{J}$ during time $t$. On the other hand, to describe the reorientation step of the molecular axis, we assume that we have tilted the molecular axis through polar angle $\theta$ with respect to the external reference system because we have tilted $\mathbf{J}$ (or changed its magnitude, or both). This yields the orientational autocorrelation function of the considered axis direction,

$$\langle \cos\theta(0)\cos\theta(t) \rangle,$$

and its correlation time

$$\tau_\theta = \int_0^\infty dt \, \langle \cos\theta(0)\cos\theta(t) \rangle. \tag{2.7a}$$

Correlation time $\tau_\theta$ is usually longer than the correlation time of $\langle \eta(0)\eta(t) \rangle$,

$$\tau_\eta = \int_0^\infty dt \, \langle \eta(0)\eta(t) \rangle. \tag{2.7b}$$

In other words, the memory of the original angular position $\eta(0)$—a position from which the molecule is moving away by rotating around a temporarily constant rotation axis—generally fades on a faster time scale than the memory of the original axis direction $\theta(0)$—a direction that is successively tilted by the collisional events (Fig. 1.3, Chapter 1). In terms of our physical picture of free rotation steps and collisional events, we may say that for sufficiently short times we ought to observe free rotor motion since the collisions have not yet intervened significantly. (Of course, for Raman and Rayleigh spectra the relations 2.7 are written in terms of second-order polynomials—see Sections 3, Chapter 1.)

We see that the use of positional and angular momentum coordinates offers indeed a flexible description of orientational motion. The set of angular momentum coordinates characterizes the purely kinetic (inertial) aspects of the orientational motion. The set of spatial coordinates characterizes the collisional-frictional (irreversible) aspects of the orientational motion. As we discuss later in this chapter, this concept can be refined by supposing orientational motion in terms of alternating events of free rotation steps of arbitrary duration and periods of motional arrest of arbitrary nature and duration. For instance, a molecule can perform librational motions during periods of orientational arrest.

It is apparent that the nature of the liquid-phase collision dynamics determines the character of a rotational relaxation process. This has led us to divide Chapter 2 into two parts. The first deals with orientational phenomena in which molecular collisions occur "at random." The second treats the more realistic regime of partly "coherent" sequences of collisions. We demonstrate that the assumption of a random collision sequence satisfactorily explains the orientational dynamics of many molecular liquids, but that a nonrandom distribution of times between collisions must be invoked to describe the more remarkable manifestations of rotational relaxation processes.

In the sections that follow immediately we discuss a theory of orientational motion based on randomly occurring collision events that affect the rotational angular momentum coordinates of the molecule.

## 2. THEORIES OF THE *M-* AND *J-*EXTENDED ROTATIONAL DIFFUSION MODELS

The simplest theoretical models that describe rotational relaxation in terms of sequential steps are the so-called *M-* and *J-*diffusion models introduced in the early 1960s.[53] In the following paragraphs we discuss this in detail for a symmetric top molecule. This furnishes a sufficiently general case (it leads easily to the linear and spherical tops) yet contains the minimum molecular symmetry to permit ready experimental verification and exploitation.

There is little doubt that newer theories of orientational motion already supersede—and certainly are going to—the *J-* and *M-*extended diffusion models in their level of approximation. Nevertheless, the *J-* and *M-*models present a tremendous jump in our understanding of rotational relaxation on the molecular level and can be considered as "basis functions" of many, more sophisticated, stochastic theories of orientational motion.

The assumptions that form the basis of the orientational motion described by the *J-* and *M-*extended diffusion models are the following:

1. The molecules undergo periods of free rotation during an arbitrary average time interval $\tau_J$ (see Eqs. 2.6 and 2.7*b*).

2. Each free rotation period is terminated by (1) a soft collision that randomizes the direction of the angular momentum vector (*M-*diffusion) or (2) by a hard collision that, in addition, randomizes the magnitude $J$ of the angular momentum vector (*J-*diffusion). The orientation of the molecule (with respect to a laboratory-fixed frame) is not changed during a collision. Note that during a collision the *M-*diffusion model transfers rotational energy for a symmetric top but only angular momentum for linear and spherical tops.[54]

3. Each collisional step (the event that terminates a free rotation period) is instantaneous (of negligible duration compared to the time between collisions).

4. The probability for a collision to occur depends only on the time elapsed since the previous collision; no memory is retained from collisional events prior to the last collision (Markov process).[55]

These models therefore impose no restrictions on the size of the angle through which the molecules rotate freely between collisions. The byname of "extended diffusion" implies that we do not deal exclusively with rotation through small angles—a condition usually called "rotational diffusion" or "Debye (rotational) diffusion."[56]

The assumptions on which these models are based are approximations because free rotation and instantaneous and binary collisions are not realistic concepts for a condensed phase; molecules in their liquid state nearly "touch." Nevertheless, these simple models describe well some important and experimentally accessible quantities of orientational motion.

To describe the theory quantitatively, we define the following:

1. A laboratory-fixed ("space-fixed") coordinate system $X, Y, Z$.
2. A molecule-fixed coordinate system $x, y, z$ congruent with the inertial axis system of the molecule (diagonal moments of inertia $I_x = I_y, I_z$) with the $z$-axis along the major $C_n$-symmetry axis; $n \geq 3$.
3. We assume that at $t = 0$ coordinates $x, y, z$ are related to coordinates $X, Y, Z$ by Euler angles $\alpha, \beta, \gamma$.

The projection of the angular momentum vector $\mathbf{J}$ ($\mathbf{J}$ is fixed in space) onto the major symmetry axis $z$ of the molecule is described by polar angle $\theta$ and azimuthal angle $\varphi$. For the free symmetric top the component of $\mathbf{J}$ along $z$, $\mathbf{J}_z = \mathbf{J}\cos\theta$, is a constant of motion. During the motion of the molecule the total angular momentum $\mathbf{J}$ precesses in the molecular frame about molecular axis $z$ (equal to the axis of $\mathbf{J}_z$) with angular frequency $\xi J_z I_x^{-1}$, where $\xi = I_x I_z^{-1} - 1$. Simultaneously, $\mathbf{J}$ nutates about $\mathbf{J}_z$ with angular frequency $J I_x^{-1}$ (see Note 1 at end of this chapter).

Hence defining at $t = 0$ the direction of $\mathbf{J}$ in $x, y, z$ by $\theta$ and $\varphi$, the transformation from the laboratory-fixed ($X, Y, Z$) to the molecule-fixed ($x, y, z$) systems at time $t$ is accomplished by the three successive rotational transformations in the sets of Euler angles $(\alpha, \beta, \gamma)$, $(\varphi, \theta, 0)$, and $(JtI_x^{-1}, -\theta, -\varphi + \xi I_x^{-1} Jt\cos\theta)$. [The reason that the angles $-\theta$ and $-\varphi$ appear in the third set is easy to see: By hypothesis (see under the preceding definition 2), at $t = 0$ frame $X, Y, Z$ is transformed into frame $x, y, z$ by rotation through the set $(\alpha, \beta, \gamma)$. To undo the rotation by $(\varphi, \theta, 0)$, we rotate by $(0, -\theta, -\varphi)$.[57]]

Designating a Wigner rotation matrix of order (or rank) $j$ by $D_{mn}^{(j)}(\alpha, \beta, \gamma)$ and its inverse by $D_{mn}^{(j)-1}(\alpha, \beta, \gamma) = D_{mn}^{(j)}(-\gamma, -\beta, -\alpha) = D_{nm}^{(j)*}(\alpha, \beta, \gamma)$, the transformation from the $X, Y, Z$ to the $x, y, z$ coordinate frame for the molecule in its first free rotation step at time $t - t_0 = t$ ($t_0 = 0$) is accomplished by the stochastic rotation operator (we omit rank index $j$)

$$\mathbf{U}(1,t) = \mathbf{D}(\alpha,\beta,\gamma)\mathbf{D}(\varphi_1,\theta_1,0)$$

$$\times \mathbf{D}\left(I_x^{-1}J_1 t, -\theta_1, -\varphi_1 + I_x^{-1}\xi J_1 t\cos\theta_1\right)$$

$$= \mathbf{D}(\alpha,\beta,\gamma)\mathbf{D}(\varphi_1,\theta_1,0)\mathbf{D}\left(I_x^{-1}J_1 t,0,0\right)$$

$$\times \mathbf{D}^{-1}(\varphi_1,\theta_1,0)\mathbf{D}\left(I_x^{-1}\xi J_1 t\cos\theta_1,0,0\right). \tag{2.8}$$

We assume that at $t = t_1$ this first free rotation step (loosely called the first diffusion step) is terminated by a collision that changes the magnitude of the angular momentum to $J_2$ and its orientation to $\theta_2, \varphi_2$ (with respect to $x, y, z$). Hence

$$\mathbf{U}(2,t) = \mathbf{D}(\alpha,\beta,\gamma)\mathbf{D}(\varphi_1,\theta_1,0)$$

$$\times \mathbf{D}\left(I_x^{-1}J_1 t_1,0,0\right)\mathbf{D}^{-1}(\varphi_1,\theta_1,0)$$

$$\times \mathbf{D}\left(I_x^{-1}\xi J_1 t_1\cos\theta_1,0,0\right)\mathbf{D}(\varphi_2,\theta_2,0)$$

$$\times \mathbf{D}\left(I_x^{-1}J_2(t - t_1),0,0\right)\mathbf{D}^{-1}(\varphi_2,\theta_2,0)$$

$$\times \mathbf{D}\left(I_x^{-1}\xi J_2(t - t_1)\cos\theta_2,0,0\right).$$

In this way we continue until the $i$th diffusion step, which commences at $t = t_{i-1}$ and ends at $t = t_i$. We then write the coordinate transformation which relates $x, y, z$ and $X, Y, Z$ during the $n$th diffusion step as follows:

$$\mathbf{U}(n,t) = \mathbf{D}(\alpha,\beta,\gamma)\prod_{i=1}^{n}\left\{\mathbf{D}(\varphi_i,\theta_i,0)\right.$$

$$\times \mathbf{D}\left(I_x^{-1}J_i(t_i - t_{i-1}),0,0\right)\mathbf{D}^{-1}(\varphi_i,\theta_i,0)$$

$$\left.\times \mathbf{D}\left(I_x^{-1}\xi J_i(t_i - t_{i-1})\cos\theta_i,0,0\right)\right\}, \tag{2.9}$$

where $J_i$, $\theta_i$, and $\varphi_i$ are the magnitude and orientation of the angular momentum vector with respect to the $x, y, z$ system at time $t_{i-1}$. The symbol $\prod$ signifies the product over all $i$.

To obtain the orientational correlation function we multiply $U(n,t)$ into $U(1,0)$, prescribe a probability distribution $\not{p}(n,t)$ of finding a molecule in its $n$th rotational diffusion step at time $t$, sum over all $n$ diffusion steps and intermediate time $t_i$, and finally take the ensemble average over original positions $\alpha, \beta, \gamma$ and angular momenta $\mathbf{J}(|\mathbf{J}|, \theta, \varphi)$.

Since we assume instantaneous, uncorrelated collisional events, $\not{p}(n, t)$ is a Poisson distribution

$$\not{p}(n, t) = \left[ \frac{(t/\tau_J)^{n-1}}{(n-1)!} \right] \exp\left( -\frac{t}{\tau_J} \right), \tag{2.10}$$

where $\tau_J$, the average time between collisions, is set equal to the angular momentum correlation time defined by

$$\langle \mathbf{J}(0) \cdot \mathbf{J}(t) \rangle \langle |\mathbf{J}(0)|^2 \rangle^{-1} = \exp\left( -\frac{t}{\tau_J} \right). \tag{2.11}$$

The exponential form of Eq. 2.11 signifies that the memory of the initial coherence of the autocorrelation of $\mathbf{J}(t)$ is instantaneously lost—the prerequisite for the assumed instantaneous forces during a collision (see Note 2).

From Eqs. 2.9 and 2.10 we obtain a general orientational correlation function for the $k$th component of a $j$th order tensor in spherical notation,

$$\hat{G}_k^{(j)}(t) = \sum_{n=1}^{\infty} \not{p}(n, t) \langle U_k^{(j)}(n, t) U_k^{(j)*}(1, 0) \rangle \langle |U_k^{(j)}(1, 0)|^2 \rangle^{-1}. \tag{2.12}$$

Inserting the matrix elements of the $\mathbf{D}^{(j)}$ (Eq. E.4$b$, Appendix E), we obtain the normalized orientational correlation function

$$\hat{G}_k^{(j)}(t) = \sum_{n=1}^{\infty} \not{p}(n, t) \langle D_{ma}^{(j)}(\alpha, \beta, \gamma) D_{ab}^{(j)}(\varphi_1, \theta_1, 0)$$

$$\times D_{cb}^{(j)*}(\varphi_1, \theta_1, 0) \exp\left[ -iI_x^{-1} J_1 t_1 (b + c\xi\cos\theta_1) \right]$$

$$\times D_{cd}^{(j)}(\varphi_2, \theta_2, 0) D_{ed}^{(j)*}(\varphi_2, \theta_2, 0)$$

$$\times \exp\left[ -iI_x^{-1} J_2 (t_2 - t_1)(d + e\xi\cos\theta_2) \right] \cdots$$

$$\times D_{yz}^{(j)}(\varphi_n, \theta_n, 0) D_{kz}^{(j)*}(\varphi_n, \theta_n, 0)$$

$$\times \exp\left[ -iI_x^{-1} J_n (t - t_{n-1})(z + k\xi\cos\theta_n) \right]$$

$$\times D_{mk}^{(j)*}(\alpha, \beta, \gamma) \rangle \langle |D_{mk}^{(j)}(\alpha, \beta, \gamma)|^2 \rangle^{-1}, \tag{2.13}$$

where summation over repeated indices is understood.

First, we average isotropically over all original positions (see also Section 3.4 of Chapter 1),[41]

$$\langle D_{ma}^{(j)}(\alpha, \beta, \gamma) D_{mk}^{(j)*}(\alpha, \beta, \gamma) \rangle_{\alpha, \beta, \gamma} = (2j+1)^{-1} \delta_{ak}.$$

Since the rotational energy is independent of azimuthal angle $\varphi$, we average isotropically over $\varphi$ (Eq. E.4b, Appendix E; Eq. 1.24, Chapter 1),

$$\langle D_{ab}^{(j)}(\varphi,\theta,0)D_{cb}^{(j)*}(\varphi,\theta,0)\rangle_\varphi = \delta_{ac}\left[d_{ab}^{(j)}(\theta)\right]^2.$$

Subsequently, we average over all intermediate times $t_{n-1}$. [Recall that instance $t_{n-1}$ is a random event, governed by $\not\!\!\rho(n,t)$.] This yields

$$\hat{G}_k^{(j)}(t) = \sum_{n=1}^{\infty} \not\!\!\rho(n,t)\left[\frac{(n-1)!}{t^{n-1}}\right]\int_0^t dt_{n-1}\int_0^{t_{n-1}} dt_{n-2}\cdots\int_0^{t_2} dt_1$$

$$\times\left\langle\prod_{l=1}^{n}\left\{\sum_{a=-j}^{j}\left[d_{ka}^{(j)}(\theta_l)\right]^2\exp\left[-iI_x^{-1}(t_l-t_{l-1})J_l(a+k\xi\cos\theta_l)\right]\right\}\right\rangle,$$

$$(2.14)$$

since there are $(n-1)!$ possible ways of writing the stochastic operator $U(n,t)$. The factor

$$\frac{(n-1)!}{t^{n-1}}dt_1\,dt_2\cdots dt_{n-1}$$

gives the product probability that a molecule at time $t$ has undergone $n-1$ collisions (finds itself in the $n$th diffusion step) during the time intervals $t_1$ to $t_1+dt_1\cdots t_{n-1}$ to $t_{n-1}+dt_{n-1}$.[52] Then

$$\hat{G}_k^{(j)}(t) = \exp\left(-\frac{t}{\tau_J}\right)\sum_{n=1}^{\infty}\left(\frac{1}{\tau_J}\right)^{n-1}\int_0^t dt_{n-1}\int_0^{t_{n-1}} dt_{n-2}\cdots\int_0^{t_2} dt_1$$

$$\times\left\langle\prod_{l=1}^{n}\left\{\sum_{a=-j}^{j}\left[d_{ka}^{(j)}(\theta_l)\right]^2\exp\left[-iI_x^{-1}(t_l-t_{l-1})J_l(a+k\xi\cos\theta_l)\right]\right\}\right\rangle_{\theta_l,J_l}$$

$$(2.15)$$

where averaging over $J_l$, $\theta_l$ still must be done. In other words, we have not yet made use of the specific *M*- and *J*-model prescriptions for averaging over **J**. Before we proceed with this, it is useful to introduce the concept of a memory function. This not only helps us to appreciate the general nature of the *M*- and *J*-extended diffusion model formalism but also simplifies the theoretical and numerical application of these models.

## 2.1.    Extended Rotational Diffusion Models: The Memory Function Approach, Langevin Equation, and Projection Operators.

Before we employ the concept of a memory function for a deeper discussion of the $M$- and $J$-extended rotational diffusion models, we need more preparation.

To this effect we recall the theory of Brownian diffusion. It describes the perpetual irregular velocity $v$ of colloidal particles immersed in a fluid by the so-called (classical) Langevin equation[58]

$$\frac{dv}{dt} = -\Gamma v(t) + \mathscr{F}(t).$$    (2.16)

The equation relates the acceleration of the particle to a general friction coefficient $\Gamma$ and to fluctuating forces $\mathscr{F}(t)$ of the medium. The terms in Eq. 2.16 have been given names. The dynamic friction $-\Gamma v$ is the "systematic" or "secular" part, and the force term $\mathscr{F}(t)$ is called the "fluctuating" or "nonlinear" part.

The secular term is expressed by the well-known Stokes law,

$$\Gamma = 6\pi \frac{d}{2} \eta_m M,$$    (2.17)

where $\eta_m$ is the macroscopic viscosity of the medium and $d$, $M$ represent diameter and mass of the (spherical) particle.

The fluctuating part $\mathscr{F}(t)$, characteristic of the Brownian motion, arises from the random changes of orientations and positions of the molecules of the fluid. $\mathscr{F}(t)$ is therefore only defined within the corresponding statistical nature of these processes. Furthermore, $\mathscr{F}(t)$ varies rapidly with respect to $v(t)$ of the suspended particle.

We now assume that Eq. 2.16 is applicable to a pure liquid. In other words, we no longer distinguish between the suspended particles and the suspending fluid. Furthermore, we replace $v(t)$ by a general stochastic system variable $A(t)$ and define $\Gamma$ to be a frequency (or time) dependent quantity $K$, writing[2b, 59]

$$\frac{d}{dt} A(t) = -\int_0^t dt' K(t') A(t - t') + \mathscr{F}(t).$$    (2.18)

(We follow through with the derivation of Eq. 2.18 later.) To solve for $K(t)$, we take the Laplace transform $\mathscr{L}$ of Eq. 2.18, obtaining[60]

$$s\tilde{A}(s) - A(0) = -\tilde{K}(s)\tilde{A}(s) + \tilde{\mathscr{F}}(s).$$

[The tilde sign denotes the Laplace-transformed function, $s$ is the Laplace variable, and $A(0) = A(t)$ at $t = 0$.] Reforming gives

$$\tilde{A}(s) = \left[s + \tilde{K}(s)\right]^{-1} A(0) + \left[s + \tilde{K}(s)\right]^{-1} \tilde{\mathscr{F}}(s)$$    (2.19)

and inverting into the time domain leads to[61]

$$A(t) = Z(t)A(0) + \int_0^t dt' \, Z(t - t') \mathcal{F}(t'), \qquad (2.20)$$

where $Z(t)$ is defined by its Laplace transform

$$\tilde{Z}(s) = \left[ s + \tilde{K}(s) \right]^{-1}. \qquad (2.21)$$

Equation 2.20 shows that the stochastic variable $A(t)$ is split into two terms. One term,

$$A'(t) = \langle Z(t)A(0) \rangle, \qquad (2.22a)$$

has a time evolution determined by $Z(t)$. The time development of term

$$A''(t) = \int_0^t dt' \, Z(t - t') \mathcal{F}(t') \qquad (2.22b)$$

is more complicated: Obviously, $A''(t)$ is nonlinear. (We remind the reader that a primed symbol is never used by us to signify differentiation.)

To obtain from Eq. 2.18 the (classical) correlation function of the stochastic variable $A(t)$,

$$\langle A(t)A(0) \rangle = \langle A(0)A(t) \rangle,$$

we employ the principle of a "projection operator." A projection operator is an operator that distributes the operand within two orthogonal (that is, nonoverlapping) subspaces. One result of this separation is termed to be "relevant," the other is "nonrelevant."[62] To clarify this,[63] consider that any arbitrary ket $|a\rangle$ possesses a projection $|a_s\rangle$ in a chosen subspace $S$ of Hilbert space $\mathcal{H}$ and a projection $|a_s^x\rangle$ in a complementary (that is, nonoverlapping) subspace $S^x$ of $\mathcal{H}$:

$$|a\rangle = |a_s\rangle + |a_s^x\rangle.$$

Let now $|s\rangle$ be a normalized vector that spans $S$ of $\mathcal{H}$. By construction,

$$\langle s|a_s^x\rangle = 0; \qquad |a_s\rangle = c|s\rangle.$$

Multiplying with $|s\rangle$ from the left gives

$$\langle s|a_s\rangle = c.$$

Therefore

$$|a_s\rangle = |s\rangle\langle s|a_s\rangle$$

and

$$\bar{P}_s \equiv |s\rangle\langle s|. \tag{2.23}$$

We see that $\bar{P}_s$ is indeed a projection operator:

1. $\bar{P}_s|a\rangle = |s\rangle\langle s|a_s\rangle + |s\rangle\langle s|a_s^x\rangle = |a_s\rangle.$
2. $(1 - \bar{P}_s)|a\rangle = |a_s\rangle + |a_s^x\rangle - |a_s\rangle = |a_s^x\rangle,$ or $(1 - \bar{P}_s)|a_s\rangle = 0.$

This also follows at once from $(1 - |s\rangle\langle s|)|a_s\rangle = |a_s\rangle - |a_s\rangle$ (see Note 3).

To get the desired correlation function

$$A'(t) = \langle A(t)A(0)\rangle \tag{2.24}$$

from $A(t)$, we must project $A(t)$ of Eq. 2.20 such that its linear part, $A'(t)$, is the projection of $A(t)$ on the $A(0) = A(t = 0)$ axis. Thus writing the identity

$$A(t) = \bar{P}_A|A(t)\rangle + (1 - \bar{P}_A)|A(t)\rangle = A'(t) + A''(t)$$

we obtain the relevant part

$$\bar{P}_A|A(t)\rangle = Z(t)A(0), \tag{2.25}$$

and the irrelevant part

$$(1 - \bar{P}_A)|A(t)\rangle = \int_0^t dt' \, Z(t - t')\mathscr{F}(t').$$

Inserting $\bar{P}_A = |A\rangle\langle A|$ (Eq. 2.23) and normalizing gives

$$Z(t) = \langle A(0)|\bar{P}_A A(t)\rangle \langle A(0)A(0)\rangle^{-1}$$

$$= \langle A(0)|A\rangle\langle A|A(t)\rangle \langle A(0)A(0)\rangle^{-1}$$

$$= \langle A(0)A(t)\rangle \langle A(0)A(0)\rangle^{-1} \equiv \hat{A}'(t). \tag{2.26}$$

Now, inspection of Eq. 2.18 shows that

$$\mathscr{F}(0) = \frac{d}{dt}A(t)\bigg|_{t=0} \equiv \dot{A}(0).$$

Consequently,

$$\langle \mathscr{F}(0)\mathscr{F}(t)\rangle = \langle \dot{A}(0)\dot{A}(t)\rangle + \int_0^t dt' \, K(t - t')\langle \dot{A}(0)A(t')\rangle.$$

In Appendix B we prove that

$$\langle \dot{A}(0)A(t)\rangle = -\left(\frac{d}{dt}\right)\langle A(0)A(t)\rangle \equiv -\dot{A}'(t) \qquad (2.27)$$

and that

$$\langle \dot{A}(0)\dot{A}(t)\rangle = -\frac{d^2}{dt^2}\langle A(0)A(t)\rangle \equiv -\ddot{A}'(t). \qquad (2.28)$$

Inserting these two relations gives

$$\langle \mathscr{F}(0)\mathscr{F}(t)\rangle = -\ddot{A}'(t) - \int_0^t dt'\, K(t-t')\dot{A}'(t).$$

Taking again the Laplace transform leads to[60] (Appendix B)

$$\mathscr{L}\{\langle \mathscr{F}(0)\mathscr{F}(t)\rangle\} = -s^2\tilde{A}'(s) + sA'(0) - \tilde{K}(s)\left[s\tilde{A}'(s) - A'(0)\right]$$

$$= -\left[s + \tilde{K}(s)\right]\left[s\tilde{A}'(s) - A'(0)\right]. \qquad (2.29)$$

We know already the Laplace transform of the correlation function of $A'(t)$: It is given by Eq. 2.21. Hence Eq. 2.29 reads

$$\mathscr{L}\{\langle \mathscr{F}(0)\mathscr{F}(t)\rangle\}\langle A(0)A(0)\rangle^{-1} = -\left[s + \tilde{K}(s)\right]\left[s\tilde{A}'(s) - 1\right]$$

$$= \left[s + \tilde{K}(s)\right]\tilde{K}(s)\left[s + \tilde{K}(s)\right]^{-1} = \tilde{K}(s),$$

or

$$\langle \mathscr{F}(0)\mathscr{F}(t)\rangle\langle A(0)A(0)\rangle^{-1} = K(t). \qquad (2.30)$$

In summary, we have obtained the two relations

$$\tilde{Z}(s) = \mathscr{L}\{\langle A(0)A(t)\rangle\} = \left[s + \tilde{K}(s)\right]^{-1}$$

and

$$K(t) = \langle \mathscr{F}(0)\mathscr{F}(t)\rangle\langle A(0)A(0)\rangle^{-1}.$$

First, they express that $K(t)$ and correlation function $Z(t) = A'(t) = \langle A(0)A(t)\rangle$ of the stochastic dynamic variable $A(t)$ are connected. Second, they show that $K(t)$ is proportional to the correlation function of the random forces in the medium.

We now demonstrate the physical meaning of $K(t)$ and the different time scales of the evolution of correlation function $Z(t)$ and of memory function

$K(t)$. We multiply Eq. 2.18 from the left by $|A(0)\rangle$.[2c] This leads to

$$\frac{d}{dt}\langle A(0)A(t)\rangle = -\int_0^t dt' K(t-t')\langle A(0)A(t')\rangle + \langle A(0)\mathcal{F}(t)\rangle. \quad (2.31)$$

On the other hand, $\langle A(0)\mathcal{F}(t)\rangle = 0$ since $\mathcal{F}(t)$ does not exist in the subspace of $A(0)$ [see Eqs. 2.22$b$, 2.25]. Consequently

$$\frac{d}{dt}\langle A(0)A(t)\rangle = -\int_0^t dt' K(t-t')\langle A(0)A(t')\rangle. \quad (2.32)$$

Now the meaning of $K(t)$ as a memory function in relation to $A'(t)$ is clear: The rate of evolution of the correlation function $\langle A(0)A(t)\rangle$ at $t=t$ depends on the characteristics of the function $K(t)$ for all $t' \leq t$. Usually the coherence of $K(t)$ has long decayed before that of $\langle A(0)A(t)\rangle$ has undergone a comparable change.

By a clever guess of $K(t)$ we can use Eq. 2.32 to arrive at an acceptable theoretical relation for $\langle A(0)A(t)\rangle$. The following application to the $M$- and $J$-extended rotational diffusion models will bear this out.

We start off with Eq. 2.32, writing it in our notation

$$\frac{d}{dt}G(t) = -\int_0^t dt' K(t')G(t-t') \quad (2.33)$$

and assume for $K(t)$

$$K(t) = K_F(t)\exp\left(-\frac{t}{\tau}\right), \quad (2.34)$$

where $K_F(t)$ denotes the memory function for free rotors and $\tau$ the time between instantaneous collisions.[64] Therefore

$$\frac{d}{dt}F(t) = -\int_0^t dt' K_F(t')F(t-t'), \quad (2.35)$$

where $F(t)$ is the orientational correlation function for the free rotors (diatomic, spherical, and so forth).

How good is our choice of $K(t)$? First, it is more realistic than an impulsive (delta function) form since our choice leads to zero-slope for $G(0)$ [see Appendix B]; a delta function does not. At $t=0$ collisions have not yet perturbed the orientational motion; the correlation function should therefore decrease parabolically during initial times owing to kinetic dephasing effects (Eq. 2.4). [We note that $K(t)$ has not zero-slope at the origin; we show later that this is responsible for instantaneous torques.] Second, the choice of $K(t)$ leads to the correct limiting behavior of $G(t)$. If $\tau \to 0$, the collision frequency $1/\tau$ is larger than frequencies of angular molecular motion. $K(t)$ therefore approximates a delta pulse (memory is immediately lost), and $G(t)$ approximates an exponential (see Eq. 2.33). If $\tau \to \infty$, $G(t)$ approximates the free rotor correlation function $F(t)$.

Finally, we demonstrate how readily this memory function approach lets us construct the *J*- and *M*-correlations of the actual (hindered) rotors. Using again the properties of Laplace transforms, we obtain from Eqs. 2.33 and 2.34[60,65]

$$s\tilde{G}(s) - G(0) = -\tilde{K}_F(s + \tau^{-1})\tilde{G}(s) \qquad (2.36a)$$

and from Eq. 2.35

$$s\tilde{F}(s) - F(0) = -\tilde{K}_F(s)\tilde{F}(s). \qquad (2.36b)$$

Inserting $G(0) = F(0) = 1$, replacing variable $s$ by $s - \tau^{-1}$ in Eq. 2.36a, and solving for $\tilde{G}(s - \tau^{-1})$, gives

$$\tilde{G}(s - \tau^{-1}) = \tilde{F}(s)\left[1 - \tau^{-1}\tilde{F}(s)\right]^{-1}$$

$$= \tilde{F}(s) \sum_{n=0}^{\infty} \tau^{-n}\tilde{F}(s)^n.$$

[Note that we have eliminated $K_F(t)$, the free rotor memory function.] Transforming into the time domain yields the convolution product series (we identify now $\tau$ with $\tau_J$ of Eq. 2.11)[61,65]

$$G(t) = \exp\left(-\frac{t}{\tau_J}\right)\left\{F(t) + \sum_{n=1}^{\infty} \tau_J^{-n}F_n(t)\right\}, \qquad (2.37)$$

with

$$F_n(t) = \int_0^t dt_1 \, F(t - t_1) \int_0^{t_1} dt_2 \, F(t_1 - t_2) \cdots \int_0^{t_{n-1}} dt_n \, F(t_{n-1} - t_n)F(t_n)$$

$$= \int_0^t dt_n \, F(t - t_n) \int_0^{t_n} dt_{n-1} \, F(t_n - t_{n-1}) \cdots \int_0^{t_2} dt_1 \, F(t_2 - t_1)F(t_1),$$

if $t_n$ is replaced by $t_1$, etc. Note that the $n$th term represents the contribution from molecules that are in their $(n + 1)$th diffusion step at $t(= t_{n+1})$. This is a slight but immaterial difference from the counting used for Eq. 2.15.

We now average over **J** according to the definitions of *J*- and *M*-diffusion. We recall that in the *J*-diffusion model **J** is randomized at each collision. We perform the average over all statistically independent pairs $J_i$, $\theta_i$ with normalized Boltzmann distributions[53,66] of all $\theta$ —at constant $J$—using

$$P_1(x, L) = \exp\left(-\tfrac{1}{2}\xi x^2 L^2\right)N^{-1}, \qquad (2.38)$$

$$P_2(L) = 2\left(\frac{1 + \xi}{\pi\xi}\right)^{1/2} \text{erf}\left\{L\left(\frac{\xi}{2}\right)^{1/2}\right\}L\exp\left(-\frac{L^2}{2}\right). \qquad (2.39)$$

$N$ is a normalization factor

$$\int_{-1}^{1} dx \exp\left(-\frac{\xi x^2 L^2}{2}\right) = \left(\frac{2}{L}\right)\left(\frac{2}{\xi}\right)^{1/2} \mathrm{erf}\left\{L\left(\frac{\xi}{2}\right)^{1/2}\right\},$$

$\mathrm{erf}(X)$ is the error function,

$$\mathrm{erf}(X) = \int_{0}^{X} dy \exp(-y^2),$$

and all variables are expressed in reduced (dimensionless) units: $L = J(I_x k_B T)^{-1/2}$, $x = \cos\theta$, $t^* = t(I_x/k_B T)^{-1/2}$, and $\omega^* = \omega(k_B T/I_x)^{-1/2}$. (To avoid confusion with index $k$, we attach subscript $B$ to the Boltzmann constant.)

Hence, for $J$-diffusion the overall probability function is given by the product of $P_1$ and $P_2$ (Eqs. 2.38, 2.39),

$$P_{12}(x, L) = \left(\frac{1+\xi}{2\pi}\right)^{1/2} L^2 \exp\left[-\frac{L^2(1+\xi x^2)}{2}\right]. \qquad (2.40)$$

Inserting this result into Eq. 2.15, we get

$$G_J^{(j,k)}(t^*) = \exp\left(-\frac{t^*}{\tau_J^*}\right) \sum_{n=1}^{\infty} \left(\frac{1}{\tau_J^*}\right)^{n-1} \int_0^{t^*} dt_{n-1}^* \int_0^{t_{n-1}^*} dt_{n-2}^* \cdots \int_0^{t_2^*} dt_1^*$$

$$\times \prod_{l=1}^{n} \left\{ \int_0^{\infty} dL_l \int_{-1}^{1} dx_l\, P_{12}(x_l, L_l) \sum_{a=-j}^{j} \left[d_{ka}^{(j)}(x_l)\right]^2 \right.$$

$$\left. \times \exp\left[-i(t_l^* - t_{l-1}^*)L_l(a + k\xi x_l)\right]\right\}$$

$$= \exp\left(-\frac{t^*}{\tau_J^*}\right) \sum_{n=1}^{\infty} \left(\frac{1}{\tau_J^*}\right)^{n-1} \left(\frac{1+\xi}{2\pi}\right)^{1/2} \int_0^{t^*} dt_{n-1}^* \int_0^{t_{n-1}^*} dt_{n-2}^* \cdots$$

$$\times \int_0^{t_2^*} dt_1^* \prod_{l=1}^{n} \left\{ \int_0^{\infty} dL_l \sum_{a=-j}^{j} \left[d_{ka}^{(j)}(x_l)\right]^2 \right.$$

$$\left. \times L_l^2 \exp\left[-\tfrac{1}{2}L_l^2(1+\xi x_l^2) - i(t_l^* - t_{l-1}^*)L_l(a + k\xi x_l)\right]\right\}. \qquad (2.41)$$

We now compare Eq. 2.41 with Eq. 2.37 and we note that they are identical. The simple memory function of Eq. 2.34 directly leads to the extended rotational *J*-diffusion model. Therefore, to obtain the *J*-model for rotors of arbitrary shape we only need to construct their free rotor correlation function $F(t)$; this has recently been accomplished for asymmetric top molecules.[67]

For *M*-diffusion, we average with Eq. 2.38 over $\theta$ at each collision but over $L$, with Eq. 2.39, after the last collision. Writing this out (Eq. 2.15) yields

$$G_M^{(j,k)}(t^*) = \exp\left(-\frac{t^*}{\tau_J^*}\right) \sum_{n=1}^{\infty} \left(\frac{1}{\tau_J^*}\right)^{n-1} \int_0^{t^*} dt_{n-1}^* \int_0^{t_{n-1}^*} dt_{n-2}^* \cdots \int_0^{t_2^*} dt_1^*$$

$$\times \int_0^{\infty} dL\, P_2(L) \prod_{l=1}^{n} \left\{ \int_{-1}^{1} dx_l\, P_1(x_l, L) \sum_{a=-j}^{j} [d_{ka}^{(j)}(x_l)]^2 \right.$$

$$\left. \times \exp\left[-i(t_l^* - t_{l-1}^*)L(a + k\xi x_l)\right] \right\}$$

$$= \exp\left(-\frac{t^*}{\tau_J^*}\right) \sum_{n=1}^{\infty} \left(\frac{1}{\tau_J^*}\right)^{n-1} \left(\frac{1+\xi}{2\pi}\right)^{1/2} \int_0^{t^*} dt_{n-1}^* \int_0^{t_{n-1}^*} dt_{n-2}^* \cdots$$

$$\times \int_0^{t_2^*} dt_1^* \int_0^{\infty} dL \prod_{l=1}^{n} \left\{ \int_{-1}^{1} dx_l \sum_{a=-j}^{j} [d_{ka}^{(j)}(x_l)]^2 \right.$$

$$\left. \times L^2 \exp\left[-\tfrac{1}{2}L^2(1+\xi x_l^2) - i(t_l^* - t_{l-1}^*)L(a + k\xi x_l)\right] \right\}. \quad (2.42)$$

Comparison of Eq. 2.42 with Eqs. 2.41 and 2.37 shows that we must replace in Eq. 2.36 and 2.37 $\tilde{G}(s)$ and $\tilde{F}(s)$ by "partial" quantities which have not yet been averaged over $\omega = J/I_x$.[68] We write this as

$$G_M(t) = \exp\left(-\frac{t}{\tau_J}\right) \int_0^{\infty} d\omega\, P_2(\omega) \left\{ F_{\omega}(t) + \sum_{n=1}^{\infty} \tau^{-n} F_{n\omega}(t) \right\}, \quad (2.43)$$

where $P_2(\omega)$ corresponds to Eq. 2.39. Accordingly, the memory function for the *M*-diffusion model is (compare with Eq. 2.34)

$$K_{\omega}(t) = K_{F\omega}(t) \exp\left(-\frac{t}{\tau}\right).$$

Let us emphasize why the memory function of Eq. 2.34 leads to the *J*- and *M*-diffusion models.

1.  The molecules rotate freely before a collision has occurred. Indeed,

$$K(t) \approx K_F(t) \qquad \text{for } t \ll \tau. \quad (2.44a)$$

2. The collisional event erases the memory of the orientational position. Indeed, $K(t)$, which is proportional to the correlation function of the intermolecular random forces of the medium (Eq. 2.30), obeys

$$K(t) \sim 0 \qquad \text{for } t \geq \tau. \qquad (2.44b)$$

## 2.2. J- and M-Extended Rotational Model Simulations, Exponential Decay, and Debye Rotational Diffusion

It is useful to discuss first well-defined simulations of the $J$- and $M$-diffusion models to underline their differences. This will help us when we, subsequently, discuss real liquid systems.

Figure 2.1 displays the numerical evaluation of the correlation function $G_k^{(j)}(t)$ (Eq. 2.15) for the component $k = 0$ of a second-order tensor ($j = 2$) with $\xi = 2.833$ (which corresponds to ethane, $CH_3CH_3$).[66] We recall from Section 3.4.2 of Chapter 1 that $G_0^{(2)}(t)$ corresponds to Eq. 1.81,

$$\hat{G}(t) = \left(\tfrac{1}{2}\right)\left\langle 3[\hat{\mathbf{m}}^z(0)\cdot\hat{\mathbf{m}}^z(t)]^2 - 1 \right\rangle = \left\langle P_2[\cos^z(t)] \right\rangle$$

$$= \left\langle P_2[\hat{\mathbf{m}}^z(0)\cdot\hat{\mathbf{m}}^z(t)] \right\rangle.$$

In other words, if we had contemplated a data comparison of these $J$- and $M$-model computations, we would have to perform a VV-VH Raman scattering experiment (Section 3.2, Chapter 1) on a totally symmetric vibrational mode of ethane.

The results of the simulations are plotted as a function of the reduced collision frequency $\tau_J^{*-1} \equiv \beta^*$ for the $M$-diffusion model in Fig. 2.1a (Eq. 2.42) and for the $J$-model (Eq. 2.41) in Fig. 2.1b. The free rotor correlation function is also displayed. We notice several important and general aspects of the computed orientational correlation functions.

1. For sufficiently short times both models show identical $G(t)$ irrespective of the value of $\beta^*$. The system is still in a free rotor regime where collisions had not yet time to significantly slow the decay rate of $G(t)$.

2. For longer times we see that $G(t)$ decays slower the larger the collision frequency $\beta$. Intermolecular torques are beginning to slow the rate of coherence loss of the orientation of the vibrational transition moment tensor.

3. We notice that $G_M(t)$ decreases slower than $G_J(t)$ at longer times—all other conditions being the same. This effect arises from the different averaging procedures over angular momentum $\mathbf{J}$. To show the onset of this difference, we develop the convolution integrals for $F_n(t)$ [Eq. 2.37] and for $F_{n\omega}(t)$ [Eq. 2.43] up to $n = 1$. We take the simple example of a linear rotor (Eq. 2.4) and obtain

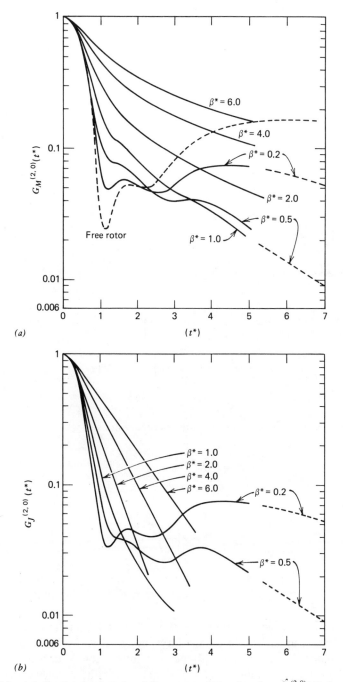

**Figure 2.1.** (a) Simulated orientational Raman correlation functions $\hat{G}_M^{(2,0)}(t^*)$ from Eq. 2.42, using $\xi = 2.833$ and several values of the reduced collision frequency $1/\tau_J^* = \beta^*$. The straight dash lines show the limiting decay ($t^*$ large) for some of the curves with low $\beta^*$. (b) Simulated orientational Raman correlation functions $\hat{G}_J^{(2,0)}(t^*)$ from Eq. 2.41, using $\xi = 2.833$ and several values of the reduced collision frequency and displaying the limiting decay curves as for Fig. 2.1a. [From A. G. St. Pierre and W. A. Steele, *J. Chem. Phys.* **57**, 4638 (1972).]

for $M$-diffusion

$$\tau_J^{-1} \int_0^t dt_1 \, F_\omega(t - t_1) F_\omega(t_1)$$

$$= \tau_J^{-1} \left\langle \int_0^t dt_1 \left[ 1 - \frac{\omega^2(t - t_1)^2}{2} \pm \cdots \right] \left[ 1 - \frac{\omega^2 t_1^2}{2} \pm \cdots \right] \right\rangle$$

$$= \tau_J^{-1} \int_0^t dt_1 \left\{ 1 - \frac{\langle \omega^2 \rangle \left[ (t - t_1)^2 + t_1^2 \right]}{2} + \frac{\langle \omega^4 \rangle (t - t_1)^2 t_1^2}{4} \pm \cdots \right\}$$

and for $J$-diffusion

$$\tau_J^{-1} \int_0^t dt_1 \, F(t - t_1) F(t_1)$$

$$= \tau_J^{-1} \int_0^t dt_1 \left[ 1 - \frac{\langle \omega^2 \rangle (t - t_1)^2}{2} \pm \cdots \right] \left[ 1 - \frac{\langle \omega^2 \rangle t_1^2}{2} \pm \cdots \right]$$

$$= \tau_J^{-1} \int_0^t dt_1 \left\{ 1 - \frac{\langle \omega^2 \rangle \left[ (t - t_1)^2 + t_1^2 \right]}{2} + \frac{\langle \omega^2 \rangle^2 (t - t_1)^2 t_1^2}{4} \pm \cdots \right\}.$$

With the help of Eq. 2.2 we find the average values

$$\langle \omega^2 \rangle = \int d\omega \, P(\omega) \omega^2 = \frac{2 k_B T}{I}$$

and

$$\langle \omega^4 \rangle = 8 \left( \frac{k_B T}{I} \right)^2. \tag{2.45}$$

Inserting these results into the preceding series development shows that the correlation function for $M$-diffusion begins to rise above that of $J$-diffusion at order $t^5$.

In terms of a physical picture, we rationalize this as follows: $J$-diffusion averages over $|\mathbf{J}|$ at each collision, $M$-diffusion averages over $|\mathbf{J}|$ after the last collision. Hence the memory of $|\mathbf{J}|$ persists longer for $M$-diffusion. Consequently, after a certain number of collisional events the correlation function decays perceptibly more slowly for $M$- than for $J$-diffusion.

4. The further decay of $G(t)$ for either rotational diffusion model becomes exponential because the original orientational coherence is erased by the large number of complicated dynamic events (rotation-collision steps) that have

occurred since $t = 0$.[13] At each $t$, the further development of the system does not depend on its time evolution prior to $t$ (Markov process).[69] In fact, we can observe similar exponential long-time behavior in almost all experimental correlation functions we have opportunity to discuss.

The exponential tail of a correlation function is frequently taken to indicate that the system is in the so-called Debye rotational diffusion regime—a model of orientational motion where it takes a large number of small angular random steps to reorient the body through an appreciable angle. However, to deduce such definite motional characteristics solely from an observed exponential decay of an orientational correlation function is, at best, audacious and usually wrong for the systems we are concerned with. Since we think that it is important to clear up this misunderstanding, we discuss this here in some detail. First, we recall the principles of the Debye rotational diffusion model and then show the fallacy of taking exponential decay as criterion for an identification of this model.

The formalism of the Debye model is based on the macroscopic rotational diffusion equation[56, 70]

$$\frac{\partial}{\partial t}P(\Omega, t) = D\nabla_\Omega^2 P(\Omega, t),  \tag{2.46a}$$

where $P(\Omega, t)$ is the probability of finding the (spheroidal) molecule at time $t$ in orientation $\Omega = (\alpha, \beta, \gamma)$ (see also Eq. 1.91 of Chapter 1), $\nabla_\Omega^2$ is the Laplacian operator on the surface of a sphere, and $D$ is the rotational diffusion coefficient

$$D = \frac{kT}{6V_M\eta_m}  \tag{2.46b}$$

related to the molecular volume $V_M$ of the liquid and its macroscopic viscosity $\eta_m$ (see also Eq. 2.17 for the translational analogue).

The solution of Eq. 2.46a is achieved by expanding $P(\Omega, t)$ in terms of spherical harmonics. The result is

$$f_k^{(j)}(t) = f_k^{(j)}(0)\exp\left(-\frac{t}{\tau^{(j)}}\right),$$

with

$$\frac{1}{\tau^{(j)}} = (j+1)jD.  \tag{2.47}$$

(We have changed the order index $l$ of Eq. 1.91, Chapter 1, to $j$.)

Note that Eq. 2.47 is valid for the spherical top. For a symmetric top molecule, there are two independent principal rotational diffusion coefficients, $D_\perp$ and $D_\parallel$. $D_\perp$ describes the orientational motion of the $C_3$ axis (due to rotational diffusion around the two perpendicular axes) and $D_\parallel$ characterizes the orientational motion of an axis direction in the degenerate plane of the symmetric top (due to rotational diffusion around the $C_3$ axis and around one of the perpendicular axes). (See Section 2.1, Chapter 1.)

The orientational correlation time of Eq. 2.47 is related to $D_\parallel$ and $D_\perp$ by[71]

$$\frac{1}{\tau^{(j,k)}} = j(j+1)D_\perp + (D_\parallel - D_\perp)k^2. \tag{2.48}$$

Thus for the Raman tensor of a parallel band of ethane we find

$$\hat{G}(t) = \exp(-6D_\perp t) \tag{2.49}$$

and for the Raman tensor of a perpendicular band (Eq. 1.95, Chapter 1) we get

$$\hat{G}(t) = (a^2 + b^2)^{-1}\{a^2\exp[-(2D_\perp + 4D_\parallel)t]$$

$$+ b^2\exp[-(5D_\perp + D_\parallel)t]\}. \tag{2.50}$$

The possible misuse of this model may enter in two ways, one very obvious, the other quite surreptitious and not yet widely appreciated. We first discuss the obvious misuse of the Debye model.

We recall that the spectral density $\mathscr{I}(\omega)$ associated with an exponential correlation function $\exp(-t/\tau)$ (Eq. A.14, Appendix A; Appendix J),

$$\mathscr{I}(\omega) = 2\int_0^\infty dt\cos(\omega t)\exp\left(-\frac{|t|}{\tau}\right), \tag{2.51}$$

is a Lorentzian

$$\mathscr{I}(\omega) = 2\tau(1 + \tau^2\omega^2)^{-1}.$$

Now, the conceptual difficulty with a Lorentzian spectrum is that $\mathscr{I}(\omega)$ does not decrease to zero as $\omega \to \infty$. This means that the correlation function diverges at short times. On the other hand, we know that for short times collisions have not yet intervened and, consequently, molecular motion is purely kinetic. Because of this, the correlation function has no first-order term[72] in $t$ (zero-slope at $t = 0$; see Appendix B). Consequently, the Debye model fails with the first-order term. For our purposes it cannot be used to give a credible description of the details of molecular motion since it is invalid for initial time intervals where the correlation function is the least "scrambled" by complicated orientation-collision events. We summarize: Since the correlation functions are exponential at longer times (Fig. 2.1), and since this has nothing to do with Debye-type orientational motion, and since such exponential tail leads to a spectral density whose *central portion* is Lorentzian, a Lorentzian band contour does not prove small-angle, Debye-type rotational diffusion.

Second, we outline a more subtle but equally important point about exponential orientational relaxation. We show that long-time exponential decay of an orientational correlation is merely a consequence of the long-range isotropy of the liquid and has nothing to do with the size of the orientational steps of the reorienting molecules.[38, 71a] We start with a so-called master equation[73] for the conditional probability $P(\Omega, t; \Omega_0)$ that the body has orientation $\Omega$ at $t$ when it had orientation $\Omega_0$ at $t = 0$,

$$\frac{\partial}{\partial t} P(\Omega, t; \Omega_0) = \int d\Omega' \, B(\Omega, \Omega') P(\Omega', t; \Omega_0)$$

$$- \int d\Omega' \, B(\Omega', \Omega) P(\Omega, t; \Omega_0). \tag{2.52}$$

Here, $B(\Omega, \Omega')$ is the time-independent rate for reorientation from $\Omega'$ to $\Omega$ ("transition kernel"). Equation 2.52 describes the rate of change of a quantity through balancing its gain (first term) and loss. (We use this important formalism again at a later occasion.) The restrictions imposed on $P$ and $B$ are as follows:

1. The probability distribution is positive and normalized,

$$P(\Omega, t; \Omega_0) > 0; \qquad \int d\Omega \, P(\Omega, t; \Omega_0) = 1. \tag{2.53}$$

2. The transition kernel obeys the detailed balance condition (see also Eq. A.3 in Appendix A),

$$B(\Omega, \Omega') P_e(\Omega') = B(\Omega', \Omega) P_e(\Omega), \tag{2.54}$$

   where $P_e(\Omega)$ corresponds to the equilibrium state.
3. Equation 2.52 is valid for times longer than inverse transition rates of $B(\Omega, \Omega')$.
4. Inertial effects are excluded; the process is Markovian.
5. There are no restrictions on the size of reorientation angle $\Omega - \Omega'$.

Defining[71a]

$$A(\Omega, \Omega') \equiv B(\Omega, \Omega') - V(\Omega)\delta(\Omega - \Omega'),$$

where $V(\Omega) = \int d\Omega' \, B(\Omega', \Omega)$ is the "leaving rate," we can write an equivalent, more concise form of Eq. 2.52 (compare with Eq. 2.46a)

$$\frac{\partial}{\partial t} P(\Omega, t; \Omega_0) = \int d\Omega' \, A(\Omega, \Omega') P(\Omega', t; \Omega_0). \tag{2.55}$$

In case of isotropy, the transition kernel $A(\Omega, \Omega')$ simplifies to

$$A(\Omega, \Omega') = A(\Omega - \Omega') = A(\Omega' - \Omega). \tag{2.56}$$

In other words, we need only consider angle differences and not absolute values (see also Eqs. 1.90–1.91 in Chapter 1) because there is an equal probability for finding a molecule in any orientation.

To accomplish our purpose, we first solve Eq. 2.52 or 2.55 for $P(\Omega, t; \Omega_0)$ and then construct with $P(\Omega, t; \Omega_0)$ an orientational correlation function as done in Sections 2.1, 2.2, and 3.4.2 of Chapter 1. The result will tell us which characteristics of the transition kernel $A(\Omega, \Omega')$ are fundamental to exponential decay of the correlation function.

We begin by expressing $P(\Omega, t; \Omega_0)$ in terms of the eigenvalues $\lambda_n$ and eigenfunctions $\psi_n$ of the transition operator $A(\Omega, \Omega')$,

$$P(\Omega, t; \Omega_0) = \sum_n^\infty \psi_n^*(\Omega_0) \psi_n(\Omega) \exp(\lambda_n t) \qquad (2.57)$$

with

$$\int d\Omega' A(\Omega, \Omega') \psi_n(\Omega') = \lambda_n \psi_n(\Omega).$$

This relation is easily verified by inserting Eq. 2.57 into Eq. 2.55.

Notice that this expansion of the general transition kernel $A(\Omega, \Omega')$ depends on both $\Omega$ and $\Omega'$. As a consequence a correlation function of its Wigner rotation matrix expansion (as, for instance, Eq. 1.92 of Chapter 1),

$$\left\langle D_{nm}^{(j)*}[\Omega(0)] D_{n'm'}^{(j')}[\Omega(t)] \right\rangle \equiv \mathscr{D}_{nm,n'm'}^{(jj')}(t)$$

$$= \int d\Omega_0 P(\Omega_0) D_{nm}^{(j)*}(\Omega_0) \int d\Omega P(\Omega, t; \Omega_0) D_{n'm'}^{(j')}(\Omega)$$

$$[\Omega_0 \equiv \Omega(0)], \qquad (2.58a)$$

yields an infinite sum of exponentials,

$$\mathscr{D}_{nm,n'm'}^{(jj')}(t) = \sum_k^\infty \int d\Omega_0 P(\Omega_0) D_{nm}^{(j)*}(\Omega_0) \psi_k^*(\Omega_0)$$

$$\times \int d\Omega D_{n'm'}^{(j')}(\Omega) \psi_k(\Omega) \exp(\lambda_k t), \qquad (2.58b)$$

with $\lambda_k \leq 0$ since $P(\Omega, t; \Omega_0) \leq 1$.

Depending on the form of $A(\Omega, \Omega')$, Eq. 2.58b may lead to any type of spectral density.[71a] However, in situations where we can assume (macroscopic) isotropy in the liquid—which is of sole interest to us—we show in Appendix C that the eigenfunctions $\psi_n(\Omega)$ of the operator $\int d\Omega' A(\Omega - \Omega')$ are a finite linear expansion of the Wigner rotation matrices,

$$\psi_m^{(j)}(\Omega) = \sum_n b_n^{(j)} D_{nm}^{(j)*}(\Omega),$$

with

$$\int d\Omega' A(\Omega - \Omega') D_{nm}^{(j)*}(\Omega') = \lambda_m^{(j)} D_{nm}^{(j)*}(\Omega) \qquad (2.58c)$$

and eigenvalues

$$\lambda_m^{(j)} = \left(\frac{8\pi^2}{2j+1}\right) c_{nm}^{(j)}.$$

Therefore, in terms of the expansion coefficient $c_{nm}^{(j)}$, we obtain

$$A(\Omega) = \sum_{j=0}^{\infty} \sum_{n=-j}^{j} \sum_{m=-j}^{j} c_{nm}^{(j)} D_{nm}^{(j)}(\Omega).$$

Although the conditional probability $P(\Omega, t; \Omega_0)$ is an infinite series expansion of the $D$ (Eq. 1.91, Chapter 1), insertion of the finite rotation matrix development of the eigenfunctions $\psi_m^{(j)}(\Omega)$ into Eq. 2.58b leads to a correlation function with a finite number of exponentials. (Refer to Eq. 1.94 for the special case $j = 2$.) Since there is no restriction on $A(\Omega - \Omega')$ with regard to the size of the rotational diffusion step $\Omega - \Omega'$, exponential decay evidently does *not* a priori mean Debye-type rotational motion. We summarize:

1.  Long-time exponential decay of the orientational correlation function is to be expected. It expresses the Markovian nature of the rotational relaxation process.

2.  A Lorentzian band contour cannot, a priori, be interpreted to reflect Debye-type (very small angle) orientational motion. Indeed, for noninertial motion a Lorentzian band contour is the consequence of the macroscopic isotropy of the medium—an assumption we make for all systems discussed here.

3.  It is difficult to draw detailed conclusions about the nature of a reorientational process by evaluating solely the exponential portion of the correlation function. If the correlation function is essentially exponential, this still does not give unambiguous evidence about the nature of the reorientation process (such as the size of the reorientational diffusion steps).

In the remainder of this section we discuss the question whether the $J$- and $M$-extended rotational diffusion models approach Debye-type orientational motion if the collision frequency $\tau_J^{-1} \equiv \beta$ greatly exceeds the angular velocity of the freely rotating molecule. This point is not only of inherent interest but also of practical significance for comparisons with relaxation data obtained by dipolar and nuclear magnetic relaxation methods. We investigate this question by setting the reduced angular momentum correlation time smaller than unity, or the absolute correlation time inferior to the inverse root-mean-square angular velocity (Section 2.1) $\tau_J^* \ll 1$, $\tau_J \ll (I/k_B T)^{1/2}$. For our purposes it is more convenient to compute the spectral density $\mathscr{J}(\omega)$ rather than $G(t)$.[53] We therefore take the Fourier transform of Eq. 2.15 (see Eq. A.14, Appendix A) and obtain

$$2\mathcal{I}^{(j,k)}(\omega) = \mathrm{Re}\int_0^\infty dt\exp(i\omega t)\hat{G}(t)$$

$$= \mathrm{Re}\sum_{n=1}^\infty \left(\frac{1}{\tau_J}\right)^{n-1}\left\langle \int_0^\infty dt_1\exp\left(-\frac{t_1}{\tau_J}\right)\int_0^\infty d(t_2-t_1)\exp\left(-\frac{t_2-t_1}{\tau_J}\right)\right.$$

$$\times\cdots\int_0^\infty d(t-t_{n-1})\exp\left(-\frac{t-t_{n-1}}{\tau_J}\right)\sum_{a=-j}^{j}\left[d_{ka}^{(j)}(x_1)\right]^2$$

$$\times\exp\left[-i\omega-i(a+k\xi x_1)L_1 t_1\right]\sum_{a=-j}^{j}\left[d_{ka}^{(j)}(x_2)\right]^2$$

$$\times\exp\left[-i\omega-i(a+k\xi x_2)L_2(t_2-t_1)\right]\cdots\sum_{a=-j}^{j}\left[d_{ka}^{(j)}(x_l)\right]^2$$

$$\left.\times\exp\left[-i\omega-i(a+k\xi x_l)L_l(t_l-t_{l-1})\right]\right\rangle_{L_l,x_l}$$

$$= \mathrm{Re}\sum_{n=1}^\infty \left(\frac{1}{\tau_J}\right)^{n-1}\left\langle \prod_{l=1}^{n}\int_0^\infty d(t_l-t_{l-1})\sum_{a=-j}^{j}\left[d_{ka}^{(j)}(x_l)\right]^2\right.$$

$$\left.\times\exp\left[-(t_l-t_{l-1})(\tau_J^{-1}+i\omega+iaL_l+ik\xi x_l L_l)\right]\right\rangle_{L_l,x_l}.$$

For $J$-diffusion we get

$$2\mathcal{I}_j^{(j,k)}(\omega) = \mathrm{Re}\sum_{n=1}^\infty \left(\frac{1}{\tau_J}\right)^{n-1}\left\{\int_0^\infty dt\int_0^\infty dL\int_{-1}^{1}dx\,P_{12}(x,L)\sum_{a=-j}^{j}\left[d_{ka}^{(j)}(x)\right]^2\right.$$

$$\left.\times\exp\left[-t(\tau_J^{-1}+i\omega+iaL+ik\xi xL)\right]\right\}^n$$

$$= \mathrm{Re}\sum_{n=1}^\infty \tau_J\left\{\int_0^\infty dL\int_{-1}^{1}dx\,P_{12}(x,L)\right.$$

$$\left.\times\sum_{a=-j}^{j}\left[d_{ka}^{(j)}(x)\right]^2\left[1+i(\omega+aL+k\xi xL)\tau_J\right]^{-1}\right\}^n.$$

$$(2.59)$$

Equation 2.59 represents a convergent geometric series. Abbreviating the argument of the $n$th-power term by $B_J^{(j,k)}(\omega)$ gives

$$2\mathscr{I}_J^{(j,k)}(\omega) = \tau_J \mathrm{Re}\{B_J^{(j,k)}(\omega)[1 - B_J^{(j,k)}(\omega)]^{-1}\}. \tag{2.60}$$

Since $\tau_J^* \ll 1$, we can expand the denominator of Eq. 2.59 into a power series that we break off with the quadratic term. The denominator of Eq. 2.59 is then

$$\approx [1 - i(a + k\xi x)L\tau_J - (a + k\xi x)^2 L^2\tau_J^2](1 + i\omega\tau_J)^{-1}.$$

Averaging over $L$ and $x$ (Appendix E) leads to the following results: (1) Terms linear in $x$ vanish. (2) The average of term $x^2$ (Eq. E.7) is to be multiplied by $\tau_J^2$. (3) The constant term gives unity because $P_{12}(x, L)$ is normalized (Eq. 2.40), and the $d_{ka}^{(j)}$ are orthogonal. Therefore

$$B_J^{(j,k)}(\omega) \approx \left(1 - k_B T\tau_J^2\{[j(j+1) - k^2]I_x^{-1} + k^2I_z^{-1}\}\right)(1 + i\omega\tau_J)^{-1}$$

$$= (1 - k_B T\tau_J^2\mathscr{B})(1 + i\omega\tau_J)^{-1},$$

with

$$[j(j+1) - k^2]I_x^{-1} + k^2I_z^{-1} \equiv \mathscr{B}.$$

Equation 2.60 reads now

$$2\mathscr{I}_J^{(j,k)}(\omega) \approx k_B T\tau_J\mathscr{B}\{\omega^2 + (k_B T\tau_J\mathscr{B})^2\}^{-1}, \tag{2.61}$$

where we left out the higher-order term $(k_B T\tau_J^2\mathscr{B})^2$ in the numerator.

We recall that $\mathscr{I}(0)$ is proportional to the correlation time of the associated correlation function (see Eq. 1.68 in Chapter 1). Setting in Eq. 2.61 $\omega = 0$, we find

$$\{k_B T\tau_J([j(j+1) - k^2]I_x^{-1} + k^2I_z^{-1})\}^{-1} \equiv \tau_R^{(j,k)}, \tag{2.62}$$

which gives the Lorentzian spectral density

$$\mathscr{I}_J^{(j,k)}(\omega) \propto \tau_R^{(j,k)}\{1 + (\omega\tau_R^{(j,k)})^2\}^{-1}. \tag{2.63}$$

On first sight it appears that the Debye limit is reached, since we have constructed a "Debye-type" condition $\tau_J^* \ll 1$ of a symmetric top molecule and have obtained a Lorentzian band contour (exponential correlation function) as result. On the other hand, if we compare Eq. 2.62 with Eq. 2.48 we notice that agreement would require the equivalence of

$$\left(\frac{k_B T}{I_x}\right)\tau_J \quad \text{with } D_\perp \tag{2.64}$$

and of

$$\left(\frac{k_B T}{I_z}\right)\tau_J \quad \text{with } D_{\parallel}.$$

Evidently this means that $D_{\parallel}$ and $D_{\perp}$ are linearly dependent. This, of course, is a nonsupportable restriction on the rotational diffusion model[71b]—in particular since it is unlikely that the anisotropy of orientational motion is solely due to mass effects $(I_x, I_z)$.[74] Hence the Debye model for anisotropic rotational diffusion is not the limiting case of the $J$-extended rotational diffusion model for the symmetric top if the collision frequency becomes large. The $M$-model does not even formally approach Debye behavior since $|\mathbf{J}|$ is not randomized at each collision; the $M$-model retains "too much" memory.

Setting in Eq. 2.64 $I_x = I_z = I$ gives the well-known Hubbard relation for a spherical top,[75]

$$\tau_R^{(j)}\tau_J = \frac{I}{k_B T}\{j(j+1)\}^{-1}, \tag{2.65}$$

which is valid for $\tau_J \ll \tau_R$. In this exceptional example the $J$-extended rotational diffusion model for $\tau_J^* \ll 1$ appears to approach the corresponding Debye model. We note that the rotational diffusion coefficient $D$ is related to the angular momentum correlation time $\tau_J$ (Eq. 2.11; Eq. 1.67, Chapter 1) by[76]

$$D \equiv \frac{k_B T}{I}\int_0^\infty \langle \mathbf{J}(0)\cdot\mathbf{J}(t)\rangle\langle \mathbf{J}^2\rangle^{-1}dt$$

$$= \left(\frac{k_B T}{I}\right)\tau_J. \tag{2.66}$$

## 3.  APPLICATION OF THE $J$- AND $M$-EXTENDED DIFFUSION MODELS TO THE ANALYSIS OF VIBRATION-ROTATION CONTOURS OF CONDENSED-PHASE SYSTEMS

In the following three sections we compare orientational correlation functions from observed vibration-rotation spectra with the $J$- and $M$-diffusion models. We recall that these models dispose of one adjustable parameter $\tau_J$, the angular momentum correlation time (Eqs. 2.11 and 2.66).

We either assume or—better yet—prove experimentally that vibrational relaxation is not significant during the time scales of rotational relaxation.

### 3.1.  Carbon Monoxide Dissolved in Liquid Nitrogen and Oxygen

The infrared spectrum of the fundamental of carbon monoxide (CO) in solution with methane ($CH_4$), oxygen, nitrogen, argon, and xenon has been evaluated in terms of the $M$- and $J$-diffusion models.[77] We discuss here the CO-$N_2$ and CO-$O_2$ systems.

The CO fundamental is near 2075 cm$^{-1}$. We recall that we obtain the experimental correlation function from the numerical evaluation of the infrared spectral density (Eq. 1.35, Chapter 1). We set the factor correlation function $\langle m(0)m(t) \rangle = 1$ according to our assumption on the relative insignificance of pure vibrational relaxation. We obtain the model correlation functions $G^{(1,0)}(t)$ by evaluation of Eq. 2.37 (for $J$-) and Eq. 2.43 (for $M$-diffusion), remembering that the free rotor correlation function $F(t)$ is given by Eq. 2.3.

Figure 2.2$a$ displays the fit of the $J$-model $G_J^{(1,0)}(t^*)$ to the experimental correlation function with $\tau_J^* = 0.5$ ($\tau_J = 0.20 \times 10^{-12}$ sec). The experimental purely imaginary autocorrelation function is also shown; it is not significant. Therefore quantal effects are not important (Appendix A). Agreement between experiment and theory is good; this is not the case for the $M$-model (not shown here).

On the other hand, for CO dissolved in liquid $O_2$ neither $J$- nor $M$-diffusion models fit. Figure 2.2$b$ shows the application of the $M$-diffusion model with $\tau_J^* = 0.4$.

We see that the value of the absolute average collision frequency is within 10% for both solvents (we have made a correction for the different ambient conditions). Second, we notice that rotational correlation of CO decays faster in liquid $N_2$ than in liquid $O_2$. Hence we conclude that the $O_2$ molecules exert a larger torque on the dissolved CO molecules than do the $N_2$ molecules—as we could reasonably expect from the physical nature of these solvents.

Third, since the $M$-diffusion model for a linear molecule does not admit collision-induced transfer of rotational energy, it is not surprising that the simulated decay of the rotational coherence in $M$-diffusion is slower than that

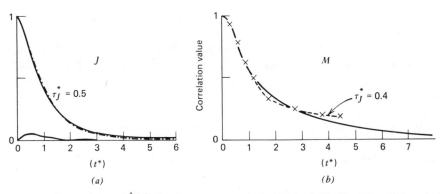

**Figure 2.2.**  Theoretical ($\hat{G}^{(1,0)}$; broken curves) and experimental ($\langle \mathbf{m}(0)\mathbf{m}(t) \rangle$; solid curves) infrared correlation functions of the vibration-rotation band of the CO fundamental ($a$) dissolved in liquid $N_2$ at 75°K ($J$-model); ($b$) dissolved in liquid $O_2$ at 55°K ($M$-model). [From G. Lévi, M. Chalaye, F. Marsault-Hérail, and J. P. Marsault, *Mol. Phys.* **24**, 1217 (1972).]

observed: The elastic collision-induced interaction defined for $M$-diffusion simulates a longer memory than that of the real system.

The intermolecular torque $\mathbf{O}H_1$ is a useful quantity for characterizing the frictional forces in liquids. Here $\mathbf{O}$ is the gradient operator with respect to rotation about the principal axes of inertia, and $H_1$ is the angle-dependent intermolecular potential.[28] The observable quantity $\langle (\mathbf{O}H_1)^2 \rangle$ is obtained from the fourth rotational spectral moment of the band contour,

$$M(4) \equiv \int_{\text{band}} d\omega \, \hat{\mathcal{I}}(\omega)(\omega - \omega_0)^4.$$

(We take this up in Section 4.) Since the contributions to the integral are appreciable at $\omega$ far off the center frequency $\omega_0$, a meaningful determination of $M(4)$ requires sophisticated data acquisition and smoothing capabilities. Little work along such lines can be reported for liquid-phase systems; compressed gases have received considerably more attention (but such systems are not within the scope of our interests).

With respect to the systems just discussed, it had been suggested that the exponential tail in the correlation function (Fig. 2.2) reflects the Debye limit $\tau_J^* \ll 1$: It was believed that the intermolecular torques gradually decreased the size of the angular steps of the reorienting CO molecules during the evolution of the correlation function.[77] That such an idea is wrong is easily understood by recalling that $\tau_J$, the average time between collisions, is a constant in the $J$- and $M$-rotational diffusion models (Section 2.2). Time $\tau_J$ depends on the fluctuating forces in the medium (Section 2.1), considered to be an equilibrium system.

### 3.2. Methane and Perdeuteromethane and Their Solutions in Liquefied Argon: Coriolis Interaction in Spherical Top Molecules

We discuss in this section the application of the $J$- and $M$-diffusion models to the observed band contours of the triply degenerate $\nu_3$ and $\nu_4$ infrared fundamentals of methane.[78] We recall that the direction of the vibrational transition moment lies along any axis of the molecule. First we deal with the $\nu_3$ fundamental, a C-H (3020 cm$^{-1}$) and C-D (2258 cm$^{-1}$) stretch. Figure 2.3 shows the comparison for neat liquid CH$_4$ ($a$) and for 0.15% CH$_4$ dissolved in liquid Ar ($b$). Figure 2.4$a$, and $b$ shows the same for CD$_4$.

We see that rotational relaxation of the neat methane species fits the $J$-diffusion model whereas rotational relaxation of methane in dilute Ar solutions obeys $M$-diffusion. Attempts to fit these systems to the respective opposite diffusion model failed.

Within each model description, we notice that the angular momentum correlation time of the neat deuterated species and its Ar solution is only 60% of that of the protonated compound. Although a precise comparison is impossible because of different experimental temperatures, closer scrutiny of

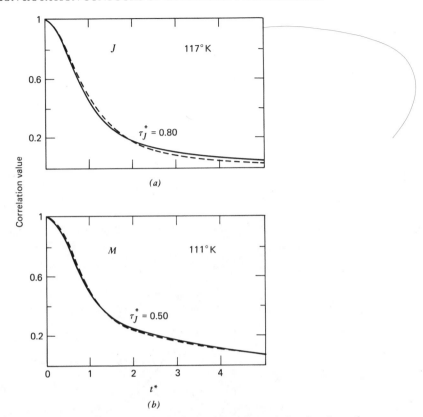

**Figure 2.3.** Experimental (solid) and theoretical (dash curves) infrared correlation functions of the triply degenerate C-H stretching fundamental of $CH_4$. (a) Neat liquid, $J$-model; (b) dissolved in liquefied Ar, $M$-model. [From R. E. D. McClung, *J. Chem. Phys.* **55**, 3459 (1971).]

the results reveals that the collision frequency $1/\tau_J^* = \beta^*$ appears to be greater for neat or dissolved $CD_4$ than for the lighter $CH_4$ systems.

Another dynamic aspect of the liquefied methane system is puzzling: Although the orientational correlation functions of all fundamentals of the methane molecule should be identical (the same transition moment vector is observed—see Section 2.1, Chapter 1), the solid curves in Fig. 2.5 show that the experimental $G(t)$ of the $\nu_3$ and the $\nu_4$ modes of $CD_4$ (in Ar) are not coincident.

These discrepancies essentially disappear by considering the different degree of Coriolis coupling in these modes.[79] We remember that Coriolis interaction is a type of coupling of rotational and vibrational motion. It is described classically by the term $\boldsymbol{\omega} \cdot \sum M_i(\mathbf{r}_i \times \mathbf{v}_i)$, where $r$ is the coordinate of mass $M$ in the moving frame of molecular axes and $v$ is its velocity with respect to a space-fixed frame. As it turns out, at certain molecular orientations the Coriolis force-induced atomic displacement $q_i$ of mass $M_i$ in vibrational mode $i$ is

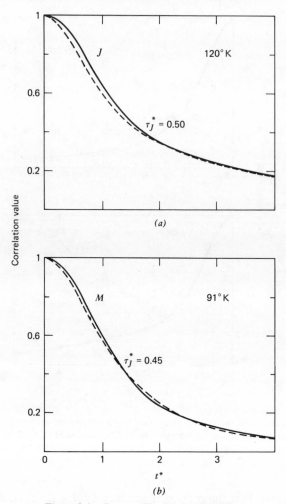

**Figure 2.4.**   Same as Fig. 2.3, but for $CD_4$.

equivalent to the normal displacement of mode $j$. The Coriolis force-induced vibration on nucleus $j$ moves then with the frequency of $q_i$. It is obvious that the influence on the other mode is the strongest when $i$ and $j$ belong to a degenerate vibrational mode of the molecule; the Coriolis-induced motion acts on resonant frequencies.[80] As a consequence, the vibrational degeneracy is lifted and terms of vibrational angular momentum $\mathbf{p}$, $|\mathbf{p}| = \zeta\hbar$, where $\zeta$,

$$-1 \le \zeta \le 1$$

is the Coriolis coupling constant, appear in the rotational Hamiltonian. For $\zeta > 0$ ($\zeta < 0$) the Coriolis effect decreases (increases) the rotational level spacing of infrared vibration-rotation levels. In other words, the band contour narrows (broadens).[81] As a consequence, for $\zeta > 0$ the correlation values $G(t)$ undergo an apparent time scale dilation. This is the case of interest here since $\zeta(\nu_4, CD_4) = 0.34$, $\zeta(\nu_3, CD_4) = 0.16$.

We present the computations of the first-order tensor free rotor correlation function for a spherical top molecule with Coriolis interaction in Appendix D. We include in Fig. 2.5 the application of the computations[79] by two dashed curves: They represent the $M$-model with $\tau_J^* = 0.70$. We observe now good agreement between theory and experiment for the $\nu_3$ mode, whereas agreement is less satisfactory for the $\nu_4$ mode. However, this is not surprising since the $\nu_4$ mode has the largest degree of Coriolis coupling of all $CD_4$ fundamentals.

As in the case of the linear molecules discussed in the preceding section, we realize that the extended $J$- and $M$-rotational diffusion models do not give a perfect description of the characteristics of the orientational motion for the spherical tops discussed here. Yet they offer explanations on some of the essential principles of rotational relaxation with the help of one adjustable parameter $\tau_J$. In addition, the memory function development for these models (Section 2.1) not only enables speedy numerical evaluation but also makes it easy to include additional or perturbative effects into the formalism. In neat liquids, application of a $J$-diffusion regime is consistent with an effective transfer of angular energy (every collision randomizes $\mathbf{J}$). In the rare-gas solutions of the methanes, the $M$-diffusion model (only the direction of $\mathbf{J}$ is randomized at each collision) is indeed better suited to describe the weaker interactions of a polyatomic molecule with an atom (devoid of rotational

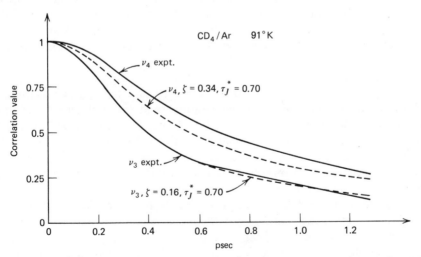

**Figure 2.5.** Experimental infrared correlation function $\langle \mathbf{m}(0)\mathbf{m}(t) \rangle$ of the contour of the triply degenerate $\nu_3$ and $\nu_4$ modes of $CD_4$ dissolved in liquefied Ar (solid curves) and the $M$-diffusion model simulations with Coriolis interaction (dashed curves). [From K. Müller and F. Kneubühl, *Chem. Phys.* **8**, 468 (1975).]

angular momentum). We note that the models do not permit partial energy transfer.

We now realize that the values of $\tau_J$ for the $\nu_3$ mode of neat $CD_4$ and $CD_4$ in Ar solution (reported in Fig. 2.4$a$ and $b$) are not correct. They are lower limits since the neglect to correct for Coriolis interaction simulates $\tau_J$ that are too short (the collision frequency is too high). There is no such caveat for the $\nu_3$ mode of $CH_4$ since its $\zeta \sim 0.04$.

Finally, we remark that it would be desirable to obtain an idea of the significance of vibrational relaxation in these systems. Since a VV-VH Raman experiment would not be fruitful (Section 3.2, Chapter 1), refuge to picosecond pulsing techniques (Section 4.3, Chapter 1) may be envisaged.

### 3.3. Neopentane, $C(CH_3)_4$: Rotational Motion in an Orientationally Disordered Crystal Structure

Neopentane is one of a large group of organic molecules of spherical or near-spherical shape that undergo a solid-state phase transition below the melting point. The structure changes from a less-symmetric (hexagonal) ordered to a high-symmetric (cubic) "disordered" crystalline phase. This transition is generally associated with a large entropy change comparable to or even greater than that during the solid-liquid transition. A waxy appearance or plastic consistency of the high-temperature solid phase has led to name such compounds plastic crystals.

The main point of interest for us lies in the plastic phase, where the crystal is *orientationally* disordered: Although the radial positions of the molecules remain in a long-range ordered state, their orientational positions are "more or less" randomly distributed. The question of importance to us concerns the dynamic nature of this orientational disorder; we would like to know how the molecules reorient. For instance, we inquire whether they perform orientational jumps between certain preferred and distinguishable orientations or whether rotation is liquidlike (all orientations are distinguishable).

We first give some background on relevant characteristics of neopentane. The low-temperature solid–high-temperature solid (plastic) phase transition occurs at 140°K and the plastic-phase–liquid-phase transition at 253°K. It had been demonstrated that orientational motion of the whole $C(CH_3)_4$ molecule in the plastic phase takes place with an apparent activation energy of about 4 kJ/mole or 330 cm$^{-1}$. However, the nature of this rotational freedom was not clear, for instance, whether uncorrelated large-angle jumps by 120° (around each of the four threefold symmetry axes) took place or whether small-angle molecular rotation (Section 2.2) was the predominant orientational process.[82]

A study of the dynamic orientational behavior of neopentane and similar systems by fluctuation spectroscopy is challenging. It touches upon important and generally interesting processes such as molecular motion and phase transitions, melting (the rotational degrees of freedom melt at the transition from the ordered crystalline to the disordered crystalline structure), and so forth. Furthermore, we entertain the interesting condition of translational rigidity of the molecular environment but may encounter the simplifications arising from the macroscopic orientational isotropy of the medium (Section 2.2).

The fundamental mode of neopentane, investigated in the first reported study of its dynamics based on infrared and Raman fluctuation spectroscopy, was $\nu_{18}$ at 923 cm$^{-1}$.[83] This fundamental (a C-C skeletal stretch) is the only mode free of overlap. We recall that only triply degenerate modes of spherical molecules are infrared active, whereas totally symmetric, doubly degenerate, and triply degenerate modes are active in the Raman. Early spectroscopic work had assigned the 923-cm$^{-1}$ band to a superposition of a triply degenerate $F_2$ species $\nu_{18}$ and a doubly degenerate $E$ species $\nu_7$ of the $T_d$ point group of the isolated neopentane molecule. Moreover, a (incorrect) calculation of the theoretical second Raman spectral moments for $F_2$ and $E$ modes[83] led to an assignment of the observed 923-cm$^{-1}$ Raman fundamental to the doubly degenerate $\nu_7$ mode. It is now clear that this Raman assignment was wrong; according to more recent studies the 923-cm$^{-1}$ band represents the $\nu_{18}(F_2)$ fundamental in both the infrared and the Raman.[84a] The $\nu_7$ $E$-mode is located elsewhere in the Raman spectrum and, apparently, weak. In other words, we can proceed in a straightforward manner with the numerical contour analysis of the 923-cm$^{-1}$ profiles in both infrared and Raman since we meet the necessary condition of dealing with the same oscillator of the neopentane molecule.

As a $F_2$ Raman mode is completely depolarized, the purely vibrational correlation function cannot be determined by the VV-VH polarization experiment (Section 3.2, Chapter 1). We then assume that $\langle m(0)m(t) \rangle \sim 1$ in Eqs. 1.36 and 1.48. We also neglect the Coriolis effect $\zeta(\nu_{18}) = 0.08$.

The theoretical correlation functions for the $J$- and $M$-models are obtained from the symmetric-top formulations (Eqs. 2.41, 2.42) by setting $\xi = (I_x/I_z) - 1 = 0$ and the indices $j, k = 1,0$ (infrared) and $2,0$ (Raman). (See Note 4.) Figure 2.6$a$ and $b$ displays the quantitative evaluation of the experimental spectral densities of the 923-cm$^{-1}$ band for the infrared (Eqs. 1.25, 1.26, 1.35, Chapter 1) and for the Raman (Eq. 1.79). [● = infrared, ○ = Raman.]

Figure 2.6$a$ shows the comparison of the experimental and theoretical correlation functions for the plastic phase of neopentane, whereas Fig. 2.6$b$ shows the same for the liquid phase. We note that the observed vibrational-rotational correlation lies between the $M$- and $J$-diffusion limits for a common best value of $\tau_J$ (computer fitted to the models). This seems realistic in view of the simulated strong ($J$-model) and weak ($M$-model) collisions. Recall that these models do not permit partial randomization of **J**.

Within the simplifications and assumptions of the theoretical models, re-orientational motion of C(CH$_3$)$_4$ in either plastic or liquid phase consists of free random rotation steps that are interrupted by randomly occurring collisions. We observe that the collision frequency $1/\tau_J$ is higher in the lower-temperature (higher-density) plastic than in the higher-temperature (lower-den-

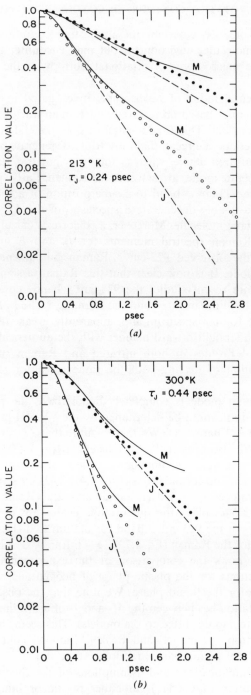

**Figure 2.6.** (*a*) Experimental infrared (solid) and Raman (open points) correlation values from the 923-cm$^{-1}$ mode of neopentane in its plastic phase. The solid (*M*-model) and dashed curves (*J*-model) are the best-fit simulations with a common $\tau_J$. (*b*) Same as Fig. 2.6*a*, but for liquid neopentane at 300°K. [R. C. Livingston, W. G. Rothschild, and J. J. Rush, *J. Chem. Phys.* **59**, 2498 (1973).]

sity) liquid phase. This may signify that proximity or steric effects outweigh kinetic effects. A clear distinction could be made only on the basis of constant density versus temperature data.

It is interesting to examine whether the orientational characteristics undergo a perceptible discontinuity at the plastic-phase–liquid-phase transition at 253°K. The semilogarithmic plot of the experimental infrared and Raman correlation time $\tau$ (Eq. 2.7a) versus reciprocal temperature in Fig. 2.7 shows that no such discontinuity is apparent for either tensor correlation time.

An activation energy of 4.1 kJ/mole or 340 cm$^{-1}$ in the range 170–260°K is computed from these plots. The value agrees with corresponding results from cold neutron scattering[82a] and proton spin relaxation measurements.[85] We do not persue this further.

Since our theoretical model assumes that the orientational motion of neopentane in its plastic and liquid phases is characterized by random angular steps, we can estimate the average angle $\theta_{\text{free}}$ of such free rotation steps from the time intervals $\tau_{\text{free}}$ where free rotor and experimental correlation functions coincide. As an example, for the plastic phase at 214°K we find $\tau_{\text{free}} \sim 0.2 \times 10^{-12}$ sec from Fig. 2.8. Using Eqs. 1.36 and 1.86b, Chapter 1, we then get

$$\langle \hat{\mathbf{m}}(0)\cdot\hat{\mathbf{m}}(t)\rangle = \langle\cos\theta\rangle \sim 0.95$$

$$\langle\cos^2\theta\rangle \sim 0.90, \tag{2.67}$$

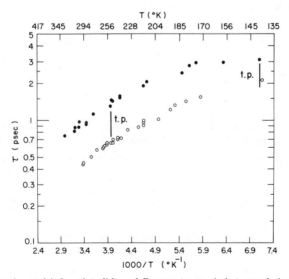

**Figure 2.7.** Experimental infrared (solid) and Raman (open circles) correlation times from the correlation functions of Fig. 2.6 (and others not shown here) as a function of temperature. The phase transitions, ordered-crystal–disordered-crystal (plastic) and plastic crystal–liquid, are indicated by vertical lines (t.p. = 140 and 253°K, respectively). [R. C. Livingston, W. G. Rothschild, and J. J. Rush, *J. Chem. Phys.* **59**, 2498 (1973).]

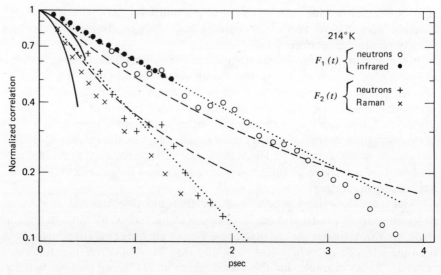

**Figure 2.8.** Comparison between the experimental first- and second-order orientational correlation values (from infrared absorption, Raman, and neutron scattering data) describing the orientational motion of neopentane in its plastic phase at 214°K. The solid curves show the initial time evolution of the first- (upper curve) and second-order free rotor correlation functions of the molecule. The dashed curves represent the orientational autocorrelation functions of the first- and second-order tensors based on the jump model of the configuration shown in Fig. 2.9a. The dotted lines represent a fit of the neutron data to purely exponential decay using the ratio of second- to first-order correlation times of 0.5—a value also read from Fig. 2.7. [Chr. Steenbergen and L. A. de Graaf, *Physica* **96B**, 1 (1979).]

which gives

$$\theta_{\text{free}} \sim 0.32 \text{ rad or 18 deg.}$$

It is pleasing that $\tau = 0.2 \times 10^{-12}$ sec (Fig. 2.8) agrees with $\tau_J = 0.24 \times 10^{-12}$ sec (Fig. 2.6a); recall that $\tau_J$ is defined as the time interval between free rotation steps. Hence the results of the experiment and the assumptions of the theoretical model agree.

On the other hand, the question comes to mind whether a motion consisting of random-size, free orientational steps can truly characterize the orientational motion of a molecule in a solid; we should expect instead molecular jumps between a certain number of preferred (minimal energy) orientations within the crystal lattice. Let us then propose that in the plastic phase the neopentane molecule rests for a relatively long interval in a certain orientation and performs whole-molecule librations, of high frequency and small amplitude, about this momentary equilibrium orientation.

Thereafter the molecule jumps rapidly by an appreciable angle—the size of which depends on the distribution of orientational potential wells—into another equilibrium orientation where it performs librational motions, and so forth. This model (we deal with it quantitatively at a later opportunity) has an orientational correlation function with pronounced oscillatory components at shorter time and a smoother oscillatory decay at longer times. At longer times the exponential Markovian tail makes the observations of the oscillatory motion difficult; it is "smeared out." However, even at shorter times the frequency and the amplitude of the librational motion may be difficult to detect for the following reason: The evolution of a correlation function at short times is determined by the intensity profile in the wings of the band ($|\omega - \omega_0|$ large) where it is difficult to perform precise absorption or scattering measurements because of the spectral background noise; usually, this leads to an underestimate of the experimental frequency range of the band profile. In other words, the profile $I(\omega)$ is cut at a certain frequency $\omega'$ where $I(\omega') \neq 0$. We then write the resulting experimental correlation function

$$G_{exp}(t) = \int_{-\infty}^{\infty} d\omega \, y(\omega) \mathscr{I}(\omega) \exp(-i\omega t),$$

with

$$y(\omega) \begin{cases} 1, & -\omega' \leq \omega \leq +\omega' \\ 0, & \text{otherwise.} \end{cases} \tag{2.68}$$

The rectangular function $y(\omega)$ is called window function.[86] Using the convolution theorem, we rewrite Eq. 2.68 (for a symmetric band profile)

$$G_{exp}(t) = \int_{-\infty}^{\infty} dt' G(t') g(t - t'), \tag{2.69}$$

with

$$G(t') = \int d\omega \, \mathscr{I}(\omega) \cos(\omega t') \tag{2.70a}$$

$$g(t') = \int d\omega \, y(\omega) \cos(\omega t'). \tag{2.70b}$$

Equation 2.70b can be solved to give

$$g(t) = \frac{2\omega_{max} \sin(\omega_{max} t)}{\omega_{max} t}, \tag{2.71}$$

where $(\sin x)/x$ is often designated by $\text{sinc}(x)$.

Clearly, sophisticated data sampling and smoothing procedures are required to distinguish oscillatory behavior of $G(t)$ simulated by insufficient integration of the frequency range from that induced by a damped-oscillatory rotational relaxation process as defined previously.

In such a situation recourse to other relaxation techniques is often helpful; let us consider results of a study of the dynamics of molecular orientation of the neopentane molecule by quasi-elastic neutron scattering on its protons.[87] Although this technique does not fall into the frame of our book, its application to studies of the dynamics of neopentane has been fruitful.

The neutron scattering technique shows superior resolution at times considerably longer than those reported in Fig. 2.6,[88] since the momentum transfer $\kappa[\text{Å}^{-1}]$ between neutron and proton can be varied to a value that is inferior to the inverse jump length of a proton as the molecule reorients from one to another position in the lattice of the crystal. (Recall that momentum transfer is essentially zero in optical spectroscopy.) The measurements therefore yield a characteristic time and a jump distance. If the jump distance is prescribed, they yield a characteristic time and a second adjustable parameter.

In practice a model of reorientational motion of the neopentane molecule between equilibrium orientations is constructed on the basis of a master or rate equation (Eq. 2.52) of probabilities $P(\mathbf{r}, t)$ for finding the proton at site $\mathbf{r}_j(t)$ at time $t$ if it was at site $\mathbf{r}_i(0)$ at $t = 0$. Depending on the prescribed geometrical model of angular motion,[89] the number of sites occupied by proton jumps can be accounted for and the master equation can be solved. Assuming that the jumps are "instantaneous," the transition kernel (Eq. 2.54) is simply the inverse of the average time $\tau$ between jumps. Subsequently, the orientational autocorrelation functions (Note 5) for the proton positions are generated and Fourier inverted to give the so-called orientational intermediate scattering function $I_{\text{rot}}(\kappa, t)$, which is then compared with the experimental scattering data.

A comparison of the first- and second-order tensor correlation functions obtained from the experimental vibration-rotation (Fig. 2.6$a$, and $b$) and from neutron data[88] is shown in Fig. 2.8. We see that agreement is good.

The geometry of the orientational jump model is shown by the configuration in Fig. 2.9$a$. The neopentane molecule performs 90-degree jumps around the fourfold vertical axis $z$ and 180-degree jumps around the two twofold axes $x$, $y$. The 90-degree jumps occur one order of magnitude more frequently than the 180-degree jumps. This ratio is taken as one of the two adjustable parameters since there is no a priori reason why the jump frequency around axis $z$ and axes $x$, $y$ should be the same.

The residence time (time between orientational jumps) is chosen as the second adjustable parameter; it amounts to $1.0 \times 10^{-12}$ sec at 214°K. Note that the configuration of Fig. 2.9$a$ of the tetrahedral neopentane molecule within the crystal cube has six distinguishable orientations (degree of orientational disorder). Although other models of molecular reorientation are possible, such as the configuration of Figs. 2.9$b$ and 2.10, they do not account as well for the neutron data as the configuration of Fig. 2.9$a$.[88]

The striking aspect in Figs. 2.6$a$ and 2.8 is that basically different model concepts of orientational motion of the neopentane molecule in its plastic phase—isotropic $M$-$J$-diffusion and jump diffusion about specific equilibrium

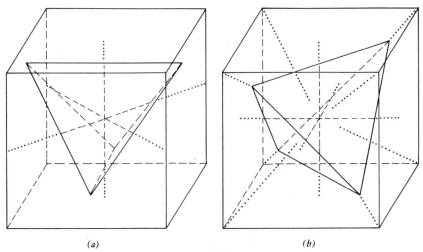

*(a)*                                                    *(b)*

**Figure 2.9.** Carbon skeleton of the tetrahedron of neopentane in two typical equilibrium positions within the plastic phase of the cubic crystal lattice (not identical to the unit cell). Common symmetry elements of the tetrahedron and the cube are indicated. The dotted lines from the corners of the cube are threefold rotation axes (configuration *b*), the dot-dash lines from the middle of a cube edge and a cube face are twofold symmetry axes (configurations *a* and *b*). The vertical axis (equal *z*) for configuration *a* is special since it represents a fourfold rotation axis of the lattice. Configuration *a* has the symmetry elements of the $D_{2d}$, configuration *b* of the $T_d$ point groups. There are six distinguishable orientations of the tetrahedron for *a*, two for configuration *b*. [From Chr. Steenbergen and L. A. de Graaf, *Physica* **96B**, 1 (1979).]

orientations in the crystal—yield orientational correlation functions that resemble each other. We necessarily conclude that orientational correlation functions may turn out to be model insensitive to a first approximation. (The reader will encounter other examples in the current literature.) This suggests that sophisticated experimental methods are needed to distinguish between theoretical models.

Another important point is profitably brought up at this occasion and is of far wider significance than has been generally realized since it involves what may be called the "principle of the isolated oscillator" in vibrational-rotational fluctuation spectroscopy (Section 1.5, Chapter 1). As a simple example, let us assume that we intend to Fourier invert an observed band profile of a C-Cl mode of $CHCl_3$. $CHCl_3$ is an isotopic mixture of $CH^{35}Cl_3$, $CH^{35}Cl_2{}^{37}Cl$, $CH^{35}Cl^{37}Cl_2$, and $CH^{37}Cl_3$ in their natural abundance ratio of about $1 : 1 : \frac{1}{3} : \frac{1}{9}$. Clearly, each isotopically different C-Cl oscillator has a slightly different resonance frequency due to the Cl mass effect.[90] Therefore the observed profile is inherently broadened since it is a superposition of four, slightly frequency-shifted profiles. Obviously, this broadening has nothing to do with any relaxation process. To utilize such modes for our purposes, the profile must be "renormalized" to reduce it to that of *one* oscillator.[10]

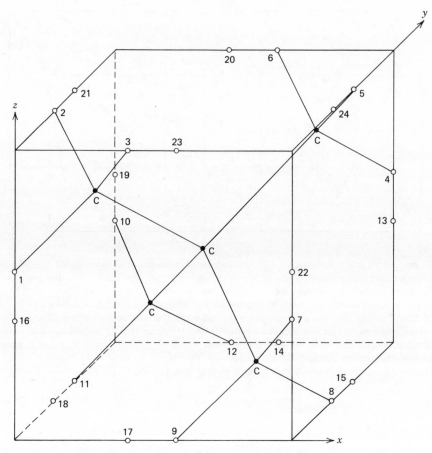

**Figure 2.10.** Position of the H and C atoms of the neopentane molecule for the configuration of Fig. 2.9*b* and its reachable H atom jump sites. By 90°-jumps around twofold axes, the H atom at site 1 can reach sites 17, 18, 19, 21, 22, and 23 in one, sites 2–12 in two, all others in three jumps. By 120°-jumps around threefold axes, the reachable positions are 2, 3, 5, 6, 8, 9, 11, and 12 (one jump) and 4, 7, and 10 (two jumps). By 180°-jumps around twofold rotation axes, H atom 1 can reach positions 4, 7, and 10. It does not reach other sites. [From Chr. Steenbergen and L. A. de Graaf, *Physica* **96B**, 1 (1979)].

Let us now turn our attention to neopentane and inspect Fig. 2.11. It displays the infrared profile of the 923-cm$^{-1}$ mode in a temperature region that covers the low-temperature ordered crystal phase and the disordered crystal (plastic) phase. (We neglect a side band, of unknown origin, at 943 cm$^{-1}$.) We notice that the band contour in the ordered solid phase has split into four overlapping peaks, which probably originate from a factor group splitting[91] (four molecules per unit cell). At any rate, the important aspect is that the band splitting persists well into the plastic phase of the crystal, although the components are no longer resolvable because of the onset of orientational broadening. (The side band at 943 cm$^{-1}$ shows similar behavior.) Consequently the band contour is broadened by the presence of (at least) four, slightly frequency-shifted, oscillators. We would not have suspected this broadening without being aware of the splitting of the entire contour at lower temperatures (see, in Fig. 2.11, the contour at 164°K). In other words, we encounter here an analogous situation as in the example of isotope-induced broadening discussed before: We find in the spectrum of neopentane a certain degree of multiple-oscillator-induced, nonrelaxational band broadening.

It is now obvious that the reported evaluation of the time development of the orientational correlation functions from the neopentane data, which was performed without due recognition of this nonrelaxational broadening,[83] represents an upper limit of the rate of orientational motion; the actual decay of the orientational correlation decreases more slowly than presented by Fig. 2.6$a$ and $b$.[92] (In this context, it is conceivable that remnants of factor or site group splittings persist, at least for short times, in strongly associated or densely packed liquids. Unless recognized, such effect leads to errors in the estimate of the rate of orientational and vibrational relaxation.)

We now realize that the neopentane system presents an example of the limit to which the extended $M$- and $J$-rotational diffusion models can be stretched. Obviously, they fail at low temperatures. On the other hand, at higher temperatures (where the energy difference of the potential wells of the crystal is

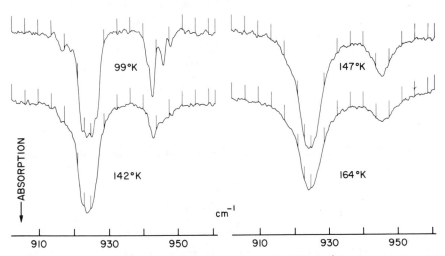

**Figure 2.11.** Photograph of several infrared spectra of the 923-cm$^{-1}$ fundamental ($\nu_{18}$, C-C stretch) of neopentane between 99–164°K, covering the range of the low-temperature hexagonal ordered crystal and the high-temperature cubic disordered crystal (plastic) phases. The transition temperature between the two phases is 140°K. The weaker band at 943 cm$^{-1}$ is of unknown origin; the weak side peak to the main absorption, at 915 cm$^{-1}$, represents the $^{13}$C isotopic species. Spectral slitwidth = 0.5 cm$^{-1}$. [R. C. Livingston, W. G. Rothschild, and J. J. Rush, *J. Chem. Phys.* **59**, 2498 (1973).]

overridden owing to the population of higher energy states) the $M$- and $J$-models may well turn out to better reflect the orientational characteristics of the molecule than do simple orientational jump models.[88]

The greatest uncertainty in the evaluation of the neopentane data, however, is introduced by our lack of precise information on concomitant vibrational relaxation. Raman studies of monocrystalline neopentane have shown that vibrational relaxation is, indeed, significant.[84b] We discuss this in Section 2.6 of Chapter 3.

## 4.   ROTATIONAL SPECTRAL MOMENTS: DEFINITIONS AND SIGNIFICANCE

A correlation function can be developed into a power series of its independent variable, time $t$. Expansions of this sort are often useful, particularly for short times, since the series converges sufficiently well with a few terms. Especially for orientational correlation functions, the expansion coefficients of the series development yield a few measurable and model-independent equilibrium quantities of the system that can be given physical meaning—assuming that molecules in their condensed phase keep their gas-phase identity (moments of inertia, vibrational force constants, Coriolis parameters, etc.).[93]

In the following we first describe how these quantities are defined and measured and thereafter discuss how to compute them theoretically. Let us write the $n$th time derivative of a general Fourier transform $X(t) = \int d\omega\, Y(\omega)\exp(i\omega t)$,

$$\frac{d^n}{dt^n} X(t) = \int_{-\infty}^{\infty} d\omega\, Y(\omega)(i\omega)^n \exp(i\omega t)$$

or

$$\left[ \frac{d^n}{dt^n} X(t) \right]_{t=0} = \int_{-\infty}^{\infty} d\omega\, Y(\omega)(i\omega)^n. \tag{2.72}$$

We see that the Taylor series time development of a correlation function (Eqs. A.7, Appendix A) is then

$$1 + it \int_{-\infty}^{\infty} d\omega\, \hat{\mathscr{I}}(\omega)(\omega - \omega_0) - \frac{t^2}{2} \int_{-\infty}^{\infty} d\omega\, \hat{\mathscr{I}}(\omega)(\omega - \omega_0)^2$$

$$- i\frac{t^3}{6} \int_{-\infty}^{\infty} d\omega\, \hat{\mathscr{I}}(\omega)(\omega - \omega_0)^3 + \frac{t^4}{24} \int_{-\infty}^{\infty} d\omega\, \hat{\mathscr{I}}(\omega)(\omega - \omega_0)^4 \pm \cdots .$$

$$\tag{2.73}$$

The integrals appearing in Eq. 2.73 are called "moments" (zeroth, first, second,...) of the spectral density; they are frequently taken about $\omega_0 = \omega_0^{\nu}$, the free molecule pure vibrational (internal) transition frequency $j = 0$ to $j = 0$. $j$ is the angular momentum quantum number of vibration-rotation mode $v$. Recalling that symbol $\hat{\phantom{x}}$ denotes normalized quantities, we write

$$M(n) = \int_{-\infty}^{\infty} d\omega \, \hat{\mathscr{I}}(\omega)(\omega - \omega_0)^n$$

$$= \int_{-\infty}^{\infty} d\omega \, \mathscr{I}(\omega)(\omega - \omega_0)^n \left\{ \int_{-\infty}^{\infty} d\omega \, \mathscr{I}(\omega) \right\}^{-1}. \qquad (2.74)$$

Notice that $M(n)$, except for $M(0)$, depends on the value of $\omega_0$. We can show that it is more useful to compute the moments about the "solvent-shifted" band origin $\omega_s^v$, defined by[93]

$$\omega_s^v = \omega_0^v + \hbar^{-1} \langle i|(\boldsymbol{\varepsilon} \cdot \hat{\mathbf{m}})^2 (H_1^v - H_1^0)|i \rangle, \qquad (2.75)$$

where $H_1^v$, $H_1^0$ is the intermolecular perturbation Hamiltonian in the excited and ground vibrational state, respectively, $\hat{\mathbf{m}}$ is the direction of the vibrational transition moment vector in the molecule with respect to the inertial axis coordinate system, and unit vector $\boldsymbol{\varepsilon}$ indicates the direction of the incident radiation field. We average over initial states $|i\rangle$. (See Eq. 1.33, Chapter 1.)

We have omitted in Eq. 2.75 presently uninteresting (but not necessarily insignificant) effects of local electric fields.[93] We also assume that vibrational and orientational (external) coordinates are statistically uncorrelated.

If we consider a spectral density $\hat{\mathscr{I}}(\omega)$ as probability density of finding the stochastic frequency between $\omega$ and $\omega + d\omega$, we get

$$M(n) = \langle (\omega - \omega')^n \rangle = \int d\omega \, \hat{\mathscr{I}}(\omega)(\omega - \omega')^n, \qquad (2.76)$$

with $\langle \omega - \omega' \rangle = 0$. $\omega'$ designates a chosen, convenient band center; clearly, if $\hat{\mathscr{I}}(\omega)$ is symmetric about $\omega'$, the odd moments $M(3)$, $M(5)$,... vanish (the correlation function is real—see Appendix A).

The physical meaning of the lowest-order $M(n)$ is as follows:[93]

1. The nonnormalized $M(0)$ gives the total intensity of the spectral density.
2. $M(1)$ yields the band center. $M(1)$ is essentially independent of intermolecular forces. It is convenient to set $M(1) = 0$.
3. $M(2)$ is a measure of the average molecular rotational kinetic energy (Eq. 2.2). $M(2)$ is also near-independent of the intermolecular forces.
4. $M(3)$ describes the asymmetry of the band contour. $M(3)$ is principally a quantum-mechanical effect.[72] (See Eq. A.3, Appendix A.) $M(3)$ is the lowest-order moment that depends on the molecular pair distribution functions; we assume that the intermolecular perturbation potential is a sum of pair interactions.

5. $M(4)$ is related to the mean-squared intermolecular torque hindering the orientational motion of the molecules (see end of Section 3.1). $M(4)$ also depends only on the pair (and not on higher) distribution functions. For instance, $M(4)$ of a parallel-polarized infrared vibration-rotation or far-infrared pure rotational transition of a linear molecule, is given by

$$M(4) \approx 8\left(\frac{k_B T}{I}\right)^2 + I^{-2}\langle i|(OH_1)^2|i\rangle.$$

(For the leading term in this relation, see Eq. 2.45.)

6. $M(n)$, with $n \geq 5$, are too complicated and not of interest.

### 4.1. Rotational Second Spectral Moment: Theory

The second pure rotational moment ($n = 2$), $M(2) = \langle(\omega - \omega_s^\nu)^2\rangle$, is of particular interest to us since, as we mentioned before, it is essentially related to the rotational thermal kinetic energy. We then expect that it characterizes the initial non-Markovian decay of the orientational correlation, a decay that is caused by pure dephasing of the angular velocities. (Refer to Appendix B for the special case of the $J$- and $M$-extended diffusion models and to Section 1 for an introductory discussion on linear molecules.) To compute $M(2)$ we replace $\hat{\mathscr{I}}(\omega)$ in Eq. 2.74 by its quantum-mechanical expression (see Eq. 1.63, Chapter 1, and Note 6)[93]:

$$M(2) = \frac{\sum_{if}|\langle i|\boldsymbol{\varepsilon}\cdot\mathbf{m}|f\rangle|^2 (E_f - E_i - \hbar\omega_s^\nu)^2}{\hbar^2\sum_{if}|\langle i|\boldsymbol{\varepsilon}\cdot\mathbf{m}|f\rangle|^2}. \qquad (2.78)$$

Here, we have used the property of the delta function $\delta(\omega + \omega_{if}) = \delta(\omega + \omega_i - \omega_f)$ to express $\omega$ in Eq. 2.74 by $(E_f - E_i)/\hbar$. The quantities $E_i, E_f$ are the initial and final external energy levels, and $\omega_s^\nu$ is the solvent-shifted band center (Eq. 2.75) for the internal vibrational transition $v \to v'$ of an appropriate vibration-rotation band $\nu$.

We recall that the transition moment $\mathbf{m}$ is averaged over the internal states, $\mathbf{m} \propto \langle v|\boldsymbol{\mu}|v'\rangle$ (see Eq. 1.33 and 1.34 in Chapter 1) and is understood to be a single-molecule-fixed vector of constant amplitude. ("Single-molecule-fixed" because we omit in Eq. 2.78 any induced absorption due to a cluster of molecules possessing a "combined" transition moment vector. "Constant amplitude" because we consider the vibrational relaxation to be known.)

To average Eq. 2.78 over all external final states $|f\rangle$ we replace the energy eigenvalues by their operators

$$E_i\|H^0 + H_1^0 + \hbar\omega_0^0; \left(\hbar\omega_0^0 \equiv 0\right)$$

$$E_f\|H^\nu + H_1^\nu + \hbar\omega_0^\nu. \qquad (2.79)$$

In Eq. 2.79 we account for the possibility that the total Hamiltonian of the unperturbed external motion $H$ and the intermolecular potential operator $H_1$ depend parametrically on the ground and excited internal states 0 and $v$. The constant energy terms $\hbar\omega_0^v$ and $\hbar\omega_0^0$ are added for convenience: We remember that the potential energy in the Schrödinger equation of the external motion is the eigenvalue of the Schrödinger equation of the internal (vibrational) problem,

$$H_{\text{vib}}|v) = [H_1(r,\Omega) + \hbar\omega_0^v]|v).$$

Indeed, if the molecules are far apart, $H_1(r,\Omega)$ tends to zero and we should be left with the Schrödinger equation of the isolated oscillator $H_{\text{vib}}|v) = \hbar\omega_0^v|v)$.

Multiplying Eq. 2.78 out, we write

$$\hbar^2\langle i|(\boldsymbol{\varepsilon}\cdot\mathbf{m})^2|i\rangle M(2)$$

$$= \sum_f \left\{ (E_f - E_i)^2 - 2(E_f - E_i)\hbar\omega_s^v + (\hbar\omega_s^v)^2 \right\} |\langle i|\boldsymbol{\varepsilon}\cdot\mathbf{m}|f\rangle|^2$$

$$= \sum_f (E_f - E_i)\langle i|\boldsymbol{\varepsilon}\cdot\mathbf{m}|f\rangle(E_f - E_i)\langle f|\boldsymbol{\varepsilon}\cdot\mathbf{m}|i\rangle$$

$$- \sum_f \left\{ (E_f - E_i)\langle i|\boldsymbol{\varepsilon}\cdot\mathbf{m}|f\rangle\langle f|\boldsymbol{\varepsilon}\cdot\mathbf{m}|i\rangle \right.$$

$$\left. + \langle i|\boldsymbol{\varepsilon}\cdot\mathbf{m}|f\rangle(E_f - E_i)\langle f|\boldsymbol{\varepsilon}\cdot\mathbf{m}|i\rangle \right\} + (\hbar\omega_s^v)^2\langle i|(\boldsymbol{\varepsilon}\cdot\mathbf{m})^2|i\rangle.$$

Multiplying the energy terms into the matrix elements gives

$$\sum_f \left[ \langle i|\boldsymbol{\varepsilon}\cdot\mathbf{m}|E_f f\rangle - \langle E_i i|\boldsymbol{\varepsilon}\cdot\mathbf{m}|f\rangle \right]\left[ \langle E_f f|\boldsymbol{\varepsilon}\cdot\mathbf{m}|i\rangle - \langle f|\boldsymbol{\varepsilon}\cdot\mathbf{m}|E_i i\rangle \right]$$

$$- \sum_f \left\{ \left[ \langle i|\boldsymbol{\varepsilon}\cdot\mathbf{m}|E_f f\rangle - \langle E_i i|\boldsymbol{\varepsilon}\cdot\mathbf{m}|f\rangle \right]\langle f|\boldsymbol{\varepsilon}\cdot\mathbf{m}|i\rangle \right.$$

$$\left. + \langle i|\boldsymbol{\varepsilon}\cdot\mathbf{m}|f\rangle\left[ \langle E_f f|\boldsymbol{\varepsilon}\cdot\mathbf{m}|i\rangle - \langle f|\boldsymbol{\varepsilon}\cdot\mathbf{m}|E_i i\rangle \right] \right\}$$

$$+ (\hbar\omega_s^v)^2\langle i|(\boldsymbol{\varepsilon}\cdot\mathbf{m})^2|i\rangle.$$

We then replace $E_i$ and $E_f$ by their operators from Eq. 2.79 and, in the subsequent algebra, make use of the "trick" of combining appropriate terms to commutators. For instance, in each of the four square-bracketed difference terms of the foregoing development, we add and subtract $H^v\boldsymbol{\varepsilon}\cdot\mathbf{m}$. This leads to the commutators $[\boldsymbol{\varepsilon}\cdot\mathbf{m}, H^v]$ or $[H^v, \boldsymbol{\varepsilon}\cdot\mathbf{m}]$, respectively. We also replace all terms containing the factor $\hbar\omega_0^v$ by $\hbar\omega_s^v$ according to Eq. 2.75

$$\hbar\omega_0^v = \hbar\omega_s^v - \langle i|(\boldsymbol{\varepsilon}\cdot\mathbf{m})^2(H_1^v - H_1^0)|i\rangle\langle i|(\boldsymbol{\varepsilon}\cdot\mathbf{m})^2|i\rangle^{-1}.$$

It turns out that this procedure causes all terms in $\omega_s''$ or its higher powers to cancel (which justifies this choice of $\omega'$).

The resulting expression, after lengthy algebra, reads

$$\hbar^2 \langle (\boldsymbol{\varepsilon}\cdot\mathbf{m})^2 \rangle M(2)$$

$$= \left\{ \left\langle (\boldsymbol{\varepsilon}\cdot\mathbf{m})^2 (H_1^v - H_1^0)^2 \right\rangle - \left\langle (\boldsymbol{\varepsilon}\cdot\mathbf{m})^2 (H_1^v - H_1^0) \right\rangle^2 \left\langle (\boldsymbol{\varepsilon}\cdot\mathbf{m})^2 \right\rangle^{-1} \right\}$$

$$- \left\langle [\boldsymbol{\varepsilon}\cdot\mathbf{m}, H^v]^2 \right\rangle + \left\langle [H^v, \boldsymbol{\varepsilon}\cdot\mathbf{m}][H^v - H^0, \boldsymbol{\varepsilon}\cdot\mathbf{m}] \right\rangle$$

$$+ \left\langle [(H^v - H^0)\boldsymbol{\varepsilon}\cdot\mathbf{m}, [H^v, \boldsymbol{\varepsilon}\cdot\mathbf{m}]] \right\rangle$$

$$+ \left\langle (H^v - H^0)(\boldsymbol{\varepsilon}\cdot\mathbf{m})^2 (H^v - H^0) \right\rangle + \cdots \qquad (2.80)$$

The dots indicate that we have omitted several higher-order terms, for instance

$$\left\langle \boldsymbol{\varepsilon}\cdot\mathbf{m}[H_1^v - H_1^0, [H^v, \boldsymbol{\varepsilon}\cdot\mathbf{m}]] \right\rangle,$$

since their contribution is negligible.[93]

Equation 2.80 gives the desired result. It represents a model-independent expression for the second purely rotational spectral moment taken with respect to the solvent-shifted band center frequency $\omega_s''$. Let us now inspect the various terms in Eq. 2.80.

1.   The term in the curly brackets represents the mean-squared fluctuation of the difference of the intermolecular perturbation operator $H_1$ between the lower and upper internal levels involved in the observed vibrational transition.

2.   The last three terms describe effects due to the difference of the external, zero-order Hamiltonian $H$ with respect to the internal levels; they account for the difference of the moments of inertia of the lower and upper vibrational state.

3.   The commutator

$$- \left\langle [\boldsymbol{\varepsilon}\cdot\mathbf{m}, H^v]^2 \right\rangle \qquad (2.81)$$

represents the most significant term in Eq. 2.80. To compute it, we must specify $H^v$,

$$H^v \equiv \left(\frac{1}{2}\right)\frac{\mathbf{P}^2}{M} + \left(\frac{1}{2}\right)\mathbf{J}\cdot\bar{\mathbf{I}}^v\cdot\mathbf{J}, \qquad (2.82)$$

where $\mathbf{P}, \mathbf{J}$ are the linear and angular momentum operators, respectively, $\bar{\mathbf{I}}$ is the reciprocal moment of inertia tensor, and $M$ the mass of the molecule. Then, Eq. 2.81 reads

$$- \left\langle [\boldsymbol{\varepsilon}\cdot\mathbf{m}, \tfrac{1}{2}\mathbf{J}\cdot\bar{\mathbf{I}}^v\cdot\mathbf{J}]^2 \right\rangle$$

since the coordinate terms commute. The commutator is solved using the well-known commutator relation between the Cartesian components of angular momentum and coordinates, $[J_y, y] = 0$, $[J_x, y] = i\hbar z$, and so on. A typical term of the six nonzero terms is

$$\left[ m_x, \tfrac{1}{2} J_y \bar{I}_y^\nu J_y \right] = -\frac{i}{\hbar} \left\{ m_z \bar{I}_y^\nu J_y + J_y \bar{I}_y^\nu m_z \right\},$$

where we twice made use of the commutator relation $[A, BC] = [A, B]C + B[A, C]$ and denoted by $x, y, z$ the inertial axes of the molecule. Upon squaring and averaging, terms linear in $J_i$ vanish. Hence we are left with the expression

$$\left\langle [\boldsymbol{\varepsilon} \cdot \mathbf{m}, H^\nu]^2 \right\rangle \hbar^{-2} = \left( m_x^2 + m_z^2 \right) \left\langle \left( \bar{I}_y^\nu J_y \right)^2 \right\rangle + \left( m_y^2 + m_z^2 \right) \left\langle \left( \bar{I}_x^\nu J_x \right)^2 \right\rangle$$

$$+ \left( m_x^2 + m_y^2 \right) \left\langle \left( \bar{I}_z^\nu J_z \right)^2 \right\rangle. \tag{2.83}$$

Since we are interested in systems of classical rotors (see Section 1) we express $\left\langle (J_i)^2 \right\rangle$ by the equipartition theorem

$$\frac{\left\langle (J_i)^2 \right\rangle}{2 I_i^0} = \frac{1}{2} \left\langle (J_i)^2 \right\rangle \bar{I}_i^0 = \frac{1}{2} k_B T. \tag{2.84}$$

($I_i^0$ is the moment of inertia in the ground vibrational state.) In terms of the rotational constants $B_i = (\hbar / 4\pi c) \bar{I}_i$, this gives

$$\left\langle \left( J_i \bar{I}_i^\nu \right)^2 \right\rangle = \frac{\left( 8\pi^2 c / h \right) k_B T \left( B_i^\nu \right)^2}{B_i^0}$$

in units of radian$^2$ sec$^{-2}$ (with $k_B = 1.38 \times 10^{-16}$ erg degree$^{-1}$) or

$$= \frac{2 k_B T \left( B_i^\nu \right)^2}{B_i^0} \qquad [\text{cm}^{-2}]$$

with $k_B = 0.695$ cm$^{-1}$ degree$^{-1}$. ($B$ is in cm$^{-1}$.) Therefore

$$M(2) = 2 k_B T \left\{ \frac{\left( \hat{m}_x^2 + \hat{m}_z^2 \right) \left( B_y^\nu \right)^2}{B_y^0} + \frac{\left( \hat{m}_y^2 + \hat{m}_z^2 \right) \left( B_x^\nu \right)^2}{B_x^0} \right.$$

$$\left. + \frac{\left( \hat{m}_x^2 + \hat{m}_y^2 \right) \left( B_z^\nu \right)^2}{B_z^0} \right\} \qquad [\text{cm}^{-2}], \tag{2.85}$$

where $\hat{m}_x$ is the direction cosine between the transition dipole moment and inertial axis $x$. $B_x$ is the rotational constant.

In most applications we do not differentiate between $B^v$ and $B^0$ and thus write

$$M(2) = 2k_B T \left\{ \left( \hat{m}_x^2 + \hat{m}_z^2 \right) B_y^0 + \left( \hat{m}_y^2 + \hat{m}_z^2 \right) B_x^0 + \left( \hat{m}_x^2 + \hat{m}_y^2 \right) B_z^0 \right\} \qquad [\text{cm}^{-2}].$$

$$(2.86)$$

We summarize: The rotational second moment $M(2)$ of a band contour belonging to a first-order tensor vibrational transition moment, taken about the solvent-shifted band center, is a measure of the rotational kinetic energy. It depends only on the temperature, the rotational constants, and the direction of the vibrational transition moment with respect to the inertial axis system of the molecule. The assumptions are (1) classical rotational motion, (2) negligible fluctuations in the difference of the intermolecular potential, (3) negligible difference of the rotational constants between upper and lower vibrational levels, (4) absence of appreciable statistical correlation between orientational and vibrational relaxation processes, (5) absence of significant induced absorption ("collective effects"), and (6) absence of effects of local electric fields.

It is not difficult to see how Eq. 2.86 is to be modified for far-infrared, pure rotational profiles or for the higher-order tensors of Raman vibration-rotation and Rayleigh pure rotation contours. Obviously, points 3 and 4 are irrelevant for far-infrared band contours; on the other hand, points (5) and (6) are particularly important. For Raman and Rayleigh contours, Eq. 2.78 is more complicated.[94] However, the resulting expressions are readily obtained by twice differentiating the appropriate free rotor correlation function and setting in the result $t = 0$ (Eq. 2.72). This process is also easily carried out in the presence of Coriolis interaction in doubly degenerate vibrational modes of symmetric top molecules; the relevant calculations and results are shown in Appendix E. Note that we neglect quantum-mechanical corrections to the $M(2)$.[72,94] The labor is not worthwhile in view of present experimental uncertainties.

Finally, it is useful to show that the solvent-shifted band center $\omega_s^v$ is a more meaningful quantity than the free molecule band center $\omega_0^0 \equiv \omega^0$ of the vibration-rotation contour. For instance, if the perturbation Hamiltonian is different for the ground and excited vibrational levels 0 and $v$, the stochastic transition dipole moment matrix element is[13] (Eq. 1.27, Chapter 1)

$$\left( n | \exp(iH^v t) \mu^v(0) \exp(-iH^0 t) | m \right) = \left( n | \mu^v(t) | m \right). \qquad (2.87)$$

In other words, the Hamiltonian is not the same in the arguments of the exponentials; the physical meaning of a correlation function $\langle \mu^v(0) \cdot \mu^v(t) \rangle$ is "muddled." Nevertheless, when we refer all $M(n), n \geq 2$, to the solvent-shifted band center $\omega_s^v$,[95] only the fluctuations of the solvent shift and not the solvent shift itself distort the band shape (Eq. 2.80).

## 4.2.   Rotational Second Spectral Moment: Applications

In the following sections we discuss the usefulness of measuring the rotational second spectral moment for the evaluation of band shapes; the discussion touches experimental as well as theoretical points.

We dispense first with a frequently occurring situation which, from a fundamental point, is trivial: A gross underestimate of the experimental value of $M(2)$ relative to its expected theoretical value due to a low spectral signal to noise ratio, to overlap of adjacent band contours, or to an incorrect baseline, in other words, whenever the operational integration over all frequencies $|\omega - \omega_s^v|$ was stopped significantly short of the true band limit (Eq. 2.68).

Frequently this experimental shortcoming is accompanied by faulty logic. Since the *central* portion of an orientational band contour is usually Lorentzian or nearly so (Note 7), an underestimate of the true band limits tends to accentuate the Lorentzian portion and thereby to exaggerate exponential decay of the orientational correlation function. Even worse, this may lead to the assumption that orientational motion obeys the Debye model—an assumption which is, as we know, incorrect (Section 2.2).[96]

### 4.2.1.   Rotational Second Spectral Moment and Induced Absorption in Polar Systems.

Occasionally the observed experimental value of the rotational second spectral moment $M(2)_{expt}$ greatly exceeds the theoretical value $M(2)_{theory}$. An example of this is $M(2)_{expt}$ obtained from the Raman profile of the parallel-polarized (totally symmetric) carbon–chlorine deformation fundamental $(\nu_3, A_1)$ of liquid chloroform near 365 cm$^{-1}$.[10b] Since the vibrational relaxation of this mode is relatively small, we can safely assume that rotational and vibrational relaxation is uncorrelated and therefore consider the total second spectral moment the sum of the purely orientational and purely vibrational contributions.[97] Second, there is no short-time quasi-crystalline broadening in the room-temperature liquid since factor or site group splittings are lacking[98] in the crystal spectrum of $\nu_3$ (Section 3.3). The profile of $\nu_3$ need only be corrected for the Cl-isotope effect and for two hot bands; either correction does not require precision since only the central portion of the contour is perturbed by these effects.

Therefore it is easy to obtain the precise value of the purely rotational second spectral moment $M(2)_{expt}$ for the $\nu_3$ mode. It amounts to 404 cm$^{-2}$.[10b] This compares to $M(2)_{expt} \sim 460$ cm$^{-2}$ obtained from the depolarized Rayleigh scattering intensity of $CHCl_3$.[10a, c] (Recall that through either method the orientational motion of the $C_3$ axis of $CHCl_3$ is observed; see Eqs. 1.81 and Section 3.5, Chapter 1.) Leaving aside the deviation between the two experimental results, we note that both $M(2)_{expt}$ are significantly above $M(2)_{theory} = 270$ cm$^{-2}$ (Eq. E.8, Appendix E).

In this situation the only phenomenon that explains the inequality $M(2)_{expt} \gg M(2)_{theory}$ is a so-called induced absorption or induced scattering intensity. It arises from a temporary transition moment generated between two or more molecules during the time interval of their close encounters. Whether such encounters are called "collision" and the effect "collision-induced absorption"[99] or scattering" or "interaction-induced absorption or scattering,"[100] matters little to our discussion at present and is interesting only if the particular nature of the interaction process is to be considered.[12] It is, however, important to realize that the induced absorption or scattering intensity persists only during binary or multiple-molecular interaction times. In other words, the corresponding band profile reflects processes that are characterized by the time *of* and not by the time *between* molecular encounters.

To estimate for $CHCl_3$ the collisional contribution to the overall $M(2)$, we consider the depolarized Rayleigh scattering data.[10a] In general, it is found that the total Rayleigh spectrum of a polar molecule can be decomposed into two overlapping contours: a low-frequency, essentially Lorentzian, and a high-frequency, essentially exponential scattering profile (Fig. 2.12). The low-frequency Lorentzian arises from rotational relaxation processes that are characterized by orientational correlation times of the rotors $\tau_R^{(2,0)}$, usually of the order of $1 \times 10^{-12}$ to $1 \times 10^{-11}$ sec. Consequently, the full bandwidth at half-peak height of the Lorentzian component is (Eq. 2.51)

$$\Delta \nu_{1/2} = \frac{1}{\pi c \tau_R}$$

$$= 1\text{--}10 \text{ cm}^{-1}. \tag{2.88}$$

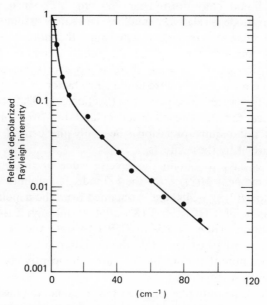

**Figure 2.12.** Depolarized Rayleigh wing spectrum of liquid $CHCl_3$ at room temperature. [P. van Konynenburg and W. A. Steele, *J. Chem. Phys.* **56**, 4776 (1972).]

On the other hand, the far-wing high-frequency exponential component represents the profile from collision-induced phenomena, $\mathscr{S}_{coll}(\omega)$. Setting [10a]

$$\hat{\mathscr{S}}_{coll}(\nu) \approx \exp\left(-\frac{\nu}{\Delta}\right), \tag{2.89}$$

where $\Delta$ is the inverse slope, we find $\Delta \sim 30 \text{ cm}^{-1}$ (see Fig. 2.12). Changing this into a lifetime of a collision $\tau_{coll}$, we obtain (Section 1 of Chapter 1)

$$\tau_{coll} \approx \frac{1}{2\pi c\Delta} \sim 2\times 10^{-13} \text{ sec.}$$

From Eqs. 2.74 and 2.89 we compute the second moment of the collisional spectral density. We find ($\omega_0 \equiv 0$)

$$M(2)_{coll} = 2\Delta^2 \sim 1800 \text{ cm}^{-2}.$$

Now we set

$$M(2)_{expt} = fM(2)_{theory} + (1-f)M(2)_{coll} \tag{2.90}$$

and therefore get the fractional contribution from $M(2)_{coll}$

$$1-f \sim 0.11.$$

We see from this estimate that a small contribution of collision-induced scattering or absorption can cause the experimental second spectral moment to rise appreciably above its theoretical value. It is obvious that this can complicate the evaluation of band contours in terms of molecular orientational motion. Even if $M(2)_{expt}$ is approximately $M(2)_{theory}$, Eq. 2.90 warns us that this agreement may be simulated by a significant value of $(1-f)M(2)_{coll}$.

In situations where $M(2)_{expt} \gg M(2)_{theory}$, there is a useful expedient for permitting us to "subtract" $M(2)_{excess}$. We simply "take in" the spectral baseline until $M(2)_{expt} = M(2)_{theory}$. In other words, we adjust the window function $y(\omega)$ of Eq. 2.68. The procedure has been attacked in the literature, [10a, 101] but in the absence of useful theories on collision-induced band shapes of molecular systems, there seems to be no better alternative.[10b, 101]

Figure 2.13 shows an application of this procedure for the $\nu_3$ mode of neat chloroform. Notice that the experimental, $M(2)$-adjusted second-order tensor orientational correlation function (Eqs. 1.81, Chapter 1) for neat $CHCl_3$ passes near the orientational correlation time $\tau_R^{(2,0)}$ (quad), the nuclear magnetic (deuteron) relaxation time of neat $CDCl_3$. We have also displayed a comparison of the first-order tensor orientational correlation function of $\nu_3$ of neat $CDCl_3$ with the correlation time $\tau_R^{(1,0)}$ from dielectric measurements of $CHCl_3$ in hydrocarbon solution. Of course, this comparison is less relevant since the two media are not identical. (Variations in the orientational characteristics of

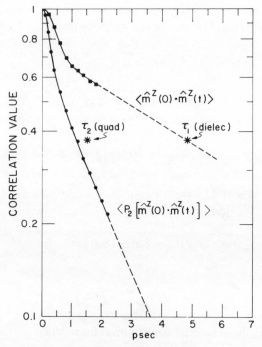

**Figure 2.13.** Second- and first-order purely orientational correlation functions of $CHCl_3$ (Raman) and $CDCl_3$ (infrared) at room temperature, obtained from the $\nu_3$ C–Cl deformation mode. The frequency range of $I_{VH}(\omega)$ was downsized to yield the theoretical purely orientational $M(2)$ as described in the text. The corresponding $\tau_2$ (quad) and $\tau_1$ (dielec) are placed at the correlation value $1/e$. [W. G. Rothschild, G. J. Rosasco, and R. C. Livingston, *J. Chem. Phys.* **62**, 1253 (1975).]

the $C_3$ axis between the neat isotopic chloroform species are insignificant for our purpose here.)

### 4.2.2. Rotational Second Spectral Moment and Presence of Weakly Coupled Vibrational Relaxation Processes.

We recall that it is difficult to determine the relative contribution of the orientational relaxation of perpendicular-polarized vibration–rotation contours of symmetric top molecules. First, for Raman contours a VV–VH scattering experiment does not lead to the desired separation of vibrational and rotational relaxation (Section 3.2, Chapter 1). Second, for infrared contours extraneous techniques must be employed to obtain the contribution of the vibrational relaxation to the infrared correlation function (see beginning of Section 4.2, Chapter 1).

Now, in these situations it is sometimes useful to measure the second spectral moment of the vibration–rotation contour and to compare the result with the expected purely rotational $M(2)_{\text{theory}}$. If $M(2)_{\text{expt}} \gg M(2)_{\text{theory}}$, we can be reasonably sure that the excess second spectral moment is mainly determined by vibrational relaxation. However, there are three caveats: (1) We must be sure that induced absorption is insignificant, (2) we do not observe entrained orientational motion (discussed in the next section), and (3) vibrational and rotational relaxation processes are statistically independent.

As an example we take the $\nu_4$ mode of neat liquid chloroform (at room temperature), a doubly degenerate carbon–hydrogen deformation fundamental at 1218 cm$^{-1}$.[10b] This mode is convenient in both infrared and Raman since the contour is intense, symmetric, little broadened by the Cl isotope effect, its wings are free of adjacent profiles, and the baselines are horizontal. On the other hand, both infrared and Raman profiles are perturbed by Coriolis interaction ($\zeta \sim +1$).

With the help of the estimated ratio $a/b \sim \sqrt{2}$[10b] of the Raman tensor elements $a$ and $b$ (Section 3.4, Chapter 1), using $\zeta = +1$, $B_x = 0.108$, $B_z = 0.054$ cm$^{-1}$, we obtain from Eqs. E.9 of Appendix E the theoretical values ($IR =$ infrared, $R =$ Raman)

$$M(2, IR) = M(2)_{\text{theory}}^{(1,1)} \approx 2k_B T B_x = 43 \text{ cm}^{-2},$$

$$M(2, R) = (a^2 + b^2)^{-1}\{a^2 M(2)_{\text{theory}}^{(2,2)} + b^2 M(2)_{\text{theory}}^{(2,1)}\}$$

$$\approx 2k_B T\{3B_x + 6B_z\} = 272 \text{ cm}^{-2}. \tag{2.91}$$

The experimental values are $M(2, IR) = 208$ and $M(2, R) = 459$ cm$^{-2}$.[102] Since uncorrelated contributions of $M(2)$ add, we find

$$M(2, IR)_{\text{excess}} = M(2, IR)_{\text{expt}} - M(2, IR)_{\text{theory}} \sim 170 \text{ cm}^{-2}. \tag{2.92}$$

Similarly,

$$M(2, R)_{\text{excess}} \sim 187 \text{ cm}^{-2}.$$

It is clear that the infrared contour—where $M(2)_{\text{excess}}$ is about four times $M(2)_{\text{theory}}$—cannot be evaluated in terms of pure orientational relaxation processes since the nonorientational relaxation processes significantly overshadow the orientational relaxation. Partly, this is caused by the Coriolis interaction which diminishes in mode $\nu_4$ the effective rate of the rotational motion around the $C_3$ axis by a factor of about 4. In fact, the short-time orientational motion of the infrared transition moment of the degenerate $\nu_4$ mode is essentially the same as that of a perpendicular infrared band of a *linear* polyatomic molecule with the same moments of inertia $I_x = I_y$, $I_z = 0$.[103]

On the other hand, for the $\nu_4$ Raman mode the apparent kinetic orientational motion around the $C_3$ axis is augmented by the Coriolis interaction: From Eq. E.8 (or Eq. E.9 with $\zeta = 0$) of Appendix E we find

$$M(2, R) = 2k_B T\{3B_x + 3B_z\} = 200 \text{ cm}^{-2} \tag{2.93}$$

in the absence of Coriolis interaction compared to 272 cm$^{-2}$ in its presence (Eq. 2.91).

It is interesting that $M(2)_{\text{excess}}$ is about the same for both tensor correlations, namely 170 cm$^{-2}$ for the infrared and 187 cm$^{-2}$ for the Raman (Eq. 2.92). This corroborates the assumption about the equality of the infrared and Raman pure vibrational correlation functions for the same vibration–rotation mode (Eq. 1.39, Chapter 1).

### 4.2.3. Rotational Second Spectral Moment and Entrained Orientational Motion.

A survey of the literature on second spectral rotational moments shows that the condition

$$M(2)_{\text{expt}} \ll M(2)_{\text{theory}}$$

prevails. In this section we therefore discuss one of the two more likely and generally significant reasons causing this inequality. Since there is no generally accepted theory that explains the inequality, our discussion is empirical. Nevertheless, our ideas are reasonable and in agreement with data.

We propose that the orientational motion of the "active" or "probe" molecule sterically entrains to a certain extent the orientational motion of its neighbor(s).[104] Evidently, the effective moment of inertia of such a couple or cluster is larger than that of the probe molecule (introduced by way of the equipartition principle into Eq. 2.86). Note that the cluster need not be rigid.

There are data that support this assumption: A study on the pressure dependence of the second rotational spectral moment of the C–I stretching fundamental of $CH_3I$ ($\nu_3$) established that $M(2)_{\text{expt}}$ decreased with increasing pressure.[105] According to Eq. 2.86, however, $M(2)$ should be pressure independent. On the other hand, higher pressure brings the molecules into closer distances and thereby increases the "compactness" of a molecular cluster or augments the number of its participants. In fact, these considerations touch upon a question that has been of considerable interest, namely, the nature of the orientational motion of liquid–phase complexes. (We discuss this in Section 5.1.)

Finally, let us make it clear that this phenomenon is not a cross-correlation effect in the sense of Eq. 1.41 (Chapter 1). We still consider only the autocorrelation of the orientational coordinate of the transition moment of the one excited molecule within the cluster.

### 4.2.4. Rotational Second Spectral Moment and Presence of Strongly Coupled Vibrational Relaxation.

In the preceding section we discussed that entrained orientational motion can lead to a value of the experimental rotational second spectral moment that falls appreciably short of that predicted by theory. In this section we consider another phenomenon that may lead to $M(2)_{\text{expt}} \ll M(2)_{\text{theory}}$, namely, strong vibration–rotation interaction. Such an effect can be considered to change orientational coordinates into vibratory coordinates.

At present, no useful theory is available that permits ensemble averaging if the density matrix $\rho_{vr}$ can no longer be written as a product of a purely vibrational $\rho_v$ and purely orientational $\rho_r$ factor (refer to Eq. 1.33, Chapter 1). Nevertheless, two published theories are of interest since they predict a decreased rotational spectral second moment in the presence of appreciable vibration–rotation coupling. On closer scrutiny, one of these studies contains a sign error.[106] Therefore the theory predicts $M(2)_{\text{vib-rot}} \geq M(2)_{\text{rot}}$ (see also Section 4.2.2). The other theory yields a negative purely vibrational second moment,[107] a result that is not helpful. However, if we reinterpret the negative moment as an interference term from coupled vibrational-orientational coordinates (see end of Section 3.2, Chapter 1), the suggested approach merits reconsideration.

To give an example, we consider neat liquid methyl iodide, the system most extensively studied by fluctuation spectroscopy (and still one of the least understood). The published experimental values of its rotational second spectral moment of a parallel-polarized band ($\nu_3$) at 525 cm$^{-1}$ in the Raman spectrum vary from 27 to 610 cm$^{-2}$;[108] the theoretical value is $M(2)^{(2,0)}_{\text{theory}} = 600$ cm$^{-2}$ at 30°C (Eq. E.8, Appendix E). Undoubtedly, a precise experimental value depends on precise measurements of $I_{\text{VH}}(\omega)$ and $I_{\text{VV}}(\omega)$; we recall that

$$M(2)^{(2,0)}_{\text{VH}} = \int_{-\infty}^{\infty} d\omega \, \hat{I}_{\text{VH}}(\omega)(\omega - \omega')^2$$

$$= M(2)^{(2,0)}_{\text{aniso}} + M(2)^{(2,0)}_{\text{iso}}$$

$$\equiv M(2)^{(2,0)}_{\text{rot}} + M(2)^{(2,0)}_{\text{iso}} \tag{2.94}$$

(Eq. 1.50, Chapter 1; Eq. 2.73) and

$$M(2)^{(2,0)}_{\text{iso}} = \int_{-\infty}^{\infty} d\omega \left\{ \overline{I_{\text{VV}}(\omega) - \left(\tfrac{4}{3}\right) I_{\text{VH}}(\omega)} \right\}(\omega - \omega')^2 \tag{2.95}$$

(Eq. 1.47, Chapter 1), with

$$\hat{I}(\omega) = I(\omega) \left\{ \int_{-\infty}^{\infty} d\omega \, I(\omega) \right\}^{-1}.$$

A reliable experimental value of the pure rotational second spectral moment $M(2)_{\text{rot}} = M(2)_{\text{aniso}}$ therefore requires a reliable value of the experimental purely vibrational second spectral moment, $M(2)_{\text{iso}}$.

The experimental literature values of $M(2)_{\text{rot}}$ which reach the theoretical value of 600 cm$^{-2}$ are suspect. One study reported a theoretically questionable extrapolation procedure of the far wings of the $\nu_3$ mode,[108] whereas another study underestimated the vibrational spectral moment $M(2)^{(2,0)}_{\text{iso}}$ by a factor of about 3—hence overestimated $M(2)_{\text{rot}}$ (Eq. 2.94).[109]

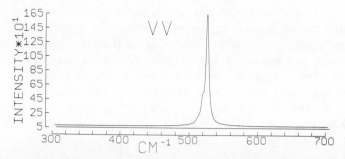

**Figure 2.14.** VV-Raman spectrum of the $\nu_3$ fundamental of liquid $CH_3I$ at ambient temperature and pressure. Note the hot bands $v, v' = 1,2$ and $2,3$ as low-frequency shoulders to the fundamental. The spectrum is truthfully reconstructed in steps of 0.2 cm$^{-1}$ from the digital average of 49 successive scans. Ordinate: 2048 units = 1 volt. Slit widths: 0.5 cm$^{-1}$. [Based on W. G. Rothschild, J. Devaure, R. Cavagnat, and J. Lascombe, Lab. Spectrosc. Infrarouge, Université Bordeaux I, 1976.]

The spectral experiments are indeed difficult to perform. First, there are two hot bands at lower frequencies of the $\nu_3$ fundamental, the $v = 1 \to 2$ mode (its intensity is ~ 16% of the fundamental) and the $v = 2 \to 3$ ( ~ 1%) mode. They require careful correction.[10a, 11, 105] Second, the Rayleigh wing—the Raman baseline—rises appreciably below 375 cm$^{-1}$. If not realized, this results in an underestimate of $M(2)_{expt}$. Furthermore, the vibrational spectral second moment (Eq. 2.95) amounts to 170 cm$^{-2}$, or about 0.3 $M(2)_{aniso}$.

At any rate, a careful redetermination of $M(2)_{aniso}^{(2,0)}$ of the mode yielded 370 cm$^{-2}$.[105] [Recall that $M(2)_{theory}^{(2,0)} = 600$ cm$^{-2}$.] The high-quality spectra, reconstructed by computer from digital data, are shown in Figs. 2.14–2.16. The lines drawn below the spectra are horizontals. Each spectrum is the average of 49 automatic scans, sampling each 0.1 cm$^{-1}$, analog-digital conversion, writing on magnetic tape, and employing the fastest scanning speeds consistent with

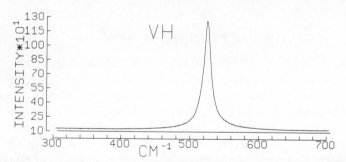

**Figure 2.15.** VH-Raman spectrum of the $\nu_3$ fundamental of liquid $CH_3I$. Note that the two hot bands are hidden. Spectral conditions are as in Fig. 2.14, but the absolute ordinate scale is different. Slits: 0.8 cm$^{-1}$. The line drawn below the spectrum is a parallel to the frequency axis.

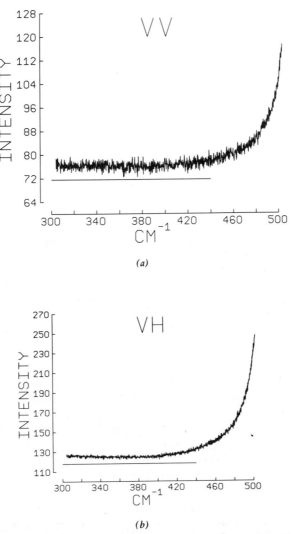

**Figure 2.16.** (*a*) Low-frequency portion of the VV spectrum of Fig. 2.14, replotted with an extended ordinate scale. (*b*) Low-frequency portion of the VH spectrum of Fig. 2.15, replotted with an extended ordinate scale.

resolution and time constants. Spectral slit widths were below 20% of the halfwidths; no numerical corrections were therefore needed. Note that the baseline is flat and that there is no fluorescence, overlap with adjacent fundamentals (the nearest fundamental is at 884 cm$^{-1}$), or other perturbations of the wings of the contour. In particular, Fig. 2.16$a$ and $b$ show, on the example of the lower frequency limits of the $I_{VV}$ and $I_{VH}$ profiles, that the frequency range of the contours can be read within a few cm$^{-1}$. Note the Rayleigh background.

At present, there is no unambiguous theoretical explanation of why the experimental pure rotational second spectral moment falls so pronouncedly below its theoretical value; it appears that it is not the fault of the data.

## 5. ROTATIONAL RELAXATION FROM VIBRATION–ROTATION AND ROTATIONAL CONTOURS: SPECIAL TOPICS

### 5.1. Rotational Motion in Liquid-Phase Complexes

In the following sections we discuss the applications of rotational relaxation data from vibrational–rotational and pure rotational spectra to some special topics.

The first topic deals with the characteristics of the rotational motion of complexes in liquids. We attempt to find out whether the two (or more) molecules that supposedly form a cluster reorient end-over-end as a whole unit of sufficient rigidity. This problem has attracted considerable interest since the concept of a liquid-phase complex is difficult to grasp except in the limiting case where the interaction between the partners ("moieties") is so strong that it is preferable to consider the association product a weak compound, or in the opposite limit where the interaction is so weak that it is more reasonable to look upon the angular interactions as "random" (no short-lived preferential relative orientations).

In our analysis we persue the following approach: In one of the molecules we choose a polarization direction of the vibrational transition moment vector $\mathbf{m}(t)$ that lies along the very axis direction that purportedly connects this ("probe") molecule with its complexing partner. In this way we observe the component of the orientational motion that tends to twist the probe molecule out of the intermolecular "bond." If complexing exists, this motional component should be hindered and should indicate, possibly, oscillatory-librational motion about the intermoiety axis. [In an additional conceivable experiment, we pick the transition moment vector $\mathbf{m}(t)$ that lies in a perpendicular direction to the intermoiety axis; by this we probe the "internal rotation" of the complex.]

We require that the interaction between the moieties does not lead to a band splitting or to unresolved broadening due to a shift of the oscillator frequency of the complexed moiety. (There will be an equilibrium distribution of "complexed" and "free" oscillators.) If this were the case, we would deal with two overlapping profiles and therefore could not use the simple theory (refer to end of Section 3.3).

As an example, we discuss the $CHCl_3$-$C_6H_6$ complex. Its purported structure is shown in Fig. 2.17. The application of these ideas is displayed in Fig. 2.18 by the infrared correlation function of the totally symmetric $\nu_3$ mode of $CDCl_3$ at 367 cm$^{-1}$.[110] (The profile of the deuterium isotope is cleaner than that of the protonated compound.) Since the vibrational relaxation of the $\nu_3$

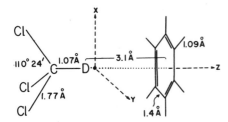

**Figure 2.17.** Molecular structure of the chloroform–benzene complex according to Ref. 111a.

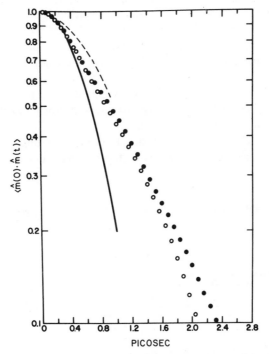

**Figure 2.18.** First-order tensor autocorrelation functions from the parallel-polarized infrared $\nu_3$ mode of $CDCl_3$ in benzene, describing the orientational motion of the $C_3$ symmetry axis of $CDCl_3$ (equal $z$) as the molecule rotates around its perpendicular axes $x$ and $y$. Note that direction $z$ contains the permanent dipole moment of $CDCl_3$ as well as the intermolecular bond of the proposed chloroform–benzene complex—see Fig. 2.17. Solid points: analytical $1:4$ mole/mole $CDCl_3:C_6H_6$ mixture. Open points: analytical equimolar mixture. Solid line: freely rotating $CDCl_3$ molecules. Dashed curve: freely rotating mixture of the complexed and uncomplexed $CDCl_3$ molecules ($0.9:1$) in an analytical $1:4$ mixture, assuming *rigid* complexation of $CDCl_3$ to $C_6H_6$ (except for internal rotation around axis $z$). [W. G. Rothschild, *Chem. Phys. Lett.* **9**, 149 (1971); see Eq. 2 of this reference.]

**127**

fundamental is insignificant,[10b] the observed infrared autocorrelation function describes the orientational motion *of* the $C_3$ symmetry axis (direction $z$, containing the permanent dipole moment and the supposed intermoiety bond to the $C_6H_6$ molecule) as the $CDCl_3$ molecules rotate *around* axes $x$ and $y$ perpendicular to the $C_3$ axis. In other words, we indeed observe that motion which twists the $CDCl_3$ moiety out of the intermoiety bond. Yet, Fig. 2.18 shows no discernible difference between the short-time behavior of complexed and freely rotating $CDCl_3$ molecules. It is unlikely that the $CDCl_3$-$C_6H_6$ complex reorients as a rigid unit.

To bring this out more clearly, we have computed the corresponding rotational correlation function for the *free* rotation of $CDCl_3$ molecules at thermodynamic equilibrium of complexed and noncomplexed molecules in a solution of analytical composition $CDCl_3 : C_6H_6 = 1 : 4$ mole/mole. The calculations (of the moments of inertia) were based on the assumption that the complexed chloroform molecules belong to a rigid $CDCl_3$-$C_6H_6$ species as defined in Fig. 2.17. Note that this species need not be rigid with respect to rotation about the intermolecular bond (internal rotation) since this motion is not observed in a parallel-polarized band (Section 2.1, Chapter 1). The result is entered into Fig. 2.18 by the dashed curve. It does not agree with the experiment.

On the basis of this information we conclude that the complex between chloroform and benzene does not prevent the chloroform moiety from individual end-over-end orientational motion.

To ascertain whether results of this type are meaningful, it is useful to consider some purely kinetic aspects of the problem. Let us write the kinetic reaction

$$CDCl_3 + C_6H_6 \underset{k_{-1}}{\overset{k_1}{\rightleftharpoons}} CDCl_3 \cdots C_6H_6, \qquad k_f = \frac{k_1}{k_{-1}},$$

and the equation for the equilibrium condition between complexed and uncomplexed chloroform molecules,

$$\frac{x/(a_0 + b_0 - x)}{(a_0 - x)(b_0 - x)/(a_0 + b_0 - x)^2} \equiv k_f,$$

where $a_0, b_0$ are the initial (analytical) number of molecules of $CDCl_3$ and $C_6H_6$, respectively, $x$ is the number of molecules of the complex, and $a_0 + b_0 - 2x + x = a_0 + b_0 - x$ is the total number of molecules in the solution. Evaluation of the expression, with the boundary condition $x = 0$ for $k_f = 0$, gives

$$\frac{x}{(a_0 + b_0 - x)} = (x)_{mf} = (1 - \sqrt{A})(1 + \sqrt{A})^{-1}$$

where mf = mole fraction and

$$A = 1 - \frac{4k_f n_0}{(1 + k_f)(n_0 + 1)^2}, \qquad n_0 = \frac{a_0}{b_0}.$$

Calculating from this expression the number ratio of complexed $CDCl_3$ to uncomplexed $CDCl_3$ molecules $(x)_c$ yields

$$\frac{x}{a_0 - x} \equiv (x)_c = \frac{1 - \sqrt{A}}{(n_0 - 1)(n_0 + 1)^{-1} + \sqrt{A}}.$$

We have plotted in Fig. 2.19 $(x)_c$ as function of the stability constant $k_f$ for $n_0 = 1$ and $\frac{1}{4}$. This information is important for our discussion since we observe in our experiments the transition moment averaged over all $CDCl_3$ molecules,

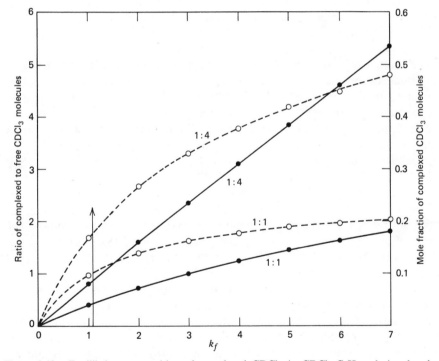

**Figure 2.19.** Equilibrium composition of complexed $CDCl_3$ in $CDCl_3$-$C_6H_6$ solution for the analytical molar ratio $CDCl_3 : C_6H_6 = 1:1$ and $1:4$ as function of the stability constant $k_f$. Left ordinate: ratio of complexed to uncomplexed chloroform molecules (solid curves). Right ordinate: mole fraction of complexed chloroform molecules (broken curves). The equilibrium stability constant is $k_f = 1.145$ as reported by J. Homer and M. C. Cooke, *J. Chem. Soc.* (A) (1969), 777.

complexed by $C_6H_6$ or not. We therefore desire that $(x)_c$, the fraction of complexed $CDCl_3$ molecules, be as large as possible: Using $k_f = 1.1$,[111a] we compute that about one of two $CDCl_3$ molecules is complexed by $C_6H_6$ in the concentration range investigated.

The "inverse" experiment, namely, the concentration dependence of the rotational bandwidth of a totally symmetric (Raman) fundamental of benzene (species $A_{1g}$, mode $\nu_2$, 990 cm$^{-1}$) dissolved in $CHCl_3$, has been reported.[111b] This mode is likewise polarized along the intermoiety axis of the complex and therefore should give corroborative evidence to the foregoing results.

However, the data are ambiguous. The observed orientational halfwidth of the $\nu_2$ mode of $C_6H_6$ exhibited a slight maximum at a solution composition of 4 : 1 mole/mole $C_6H_6 : CHCl_3 = 0.2$ mole fraction $CHCl_3$, but was flat elsewhere. The significance of this bandwidth maximum is not clear in the context of complex formation between $C_6H_6$ and $CHCl_3$. If anything, we should expect a minimum of the bandwidth since the orientational motion of axis $z$ of $C_6H_6$ should be hindered by complex formation with chloroform; we expect a larger value of $\tau_R^{(2,0)}$ of dissolved than of neat $C_6H_6$.

Furthermore, at the analytical composition $C_6H_6 : CHCl_3 = 4 : 1$ mole/mole the number ratio of complexed to uncomplexed benzene molecules is 0.13 (Fig. 2.19). Hence the reported maximum of the bandwidth of the $\nu_2$ mode of $C_6H_6$ was observed in a $C_6H_6$-$CHCl_3$ solution that contained nine times more free than complexed benzene molecules. We therefore conclude that the orientational Raman widths of the $\nu_2$ mode of $C_6H_6$ in $CHCl_3$ solution do not give any evidence for or against the supposed 1 : 1 $CHCl_3$-$C_6H_6$ complex formation. We note that the choice of $\nu_2$ for benzene was poor. First, its degree of depolarization is only 0.01; the contribution of the VH profile is possibly too insignificant to guarantee good separation of $I_{iso}$ and $I_{aniso}$ in the solution spectra. Second, the degree of vibrational relaxation of the $\nu_2$ mode (judged by the halfwidths of $I_{VV}$ and $I_{VH}$) is about 0.4 of $I_{aniso}$. The corresponding values of $\nu_3$ of $CDCl_3$ and $CHCl_3$ are about 0.1.[10b]

In conclusion, Fourier transformation of the vibrational-rotational contour of the $\nu_3$ mode of chloroform dissolved in benzene tells us that the complex does not reorient its intermoiety bond as a rigid axis, at least not during the times of our experiment (order of picoseconds, see Fig. 2.18). On the contary, the data predict that the chloroform moiety performs random reorientational motion. However, we cannot exclude a certain degree of fast librational motion (Section 3.3); the data are not accurate enough. Furthermore, we should consider that band broadening by concentration fluctuations complicates the simple description we have given here. (We discuss this in Chapter 3.) Nevertheless, application of fluctuation spectroscopy to the orientational mobility of weak solution complexes is a fascinating topic, and it would be unreasonable to get discouraged by the failure of early work to give unambiguous evidence. The majority of solute–solvent systems of interest is nonideal; it would indeed be surprising if this did not show up in their orientational dynamics.

We believe that careful experiments in the far-infrared, where the purely rotational profiles are observable, give a better chance for success than the evaluation of vibration-rotation spectra. Although vibration-rotation spectra offer the opportunity to study the internal rotation of the complex through a perpendicular-polarized mode, this advantage is effectively lost by concomitant coupling effects (Section 4.2.2). (This complication had not been realized in earlier studies.[110])

## 5.2.   Orientational Correlation Times and Model Concepts of Rotational Motion

Orientational correlation times are not as informative as their correlation functions since they integrate over the whole time domain (Eqs. 1.67, 1.68, Chapter 1). Nevertheless, they are useful for giving an idea about the contribution of rotational relaxation processes and may point at any serious or unexpected inconsistencies between experiment and theory. If carefully applied, they serve to distinguish between different model concepts of rotational relaxation (as shown later). In certain instances, only the orientational correlation time is obtainable from the experiment; it can always be estimated from the halfwidth of the band contour. We must not forget, however, that this approach accentuates the Markovian (exponential) part of the orientational correlation and that it tells us nothing about the short-time, dynamically coherent characteristics of the orientational motion.

In the following we first discuss experimental second-order tensor correlation times of the orientational motion of the $C_3$ axis of trimethylchloro- and trimethylbromomethane, $(CH_3)_3CCl$ and $(CH_3)_3CBr$.[109] These are symmetric top molecules of near-identical geometric shape and polarity but different moments of inertia about the perpendicular axes. Thereafter, we demonstrate certain aspects of orientational motion by evaluating the ratio $\tau_{aniso}^{(1)}/\tau_{aniso}^{(2)}$ of a first- and second-order orientational correlation time belonging to the same axis direction of the molecule.

The data on $(CH_3)_3CCl$ and $(CH_3)_3CBr$ were obtained from the parallel-polarized carbon–halogen stretching fundamentals through the standard VV–VH Raman scattering experiment (Section 3.2, Chapter 1). The results were computed from the correlation times, $\tau_{VH}^{(2,0)}$ and $\tau_{iso}^{(2,0)}$ —themselves measured from the full halfwidth (width at half-peak height) of $I_{VH}$ and $I_{iso}$ —assuming Lorentzian profiles (Eq. 2.88). The purely orientational correlation time is then (replace $G_{VH}(t)$ and $G_{iso}(t)$ in Eq. 1.49 by exponentials)

$$\frac{1}{\tau_{aniso}^{(2,0)}} \equiv \frac{1}{\tau_R^{(2,0)}} = \frac{1}{\tau_{VH}^{(2,0)}} - \frac{1}{\tau_{iso}^{(2,0)}}. \tag{2.96}$$

We have collected in Table II some of the $\tau_R^{(2,0)}$ of the halogen compounds measured at the same temperature and in the same solvent. Column 1 gives $\tau_R$

**Table II.** Second-order Tensor Correlation Times $\tau_R^{(2,0)}$ of the Orientational Motion of the Symmetry Axis of Trimethylchloro- and Trimethylbromomethane in Units of $10^{-12}$ sec[a]

| $(CH_3)_3CCl$ | | | | $(CH_3)_3CBr$ | | |
|---|---|---|---|---|---|---|
| | Solvent | | | | Solvent | |
| Neat Liquid | $CCl_4$ | $C_6H_{14}$ | °C | Neat Liquid | $CCl_4$ | $C_6H_{14}$ |
| | 1.14 | | 50 | | 1.58 | |
| 1.40 | 1.64 | 1.13 | 25 | 1.83 | 1.99 | 1.13 |
| | | 4.57 | −70 | | | 4.14 |

[a] From Ref. 109.

for the neat, column 2 for a 0.2 mole fraction of the halogen compound in $CCl_4$, and column 3 in hexane ($C_6H_{14}$) solution.

The data exhibit unexpected aspects. At 25°C the orientational motion of the compounds in dilute solution of $CCl_4$ — a spherical top — is slower than in their neat liquids.

Second, at −70°C, the average orientational motion in hexane solution is slower for the lighter $(CH_3)_3CCl$ than for $(CH_3)_3CBr$ (Columns 3), whereas at 25°C both compounds in hexane solution reorient at the same rate.

From this we draw the apparent conclusions that a spherical solvent molecule exerts stronger angle-dependent forces on the solute than the dipolar solute itself, and that the orientational correlation times for both solutes in hexane are identical at 25°C but longer for the lighter solute at −70°C. Neither observation follows intuitive ideas about the effects of solvent polarity and solute mass on orientational motion.

What may cause these unexpected results? We can only surmise that vibrational relaxation or concentration fluctuations are at the root of the apparent discrepancies between experiment and expectations. With respect to vibrational relaxation, the experimental purely vibrational correlation time $\tau_{iso}^{(2,0)}$ is of the same order of magnitude as those of $\tau_R^{(2,0)}$.[109] For instance, in the neat chloro and bromo compounds at 25°C, $\tau_{iso}^{(2,0)} = 0.85 \times 10^{-12}$ and $1.36 \times 10^{-12}$ sec versus $\tau_R^{(2,0)} = 1.40 \times 10^{-12}$ and $1.83 \times 10^{-12}$ sec (Table II). In such a situation, vibrational and rotational relaxation processes may be correlated, at least partially (Section 4.2.4). In fact, the temperature dependence of the data suggests that the correlation time of the major vibrational relaxation process involved here is temperature insensitive (we discuss the "which" and the "why" in Chapter 3), in contrast to the pronounced inverse temperature dependence of $\tau_R$.[112] (The molecules reorient faster at higher kinetic energies.) In other words, at lower temperatures the relative contribution of $\tau_{iso}$ becomes even more pronounced, and we can no longer trust that the experimental $\tau_R$ give an accurate account of the purely orientational motion.

Band broadening by concentration fluctuations, which occurs in solution spectra, may render a comparison between bandwidths of neat and dissolved compound more complicated than expected. We discuss this in Chapter 3.

Apart from absolute correlation times, it is useful to consider the ratio

$$\frac{\tau_R^{(1,k)}}{\tau_R^{(2,k)}}$$

of a first- and second-order tensor orientational correlation time of the same transition moment. Depending on the particular model of rotational motion, the ratio takes different values; comparison with the experiment thus yields conclusions on the nature of the orientational process.

For example, let us take the Debye rotational diffusion model and apply it to a spherical top molecule. We recall that in this particular situation the Debye rotational diffusion model is the limiting case of the $J$-extended rotational diffusion model for $\tau_J^* \ll 1$ (Eq. 2.65); we get

$$\left. \frac{\tau_R^{(1,0)}}{\tau_R^{(2,0)}} \right|_{\text{Debye}} = 3. \tag{2.97}$$

This value has been reported in several studies on liquid-phase dynamics.[113] The conclusion is therefore that the observed orientational motion obeys the Debye mechanism.

How good must the data be to allow this conclusion to be drawn? To answer this, we employ a theoretical model that includes $\tau_R^{(1)}/\tau_R^{(2)}$ as functional parameter (we omit superscript 0, considering spherical top molecules only). This is the so-called Ivanov jump model. It neglects kinetic effects and assumes a Poisson distribution of times between two successive events (Eq. 2.10) that randomly reorient the observed molecular axis by instantaneous jumps through angle $\Omega$ after a period of motional arrest. $\Omega$ fluctuates in a prescribed manner around an arbitrary average $\Omega_0$. The choice of $\Omega_0$ and its distribution function give the required flexibility to the model.[114]

Let us deal only with those aspects of the model that are of immediate use; the derivation of its correlation function is discussed later in this chapter. We wish to find the eigenvalues of the orientational transition kernel $A(\Omega - \Omega')$ (see Section 2.2). It is convenient to construct the operator

$$\lambda_{(1)}^{(j)} = \int d\Omega \, A(\Omega - \Omega')_{(1)} \mathbf{D}^{(j)*}(\Omega), \tag{2.98}$$

the expectation value of the rotational transformation $\mathbf{D}(\Omega)$ under $A$ for a single orientation step (indicated by the subscript). (See also Eqs. C.1, C.8, Appendix C.) According to the assumption of the jump model, $A(\Omega - \Omega')$ is

Table III. Ratio of the Infrared and Raman Correlation Times for the Reorientational Motion of an Arbitrary Axis of a Spherical Top Molecule as a Function of its Average Jump Angle $\Omega_0$.

| $\tau_R^{(1)}/\tau_R^{(2)}$ | $\Omega_0$ [In Degrees] |
|---|---|
| 3.000 | 1 |
| 2.996 | 5 |
| 2.982 | 10 |
| 2.956 | 15 |
| 2.928 | 20 |
| 2.839 | 30 |
| 2.649 | 45 |
| $\vdots$ | $\vdots$ |
| 2.000 | 75 |

nonzero only for discrete $\Omega - \Omega'$. We assume that $\Omega_0$ is the same for each single jump. Hence

$$A(\Omega - \Omega')_{(1)} = \delta(\Omega - \Omega_0). \tag{2.99}$$

For a spherical molecule, $\lambda_{(1)}^{(j)}$ must be invariant to any rotation $\mathscr{R}^{(j)}(\Omega) = \mathbf{D}^{(j)}(\Omega)$ of the coordinate system. Therefore $\lambda_{(1)}^{(j)}$ commutes with $\mathscr{R}^{(j)}$.[38, 114] Then, $\lambda_{(1)}^{(j)}$ is a multiple of the identity matrix (Schur's lemma):

$$\lambda_{(1)}^{(j)} = \lambda_{(1)}^{(j)} \mathbf{E}^{(j)}.$$

Taking the trace on either side of Eq. 2.98 gives[115]

$$\lambda_{(1)}^{(j)}(2j+1) = \int d\Omega \, A(\Omega - \Omega')_{(1)} \operatorname{Tr} \mathbf{D}^{(j)*}(\Omega)$$

$$= \int d\Omega \, A(\Omega - \Omega')_{(1)} \sum_{k=-j}^{j} \exp(ik\Omega)$$

$$= \int d\Omega \, \frac{A(\Omega - \Omega')_{(1)} \sin\left[(j+\frac{1}{2})\Omega\right]}{\sin(\Omega/2)}. \tag{2.100}$$

Insertion of Eq. 2.99 leads to

$$\lambda_{(1)}^{(j)} = \frac{(2j+1)^{-1} \sin\left[(j+\frac{1}{2})\Omega_0\right]}{\sin(\Omega_0/2)}. \tag{2.101}$$

As we show later, the eigenvalue for $N \to \infty$ jumps for this model is

$$\lambda_\infty^{(j)} = \frac{\left[1 - \lambda_{(1)}^{(j)}\right]}{\tau}, \tag{2.102}$$

where $\tau$ is the time between successive orientational steps $\Omega_0$ (Eq. 2.10). Note that $\lambda_\infty^{(j)}$ is the reciprocal correlation time of the orientational correlation function (Eq. C.11, Appendix C).

Table III shows computed values of $\tau_R^{(1)}/\tau_R^{(2)}$ as function of $\Omega_0$. The important aspect is obvious: Under common experimental precision, the result $\tau_R^{(1)}/\tau_R^{(2)} = 3$ covers a range of $\Omega_0$ from 1 to 20 degrees. Hence it is hardly permissable to conclude that the value $\tau_R^{(1)}/\tau_R^{(2)} = 3$ signifies Debye orientational motion.

### 5.3.  Experimental Angular Correlations in a Symmetric and a Spherical Top Molecule

As we mentioned at the end of Section 2.1 of Chapter 1, we gain additional information from orientational correlation data of vibration-rotation band profiles by constructing a linear combination of Cartesian-axis orientational correlation functions over a complete set of the vibrational symmetry species of the point group of the molecule. In the following discussion we elaborate on this and apply it to two experimental situations.[14, 116] Indeed, this approach deserves more attention than it has received in the past—were it not, in many instances, rendered rather difficult by inherent, spectral conditions. A "complete set" would require that we examine at least one band profile of each of the two active vibrational species for axial molecules (symmetric tops) and one profile each of the three allowed species for orthorhombic molecules. There are not many molecular systems where the appropriate contours are found well isolated from adjacent profiles.

We introduce a complete set of molecule-fixed orthonormal basis vectors $\hat{\mathbf{b}}_i(t)$, $i = 1, 2, 3$, along some chosen molecular axes. The transition moment vector $\hat{\mathbf{m}}(t)$ of a vibration-rotation mode $\nu$ is expanded in the form

$$\hat{\mathbf{m}}^\nu(t) = \sum_i a_i^\nu \hat{\mathbf{b}}_i(t). \tag{2.103}$$

Now, the rotation of vector component $\hat{\mathbf{b}}_i(t)$ is represented as follows: Consider two three-dimensional coordinate frames $F$ and $F'$ which coincide at $t = 0$. At $t = t$, frame $F'$ will have been rotated against $F$ by means of the rotation matrix $\mathcal{R}(t)$. Then, if a vector has components $\hat{\mathbf{b}}_i(0)$ in $F$, it will have component $\hat{\mathbf{b}}_i(t)$ in $F'$,

$$\hat{\mathbf{b}}_i(t) = \sum_j \mathcal{R}_{ij}(t) \hat{\mathbf{b}}_j(0).$$

This random rotation matrix therefore expresses the characteristics of the orientational motion of the molecule; notice that, in principle, we employ the same approach as that leading to Eq. 1.92 of Chapter 1. However, we discuss only first-order (infrared) tensors since the Cartesian Raman correlation functions are too cumbersome to be of much use (Section 3.4, Chapter 1). Assuming that vibrational relaxation is known or negligible, we obtain for the orientational correlation function associated with the spectral density of vibration-rotation mode $\nu$

$$\hat{G}^{\nu}(t) = \langle \hat{\mathbf{m}}^{\nu}(0) \cdot \hat{\mathbf{m}}^{\nu}(t) \rangle$$

$$= \sum_{ij} a_i^{\nu} \langle \mathcal{R}_{ij}(t) \rangle a_j^{\nu}. \qquad (2.104)$$

At present we make use of the effect of molecular point group symmetry $S_M$ on matrix $\mathcal{R}$.[38]

1.  For spherical point groups, $\mathcal{R}$ is isotropic.
2.  For axial symmetry, $\mathcal{R}$ is axial.
3.  For orthorhombic symmetry, $\mathcal{R}$ is diagonal.
4.  For lower symmetries, $\mathcal{R}$ is nondiagonal. This situation is too complicated and does not interest us here.

Now recall that among the irreducible representations of $S_M$ are the symmetry species that belong to an infrared vibrational transition moment. Therefore these transition moment species are connected to the corresponding autocorrelation functions $\hat{G}^{\nu}(t)$. Consequently, a complete set of $\hat{G}^{\nu}(t)$ forms a representation of the stochastic rotation matrix $\mathcal{R}$ under the molecular point group $S_M$. Some examples will make this clear.

***Example 1. Symmetric Top, $C_{3v}$.*** The irreducible representations ( = species) are $A_1$ and $E$. Assigning the $C_3$ symmetry axis to index 3 of basis vector $\hat{\mathbf{b}}_i$ and the two, equivalent perpendicular axes to indices 1 and 2, we find

$$\hat{G}(A_1, t) = \langle \mathcal{R}_{33}(t) \rangle$$

$$\hat{G}(E, t) = \langle \mathcal{R}_{11}(t) \rangle = \langle \mathcal{R}_{22}(t) \rangle$$

or

$$\hat{G}(E, t) = \left(\tfrac{1}{2}\right)\{\langle \mathcal{R}_{11}(t) + \mathcal{R}_{22}(t) \rangle\}. \qquad (2.105)$$

***Example 2. Spherical Top, $T_d$.*** The only irreducible infrared-active species is $F_2$. Therefore

$$\hat{G}(F_2, t) = \left(\tfrac{1}{3}\right)\{\langle \mathcal{R}_{11}(t) + \mathcal{R}_{22}(t) + \mathcal{R}_{33}(t) \rangle\}. \qquad (2.106)$$

Up to now our discussion has been a mere exercise in an application of group theory; what have we then achieved by this procedure? We have, in fact, gained the possibility of composing experimental correlation functions of dynamic variables that are not directly obtainable by the Fourier inversion of an infrared spectral density, and which may give considerable insight into rotational relaxation phenomena.

To see this, we note[14, 116] that we can parametrize rotation matrix $\mathcal{R}(t)$ not only in terms of Eulerian angles $\alpha$, $\beta$, $\gamma$ but also by a three-dimensional molecule-fixed vector $\Theta$ whose direction, determined by polar angles $\theta(t)$ and $\phi(t)$, serves as axis of rotation with rotation angle $\eta(t)$. (See also Eq. 2.6.)

For cubic and axial point groups, we have collected in Table IV the diagonal elements of $\mathcal{R}$, the correlation functions forming a representation of $\mathcal{R}$, and useful linear combinations derived from them. For instance, correlation function

$$\hat{G}(t) = \langle \cos \eta(t) \rangle \equiv \hat{G}_\eta(t) \tag{2.107}$$

describes the time evolution of the autocorrelation of the rotation angle around the average rotation axis. Functions

$$\hat{G}(t) = \langle \sin^2\theta(t)[1 - \cos\eta(t)] \rangle \equiv \hat{G}_\theta^{(-)}(t)$$

$$\hat{G}(t) = \langle \cos^2\theta(t)[1 - \cos\eta(t)] \rangle \equiv \hat{G}_\theta^{(+)}(t) \tag{2.108}$$

describe the time evolution of the autocorrelation of the average position of the

**Table IV.    Explicit Forms of the Diagonal Matrix Elements $\mathcal{R}_{ii}$ of the Representation of Cubic and Axial Molecular Point Groups, Correlation Functions of the Vibrational Symmetry Species, and Some Useful Linear Combinations**[a]

| Matrix Elements | | |
|---|---|---|
| $\mathcal{R}_{11}$ | $\cos\gamma\cos\alpha - \cos\beta\sin\alpha\sin\gamma$ | $\cos\eta + \sin^2\theta\cos^2\phi(1 - \cos\eta)$ |
| $\mathcal{R}_{22}$ | $-\sin\gamma\sin\alpha + \cos\beta\cos\alpha\cos\gamma$ | $\cos\eta + \sin^2\theta\sin^2\phi(1 - \cos\eta)$ |
| $\mathcal{R}_{33}$ | $\cos\beta$ | $\cos\eta + \cos^2\theta(1 - \cos\eta)$ |

| Correlation Functions | | |
|---|---|---|
| | Cubic | Axial |
| | $\hat{G}(F, t) = \frac{1}{3}\langle \mathcal{R}_{11}$ | $\hat{G}(E, t) = \frac{1}{2}\langle \mathcal{R}_{11} + \mathcal{R}_{22}\rangle$ |
| | $+ \mathcal{R}_{22} + \mathcal{R}_{33}\rangle$ | $\hat{G}(A, t) = \langle \mathcal{R}_{33}\rangle$ |
| $\langle \cos\eta(t)\rangle$ | $\frac{1}{2}\{3\hat{G}(F, t) - 1\}$ | $\frac{1}{3}\{\hat{G}(A, t) + 2\hat{G}(E, t) - 1\}$ |
| $\langle \sin^2\theta(t)[1 - \cos\eta(t)]\rangle$ | $1 - \hat{G}(F, t)$ | $1 - \hat{G}(A, t)$ |
| $\langle \cos^2\theta(t)[1 - \cos\eta(t)]\rangle$ | $\frac{1}{2}\{1 - \hat{G}(F, t)\}$ | $\frac{1}{2}\{1 + \hat{G}(A, t) - 2\hat{G}(E, t)\}$ |

[a] From B. Keller and F. Kneubühl, *Helv. Phys. Acta* **45**, 1127 (1972).

average rotation axis of the molecule. For instance, if $\hat{G}_\theta^{(+)}(t) \gg \hat{G}_\theta^{(-)}(t)$, the average rotation axis remains near the $C_3$ symmetry axis of the molecule during the time interval $t[\theta \sim 0]$.

The foregoing considerations are, by no means, academic. Let us give two examples.

***Example 1. Average Rotation Axis in a Symmetric Top.*** A parallel ($A_1$) and a perpendicular-polarized ($E$) infrared fundamental band contour of $CH_3I$ at room temperature have been Fourier inverted, and the infrared correlation functions have been reported.[117] Although these were based on early data, we can improve them in the light of more recent experiments. This means correcting for the pure vibrational relaxation for the $A_1$ band[118] and applying the needed modification for the Coriolis coupling for the $E$ mode.[46] (We assume here that its vibrational relaxation is not significant for present purposes, realizing that this assumption is usually not correct.)

The plot of Eqs. 2.107–2.108 based on the complete set of vibrational species correlation functions is displayed in Fig. 2.20. We see that $\hat{G}_\theta^{(+)} \gg \hat{G}_\theta^{(-)}$

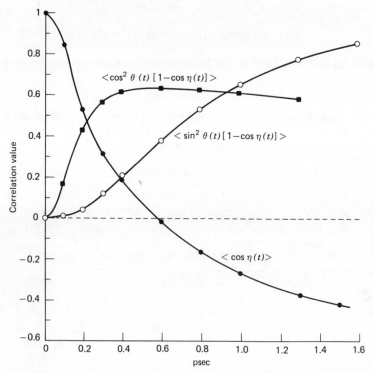

**Figure 2.20.** Liquid methyl iodide at 300°K: autocorrelation function of the average rotation angle of the average rotation axis (solid dots) and autocorrelation functions of the direction of the average rotation axis with respect to the molecular symmetry axis (squares and open circles).

for up to $t' \sim 0.4 \times 10^{-12}$ sec. This means that the average rotation axis is approximately coincident with the molecular symmetry axis during $t \sim t'$.

Rotation around the $C_3$ axis is rapid. Note that the correlation function $\langle \cos \eta(t) \rangle$, displayed in Fig. 2.20 by the solid dots, turns negative after $t' = 0.6 \times 10^{-12}$ sec. In other words, after this time interval the average angle for this rotational component ("spinning") has exceeded $\pi/2$.

Let us compare this value with the period of one free revolution around the $C_3$ axis, $[2A(1-\zeta)^2 k_B Tc^2]^{-1/2}$. It is about $0.9 \times 10^{-12}$ sec, where $A = 5.2$ cm$^{-1}$ is the appropriate rotational constant, $k_B = 0.695$ cm$^{-1}$/degree, and $\zeta = 0.19$ is the Coriolis coupling constant of the mode. We see that the molecule in its liquid phase requires on the average three times longer to rotate by $\pi/2$ around the $C_3$ axis than does the free rotor. Consequently, the frequently expressed notion that the spinning motion of liquid-phase $CH_3I$ is nearly free[119] does not seem to be corroborated by these results.

We also observe in this system that $\hat{G}_\theta^{(+)}$ is approximately $\hat{G}_\theta^{(-)}$ after $0.8 \times 10^{-12}$ sec. If rotation around the $C_3$ axis were indeed the only significant contribution to the rotational relaxation of this mode during the complete interval of observation (Fig. 1.5), then the average rotation axis should remain essentially parallel to the $C_3$ axis, $\hat{G}_\theta^{(+)} \gg \hat{G}_\theta^{(-)}$.

The second example explores further possibilities of this approach.

***Example 2.    Average Orientational Motion in*** $C(CH_3)_4$ ***and*** $CH_3I$. The correlation function $\hat{G}_\eta(t) = \langle \cos \eta(t) \rangle$ (Table IV) of a spherical and symmetric top molecule may be compared. In this way we would obtain comparative information on the orientational motion of different molecules of related shape (spheroidal).

As an example we have chosen two previously discussed systems, liquid neopentane and liquid methyl iodide. To arrive at a meaningful comparison, we apply the following two corrections:

1.  We "normalize" the absolute axis $t$ by defining a reduced time $t^* = \bar{\omega}t$ where

$$\bar{\omega} = \{ (\tfrac{2}{3}) k_B Tc^2 D \}^{1/2}.$$

(See also Eq. 2.2 and the definition of reduced units in Section 2.2.) Here $D \equiv 3A$ or $\equiv A + 2B$ for the spherical or symmetric top, respectively, is the sum of the rotational constants (remember $A \geq B \geq C$) in cm$^{-1}$. This procedure brings the thermal kinetic effects of the different molecules on a common denominator.

2.  Because there is some Coriolis interaction in both the degenerate vibrational modes [$\zeta = 0.19$ for $\nu_6$ of $CH_3I$, and $\zeta = 0.09$ for $\nu_{18}$ of $C(CH_3)_4$], we ought to modify $\bar{\omega}$ correspondingly. This entails multiply-

ing rotational constant $A$ by $(1 - \zeta)^2$. (For triply degenerate infrared modes of spherical tops, we have discussed the origin of this correction in Appendix D. For doubly degenerate modes of the symmetric top, see Appendix E.)

The numerical values of $\bar{\omega}$ are $0.12 \times 10^{12}$ sec$^{-1}$ and $0.71 \times 10^{12}$ sec$^{-1}$ for $C(CH_3)_4$ and $CH_3I$, respectively (300°K). Figure 2.21 shows the correlation function $\langle \cos \eta(t^*) \rangle$ for both molecules. It is interesting to observe from this comparison that the correlation of the average rotation angle of the average rotation axis of $C(CH_3)_4$ decreases more rapidly than that of $CH_3I$. This is striking considering the supposedly fast-spinning rotation of $CH_3I$ (around the $C_3$ axis). We then realize that rotation around the perpendicular axes in $CH_3I$ must be strongly hindered to compensate for the rate of the spinning motion.

No doubt, comparisons of this type are not only instructive in themselves but are also useful for relating the nature of the orientational motion to macroscopic quantities, such as free volume and viscosity. These are scalar parameters; it therefore makes more sense to compare them to those characteristics of rotational relaxation that reflect the average rotation around all three inertial axes of the molecule.

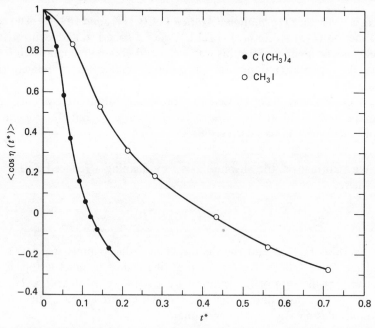

**Figure 2.21.** Liquid methyl iodide and neopentane at 300°K: comparison between the decay of the autocorrelation of the average rotation angle of the average rotation axis as a function of reduced time.

### 5.4. Energy Barriers to Free Rotation and Free Volume: Deuterochloride (DCl) in Hydrocarbon Solutions

As is well known, the vapor-phase infrared rotation-vibration profile of a diatomic molecule lacks a central branch ($Q$ branch) since vibrational-rotational transitions with the angular momentum selection rule $\Delta j = 0$ are forbidden. Therefore when such a forbidden central branch is *induced* in situations where the orientational motion of the diatomic rotors is obviously no longer free, it is clear that this phenomenon can be exploited to give us information on the perturbing intermolecular forces and the characteristics of the hindered motion.

Our concern with this effect is therefore obvious. First, we would like to know and to describe the forces responsible for the appearance of the free-molecule-forbidden $Q$ branch in a condensed medium; it is not difficult to see that the rapid fluctuation phenomena in the condensed phase—rotatory-translatory motions of the molecules—*cannot* give rise to the $Q$ branch in the hindered rotor system. Second, interesting results of experiments have been published which deal with certain characteristics of the ratio $I_Q/I_R$ of the peak intensity $I$ of the induced $Q$ and the allowed $R$ branch of the fundamental of DCl dissolved in a variety of saturated hydrocarbons.[120] These results are our incentive to delve with considerable detail into the problems posed.

Two sets of the published data are of relevance to us, namely, the dependence of $I_Q/I_R$ of the DCl mode on the length of linear hydrocarbon solvent molecules $C_nH_{2n+2}$ ($n = 5$ to 16) and on the number of methyl groups of isomers of nonane, $C_9H_{20}$ (2, 4, 5, and 6 $CH_3$ groups).

Figure 2.22 shows the first set. Figure 2.23—we have replotted the original data in a different fashion—displays the second. Inspection of these results immediately brings the following, apparently relevant, macroscopic parameter to mind: the free volume.[120] For instance, for the series of straight-chain hydrocarbons it is known that the free volume decreases with increasing chain length; from $C_5$ to $C_{16}$ it drops by about 30%. We observe (Fig. 2.22) that $I_Q/I_R$ of the DCl fundamental increases with a lengthening of the hydrocarbon chain of the solvent. This would agree with the concept of the decreasing free volume that is available to the reorienting DCl molecule.

For the series of nonane isomers, however, the free volume increases with the number of methyl groups ("branching"). Yet, $I_Q/I_R$ of the DCl fundamental does not drop with increased branching in the nonanes (Fig. 2.23); it rises. Obviously, there are either additional effects (we do not go into details)[120] or the simple notion of a macroscopic free volume as applied here is not useful.

Evidently, we will not progress until we understand the process that generates, on the molecular level, the free-molecule-forbidden $Q$ branch. We describe this as follows on the basis of a quantitative theory of wide applicability[121]: We determine the forces that induce the forbidden $Q$ branch lines in the infrared vibration-rotation band of a diatomic molecule rotating in an inert condensed medium that we consider, first, static (molecular positions and

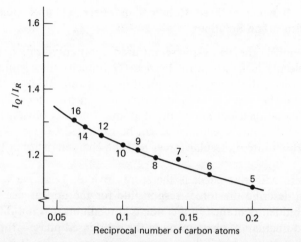

**Figure 2.22.** Observed peak intensity ratio $I_Q/I_R$ of the $Q$ and $R$ branches of the infrared fundamental of DCl dissolved in a series of straight-chain hydrocarbons of varying chain length (concentration 1.5 mole %). [From D. Richon, D. Patterson, and G. Turrell, *Chem. Phys.* **16**, 61 (1976).]

orientations of the hydrocarbon solvent molecules are frozen). Thereafter, we introduce rotational relaxation of DCl due to the actual fluctuations in the positional coordinates of the hydrocarbon solvent molecules. In other words, a quasi-static intermolecular perturbation potential generates first a rotational transition moment operator in DCl which couples states of equal angular momentum quantum numbers (the free-rotor-forbidden $Q$ branch). Then, coupling of the orientational coordinate of DCl to the *fluctuations* in the medium is invoked for causing *broadening* of the rotational lines of all vibration-rotation branches of the DCl fundamental.

The example of the perturbed diatomic rotor is so informative and useful since it demonstrates how we can solve a complicated dynamic problem by decomposing the overall phenomenon into components that occur on different time scales. We can see that, quite *generally*, the quasi-static effect produces a band splitting or generates free-molecule-forbidden transitions. [The quasi-static effect lifts a "degeneracy" of the ensemble of free (unperturbed) molecules.] On the other hand, only the "short-time" dynamic, fluctuating variables of the perturbation potential cause true relaxation (irreversibility).

We then define a nonpolar environment of configurations of relatively long lifetimes $\tau_l$ about the dipolar vib-rotor; the solvent configurations around a DCl molecule persist much longer than its orientational correlation time $\tau_R^{(1,0)}$. In turn, the fluctuations of these solvent configurations are characterized by a correlation time $\tau_c$ that reflects a faster time scale than $\tau_l$: $\tau_c \sim \tau_R$, where we drop the $j, k$ designation to $\tau_R$.

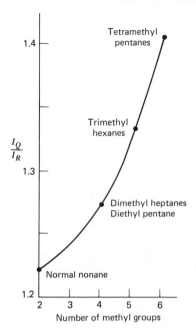

**Figure 2.23.**   Observed peak intensity ratio $I_Q/I_R$ of the $Q$ and $R$ branches of the DCl infrared fundamental dissolved in a series of nonanes ($C_9H_{20}$) of varying degrees of hydrocarbon branching. [Computed from the data of Table 2 of Ref. 120.]

Note that our theoretical approach is the "inverse" of our previous treatments: Instead of constructing a correlation function, we model the form of the frequency-dependent absorption coefficient ( = band profile).

Assuming that the index of refraction is constant over the band profile, we write Eq. 1.22 of Chapter 1—omitting uninteresting constant factors—

$$\sigma(\omega) = \omega \, \mathrm{Re} \int_0^\infty dt \exp(-i\omega t) \big\langle [m_z(0), m_z(t)] \big\rangle. \qquad (2.109)$$

We remember that the quantity within the square brackets is a commutator, here written in terms of the vibrational-rotational transition moment polarized along the bond. (See the beginning of Section 1.4, Chapter 1.) Since this is the only direction of polarization for DCl, we omit the subscript from now on.

The total system Hamiltonian is

$$H = H_a + H_b + V(t),$$

where $H_a$, $H_b$ are the zero-order Hamiltonians of the system (DCl) and the bath (or lattice)—the hydrocarbon environment—respectively. The quantity $V(t)$ represents the coupling between coordinates $a$ and $b$. We need not worry about DCl–DCl interactions since the solutions were dilute—1.5 mole % DCl.

We now designate by $\bar{V}_a$ the quasi-static part of $V(t)$ which stays constant during $\tau_l$. Hence

$$H = H_a + H_b + V(t)$$
$$= H_a + H_b + \bar{V}_a + V(t) - \bar{V}_a$$
$$= H_a + H_b + \bar{V}_a + \Delta V(t). \tag{2.110}$$

Note that $\Delta V(t)$ characterizes the part of the intermolecular potential that fluctuates about the average $\bar{V}_a$ with a time constant $\tau_c$. [We designated it previously by $H_1(t)$.] This renormalization procedure assures the proper decomposition of $V(t)$ into the quasi-static part $\bar{V}_a$ and the stochastic part $\Delta V(t) = V(t) - \bar{V}_a$ and guarantees that $\Delta V(t)$ is small. Figure 2.24 gives a schematic of the various time scales.

We must be careful in keeping track of the subscripts $a$ and $b$ which label the set of eigenfunctions $|ab\rangle = |a\rangle|b\rangle$ of the vibrational-rotational coordinates $|a\rangle = |vr\rangle$ of the active DCl oscillator and of the bath, $|b\rangle$, respectively. Hence we write

$$\{(H_a + \bar{V}_a)\}|a\rangle = E_a|a\rangle = E_{vr}|vr\rangle$$
$$H_b|b\rangle = E_b|b\rangle. \tag{2.111}$$

The first of these relations is now the renormalized zero-order problem. The bath coordinates $|b\rangle$ contain the (not further specified) external levels of the

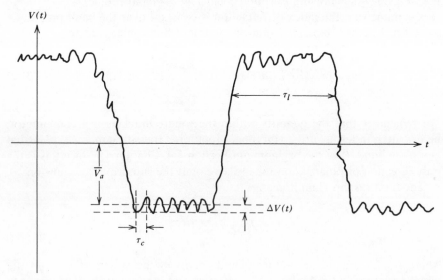

**Figure 2.24.** Time evolution of the intermolecular perturbation operator $V(t)$ coupling solvent (hydrocarbon) and solute (DCl) molecules. $\bar{V}_a$: quasi-static intermolecular potential of lifetime $\tau_l$. $\Delta V(t)$: fluctuation of $V(t)$, with respect to $\bar{V}_a$, with correlation time $\tau_c$. [D. Robert and L. Galatry, *J. Chem. Phys.* **55**, 2347 (1971).]

solvent as well as the translational motions of the active (probe) DCl molecule. As usual, Tr denotes the trace operation and $\rho_0 \equiv \rho(0)$ the statistical canonical density operator (Section 1.3, Chapter 1).

We now solve the Heisenberg equation of motion for $m(t)$. (Eq. 1.30, 1.73, or 1.100; Chapter 1.) We do this by using a "shorthand" version, making use of a Kubo operator[122] $R^\times$, defined by the operation ($R, A$ are arbitrary)

$$R^\times A = [R, A] = RA - AR.$$

Hence

$$R^\times = [R, \ ]. \tag{2.112}$$

With this, we can write the equation of motion for $m$ in the form

$$\dot{m}(t) = i\hbar^{-1}[H, m] = i\hbar^{-1}H^\times m, \tag{2.113}$$

and express its formal solution by

$$m(t) = \exp\{i\hbar^{-1}H^\times t\}m(0). \tag{2.114}$$

The exponential operator in Eq. 2.114, designated by the general symbol $U(t)$, is a time evolution operator. Its action transforms $m(0)$ to $m(t)$. (We are already familiar with other types of specific time evolution operators. See, for instance, Eq. 2.9.)

Note that the matrix elements of $U(t)$ require four, instead of two, labels (Appendix F).

We reformulate Eq. 2.114 to read

$$m(t) = \exp\{i\hbar^{-1}H^\times t\}m(0)$$

$$= \exp\{i\hbar^{-1}(H_a^\times + \bar{V}_a^\times + H_b^\times + \Delta V^\times)t\}m(0)$$

$$= U^\times(t)m(0). \tag{2.115}$$

According to plan, we first solve $U^\times$ for potential term $\bar{V}_a$; thereafter we solve $U^\times(t)$ for $\Delta V(t)$ by writing $U^\times(t)$ as (rapidly converging) power series in terms of correlation functions of the stochastic potential $\Delta V(t)$.

Since the fluctuations $\Delta V(t) = V(t) - \bar{V}_a$ are smaller than the energy level differences $\Delta E_a$ between adjacent states $|vr\rangle$, we can write $\rho(0) = \rho_a(0)\rho_b(0)$. Therefore Eq. 2.109 reads

$$\sigma(\omega) = \omega \operatorname{Re} \int_0^\infty dt \exp(-i\omega t) \operatorname{Tr}\{\rho(0)mU^\times(t)m - m\rho(0)U^\times(t)m\}$$

$$= \omega \operatorname{Re} \int_0^\infty dt \exp(-i\omega t) \operatorname{Tr}\{\rho^\times(0)mU^\times(t)m\}, \tag{2.116}$$

where we have made use of the cyclic permutation rule under the trace operation, set $m(0) = m$, and used again in an obvious fashion the Kubo operator to write

$$\rho(0)m - m\rho(0) = \rho^{\times}(0)m.$$

Employing $\rho(0) = \rho_a(0)\rho_b(0)$, the trace operation in Eq. 2.116 leads to (Note 8)

$$\text{Tr}_a\{\rho_a^{\times}(0)m\,\text{Tr}_b U^{\times}(t)\rho_b(0)m\}.$$

Writing this in matrix notation (Appendix F) and after insertion into Eq. 2.116 gives (Note 9)

$$\sigma(\omega) = \omega\,\text{Re}\int_0^{\infty} dt\exp(-i\omega t)\sum_{\substack{aa' \\ a''a'''}} ((aa'|\rho_a^{\times}(0)|aa'))$$

$$\times (a|m|a')((a'a|\langle U^{\times}(t)\rangle_b|a'''a''))(a'''|m|a''), \qquad (2.117)$$

with

$$\langle U^{\times}(t)\rangle_b \equiv \left\langle \exp\{i\hbar^{-1}(H_a^{\times} + \bar{V}_a^{\times} + H_b^{\times} + \Delta V^{\times})t\}\right\rangle_b$$

$$= \text{Tr}_b\{U^{\times}(t)\rho_b(0)\}$$

$$= \sum_{b'} ((bb|U^{\times}(t)|b'b'))(b'|\rho_b(0)|b'). \qquad (2.118)$$

The symbol $\langle \ldots \rangle_b$ means here that the average *was* taken only over coordinate $|b)$. Notice that $\rho_a(0)$ and $\rho_b(0)$ are assumed to be diagonal in their respective bases $|a)$ and $|b)$. The density operator over states $|a)$,

$$\rho_a(0) = \exp\left(-\frac{H_a + \bar{V}_a}{k_B T}\right)\left\{\text{Tr}_a\exp\left(-\frac{H_a + \bar{V}_a}{k_B T}\right)\right\}^{-1}$$

indicates that the solute molecule is prepared at $t = 0$ (when the interaction with the radiation field is turned on) in a state that "feels" the polarizing effect of the intermolecular potential $\bar{V}_a$, an effect that does not average out to zero during the relaxation time $\tau_R$ of the solute molecule. (Fig. 2.24.)

The time evolution of the system, in terms of the rapid fluctuations of small magnitude ($\tau_c^{-1} \ll \Delta E_a/\hbar$), is governed by $\langle U^{\times}(t)\rangle_b$; with $\rho = \rho_a\rho_b$, we split the time evolution operator into a part that operates on system variable $|a)$, the other on bath coordinate $|b)$. We first write $U^{\times}(t)$ in the interaction representation[26] (Appendix G),

$$U^{\times}(t) = \exp\{i\hbar^{-1}(H_a^{\times} + \bar{V}_a^{\times} + H_b^{\times})t\}\exp_0\left\{i\hbar^{-1}\int_0^t dt'\,\Delta V^{\times}(t')\right\}$$

$$= \exp\{i\hbar^{-1}(H_a^{\times} + \bar{V}_a^{\times})t\}\exp_0\left\{i\hbar^{-1}\int_0^t dt'\,\Delta V^{I\times}(t')\right\}, \qquad (2.119)$$

where

$$\Delta V^{I}(t) = \exp\left\{- i\hbar^{-1}\left(H_a + \overline{V}_a + H_b\right)^{\times} t\right\} \Delta V(t).$$

Note that $\exp_0$ denotes a "time-ordered exponential" since the perturbation power series of $\exp\{i\hbar^{-1}\int dt' \Delta V^{I \times}(t')\}$ leads to product terms $\Delta V^{I \times}(t')\Delta V^{I \times}(t'')\ldots$ of a priori noncommuting factors; the effect of $\Delta V^{I}$ at time $t'$ is not the same as that at $t = t''$. Taking in $U^{\times}$ the trace over coordinate $b$ (Eq. 2.118), we get, remembering Eq. 2.119,

$$\langle U^{\times}(t)\rangle_b = \mathrm{Tr}_b\{U^{\times}(t)\rho_b(0)\}$$

$$= \mathrm{Tr}_b\left\{U^{\times}(t)\frac{\exp(-H_b/k_B T)}{\mathrm{Tr}_b \exp(-H_b/k_B T)}\right\}$$

$$= \exp\left\{i\hbar^{-1}\left(H_a + \overline{V}_a\right)^{\times} t\right\}\left\langle \exp_0\left\{i\hbar^{-1}\int_0^t dt' \Delta V^{I \times}(t')\right\}\right\rangle_b$$

$$= \exp(i\omega_{a'a}t)\langle \Gamma^{\times}(t)\rangle_b, \tag{2.120}$$

with

$$\Gamma^{\times}(t) = \exp_0\left\{i\hbar^{-1}\int_0^t dt\, \Delta V^{I \times}(t)\right\} \tag{2.121}$$

and

$$\omega_{a'a} = \hbar^{-1}(E_{a'} - E_a).$$

Inserting this result into Eq. 2.117 gives

$$\sigma(\omega) = \sum_{\substack{aa' \\ a''a'''}} \omega((aa'|\rho_a^{\times}(0)|aa'))\mathrm{Re}(a|m|a')(a'''|m|a'')$$

$$\times \int_0^{\infty} dt \exp\{-i(\omega - \omega_{a'a})t\}((a'a|\langle \Gamma^{\times}(t)\rangle_b|a'''a'')). \tag{2.122}$$

The further development of the theory proceeds along our planned path. First, we solve for $\sigma(\omega)$ without considering line broadening; $\Gamma^{\times}(t) = 1$. Thereafter we perform the coupling of the orientational coordinate to the bath coordinates. In particular:

1. We specify convenient wave functions $|a)$ and construct $\overline{V}_a$, the average quasi-static intermolecular perturbation potential function. With this, we solve (the first of) Eq. 2.111 and thereby obtain the (perturbed) vibrational-rotational eigenfrequencies $\omega_{a'a} = \omega_{v'r'} - \omega_{vr}$ and eigenfunctions $|vr)$.

2. Subsequently, we introduce fluctuation phenomena of $V(t)$ through the solution of Eq. 2.118.

3. Finally, we compute $\sigma(\omega)$ according to Eq. 2.122.

Evidently, this detailed method can give us a great deal of insight into the characteristics of orientational motion in condensed phases. However, it is also clear that the method is too complex to be successful for most systems. We have to consider a large number of transition dipole moment matrix elements, we have to construct the line spectrum of the rotor perturbed by the quasi-static environment (tabulate all frequencies $\omega_{a'a}$), and we have to introduce the line broadening about each $\omega_{a'a}$ through $\langle \Gamma^{\times}(t) \rangle_b$. This approach may give us theory in excess of readily available experimental information or even interest.

To construct the quasi-static potential $\overline{V}_a$ (the calculations are not of concern here), we assume that the rotating DCl molecule is subjected to the combined torques of a large number of surrounding "point interaction centers" that are placed on the hydrocarbon molecules in a certain spacial configuration. The anisotropic interactions are introduced by placing the DCl molecule off-center, by distance $\vartheta$, within the spherical "solvent cage" made of these interaction centers. $\overline{V}_a$ is then obtained in the form[123]

$$\overline{V}_a = -\lambda \cos\theta, \tag{2.123}$$

where $\lambda \leq 1$ is a positive parameter denoting the strength of the solute–solvent interaction and $\theta$ is the polar angle between the dipole moment of DCl and the symmetry axis $\vartheta$ of the solvent cage anisotropy.

Equation 2.123 is inserted as perturbation operator into the eigenvalue problem of the unperturbed rotor. Choosing the quantization axis along direction $\vartheta$ and expressing by $|r\rangle = \psi_{jm}(\theta, \varphi)$ the usual linear rotor wave functions in terms of polar and azimuthal angle $\theta$ and $\varphi$, Schrödinger's equation for the perturbed rotor reads

$$\{H_r + \overline{V}_a\}\psi_{jm}(\theta, \varphi) = \left\{ -\frac{\hbar^2}{2I} \nabla_r^2 - \lambda \cos\theta \right\} \psi_{jm}(\theta, \varphi)$$

$$= E_{jm}\psi_{jm}(\theta, \varphi). \tag{2.124}$$

$\nabla_r^2$ is the rotational kinetic energy operator, and $I$ is the moment of inertia.

The Hamiltonian of Eq. 2.124 is formally identical with the Hamiltonian for a first-order Stark effect (static electric field); this effect splits the $(2j + 1)$-fold level degeneracy into $j + 1$ sublevels $|m| = 0, 1, 2, \ldots$[124]

Figure 2.25 shows results of some model computations. We have plotted the square of the matrix element of the rotational transition moment $\mu$,

$$\left| \int \psi_{j'm'} \mu \psi_{jm} \, d\theta \, d\varphi \right|^2,$$

(which is proportional to the line intensity) as function of the reduced strength parameter $\lambda^* = \lambda(\hbar^2/2I)^{-1}$ for the transitions $v, v' = 0, 1$; $jm, j'm' = 00, 00$ and $00, 10$. We notice not only the rapid and relatively strong intensity increase of the free-rotor-forbidden $\Delta j = 0$ branch with growing strength of the quasi-static rotor–environment interaction, but also a concomitant intensity decrease of the allowed $P$ and $R$ ($\Delta j = \pm 1$) transitions.

Figure 2.26 shows computed "bar" or "stick" spectra for $\lambda^* = 6$, 10, and 16. Notice again the increase of the $Q$ branch intensity with increasing $\lambda^*$ and the concomitant "smearing" of the rotational $P, R$ fine structure. (The ordinate scale of Fig. 2.26c is one-half that of Fig 2.26a and b.)

In summary, the quasi-static intermolecular potential $\overline{V}_a$ defined here causes the following changes in the infrared vibration-rotation spectrum of the fundamental of the dissolved polar diatomic molecule in the nonpolar (and nonpolarizable) medium:

1. A free-rotor-forbidden $Q$ branch is induced in the diatomic vib-rotor; its intensity increases with increasing strength of interaction parameter $\lambda$.

2. The intensity of the free-rotor-allowed $P$ and $R$ branches decreases, and their peak absorption is shifted to higher frequencies.

3. Parameter $\lambda$ can be given a clearer physical meaning by relating it to the mean-squared intermolecular torques[72]

$$\langle a | (OV_a)^2 | a \rangle$$

(Section 4), where

$$\langle (OV_a)^2 \rangle \approx \frac{M(4)_{\text{expt}} - 2M(2)^2_{\text{theory}}}{4B^2}.$$

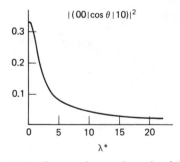
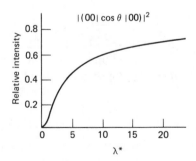

**Figure 2.25.** Computed squared rotational transition moments $|(jm|\cos\theta|j'm')|^2$ of the diatomic rotor in a perturbing medium as a function of the reduced coupling parameter $\lambda^*$ of the quasi-static intermolecular potential $\overline{V}_a$. Note that the transition 00–00 is forbidden in the vibration-rotation spectrum of the free rotors ($\lambda^* = 0$). [D. Robert and L. Galatry, *J. Chem. Phys.* **55**, 2347 (1971).]

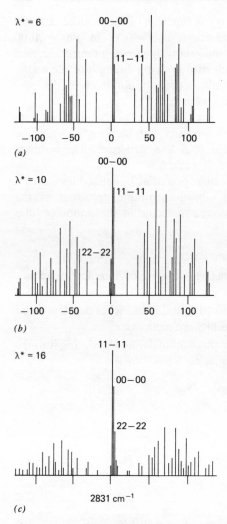

**Figure 2.26.** Computed bar ("stick") spectrum of the vibration-rotation contour of the infrared fundamental of HCl for three values of the reduced coupling parameter $\lambda^*$ of the quasi-static intermolecular potential $\bar{V}_a$. The length of each bar is proportional to the integrated absorption of the corresponding $jm - j'm'$ transition—some of which are indicated. Note the increasing importance of the induced $Q$ branch and the progressive frequency shifts of the free rotor allowed vibrational-rotational lines with increasing $\lambda^*$. The ordinate scale of spectrum $c$ is one-half that of the others. [D. Robert and L. Galatry, *J. Chem. Phys.* **55**, 2347 (1971).]

[Recall that $k_B T/I$ is $\frac{1}{2}$ the classical rotational second spectral moment of the diatomic molecule (Eq. 2.2, or Eq. E.8, Appendix E) and $B = 2/I$.] Operating with $O = \partial/\partial\theta$ on $\bar{V}_a$ gives

$$\left\{ \frac{\partial}{\partial\theta}[-\lambda\cos\theta] \right\}^2 = \lambda^2\sin^2\theta.$$

Taking the average over all $\theta$ yields

$$\langle \sin^2\theta \rangle = \tfrac{2}{3}.$$

With $M(4)_{expt} = 0.39 \times 10^8$ cm$^{-4}$ and $M(2)_{theory} = 0.44 \times 10^4$ cm$^{-2}$, we get

$$\lambda \sim 61 \text{ cm}^{-1}$$

as average over all (16) hydrocarbon solvent systems studied.

Defining $2\lambda$ as the rotational barrier to free rotation (a more meaningful quantity than an activation energy of an orientational correlation time or an equivalent bandwidth),[125] we find

$$2\lambda \sim 120 \text{ cm}^{-1} \quad \text{or} \quad 2\lambda^* = \frac{2\lambda}{4B} \sim 6$$

as a crude estimate. In other words, DCl molecules in rotational levels $j \geq 5$ would be at the top or above the barrier. We may consider the rotational barrier to express the tendency of the probability density of the rotor, $|\psi_{jm}(\theta)|^2$, to be "pulled" toward the direction of $\vartheta$. This tendency increases with $\lambda$ and is, for a given $j$, the largest if $m = j$.[121] (Of course, the effect of $\lambda$ on the rotational energies of the rotor is the greater the smaller $j$.) Hence the rotational barrier does not cause a *uniformly* slowing effect on those rotors that are in energy levels below or near $2\lambda$, but puts a "modulated friction" on the orientational motion. Not surprisingly, neither $J$- nor $M$-diffusion models fit the data.[120]

We begin to realize that rotational relaxation in condensed media with even moderately anisotropic intermolecular forces is really qualitatively and quantitatively different from that encountered in more mobile systems. Concepts such as free rotation steps interrupted by collisions must then be enlarged by a description that allows "residence" or lifetimes of certain orientations. In fact, we already discussed a special case of this in Section 3.3. We discuss further experimental and theoretical examples later on.

We now introduce line broadening through $\Gamma^\times(t)$. We develop Eq. 2.121 by an iterative method that gives a perturbation series expansion of $\Gamma^\times(t)$. We do not yet specify the exact nature of $\Delta V(t)$. In fact, we do not need such detailed information; we are aiming at *correlation functions* of $\Delta V(t)$ since these are easier to construct and to use than $\Delta V(t)$. Differentiating Eq. 2.121 gives, with $\Gamma^\times(0) = 1$,         $\dot{\Gamma}^\times = \Delta V(t)\,\Gamma^\times(t)$  (integrating $\longrightarrow$)

$$\Gamma^\times(t) = 1 + i\hbar^{-1} \int_0^t dt_1\, \Delta V'^\times(t_1)\, \Gamma^\times(t_1).$$

[Note also Eqs. G.1 and G.3, Appendix G, with $T_1^\dagger \equiv \Gamma^\times(t)$, $H_1 \equiv \Delta V^\times$.] This relation indeed allows an iteration procedure; the next step is (we omit superscript $I$ and set $\hbar = 1$)

$$\Gamma^\times(t) = 1 + i \int_0^t dt_1\, \Delta V^\times(t_1)\left\{ 1 + i \int_0^{t_1} dt_2\, \Delta V^\times(t_2)\, \Gamma^\times(t_2) \right\},$$

and so forth (recall that $\Delta V^\times(t)$ is small by construction). We then obtain the general relation

$$\Gamma^\times(t) = 1 + \sum_{n=1}^{\infty} i^n \int_0^t dt_1 \int_0^{t_1} dt_2 \cdots \int_0^{t_{n-1}} dt_n \, \Delta V^\times(t_1) \Delta V^\times(t_2) \cdots \Delta V^\times(t_n)$$

$$= 1 + \sum_{n=1}^{\infty} \frac{i^n}{n!} \int_0^t dt_1 \int_0^t dt_2 \cdots \int_0^t dt_n \, \Delta V^\times(t_1) \Delta V^\times(t_2) \cdots \Delta V^\times(t_n),$$

$$(2.125)$$

which equals

$$\exp_0\left\{ i \int_0^t dt' \, \Delta V^\times(t') \right\}. \qquad (2.126)$$

The operators $\Delta V^\times(t_1), \ldots \Delta V^\times(t_n)$ are entered in chronological order—here with increasing values of $t$ to the left: $t_n < t_{n-1} < t$. We repeat that time ordering is necessary since, a priori,

$$\cdots \iint dt_1 \, dt_2 \, \Delta V^\times(t_1) \Delta V^\times(t_2) \cdots \neq \cdots \iint dt_1 \, dt_2 \, \Delta V^\times(t_2) \Delta V^\times(t_1) \cdots$$

$$(2.127)$$

Comparing now Eq. 2.126 with Eq. 2.121, we see that we have obtained a prescription for computing $\Gamma^\times(t)$ to any order of perturbation. There is "only" one more problem to solve: We must take the ensemble average—our goal is $\langle \Gamma^\times(t) \rangle_b$. Hence we would like to transform the average of an exponential

$$\left\langle \exp_0\left\{ i \int_0^t dt \, \Delta V^\times(t) \right\} \right\rangle$$

into an exponential of an average (Eqs. 2.125, 2.126),

$$\exp_0\left\{ \sum_{n=1}^{\infty} \frac{i^n}{n!} \int_0^t dt_1 \cdots \int_0^t dt_n \langle \Delta V^\times(t_1) \cdots \Delta V^\times(t_n) \rangle^c \right\}, \qquad (2.128)$$

called cumulant average. It is designated by superscript $c$.

In most situations the series expansion converges with the second-order term (central limit theorem),[126, 127] the well-known dispersion (Eq. H.2, Appendix H)

$$\langle \Gamma^\times(t) \rangle = \exp\left\{ -\left(\frac{1}{2}\right) \int_0^t \int_0^t dt_1 \, dt_2 \langle \Delta V^\times(t_1) \Delta V^\times(t_2) \rangle^c \right\}. \qquad (2.129)$$

It becomes the ordinary average for $\int_0^t dt_1 \langle \Delta V^\times(t_1) \rangle^c = 0$ (zero average frequency shift). We generally impose this condition (Section 4.1).

We then have the result that the ensemble-averaged line broadening operator is a second-order time evolution operator; its argument is the integral over the autocorrelation function of the stochastic coupling parameter for all intermediate times $0 \leq t' \leq t$. In other words, $\langle \Gamma^\times(t) \rangle_b$ has a memory since events for all $t' \leq t$ are mapped into $t$.

We are now ready to compute the complete absorption coefficient $\sigma(\omega)$ in Eq. 2.122. First, we write the matrix elements of $\langle \Delta V^\times(t_1) \Delta V^\times(t_2) \rangle_b$ over system states $| aa' \rangle$ and bath states $| bb \rangle$,

$$((a'a|\langle \Delta V^\times(t_1) \Delta V^\times(t_2) \rangle_b | a''' a'')), \qquad (2.130)$$

using Eqs. 2.112, 2.118, and our assumption $(b| \rho_b(0)| b') = \rho_b(0) \delta_{bb'}$. This is a messy expression. We simplify it by throwing away the nondiagonal elements,

$$(a'|a''') = \delta_{a'a'''}, \qquad (a|a'') = \delta_{aa''}.$$

Physically, this means we are neglecting interference terms between the broadening effects of different spectral lines.[121] We are justified in doing this by our assumption on the relative smallness of the fluctuations of $\overline{V}_a$ compared to the difference between adjacent systems energy levels: $(a| \rho_a(0)| a') = \rho_a(0) \delta_{aa'}$. Employing the relation

$$((hi| A^\times | jk)) = (h| A | j) \delta_{ki} - (k| A | i) \delta_{hj},$$

(Eq. F.3, Appendix F) and using $a'', a'''$ as dummy indices, Eq. 2.130 reads

$$\sum_{\substack{a''a'''b \\ b'b''b'''}} ((a'a|((bb| \Delta V^\times(t_1)| a''' a''))| b'b''))$$

$$\times ((a''' a''|((b'b''| \Delta V^\times(t_2)| a'a))| b''' b'''))(b''' | \rho_b(0)| b''')$$

$$= \sum \{[(a'|(b| \Delta V(t_1)| a''')| b') \delta_{a''a} \delta_{b''b}$$

$$- (a''|(b''| \Delta V(t_1)| a)| b) \delta_{a'a'''} \delta_{bb'}]$$

$$\times [(a''' |(b'| \Delta V(t_2)| a')| b''') \delta_{aa''} \delta_{b'''b''}$$

$$- (a|(b''' | \Delta V(t_2)| a'')| b'') \delta_{a'''a'} \delta_{b'b'''}]\}(b''' | \rho_b(0)| b''').$$

$$(2.131)$$

Multiplying out yields—after some simplification in the notation of the dummy indices—

$$((a'a|\langle \Delta V^\times(t_1) \Delta V^\times(t_2)\rangle_b|a'a))$$

$$= \sum_{bb'a''} \{(a'b|\Delta V(t_1)|a''b')(a''b'|\Delta V(t_2)|a'b)$$

$$+ (ab|\Delta V(t_2)|a''b')(a''b'|\Delta V(t_1)|ab)$$

$$- (a'b|\Delta V(t_1)|a'b')(ab'|\Delta V(t_2)|ab)$$

$$- (a'b|\Delta V(t_2)|a'b')(ab'|\Delta V(t_1)|ab)\}(b|\rho_b(0)|b)$$

$$= \langle \Delta V(t_1) \Delta V(t_2)\rangle. \tag{2.132}$$

Taking only the first matrix element and recalling (Eq. 2.121) that $\Delta V^\times$ is in the interaction representation (Eq. 2.119), Eq. 2.129 reads

$$\langle \Gamma^\times(t)\rangle_b = \exp\left\{ -\frac{1}{2} \int_0^t \int_0^t dt_1 dt_2 \left\langle (a'b|\Delta V^I(t_1)|a''b')(a''b'|\Delta V^I(t_2)|a'b)\right\rangle_b\right\}$$

$$= \exp\left\{ -\frac{1}{2} \int_0^t \int_0^t dt_1 dt_2 \left\langle (a'b|\exp\{i[H_a + \overline{V}_a + H_b]^\times t_1\}\Delta V(t_1)|a''b')\right.\right.$$

$$\left.\left. \times (a''b'|\exp\{i[H_a + \overline{V}_a + H_b]^\times t_2\}\Delta V(t_2)|a'b)\right\rangle_b\right\}. \tag{2.133}$$

Therefore (see Appendix F, Eq. F.5)

$$\langle \Gamma^\times(t)\rangle_b = \exp\left\{ -\frac{1}{2} \int_0^t \int_0^t dt_1 dt_2 \exp\{i(\omega_{a'a''} + \omega_{bb'})(t_1 - t_2)\}\right.$$

$$\left. \times \left\langle (a'b|\Delta V(t_1)|a''b')(a''b'|\Delta V(t_2)|a'b)\right\rangle_b\right\}. \tag{2.134}$$

It is obvious that this complicated expression is difficult to evaluate since we must write down the matrix elements of $\Delta V(t)$ over system and bath coordinates. Out of ignorance we therefore guess a *convenient* relation and put

$$\langle \Delta V(t_1) \Delta V(t_2)\rangle \langle |\Delta V(0)|^2\rangle^{-1} = \exp\left\{ -\frac{t_1 - t_2}{\tau_c}\right\},$$

$$t_1 \geq t_2 \tag{2.135}$$

where $\tau_c$ is the lifetime of the autocorrelation of $\Delta V(t)$ (see Fig. 2.24). In other words, we can dispose of one adjustable parameter. [We assume here that $\langle \Delta V(t_1) \Delta V(t_2)\rangle$ is time-shift invariant—see Section 1.3, Chapter 1.]

Insertion of Eq. 2.135 into Eq. 2.134 and integrating yields, noting

$$\left(\frac{1}{2}\right)\int_0^t\int_0^t dt_1\, dt_2 = \int_0^t dt_1 \int_0^{t_1} dt_2,$$

$$\langle \Gamma^\times(t)\rangle_b = \exp\Bigg\{ \langle (a'b|\Delta V(0)|a''b')(a''b'|\Delta V(0)|a'b)\rangle_b$$

$$\times\Bigg( \frac{-t}{-i(\omega_{a'a''}+\omega_{bb'})+\tau_c^{-1}}$$

$$-\frac{\exp\{[i(\omega_{a'a''}+\omega_{bb'})-\tau_c^{-1}]t\}-1}{[i(\omega_{a'a''}+\omega_{bb'})-\tau_c^{-1}]^2}\Bigg)\Bigg\}. \qquad (2.136)$$

The limiting time evolutions of Eq. 2.136 allow considerable simplifications. For instance, we may neglect in Eq. 2.136 the second term in the argument of the main exponential. In other words, we would then consider the situation

$$t \gg \tau_c.$$

In terms of a physical picture, we forgo any knowledge about the coherent or purely kinetic characteristics of $\langle \Gamma^\times(t)\rangle_b$; $\langle \Gamma^\times(t)\rangle_b$ retains no memory of its history and only maps those events into time point $t$ that immediately precede $t$.

Now, what do we understand by "events"? We may think in terms of collisions between rapidly fluctuating "interaction points" of the solvent molecules ("segmental motion") with the rotatory-translatory degrees of freedom of the diatomic solute where—according to the foregoing—we disclaim all knowledge of the short-time characteristics of such segmental motion.

It is interesting and important to see what form $\langle \Gamma^\times(t)\rangle_b$ would assume in the opposite situation,

$$t \ll \tau_c.$$

In other words, we now consider only time intervals during which the autocorrelation of $\Delta V(t)$ has not yet decayed into random fluctuations. We may then develop the exponential in the second term in Eq. 2.136 into a power series up to (and including) second order in $t$. We obtain

$$\langle \Gamma^\times(t)\rangle_b = \exp\left\{ -\frac{1}{2}\langle (a'b|\Delta V(0)|a''b')(a''b'|\Delta V(0)|a'b)\rangle_b t^2\right\}, \qquad (2.137)$$

*independent* of $\tau_c$ and leading to a Gaussian profile line broadening.

The condition $t \ll \tau_c$ represents a quasi-static regime where the line broadening is caused solely by the equilibrium distribution of $\langle |\Delta V(0)|^2\rangle_b$; times are too short for effective collisional randomization of $\langle \Gamma^\times(t)\rangle_b$. We shall encounter an analogous dynamic effect in vibrational dephasing. Hence we deal here with a rather general way of looking at the temporal development of relaxation processes. At any rate, notice that starting with the simple exponential description for some autocorrelation function of a random force or a random potential (Eq. 2.135) we arrive nevertheless at a realistic time evolution operator (here $\langle \Gamma^\times(t) \rangle_b$) of some broadening effect that has the correct short-time characteristics (namely "coherence," or zero-slope at zero time, or "memory"). Contrarily, the "lower-order" potential autocorrelation function $\exp(-t/\tau_c)$ possesses a (physically unreasonable) finite zero-time slope (Section 2.1).

Which of the approximations $t \gtrless \tau_c$ is the more applicable to our immediate interest here? Clearly, it is the condition $t \gg \tau_c$ since we, eventually, integrate in Eq. 2.122 over all times. In this situation, the long-time tail of $\langle \Gamma^\times(t) \rangle_b$ exerts a larger influence on $\sigma(\omega)$ than its short-time, quasi-static development. Representative values of $\tau_c$ (we have more to say about them in Chapter 3) are usually of the order of $1 \times 10^{-12}$ to $1 \times 10^{-13}$ sec. On the other hand, the upper limit of $t$—that is to say, the time interval during which we have knowledge of the system—is of the order of several picoseconds (Eq. 1.101, Chapter 1). Consequently, we do not make a large error in omitting the regime $t \ll \tau_c$ although a full description of $\langle \Gamma^\times(t) \rangle_b$ requires consideration of both terms of Eq. 2.136.

We then truncate the argument in Eq. 2.136 after the first term and insert the result into Eq. 2.122. We obtain

$$\sigma(\omega) = \omega \sum_{aa'} \left[ (a|\rho_a(0)|a) - (a'|\rho_a(0)|a') \right] |(a|m|a')|^2$$

$$\times \frac{\gamma_{a'a}}{\left[ \omega - (\omega_{a'a} - \gamma'_{a'a}) \right]^2 + \gamma_{a'a}^2}$$

with

$$\gamma_{a'a} = \sum_{a''bb'} (b|\rho_b(0)|b) \left\{ \frac{|(a'b|\Delta V(0)|a''b')|^2 \tau_c}{1 + (\omega_{a'a''} + \omega_{bb'})^2 \tau_c^2} + \frac{|(ab|\Delta V(0)|a''b')|^2 \tau_c}{1 + (\omega_{aa''} + \omega_{bb'})^2 \tau_c^2} \right.$$

$$\left. -2 \frac{(a'b|\Delta V(0)|a'b')(ab'|\Delta V(0)|ab) \tau_c}{1 + \omega_{bb'}^2 \tau_c^2} \right\} \qquad (2.138)$$

and

$$\gamma'_{a'a} = \sum_{a''bb'} (b|\rho_b(0)|b) \left\{ \frac{|(a'b|\Delta V(0)|a''b')|^2 (\omega_{a'a''} + \omega_{bb'}) \tau_c^2}{1 + (\omega_{a'a''} + \omega_{bb'})^2 \tau_c^2} \right.$$

$$\left. - \frac{|(ab|\Delta V(0)|a''b')|^2 (\omega_{aa''} + \omega_{bb'}) \tau_c^2}{1 + (\omega_{aa''} + \omega_{bb'})^2 \tau_c^2} \right\},$$

where all matrix elements of Eq. 2.132 with the *same* $\tau_c$ (another simplification) have been included.

Numerical evaluation of Eq. 2.138 is straightforward.[121] We briefly sketch the method.

1.  For the density matrix, we use

$$(a|\rho_a(0)|a) = \frac{\kappa_m \exp(-\beta E_{jm})}{\sum_m \kappa_m \exp(-\beta E_{jm})}, \qquad \begin{cases} \kappa_m = \begin{array}{l} 1, m = 0 \\ 2, \text{otherwise} \end{array} \\ \beta^{-1} = k_B T. \end{cases}$$

Recall that $|a) = |v)|r)$. Hence the term $(a'|\rho_a(0)|a')$ is negligible since almost all solute molecules are in their vibrational ground state. Since we neglect vibrational relaxation, the energy states, damping coefficient $\gamma_{aa'}$, and so forth, pertain to rotational coordinates of the diatomic molecule.

2.  For the transition matrix element we use

$$|(a|m|a')|^2 = |(\psi_{jm}|\cos\theta|\psi_{j'm})|^2 + |(\psi_{jm}|\sin\theta\cos\varphi|\psi_{j'm'})|^2,$$

where the rotational wave functions $\psi_{jm}$ are obtained from Eq. 2.124.

3.  The matrix elements $(jm|\Delta V(\theta,0)|j'm')$ determine the root-mean-square fluctuation of $V(t)$ about its mean $\bar{V}_a$ (Fig. 2.24),

$$\Delta V(\theta,0) = \langle |V(t) - \bar{V}_a|^2 \rangle^{1/2} = \lambda' \cos\theta,$$

where $\lambda'$ is a characteristic line-broadening parameter of the order of 40 cm$^{-1}$ and the indices $j, m$ label the perturbed wave function $\psi_{jm}$. The calculation of $\lambda'$ itself is based on the assumption that the fluctuations of $V(t)$ arise from the relative radial motion of solute molecule and solvent cage.

4.  The correlation time $\tau_c$ describes the average lifetime of a fluctuation of $\bar{V}_a$ (Fig. 2.24). Estimates of $\tau_c$ lead to relatively short values, $\tau_c \lesssim 0.6 \times 10^{-13}$ sec.

5.  Finally, Fig. 2.27 displays the results of applying the foregoing to HCl in CCl$_4$ solution at 300°K. Although this system is not exactly comparable to the DCl–hydrocarbon system, the computations are based on identical principles, namely, a nonpolar, nonpolarizable solvent medium with characteristic times as depicted in Fig. 2.24.

The values of the parameters $\lambda^*$, $\lambda'$, and $\tau_c$ are indicated in the figure. An experimental determination of the absorption coefficient of the diatomic rotor in the solution is also given. In particular, contour ---- represents the combined profile of the free rotor allowed transitions, $|\Delta j| = 1$, $\Delta m = 0$ and $|\Delta j| = 1$, $\Delta m = \pm 1$. They are (as we mentioned before) frequency shifted and intensity perturbed through $\bar{V}_a$ (see also Fig. 2.26). Contour —— represents the profile of the induced transition $\Delta j = 0$, $\Delta m = \pm 1$ (resonant with respect to rotation)

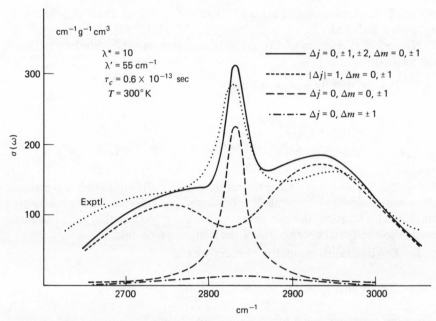

**Figure 2.27.** Computed and experimental band profile of the fundamental of HCl in dilute $CCl_4$ solution at 300°K. The components of the computed profile are labeled by their rotational level quantum numbers of the perturbed wave functions $\psi_{jm}$ of Eq. 2.124. [D. Robert, Thèse Docteur ès Sciences Physiques, Université de Besançon, Oct. 17, 1967, C.N.R.S. N° A.O. 1554; J. Lascombe, P. V. Huong, and M. L. Josien, *Bull. Soc. Chim. Fr.* (1959), p. 1175 (experimental spectrum).]

and $\Delta j = 0$, $\Delta m = 0$ (nonresonant). Contour — · — · depicts the profile due to the resonant induced transitions $\Delta j = 0$, $\Delta m = \pm 1$.

Comparison of computed and experimental profiles shows that the theory accounts reasonably well for the phenomena that perturb the orientational motion of the diatomic molecule. In fact, comparison with the stick spectra of Fig. 2.26 makes it clear that the proper choice of the potential parameter $\lambda^*$ is the most important prerequisite. The approximations that we introduced for calculating the line broadening, namely Eq. 2.135, the condition $t \gg \tau_c$, and the combination of the numerical values of $\lambda'$ and $\tau_c$, are of secondary importance.

We conclude this section with several remarks. First, it is obvious that this type of modeling allows great freedom of choosing complicated intermolecular potential functions, correlation times, and so on, and is limited only by our ignorance and paucity of *discriminating* experimental data. In many instances we therefore do not need such detailed theory and are better served by evaluating the appropriate time evolution (correlation function) of the process. Second, consideration of a peak absorption coefficient or of the ratio of the peak intensities of $Q$ and $R$ branches[120] is insufficient to disentangle the pertinent input parameters $\Delta V$, $\overline{V}_a$, and $\tau_c$. Third, we note that the far-infrared spectrum of the halide molecule in solution can be computed by the same methods; of course, the nonresonant transitions $\Delta j = 0$, $\Delta m = 0$ are now of zero intensity.

In connection with our discussion here, we mention that several empirical models that relate the orientational motion to the free volume of the medium (neat liquid or solvent) have been proposed and tested with apparent success.[18b, 128] These models are based on the concepts of binary collisions and hard-sphere diameters; they do not exceed the interpretive level of orientational models such as the $J$-extended diffusion model. In view of the expected connection of free volume and orientational mobility, the paucity of sophisticated but applicable free-volume theories is regrettable, particularly since studies of orientational motion under high pressure—the most appropriate experimental technique—are becoming routine.[105, 118a]

Another macroscopic parameter, namely, the shear viscosity of the medium, has enjoyed greater attention than the free volume. We discuss aspects of this in the next section.

### 5.5. Frictional Hindering of Free Rotation and Viscosity: Small Spherical and Ellipsoidal Molecules

Since the earliest studies on orientational relaxation phenomena by nuclear magnetic relaxation, attempts have been made to see a fundamental connection between orientational correlation times and macroscopic viscosity. This interest has been carried over into the domain of fluctuation spectroscopy of rotational relaxation as discussed in this chapter. For instance, Fig. 2.28 shows a typical plot[129] of the measured orientational correlation time for a second-order tensor (Raman anisotropic, Rayleigh depolarized) as a function of macroscopic viscosity. It is apparent that the plot is nicely linear and possesses a (extrapolated) zero-viscosity intercept.[108] On the basis of such data, the following conclusions have been drawn:[130]

1.  Linearity between the orientational correlation time and shear viscosity points to a fundamental relation between the two; it was hoped that orientational correlation times may be predicted from viscosity data.

2.  The intercept yields a correlation time that appears to equal the time interval $(2\pi/9)(I/k_B T)^{1/2}$ required for a *freely rotating* molecule to turn through an angle of 41°—the angle for the autocorrelation of a second-order Legendre polynomial (Eq. 1.81, Chapter 1) to drop to $1/e$ of its original (at $t = 0$) value[18b] (Note 10).

Let us first define some concepts and thereafter go back to these results.

Macroscopic viscosity is defined as the resistance to shear if a thin liquid layer is moved ("sheared") between parallel plates. The dimension of viscosity is force times time divided by area; the unit is the poise ($P$). This tells us that we deal with linear momentum transfer. In fact, viscosity is essentially a

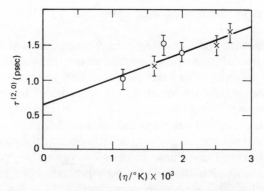

**Figure 2.28.**   Orientational relaxation time $\tau_R^{(2,0)}$ of $CH_3I$ dissolved in $C_6H_6$ and $CS_2$ as function of viscosity per degree Kelvin. $\bigcirc = 309°K$; $\times = 284°K$. The intercept amounts to $0.64 \times 10^{-12}$ sec; the value of $(2\pi/9)(I/k_BT)^{1/2}$ equals $0.36 \times 10^{-12}$ sec on the basis of the moment of inertia normal to the symmetry axis of $CH_3I$ (a motion that is observed by $\tau_R^{(2,0)}$ —see Section 3.4.2 of Chapter 1, Table I). [From D. R. Jones, H. C. Andersen, and R. Pecora, *Chem. Phys.* **9**, 339 (1975).]

phenomenon associated with the attractive forces in a liquid[131]; the momentum transfer tries to equalize the velocities of the moving layers.

We note that the hydrodynamic boundary conditions for the viscosity dependence of the orientational motion of a spherical molecule (Eq. 2.46*b*) have been defined in terms of the "rough sphere"; the surface of the reorienting molecule entrains its neighbors ("stick" boundary condition). A perfectly slipping sphere would have zero viscosity. On the other hand, we know that on the molecular level interactions take place through impulsive collisions between few neighbors. It is therefore not surprising that such a macroscopic rough sphere model overestimates $\tau_R$ (underestimates the rotational diffusion coefficient $D$).[108]

Attempts have been made to ameliorate the situation, proposing the concept of "microviscosity" by applying pure hydrodynamics "differentially" to concentric layers of finite thickness around the rotating molecule. Later, a more flexible model was proposed[132a] which admitted collisional effects in a thin boundary layer around the spherical molecule; the hydrodynamic ("collective") effects occurred only in an outer region about the molecule. The theory predicts that the rotational diffusion constant $D$ is a sum of collisional (subscript $E$) and hydrodynamic (subscript $h$) terms,

$$D = D_E + D_h,$$

where $D_E$ is the so-called Enskog collisional diffusion constant. For systems of immediate interest here, $D_E > D_h$.

The situation is simpler for systems of small molecules of ellipsoidal shape since their orientational motion (in the repulsive potential region) displaces their neighbors; the rotational motion of a perfectly slipping *ellipsoid* would show a viscosity dependence. A purely hydrodynamic theory based on this principle has been developed.[132b] It computes the friction coefficient $\zeta \propto D_h^{-1}$ for a perfectly slipping spheroid (an ellipsoid with axes $a = b \gtrsim c$) relative to its friction coefficient for stick as a function of the ratio min/max of its shorter (min) and longer (max) spheroidal semiaxes. Hence $\zeta_a = \zeta_b$, $\zeta_c = 0$ (rotation around the symmetry axis does not displace molecules). In the following we give an example:

1. Measurements of the Rayleigh depolarized correlation time of toluene ($C_6H_5 \cdot CH_3$) as a function of macroscopic viscosity (adjusted by solvents and/or temperature) give the usual linear plot (as in Fig. 2.28), with an experimental slope of $6.1 \times 10^{-12}$ sec/cP (cP = centipoise).[130]

2. A geometric consideration of the shape of the toluene molecule yields a value of min/max approximately 0.41.

3. The slope $\tau_R^{(2)}/\eta_m$ for stick boundary conditions is estimated from $\eta_m$ (Eqs. 2.46b and 2.47) to approximately $23 \times 10^{-12}$ sec/cP.[130]

4. From Fig. 2.29 we find, for min/max = 0.41, a ratio of slip/stick friction coefficient = 0.39.

5. Therefore we compute a slope $\tau_R^{(2)}/\eta_m$ for slip of $0.39 \times 23 \times 10^{-12}$ sec/cP—which is sufficiently close to the experimental result $6.1 \times 10^{-12}$ sec/cP considering the nontrivial task of computing min/max for toluene.

We should also keep in mind that the comparison of orientational correlation times from fluctuation spectroscopic methods with reorientation times from

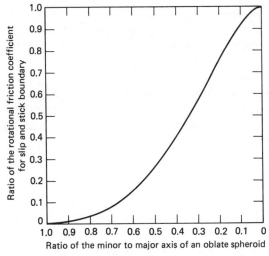

**Figure 2.29.** Ratio of the hydrodynamic rotational friction coefficient for slip and stick boundary conditions as a function of the ratio of the smaller and larger semiaxes of the same rotational oblate spheroid. [From Table I, Ref. 132b.]

macroscopic viscosity theory puts us into the worst possible situation:

1.  For large rotors in a solvent of small molecules the stick condition for relating $\tau_R$ (or $D$) is satisfied (Eq. 2.46$b$). However, there are no large rotors of interest to us since their long $\tau_R$ are difficult to extract from the other, faster relaxation processes that broaden a spectroscopic profile.

2.  If solvent and solute (probe) molecules are of comparable size, the assumptions of the simple viscosity theory fail. Relating molecular reorientation times and macroscopic viscosity is futile; a linear relationship and a zero-viscosity intercept are devoid of a theoretical basis.

3.  If the size of the rotor is inferior to that of the solvent molecules, the concept of macroscopic viscosity loses meaning since the smaller rotor sees only certain local viscosities of the larger solvent molecule. For instance, orientational motion of (a) methylene chloride ($CH_2Cl_2$) in a medium of linear polystyrene molecules[133a] and of (b) carbon oxysulfide (OCS) in $n$-alkanes ($n = 10$–$17$)[133b] was shown to be independent of the macroscopic viscosity (which strongly increased with molecular weight). Clearly, the macroscopic viscosity (momentum transfer between long-chain molecules) has little do to with the orientational motions of the small probe molecule localized somewhere along the chain.

Only theories that also take collisional effects into consideration will get us further.[132a] Much more work needs to be done.

## 5.6.  Orientational Pair Correlation in Symmetric Top Molecules

In this section we discuss some Rayleigh scattering phenomena of immediate interest to us. We recall that a comparison of correlation data from a depolarized Rayleigh spectrum (Eq. 1.51, Chapter 1) with correlation data from the anisotropic scattering intensity of a polarized Raman vibration-rotation band (Eq. 1.49) probes the so-called orientational pair correlations: Since the same molecular axis and the same Legendre polynomial is observed in both scattering experiments, any difference between Rayleigh and Raman correlation functions should arise from the distinct (cross-correlation) term of the Rayleigh correlation function between molecules $i$ and $j$,

$$\sum_{i \neq j} \langle \boldsymbol{\beta}^{0i}(0) \cdot \boldsymbol{\beta}^{0j}(t) \rangle.$$

(We know that Raman cross correlation is negligible—see end of Section 2.2, Chapter 1.)

To reach our goal we construct a general correlation function matrix $\langle \mathbf{A}_a(t)\mathbf{A}_b^*(0) \rangle$[42] in terms of a stochastic variable $\mathbf{A}_s(t)$ to which we attach tensorial rank and physical meaning. The procedure is as follows: (1) We derive[2c] the Langevin formalism (Eq. 2.18). (2) We define $\mathbf{A}(t)$ as a coordinate that describes the orientational motion of a molecule-fixed axis. (3) The orientational components of $\mathbf{A}(t)$ comprise single- and multiple-molecule motion (auto and cross correlations). (4) We apply projection operator techniques to separate components that fluctuate on different time scales. (5) We relate the theory to Raman and Rayleigh data.

The equation of motion of the Liouville-space vector $|A_a\rangle$ (Appendix F) is[5]

$$\frac{d}{dt}|A_a(t)\rangle = iL|A_a(t)\rangle, \tag{2.139}$$

where $L$ is the Liouvillian operator $L \equiv \hbar^{-1}[H, ]$. (See Eq. 1.9, Chapter 1, and note the change of sign.) Using the projection operator $\bar{P} = \sum_c |A_c\rangle\langle A_c|$ (Eq. 2.23), we write the identity

$$\frac{d}{dt}|A_b(t)\rangle = iL\bar{P}|A_b(t)\rangle + iL(1-\bar{P})|A_b(t)\rangle. \tag{2.140}$$

Multiplying from the left by $|A_a(0)\rangle\bar{P} \equiv |A_a\rangle\bar{P}$ yields

$$\frac{d}{dt}\langle A_a|\sum_c |A_c\rangle\langle A_c|A_b(t)\rangle$$

$$= \frac{d}{dt}\langle A_a|A_a\rangle\langle A_a|A_b(t)\rangle = \frac{d}{dt}\langle A_a|A_b(t)\rangle \equiv \frac{d}{dt}C_{ab}(t)$$

$$= \langle A_a|\bar{P}iL\bar{P}|A_b(t)\rangle + \langle A_a|\bar{P}iL(1-\bar{P})|A_b(t)\rangle$$

$$= \langle A_a|\sum_c |A_c\rangle\langle A_c|iL|\sum_c |A_c\rangle\langle A_c|A_b(t)\rangle$$

$$+ \langle A_a|\sum_c |A_c\rangle\langle A_c|iL(1-\bar{P})|A_b(t)\rangle$$

$$= \Omega_{ac}C_{cb}(t) + \langle iA_a|L(1-\bar{P})|A_b(t)\rangle \tag{2.141a}$$

with

$$\Omega_{ac} \equiv \langle A_a|iL|A_c\rangle.$$

To find in Eq. 2.141a the result of $L$ acting on $(1-\bar{P})|A_b(t)\rangle$, we operate with $1 - \bar{P}$ on Eq. 2.140. This gives

$$\frac{d}{dt}(1-\bar{P})|A_b(t)\rangle$$

$$= (1-\bar{P})iL\sum_c |A_c\rangle\langle A_c|A_b(t)\rangle + (1-\bar{P})iL(1-\bar{P})|A_b(t)\rangle$$

$$= \sum_c (1-\bar{P})iL|A_c\rangle C_{cb}(t) + (1-\bar{P})iL(1-\bar{P})|A_b(t)\rangle.$$

Integration yields (Note 11)

$$(1-\bar{P})|A_b(t)\rangle = \sum_c \int_0^t dt' \exp\{(1-\bar{P})iLt'\}(1-\bar{P})iL|A_c\rangle C_{cb}(t-t').$$

$$\tag{2.141b}$$

Substitution into Eq. 2.141a yields

$$\frac{d}{dt} C_{ab}(t) = \sum_c \left\{ \Omega_{ac} C_{cb}(t) - \int_0^t dt' \, K_{ac}(t') C_{cb}(t-t') \right\} \qquad (2.142)$$

with (note that $LA_m = -A_m L$ and $(1-\bar{P})|A_m\rangle = (1-\bar{P})^2|A_m\rangle$)

$$K_{mn}(t) = \langle iLA_m|(1-\bar{P})\{\exp[i(1-\bar{P})Lt]\}(1-\bar{P})|iLA_n\rangle$$

$$\equiv \langle \dot{A}_m(0)\dot{A}_n(t)\rangle^+.$$

$K_{mn}(t)$ is the memory function of the multivariate correlation function $C_{mn}(t)$ (see also Eq. 2.32).

We make two approximations in Eq. 2.142. Following assumption (4), we consider $C_{cb}(t-t')$ constant during $t \lesssim t'$. Hence, the correlation matrix $\mathbf{C}(t)$ obeys the differential equation

$$\frac{d}{dt}\langle \mathbf{A}(t)\mathbf{A}^*(0)\rangle = \Omega\langle \mathbf{A}(t)\mathbf{A}^*(0)\rangle - \Gamma\langle \mathbf{A}(t)\mathbf{A}^*(0)\rangle \qquad (2.143)$$

with

$$\Omega = \langle \dot{\mathbf{A}}(0)\mathbf{A}^*(0)\rangle \mathbf{N}^{-1}$$

$$\mathbf{N} = \langle \mathbf{A}(0)\mathbf{A}^*(0)\rangle$$

$$\Gamma = \mathbf{F}\mathbf{N}^{-1}$$

$$\mathbf{F} = \int_0^t dt \, \langle \dot{\mathbf{A}}(t)\dot{\mathbf{A}}^*(0)\rangle^+.$$

Second, we extend the upper limit of the integral to $\infty$. This is permitted since $K(t)$ has decayed long before $C(t)$ has changed perceptibly.

The solution of Eq. 2.143 is then

$$\langle \mathbf{A}(t)\mathbf{A}^*(0)\rangle\langle|\mathbf{A}(0)|^2\rangle^{-1} = \exp[(\Omega-\Gamma)t], \qquad (2.144)$$

where $\Omega$ is the frequency matrix defined previously (see also Eq. 2.141a) and $\Gamma$ is the damping coefficient or inverse correlation time matrix

$$\Gamma = \int_0^\infty dt \, \langle \dot{\mathbf{A}}(t)\dot{\mathbf{A}}^*(0)\rangle^+ \langle|\mathbf{A}(0)|^2\rangle^{-1}. \qquad (2.145)$$

According to our assumptions, $\Gamma$ represents the set of rapidly relaxing variables whose subspace is orthogonal to that of $\mathbf{A}(t)$ (see Section 2.1).

The polarizability tensor has five irreducible anisotropic components (Eq. 1.87, Chapter 1). Therefore it is convenient to express $\mathbf{A}(t)$ by[42] $A_1(t)$,

$A_2(t), \ldots, A_5(t)$, where

$$A_1(t) = \sum_i D_{00}^{(2)}(\Omega^i(t))$$

$$A_{2,3}(t) = \frac{1}{\sqrt{2}} \sum_i \{ D_{02}^{(2)}(\Omega^i(t)) \pm D_{0-2}^{(2)}(\Omega^i(t)) \}$$

$$A_{4,5}(t) = \frac{1}{\sqrt{2}} \sum_i \{ D_{01}^{(2)}(\Omega^i(t)) \pm D_{0-1}^{(2)}(\Omega^i(t)) \}. \tag{2.146}$$

$D(\Omega^i(t))$ is a Wigner rotation matrix, and $\Omega^i(t)$ is a set of Eulerian angles of molecule $i$ at time $t$; the positive (negative) sign in Eq. 2.146 belongs to the even (odd) subscript of $A_s(t)$. We label the elements of the $5 \times 5$ matrices by subscripts $mm' = 11, 12, \ldots, 55$.

We first compute the matrix elements of the overlap matrix **N**,

$$\mathbf{N} = \sum_{ij} \langle D_{0m}^{(2)}(\Omega^i(0)) D_{0m'}^{(2)*}(\Omega^j(0)) \rangle$$

$$= N_0 \{ \delta_{mm'} + (N_0 - 1) \langle D_{0m}^{(2)}(i) D_{0m'}^{(2)*}(j) \rangle \}, \tag{2.147}$$

where $N_0$ is the number of molecules and arguments $(i)$, $(j)$ are convenient abbreviations for the set of Eulerian angles of molecule $i$ and $j$ at zero time. Explicitly,

$$N_{11} = N_0 \{ 1 + (N_0 - 1) \langle D_{00}^{(2)}(i) D_{00}^{(2)*}(j) \rangle \}$$

$$N_{22} = N_0 \{ 1 + (N_0 - 1) \langle [D_{02}^{(2)}(i) + D_{0-2}^{(2)}(i)][D_{02}^{(2)*}(j) + D_{0-2}^{(2)*}(j)] \rangle \}$$

$$N_{44} = N_0 \{ 1 + (N_0 - 1) \langle [D_{01}^{(2)}(i) + D_{0-1}^{(2)}(i)][D_{01}^{(2)*}(j) + D_{0-1}^{(2)*}(j)] \rangle \}$$

$$N_{12} = N_0 (N_0 - 1) \frac{1}{\sqrt{2}} \{ \langle D_{00}^{(2)}(i)[D_{02}^{(2)*}(j) + D_{0-2}^{(2)*}(j)] \rangle \}.$$

For a system of identical molecules possessing a threefold symmetry axis, the allowed values of $m$ and $m'$ turn out to be[42]

$$m \quad \text{and} \quad m' = 0, \pm 3, \pm 6, \ldots . \tag{2.148}$$

Hence $N_{22} = N_{33} = N_{44} = N_{55} = N_0$, $N_{pq} = 0$, and

$$\mathbf{N} = N_0 g_2 \mathbf{I},$$

where **I** is the $5 \times 5$ unit matrix and

$$N_{11} \equiv g_2 = 1 + (N_0 - 1) \langle D_{00}^{(2)}(i) D_{00}^{(2)*}(j) \rangle. \tag{2.149}$$

Quantity $g_2$ is the so-called static orientational pair correlation factor or "generalized Kirkwood $g_2$ factor" (Eq. 1.97, Chapter 1). It measures, in our example, the average pairwise static (long-time) tendency of the molecular $C_3$ axes to be aligned parallel.

To compute matrix **F**, we need the time differentials of $\mathbf{A}(t)$. They are obtained from the time derivatives of the $D_{mn}^{(2)}$ which, in turn, are calculated from Eq. E.4b (Appendix E) and relations between the time derivatives of Eulerian angles and angular velocities $\omega_p$ about molecular axes $p$.[134] The resulting expressions (the calculations are of no interest to us) are collected in Table V; we see that matrix **F** contains terms of the form

$$\int_0^\infty dt \langle \omega_p(i,t) A_s(i,t) \omega_{p'}^*(j,0) A_{s'}^*(j,0) \rangle^+.$$

Since angular variables $\omega$ fluctuate more rapidly than orientational variables $A$ (Eqs. 2.7a, 2.7b), we can (1) average independently over angular and orientational coordinates and (2) suppress the time dependence in the orientational correlation function by setting

$$\langle A_s(i,t) A_{s'}^*(j,0) \rangle \approx \langle A_s(i,0) A_{s'}^*(j,0) \rangle \equiv \langle A_s(1) A_{s'}^*(2) \rangle.$$

Separating the entire integrand into auto- (same molecule, 1) and cross-correlation (different molecules, 1 and 2) terms gives

$$N_0 \int_0^\infty dt \langle \omega_p(1,t) \omega_{p'}^*(1) \rangle^+ \langle A_s(1) A_{s'}^*(1) \rangle$$

$$+ N_0(N_0 - 1) \int_0^\infty dt \langle \omega_p(1,t) \omega_{p'}^*(2) \rangle^+ \langle A_s(1) A_{s'}^*(2) \rangle. \quad (2.150)$$

Abbreviating the first integral of Eq. 2.150 by the symbol $D_{pp'}$ (rotational diffusion tensor) and the second by $J_{pp'}$ (*dynamic* orientational pair correlation —see Eq. 1.98 of Chapter 1), the matrix elements of **F** (Eq. 2.143) are

Table V.    Coefficients[a] of $(d/dt)A_s = a\omega_x + b\omega_y + c\omega_z$

| $\dot{A}_s(t) \equiv (d/dt)A_s$ | $\omega_x$ | $\omega_y$ | $\omega_z$ |
|---|---|---|---|
| $\dot{A}_1$ | $-i\sqrt{3}\,A_4$ | $\sqrt{3}\,A_5$ | $0$ |
| $\dot{A}_2$ | $-iA_4$ | $-A_5$ | $-2iA_3$ |
| $\dot{A}_3$ | $-iA_5$ | $-A_4$ | $-2iA_2$ |
| $\dot{A}_4$ | $-i(\sqrt{3}\,A_1 + A_2)$ | $A_3$ | $-iA_5$ |
| $\dot{A}_5$ | $-iA_3$ | $A_2 - \sqrt{3}\,A_1$ | $-iA_4$ |

[a]From T. D. Gierke, *J. Chem. Phys.* **65**, 3873 (1976).

computed as follows with the help of Table V:

$$F_{11} = \int_0^\infty dt \langle \dot{A}_1(t)\dot{A}_1^*(0)\rangle^+$$

$$= 3\int_0^\infty dt \{\langle \omega_x(1,t)\omega_x^*(1)\rangle^+ \langle A_4(1)A_4^*(1)\rangle$$

$$+ \langle \omega_y(1,t)\omega_y^*(1)\rangle^+ \langle A_5(1)A_5^*(1)\rangle\}$$

$$= 3(D_{xx} + D_{yy}) \equiv 6D_\perp \tag{2.151}$$

since $D_{xx} = D_{yy}$ (axial symmetry). In the same fashion we get

$$F_{22} = D_{xx} + D_{yy} + 4D_{zz} \equiv 2D_\perp + 4D_\| = F_{33},$$

$$F_{44} = F_{55} = 5D_\perp + D_\|. \tag{2.152}$$

It is easy to see that nondiagonal elements of **F** are zero. For instance,

$$F_{12} = \int_0^\infty dt \langle \dot{A}_1(t)\dot{A}_2^*(0)\rangle^+$$

$$= \int_0^\infty dt \{\langle \sqrt{3}\,\omega_x(1,t)\omega_x^*(1)\rangle^+ \langle A_4(1)A_4^*(1)\rangle$$

$$- \langle \sqrt{3}\,\omega_y(1,t)\omega_y^*(1)\rangle^+ \langle A_5(1)A_5^*(1)\rangle\}$$

$$+ \int_0^\infty dt \{\langle \sqrt{3}\,\omega_x(1,t)\omega_x^*(2)\rangle^+ \langle A_4(1)A_4^*(2)\rangle$$

$$- \langle \sqrt{3}\,\omega_y(1,t)\omega_y^*(2)\rangle^+ \langle A_5(1)A_5^*(2)\rangle\}.$$

The first integral vanishes since $D_{xx} = D_{yy}$. The second vanishes since

$$\langle A_4(1)A_4^*(2)\rangle, \langle A_5(1)A_5^*(2)\rangle = 0$$

from Eqs. 2.146 and 2.148. Hence we conclude that the effect of the dynamic orientational pair correlation (Eq. 2.150),

$$\langle \omega_p(1,t)\omega_p^*(2,0)\rangle^+,$$

is not observed for a system of symmetric top molecules since the quantity is weighted by zero elements of $\langle \mathbf{A}(1)\mathbf{A}^*(2)\rangle$. (For a divergent result, see P. G. Wolynes and J. M. Deutch, *J. Chem. Phys.* **67**, 733 (1977).)

Next we find that the frequency matrix $\boldsymbol{\Omega}$ in Eq. 2.143 vanishes:

$$\boldsymbol{\Omega} = \langle \dot{\mathbf{A}}(0)\mathbf{A}^*(0)\rangle \mathbf{N}^{-1} = 0 \tag{2.153}$$

since $\langle \dot{\mathbf{A}}(0)\mathbf{A}^*(0)\rangle$ has only nondiagonal (see Table V), $\mathbf{N}^{-1}$ only diagonal elements.

Since we are interested in the $k$ (or $m$) = 0 component of the polarizability tensor (Eqs. 2.48, 2.49; Section 3.5, Chapter 1) we use the 1,1 element of matrix $\mathbf{F}$, $F_{11} = 6D_\perp$. Inserting Eqs. 2.149, 2.151, and 2.153 into Eq. 2.144 gives, with $\Gamma_{11} = F_{11}N_{11}^{-1}$,

$$C_{11}(t) = \langle A_1(t)A_1^*(0)\rangle\langle |A_1(0)|^2\rangle^{-1}$$

$$= \exp\left(-\frac{6D_\perp t}{g_2}\right)$$

$$\equiv \exp\left[-\frac{t}{\tau_{RAY}^{(2,0)}}\right], \tag{2.154}$$

with

$$\tau_{RAY}^{(2,0)} = \frac{g_2}{6D_\perp}$$

$$= g_2\tau_R^{(2,0)}. \tag{2.155}$$

$\tau_R^{(2,0)}$ is the $k = 0$ component second-order tensor orientational correlation time of the single-molecule (no pair correlation) effect (Section 5.2).

Note that the combination of diffusion coefficients $F_{22} = 2D_\perp + 4D_\parallel$ and $F_{44} = 5D_\perp + D_\parallel$ of the symmetric top (Eq. 2.152) could be obtained from a doubly degenerate Raman ($k$, or $m = 1, 2$) vibration-rotation mode (Eq. 2.50). However, this is not interesting here since $k \neq 0$ components are not allowed in the depolarized Rayleigh spectrum of the symmetric top.

We are now ready to discuss experimental data. First, we show in Table VI a comparison of the measured orientational diffusion coefficients (Eq. 2.155)

**Table VI. Depolarized Rayleigh and Anisotropic Raman Rotational Diffusion Constants and $g_2$ Factor[a]**

| Liquid (25°C) | $D_\perp^{RAY}$ (Units of $10^{10}$ sec$^{-1}$) | $D_\perp^R$ | $g_2$ |
|---|---|---|---|
| $C_6H_6$ | 6.1 | 6.0 | 1.0 |
| $C_6D_6$ | 5.3 | 5.4 | 1.0 |
| $CH_3CN$ | 9.4 | 18.3 | 2.0 |
| $CD_3CN$ | 8.9 | 16.3 | 1.8 |
| $CH_3I$ | 7.6 | 10.6 | 1.4 |
| $CDCl_3$ | 5.7 | 9.4 | 1.7 |
| $CHBr_3$ | 1.7 | 3.7 | 2.2 |

[a]From G. D. Patterson and J. E. Griffiths, *J. Chem. Phys.* **63**, 2406 (1975).

from depolarized Rayleigh and anisotropic polarized Raman data of several symmetric tops at ambient temperature and pressure.[113a] (The $\tau_R^{(2,0)}$ were obtained from the halfwidth of the band contour (Eq. 2.88).) Since $\tau_{RAY} \geq \tau_R$, $D_\perp^{RAY} \leq D_\perp^R$; the $g_2$ values, computed from $D_\perp^R / D_\perp^{RAY}$, are collected in column 4 of Table VI.

Table VI shows that the observed orientational motion of the molecular symmetry axis is influenced by static pair correlation in all liquid systems with the notable exception of benzene and deuterobenzene. A trend with some relevant macroscopic property, for instance with the permanent dipole moment, is conceivable but is not apparent. For instance, the $g_2$ factor of $CHBr_3$ (permanent dipole moment approximately 2 debye) equals that of $CH_3CN$ (about 4 debye). (Dipole-dipole interactions depend on the second power of the dipole moment.) The benzene data tell us that there is no static parallel alignment of the symmetry axes of benzene molecules. Indeed, its liquid-state structure (by X-ray scattering) shows that although each $C_6H_6$ molecule sees 12 near neighbors in preferred orientations, their molecular planes are inclined by 41 to 79 degrees with respect to the reference molecule. Beyond, a random distribution of relative orientations prevails.[135]

It is difficult to relate the $g_2$ factor to the structure of liquids. Simple considerations are deceiving, since it is not so apparent how many near-neighbor shells contribute to $g_2$. In other words, we do not know a priori the limiting distance $r_m$ beyond which the pair distribution function $g'(r)$ cuts off the integration volume $4\pi r^2 \, dr$,

$$4\pi \int_0^{r_m} [g'(r) - 1] r^2 \, dr \sim 0. \tag{2.156}$$

It does not seem obvious how to establish $r_m$ by a spectroscopic experiment. We therefore take recourse (as often in similar situations) to a "computer experiment" by Monte Carlo[136] or molecular dynamics[2c] calculations. Although the discussions of such methods are not within the scope of this book, an application of molecular dynamics simulations to an estimate of the range of a $g_2$ factor is instructive and useful. We apply this now to the system of liquid $CS_2$.[137]

We introduce the pair distribution function $g'(r, \Omega_1, \Omega_2)$ of finding molecules 1,2 a distance $r$ apart and at orientations $\Omega_1 = (\theta_1, \varphi_1)$ and $\Omega_2 = (\theta_2, \varphi_2)$. Equation 2.149 then reads

$$g_2 = 1 + \sum_b \left\langle \frac{1}{2} [3\cos^2(\theta_{1,b}) - 1] \right\rangle$$

$$= 1 + \left(\frac{\rho}{5}\right) \int dr \, d\Omega_1 \, d\Omega_2 r^2 \, g'(r, \Omega_1, \Omega_2)$$

$$\times \sum_{m=-2}^{2} Y_2^{m*}(\Omega_1) Y_2^m(\Omega_2). \tag{2.157}$$

$\theta_{1,b}$ is the polar angle between symmetry axis of reference molecule 1 and neighbor $b$, $\rho$ is the number density, and $g'(r, \Omega_1, \Omega_2)$ is expressed by its spherical harmonics development,[138b]

$$g'(r, \Omega_1, \Omega_2) = 4\pi \sum_{ll'm}^{\infty} g'_{ll'm}(r) Y_{l'}^{m*}(\Omega_1) Y_l^m(\Omega_2). \qquad (2.158)$$

We then insert Eq. 2.158 into Eq. 2.157; we hope that only a few terms are required to give satisfactory convergence.

The results of the molecular dynamics calculations are shown in Fig. 2.30, together with a plot of the contributing $g'_{ll'm}(r)$ coefficients of Eq. 2.158. We see that $g_2$ sums about equally over two nearest-neighbor shells and has not converged with the set limit. Hence this example demonstrates that experimental values of the static orientational pair correlation factor may be misinterpreted if we rely on intuition in choosing the range of interaction. In fact, we note that the principal pair correlation coefficient $g'_{220}(r)$ changes sign after the first nearest-neighbor shell (parallel and antiparallel alignments alter); $g_2 - 1$ (Eq. 2.157) is negative for 54.7 degrees $< \theta_{1,b} < 125.3$ degrees.

In conclusion, we discuss the pressure dependence of depolarized Rayleigh correlation times $\tau_{RAY}^{(2,0)}$ since the data can be compared with results from corresponding pressure studies of Raman anisotropic spectral densities. Again, we restrict our discussion to systems of symmetric top molecules.[129, 105, 118a]

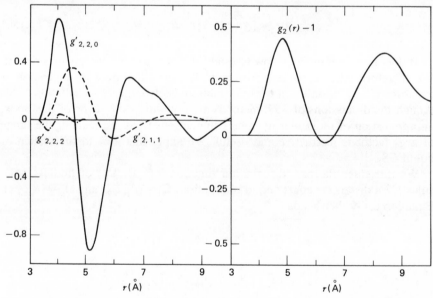

**Figure 2.30.** Distance dependence, counted from the reference molecule, of the static orientational pair correlation factor $g_2 - 1$ as obtained by molecular dynamics calculations on 512 $CS_2$ molecules. The left part of the figure displays the distance dependence of three coefficients $g'_{ll'm}(r)$ of the spherical harmonics series development of the molecular pair correlation function $g'(r, \Omega_1, \Omega_2)$ of Eq. 2.158. [O. Steinhauser and M. Neumann, *Mol. Phys.* **37**, 1921 (1979).]

The circles in Fig. 2.31 (chloroform) and Fig. 2.32 (methyl iodide) show the Rayleigh correlation times; we have added recent anisotropic Raman[105, 118a] and quadrupole nuclear magnetic relaxation data.[139] The static orientational $g_2$ factor $\tau_{RAY}^{(2,0)}/\tau_R^{(2,0)}$ of $CDCl_3$, $g_2$ about 2.0, is near pressure independent, whereas for $CH_3I$, $g_2$ approximates 2.1 at 1 bar and 1.3 at 2500 bar. Increased pressure appears to diminish the average pairwise alignment of the symmetry axes of liquid-phase $CH_3I$ molecules.

Figure 2.32 shows that the single-particle correlation times $\tau_R^{(2,0)}$ of $CH_3I$ from two different Raman studies agree. Furthermore, Raman and nuclear magnetic resonance results agree on the single-particle orientational motion of chloroform-$d$ (Fig. 2.31)—as they should since the same transition moment vector and Legendre polynomial is observed (Eq. 1.81, Chapter 1). We note that nuclear magnetic relaxation data are not influenced by orientational cross correlation or vibrational relaxation processes. Hence the agreement between the Raman and the NMR results gives us confidence that the Raman data are not perturbed by nonorientational, pressure-induced effects (for instance, vibration-rotation coupling).[129]

The multiparticle correlation times $\tau_{RAY}^{(2,0)}$ displayed in Figs. 2.31 and 2.32 disagree with comparable results compiled in Table VI (at 1 bar): They are consistently longer and thus yield a larger static orientational pair correlation factor. The reason for this discrepancy is not known; hence there is a great deal left for improvement and verification.

Interestingly, both Rayleigh studies[113a, 129] observe a vanishing *dynamic* orientational pair correlation $J_2$ for symmetric top molecules—as they should.

In conclusion, we raise two points regarding the $g$ factors. First, neutron diffraction data[140] give a more clear-cut answer to the structure of liquids than

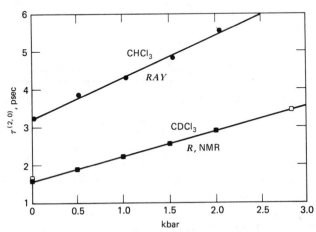

**Figure 2.31.** Pressure dependence of the second-order tensor orientational correlation time $\tau_R^{(2,0)}$ from anisotropic Raman ($R$, Ref. 105, □), quadrupolar nuclear magnetic resonance relaxation (NMR, Ref. 139, ■), and depolarized Rayleigh scattering ($RAY$, Ref. 129) experiments on liquid $CHCl_3$ and $CDCl_3$. [Adapted from S. Claesson and D. R. Jones, *Chem. Scripta* **9**, 103 (1976). By permission of the Royal Swedish Academy of Sciences.]

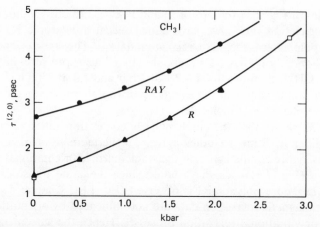

**Figure 2.32.** Pressure dependence of the second-order tensor orientational correlation time $\tau_R^{(2,0)}$ of liquid $CH_3I$ from an anisotropic Raman ($R$, Ref. 105, □; Ref. 118$a$, ▲) and depolarized Rayleigh scattering ($RAY$, Ref. 129) experiment. [Adapted from S. Claesson and D. R. Jones, *Chem. Scripta* **9**, 103 (1976). By permission of the Royal Swedish Academy of Sciences.]

the Kirkwood $g_2$ or $g_1$ factors since they yield experimental values for the distribution functions $g'_{ll'm}(r)$. [For the $g_1$ factor, in Eq. 2.149 replace $D^{(2)}$ by $D^{(1)}$.] Second, the $g_1$ factor is evidently obtainable from infrared/far-infrared (Eq. 1.41, Chapter 1) or dielectric measurements.[141] However, the necessity of local field corrections[10c] does not always assure agreement with neutron diffraction results.[140]

### 5.7.  Two-Variable Theory: Coupling of Orientational Motion and Angular Momentum

The one-variable theory discussed in the previous section is valid for a coarse division of time, namely, for intervals $t$ that exceed the orientational correlation time $1/\Gamma$. This condition allowed us the simplification of setting in Eq. 2.145 $C(t) \sim 1$ during the lifetime of the memory function $K(t)$. $C(t)$ contains the "slowly relaxing" variables of $A(t)$; this set is called a "good set of collective variables".[59]

Since an orientational correlation usually persists longer than the correlation of intermolecular torques, angular momentum, and of some general anisotropic intermolecular potential, we can extend this partitioning of effective time domains ("graining") to include additional dynamic variables of $A(t)$ in $C(t)$. The principal advantage of such an approach is that it allows us to consider the theory in accordance with experimental knowledge we have of the system. The orientational correlation functions will reflect our heightened understanding of rotational relaxation phenomena. Depending on the quality of experimental results, we need no longer maintain that, for instance, intermolecular torques act instantaneously, that angular momentum correlation decay is exponential, and so forth.

We recall from Section 5.6 the form of stochastic variable $A(t)$,

$$A(t) = \sum_l D_{mn}^{(j)}(\Omega^l(t)),$$

appropriately called "one-variable theory" since $C(t)$ contains only the slowest varying orientational coordinate. $C(t)$ is a purely exponential correlation function (Eq. 2.154); we know that it fails to describe the shorter-time evolution of the orientational motion of the system.

Let us then enlarge the set of variables of $A(t)$ included in $C(t)$.[142] We select the time derivative of $A(t)$. The reasons for this choice are easy to see. (1) The subspaces of $\dot{A}(t)$ and $A(t)$ are orthogonal to each other (see Table V),

$$\langle \dot{A}_p(0) A_p^*(t) \rangle = 0.$$

(We only consider autocorrelations.) (2) Fluctuations of $\dot{A}(t)$ are faster than fluctuations of $A(t)$; we project the slowest of $\dot{A}(t)$ into the set of $C(t)$ to keep $C(t)$ a complete collection of good ("slowly relaxing") variables. (3) Thereby we extend the validity region of $C(t)$ to shorter time intervals.

Hence we assume the set of Wigner matrices

$$D_{mn}^{(j)}(\Omega(t)), \dot{D}_{mn}^{(j)}(\Omega(t)) \equiv \frac{d}{dt} D_{mn}^{(j)}(\Omega(t)). \tag{2.159}$$

Note first that $\dot{D}(t)$ is related to the angular velocity (Table V). Therefore our choice of additional variable leads to a coupling of orientational and angular momentum coordinates. We show thereby that we introduce kinetic (inertial) effects.

The matrix equation corresponding to Eq. 2.143 is now

$$\frac{d}{dt} \begin{pmatrix} \langle A(t)A^*(0) \rangle \\ \langle B(t)A^*(0) \rangle \end{pmatrix} = \begin{pmatrix} \langle \dot{A}A^* \rangle - (\dot{A}\dot{A})^+ & \langle \dot{A}B^* \rangle - (\dot{A}\dot{B})^+ \\ \langle \dot{B}A^* \rangle - (\dot{B}\dot{A})^+ & \langle \dot{B}B^* \rangle - (\dot{B}\dot{B})^+ \end{pmatrix}$$

$$\times \begin{pmatrix} \langle AA^* \rangle^{-1} & 0 \\ 0 & \langle BB^* \rangle^{-1} \end{pmatrix} \begin{pmatrix} \langle A(t)A^*(0) \rangle \\ \langle B(t)A^*(0) \rangle \end{pmatrix} \tag{2.160}$$

with the abbreviations

$$A(t) \equiv D(t); \qquad\qquad B(t) \equiv \dot{D}(t)$$

$$\langle \dot{A}B^* \rangle \equiv \langle \dot{A}(0)B^*(0) \rangle; \qquad (\dot{A}\dot{B})^+ \equiv \int_0^\infty dt \langle \dot{A}(t)\dot{B}^*(0) \rangle^+. \tag{2.161}$$

[For $\langle \dot{A}(t)\dot{B}^*(0) \rangle^+$, see Eq. 2.142.] We have also written out the (diagonal) normalization matrix. Of course, an analogous matrix equation can be written for a vector with elements $\langle A(t)B^*(0) \rangle$, $\langle B(t)B^*(0) \rangle$. According to expectation, we see that elements $\langle \dot{A}A^* \rangle$ and $\langle \dot{B}B^* \rangle$ of the frequency matrix $\Omega$ in Eq. 2.160 vanish. (Refer to Table V and see also Eq. 2.153.) Second, elements of the form $(X\dot{A})^+$ or $(\dot{A}X)^+$ disappear since $|\dot{A}^+\rangle \equiv |\dot{D}^+\rangle$ vanishes (Note 12),

$$|\dot{D}^+\rangle \equiv (1 - \Sigma\bar{P})|\dot{D}\rangle$$

$$= (1 - |\hat{D}\rangle\langle\hat{D}| - |\hat{\dot{D}}\rangle\langle\hat{\dot{D}}|)|\dot{D}\rangle$$

$$= |\dot{D}\rangle - 0 - |\dot{D}\rangle.$$

The coefficient matrix of Eq. 2.160 then reads

$$\Omega - \Gamma \equiv \mathbf{T}^{-1} = \begin{pmatrix} 0 & \langle \dot{A}B^* \rangle\langle BB^* \rangle^{-1} \\ \langle \dot{B}A^* \rangle\langle AA^* \rangle^{-1} & -(\dot{B}\dot{B})^+\langle BB^* \rangle^{-1} \end{pmatrix}. \quad (2.162)$$

$\mathbf{T}^{-1}$ is called the transport matrix.[59]

We now determine the elements of $\mathbf{T}^{-1}$. First, consider element 2,1. Recalling $A \equiv D$, $B \equiv \dot{D}$, and Eq. 2.139, we have

$$\langle \ddot{D}D^* \rangle = \langle iLiLD| D^* \rangle = -\langle iLD| iLD^* \rangle = -\langle \dot{D}\dot{D}^* \rangle. \quad (2.163a)$$

Furthermore,

$$\langle \dot{D}\dot{D}^* \rangle\langle DD^* \rangle^{-1} = j(j+1)\omega^2 = \frac{j(j+1)k_BT}{I} \equiv \omega_j^2 \quad (2.163b)$$

for a linear or spherical classical rotor. [Remember that we decided to neglect cross correlation of the orientational motion between different molecules; refer to Eqs. B.2, B.5 of Appendix B (or Eq. 2.2) for an example with $j=1$.]

Second, we write element 2,2 of $\mathbf{T}^{-1}$,

$$(\ddot{D}\ddot{D})^+\langle \dot{D}\dot{D}^* \rangle^{-1} \equiv (\ddot{D}\ddot{D})^+\langle DD^* \rangle^{-1}\langle \dot{D}\dot{D}^* \rangle^{-1}\langle DD^* \rangle = (\hat{\ddot{D}}\hat{\ddot{D}})^+\omega_j^{-2}.$$

Since $(\ddot{D}\ddot{D})^+$ (see Eq. 2.161) has the dimension of radian$^2$ sec$^{-3}$, we define a reciprocal time

$$(\hat{\ddot{D}}\hat{\ddot{D}})^+\omega_j^{-2} \equiv \tau_\omega^{-1}, \quad (2.164)$$

where $\tau_\omega$ is a correlation time associated with angular velocities and intermolecular torques. [To see this, take from Table V

$$\ddot{D}_1 = -\left(\sqrt{3}\,D_2 + 3D_1\right)\omega_x^2 + \left(\sqrt{3}\,D_2 - 3D_1\right)\omega_y^2 - i\sqrt{3}\,D_4\dot{\omega}_x + \sqrt{3}\,D_5\dot{\omega}_y,$$

$$(2.165)$$

and note Euler's equation for the classical rigid rotor:

$$\text{torque} = \mathbf{I}\dot{\boldsymbol{\omega}} + \boldsymbol{\omega} \times (\mathbf{I}\boldsymbol{\omega}).]$$

Hence

$$\mathbf{T}^{-1} = \begin{pmatrix} 0 & 1 \\ -\omega_j^2 & -\tau_\omega^{-1} \end{pmatrix}. \tag{2.166}$$

We further assume that the autocorrelation time for the (projected) torques is short; $\langle \ddot{D}(t)\ddot{D}^*(0)\rangle^+$ is a rapidly decaying memory function with decay time $\tau_T$ ($T$ = torque). The (projected) torques are impulsive, meaning

$$\frac{\tau_T}{\tau_\omega} \equiv \omega_T^2 \tau_T^2 \ll 1, \tag{2.167a}$$

where we define a (projected) mean-squared frequency

$$\omega_T^2 = \langle \hat{\ddot{D}}^+ \hat{\ddot{D}}^+ \rangle \omega_j^{-2},$$

$$\tau_\omega^{-1} = \omega_T^2 \tau_T. \tag{2.167b}$$

Note that the definitions in Eq. 2.167b are analogous to the one-variable case (Eq. 2.150). Because of the different time scales demanded by Eq. 2.167a, we can neglect the time dependence of the projected $D\omega^2$ against that of the projected $D\dot{\omega}$. Hence

$$\tau_\omega^{-1} = (\hat{\ddot{D}}\hat{\ddot{D}})^+ \omega_j^{-2} = \omega_j^{-2} \int_0^\infty dt\, \langle \hat{\ddot{D}}(t)\hat{\ddot{D}}^*(0)\rangle^+$$

(Eq. 2.164) can be approximated by

$$\tau_\omega^{-1} = \omega_j^{-2} \int_0^\infty dt\, \langle \hat{D}(0)\hat{D}^*(0)\rangle \{\langle \hat{\omega}_x^2(0)\hat{\omega}_x^2(0)\rangle^+ + \langle \hat{\omega}_x(t)\hat{\omega}_x^*(0)\rangle^+ + \cdots \}$$

$$= \langle \hat{\ddot{D}}^+ \hat{\ddot{D}}^+ \rangle \omega_j^{-2} \int_0^\infty dt\, \langle \hat{\omega}(t)\hat{\omega}^*(0)\rangle^+$$

$$= \langle \hat{\ddot{D}}^+ \hat{\ddot{D}}^+ \rangle \omega_j^{-2} \tau_T. \tag{2.168}$$

We now solve Eq. 2.160 for $\langle B(t)A^*(0)\rangle \equiv \langle \dot{D}(t)D^*(0)\rangle$. Using Eq. 2.161, we set $(d/dt)\langle B(t)A^*(0)\rangle = \langle \ddot{D}(t)D^*(0)\rangle = (d^2/dt^2)\langle D(t)D^*(0)\rangle$ and abbreviate $\langle D(t)D^*(0)\rangle = G^{(j,0)}(t)$ by $C(t)$. With $\mathbf{T}^{-1}$ given by Eq. 2.166, the differential equation reads

$$\frac{d^2}{dt^2}C(t) = -\omega_j^2 C(t) - \tau_\omega^{-1}\frac{d}{dt}C(t). \tag{2.169}$$

Taking the Laplace transform of Eq. 2.169, designating the Laplace variable by $s$ and the Laplace-transformed function by an upper tilde sign, we obtain[143a]

$$\tilde{C}(s) = (s + \tau_\omega^{-1})[s^2 + s\tau_\omega^{-1} + \omega_j^2]^{-1}C(0)$$

$$= (s + \tau_\omega^{-1})\left[\left(s + \frac{\tau_\omega^{-1}}{2}\right)^2 - \left(\frac{\tau_\omega^{-2}}{4} - \omega_j^2\right)\right]^{-1} \tag{2.170}$$

with $C(0) = 1$. Inverting Eq. 2.170 into the time domain gives, after a little algebra,[143b]

$$C(t) = (\tau_+^{-1} - \tau_-^{-1})^{-1}\left\{\tau_+^{-1}\exp\left(-\frac{t}{\tau_-}\right) - \tau_-^{-1}\exp\left(-\frac{t}{\tau_+}\right)\right\}$$

$$= (\tau_- - \tau_+)^{-1}\left\{\tau_-\exp\left(-\frac{t}{\tau_-}\right) - \tau_+\exp\left(-\frac{t}{\tau_+}\right)\right\}$$

$$= G^{(j,0)}(t) \tag{2.171}$$

with

$$\tau_\pm^{-1} = (\tfrac{1}{2})(\tau_\omega^{-1} \pm 2\beta),$$

$$\beta = (\tfrac{1}{2})(\tau_\omega^{-2} - 4\omega_j^2)^{1/2},$$

$$\omega_j^2 = \frac{j(j+1)k_BT}{I},$$

$$4\omega_j^2\tau_\omega^2 < 1 \text{ (fast modulation).}$$

It is not difficult to see that Eq. 2.171 is an improvement over the result from the one-variable approximation (Eq. 2.144). For instance,

1.
$$\left.\frac{d}{dt}C(t)\right|_{t=0} = 0,$$

saying that $C(t)$ has the correct zero-slope at zero time.

2. The memory function $K(t)$, associated with $C(t)$, is an exponential, $K(t) = \alpha_2\exp(-t/\tau_m)$. To see this, take the Laplace transform of Eq. 2.32,

$$\frac{d}{dt}C(t) = -\int_0^\infty dt'\,\alpha_2\exp\left(-\frac{t'}{\tau_m}\right)C(t - t'),$$

which is[143c]

$$\tilde{C}(s) = (s + \tau_m^{-1})[s^2 + s\tau_m^{-1} + \alpha_2]^{-1}, \tag{2.172}$$

and compare the result with Eq. 2.170. Note that this memory function is physically more reasonable than the instantaneous memory (delta pulse) associated with Eq. 2.144 (see also Section 2.1).

In fact, we have here a segment of a "ladder approach," a hierarchy of correlation functions and associated memory functions that corresponds to the so-called Mori's continued fraction representation.[2c, 59] We can construct (by way of Eq. 2.32) the correlation function from an assumed memory function, consider the result an improved memory function to permit, in turn, generation of its (improved) correlation function, and so go on further. For instance, starting with

$$K(t) \propto \delta(t) \tag{2.173a}$$

it is obvious that the first step leads to a pure exponential $C(t)$, our result in Section 5.6 (Eq. 2.144),

$$C(t) = \exp(-\alpha_1 t). \tag{2.173b}$$

This $C(t)$ is only good for $t > 1/\alpha_1$, where $\alpha_1^{-1}$ is the orientational correlation time. The second step leads to a $C(t)$ that has the correct zero-time slope. Its time series development is given by expressions such as Eq. 2.4 or Eq. B.5 (Appendix B). This $C(t)$ therefore includes the purely kinetic effects (free rotation steps) and correctly describes orientational correlation for $t > \tau_m$, where $\tau_m$ is the correlation time of its (exponential) memory function. However, the improved $C(t)$ fails for $t \lesssim \tau_m$ since the intermolecular torques are—by assumption (Eq. 2.167a)—impulsive; their memory is a delta pulse.

Inspection of Eqs. 2.170 and 2.172 shows that we must set

$$\alpha_2 = \omega_j^2; \qquad \tau_m^{-1} = \tau_\omega^{-1}. \tag{2.173c}$$

In other words, the elements of the transport matrix $\mathbf{T}^{-1}$ are given by the zero time value and the inverse correlation time of the memory function. [The analogous relations for the one-variable approximation are $\tilde{C}(s) = (s + \Gamma)^{-1}$, with $\Gamma = \alpha_1$ defined in Eqs. 2.145 and 2.173b.]

Whereas the one-variable theory leads to the rotational diffusion model, we see that the two-variable theory (Eq. 2.171) represents a more general orientational behavior, aptly called[142b] the pseudo-GLED model. GLED symbolizes "gas-like extended diffusion"; "pseudo" reminds us that we deal here—in contrast to the GLED models discussed in Section 2—with projected quantities ("pseudo torques," for instance). This makes physical interpretation of the assumptions of the pseudo-GLED model more involved, but the pseudo-GLED model gives many of the interesting features of the GLED models in a simple, analytical manner.[142b] For instance, let us write the orientational correlation time $\tau_R = \int_0^\infty dt\, C(t)$,

$$\tau_R = \left(\omega_j^2 \tau_\omega\right)^{-1}, \tag{2.174}$$

a relation that follows readily by realizing that $\tau_R = \tau_+ + \tau_-$ with

$$\tau_\pm = 2\tau_\omega \left\{ 1 \pm \left[ 1 - 4\omega_j^2 \tau_\omega^2 \right]^{1/2} \right\}^{-1} \tag{2.175}$$

(Eq. 2.171). Imposing the fast modulation regime,

$$\tau_R \gg \tau_\omega; \qquad \omega_j^2 \tau_\omega^2 \ll 1,$$

we get from Eq. 2.175 that $\tau_- = 0$, $\tau_+ = \tau_\omega$. Hence in Eq. 2.171, $C(t) \to \exp(-t/\tau_\omega)$.

We thus have established without ado the rotational diffusion limit for $C(t) = G^{(j,0)}(t)$. (Compare with Eqs. 2.59 to 2.63.)

Another example for the usefulness of the pseudo-GLED model is its applicability to a slow modulation regime,

$$\tau_R < \tau_\omega; \qquad 4\omega_j^2 \tau_\omega^2 > 1. \tag{2.176}$$

In this situation, inversion of Eq. 2.170 gives an oscillatory correlation function:[143d]

$$C(t) = \exp(-\gamma t) \left\{ \cos(\beta t) + \frac{\gamma}{\beta} \sin(\beta t) \right\} \equiv G^{(j,0)}(t), \tag{2.177}$$

where

$$\gamma = \frac{\tau_\omega^{-1}}{2}$$

$$\beta = \left( \tfrac{1}{2} \right) \left( 4\omega_j^2 - \tau_\omega^{-2} \right)^{1/2}$$

$$\omega_j^2 = \frac{(j+1)jk_B T}{I}; \qquad 4\omega_j^2 > \tau_\omega^{-2}.$$

This interesting result depicts the situation of hindered orientational motion,[142] with relatively long correlation times $\tau_\omega$ of solidlike oscillatory rotations (SLOR; Sections 3.3 and 5.1).

It is easy to see that the two-variable model is on the interpretive level of the stochastic $J$-extended diffusion model (Section 2): (1) instantaneous torques; (2) an exponential memory function, equivalent to a Poisson distribution of times between collisions (Eq. 2.34); (3) prevalence of fast modulation, $\omega_j^2 \tau_\omega^2 < \tfrac{1}{4}$. Numerical examples abound[144]: $\omega_1^2 = M^{(1,0)}(2)$ is of the order of $10^{24}$ radian$^2$ sec$^{-2}$; $\tau_\omega = \tau_J = \{ M^{(1,0)}(2)\tau_R^{(1,0)} \}^{-1}$ (Eq. 2.174) is of the order of $0.3 \times 10^{-12}$ sec.

We recall that in the $J$- and $M$-diffusion models the direction of the angular momentum $\hat{\mathbf{J}} = \mathbf{J}/|\mathbf{J}|$ is randomized (reset to a Maxwellian distribution) at each collisional event (Eq. 2.38). On the other hand, the two-variable formalism lets us construct situations where a *selected* amount of angular momentum $\mathbf{J}$, $|\Delta\mathbf{J}|$, is transferred during a collision. Designating $\mathbf{J}$ before and after a collision by $\mathbf{J}(1)$ and $\mathbf{J}(2)$, respectively, we write

$$\mathbf{J}(2) = \mathbf{J}(1) + \Delta\mathbf{J}, \tag{2.178}$$

hence

$$|\Delta\mathbf{J}|^2 = 2|\mathbf{J}^2|(1 - \cos\theta).$$

$\theta$ is the angle between $\mathbf{J}(2)$ and $\mathbf{J}(1)$. We have set $|\mathbf{J}(1)| = |\mathbf{J}(2)| = |\mathbf{J}|$, the $M$-diffusion model.

Consider the limiting situations of Eq. 2.178. (1) For $\langle\cos\theta\rangle = 0$ we obtain the original $M$-diffusion model; upon a collision, $\hat{\mathbf{J}}$ takes all orientations with equal probability. (2) For $\langle\cos\theta\rangle = 1$, $\mathbf{J}$ never changes; the rotors are free. (3) For $\langle\cos\theta\rangle = -1$, a collision reverses the direction of $\mathbf{J}$. This limit is not included in the original $M$- or $J$-models.

The quantitative expression[145a] is obtained by multiplying $(\frac{1}{2})\tau_\omega^{-1}$ in Eq. 2.170 or Eq. 2.177 by $(1 - \cos\theta)$. This modifies $\gamma$ to

$$\gamma = \frac{1 - \cos\theta}{2\tau_\omega}$$

$$= \frac{(\frac{1}{4})|\Delta\mathbf{J}|^2}{\tau_\omega|\mathbf{J}|^2}. \tag{2.179}$$

The modified $M$-diffusion model is quite useful since we dispose of an additional adjustable parameter. Note that for

$$0 < \cos\theta < 1$$

the system may be in the slow modulation regime,

$$4\omega_j^2\tau_{\omega,\text{eff}}^2 > 1, \text{ where } \tau_{\omega,\text{eff}} = \tau_\omega(1 - \cos\theta)^{-1}.$$

As examples of the modified $M$-diffusion model,[145a] we show in Fig. 2.33 a simulation of the orientational correlation function Eq. 2.177, with $I = 91.3 \times 10^{-40}$ g cm$^2$, $\tau_\omega = 0.10 \times 10^{-12}$ sec, and $\Delta J = 9.1 \times 10^{-26}$ (uppermost curve), $\Delta J = 2.4 \times 10^{-26}$, and $\Delta J = 0.85 \times 10^{-26}$ g cm$^2$ sec$^{-1}$. The moment of inertia corresponds to $I_\perp$ of CH$_3$CN (acetonitrile)—which we consider here a "linear molecule" and accordingly set, in Eq. 2.179, $J = I_\perp\omega$. [Note that the modified $M$-model of Eq. 2.177 is still to be averaged over all frequencies (Eq. 2.43) with the distribution function Eq. 2.2.]

We see from Fig. 2.33 that $C(t)$ shows a damped oscillatory time evolution for certain $\Delta J$. A second maximum in the correlation value is indicated (which

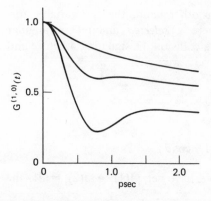

**Figure 2.33.** Simulated orientational autocorrelation function of the symmetry axis of $CH_3CN$ on the basis of the modified $M$-diffusion model. The common average time between collisions is $0.10 \times 10^{-12}$ sec, the amount of angular momentum transferred per collision is (top to bottom) $9.1 \times 10^{-26}$, $2.4 \times 10^{-26}$, and $0.85 \times 10^{-26}$ g cm$^2$ sec$^{-1}$, at 314°K. [D. Frenkel, G. H. Wegdam, and J. van der Elsken, *J. Chem. Phys.* **57**, 2691 (1972).]

is absent in the basic $M$- or $J$-diffusion model). Clearly, the smaller $\Delta J$ the closer $C(t)$ approaches the free rotor correlation function.

The reversed situation, namely, simulation of a set of $C(t)$ for fixed $\Delta J$ but different $\tau_\omega$, is shown in Fig. 2.34.[145a] We notice that the slowest decay of the orientational coherence belongs to the shortest $\tau_\omega$. The memory of the original orientation is retained longer if the collision frequency $1/\tau_\omega$ increases since this interrupts more frequently the fastest possible orientational relaxation process — free ensemble-averaged rotation. (Dephasing, see Section 1.)

It is interesting to follow the memory function $K(t)$ associated with an oscillatory correlation function $C(t)$—such as that represented by the lowest curve in Fig. 2.33. We surmise that $K(t)$ displays itself as an oscillatory time evolution. Obviously, the initial decay of $C(t)$ is paralleled by a rapid monotonic drop of $K(t)$. Nevertheless, whenever $C(t)$ curves upward or even reaches

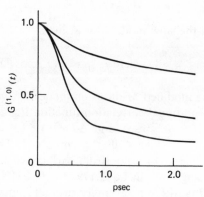

**Figure 2.34.** Simulated orientational autocorrelation function of the symmetry axis of $CH_3CN$ on the basis of the modified $M$-diffusion model. The common amount of angular momentum transferred per collision is $6.2 \times 10^{-26}$ g cm$^2$ sec$^{-1}$. The average time between collisions is (top to bottom) $0.10 \times 10^{-12}$, $0.28 \times 10^{-12}$, and $0.56 \times 10^{-12}$ sec. [D. Frenkel, G. H. Wegdam, and J. van der Elsken, *J. Chem. Phys.* **57**, 2691 (1972).]

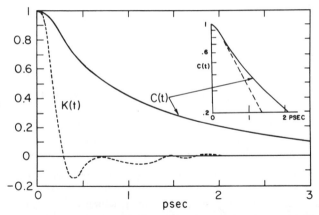

**Figure 2.35.** Experimental velocity correlation function $C(t)$ of Ar atoms in liquid Ar and its associated memory function $K(t)$ at 85°K. Inset: Correlation function replotted on a logarithmic scale, with the dashed straight line giving a long-time exponential decay of $C(t)$. [R. D. Mountain, *J. Res. Natl. Bur. Stand.* **78A**, 413 (1974).]

a second maximum, we expect a corresponding rise of $K(t)$ since the original memory of the orientational correlation is (partly) regained. Figure 2.35 shows these relations on the example of the experimental velocity correlation function of Ar atoms in liquid Ar at 85.2°K (measured by neutron diffraction); the associated memory function was numerically obtained with Eq. 2.32.[145b] We see that although $C(t)$ always stays positive, $K(t)$ becomes negative after an initial, rapid drop. This negative region of $K(t)$ indicates that $(d/dt)C(t)$ has reversed sign (Eq. 2.32); the drop of $C(t)$ is then slowed down with respect to a purely exponential (memoryless) decay.

The inset to Fig. 2.35, where we have replotted $C(t)$ on a semilogarithmic scale, shows this clearly. If the long-time decay of correlation function $C(t)$ were to follow that indicated by the dashed straight line, its memory $K(t)$ would be a pure exponential, having dropped to small values at $t \gtrsim 0.3 \times 10^{-12}$ sec.

We recall that this particular appearance of $C(t)$ and its associated $K(t)$—namely, an initial parabolic decay followed by a pure exponential time evolution for $C(t)$, and a pure exponential decay for $K(t)$—corresponds to the behavior of the $J$-extended rotational diffusion model with fast modulation. (See Figs. 2.1b and 2.6a and b.) In fact, in Chapter 3 we discuss vibrational relaxation processes that show this behavior.

### 5.8. Multiple-Variable Theory: Torques of Finite Duration and Nonrandom Collision Processes

As we mentioned in the previous section, we can extend the solution of a modified Langevin equation using three, four, ... collective variables, obtaining thereby an ever finer time graining and, consequently, an increasingly better understanding of the short-time orientational motion.

We recall that the two-variable theory attains the interpretive power of the *J*- and *M*-extended rotational diffusion models. The three-variable theory now proposes nonimpulsive torques.[142b] In terms of a physical picture, this introduces "times *of* a collision" (noninstantaneous collisions).

We realize that we may thereby exceed the concept of a Poisson distribution of times *between* collisions. For instance, a torque may last a relatively long time interval and thus trap the rotor in a librational potential well. Such, essentially solidlike, molecular environment is expected to "pulsate" with a certain coherence ("phonon mode," see end of Section 3.3). The distribution of times between collisional perturbations that the medium exerts on a rotor is therefore not random but is peaked at a time that corresponds to an inverse phonon eigenfrequency of the lattice.[145a] (Refer to Note 13.)

In a three-variable theory, the times during which the torques are switched on and off (the rate of change of intermolecular torques) are now determined by delta-pulse memory functions. To introduce gradually working torques, the four-variable approximation is required.

The theory containing one-, two-, and higher-tuple variable approximations is related to the modified Langevin equation[2c, 59]

$$\frac{d}{dt}K_n(t) = -\int_0^t dt'\, K_{n+1}(t')K_n(t-t'), \tag{2.180}$$

which formally expresses the hierarchy of correlation and associated memory functions already mentioned in Section 5.7. Here,

$$K_n(t) = \langle f_n | \exp(iL_n t) | f_n \rangle$$

$$= \langle f_n(0) f_n(t) \rangle, \tag{2.181}$$

with

$$| f_n \rangle = (1 - \bar{P}_{n-1}) i L_{n-1} | f_{n-1} \rangle \langle f_{n-1} | f_{n-1} \rangle^{-1/2},$$

$$L_n = (1 - \bar{P}_{n-1})(1 - \bar{P}_{n-2}) \cdots (1 - \bar{P}_0) iL$$

$$= \left(1 - \sum_{j=0}^{n-1} \bar{P}_j\right) iL, \tag{2.182}$$

and $L_0 \equiv L$, $f_0 \equiv A$, $\bar{P}_0 \equiv \bar{P}$. $L$ is the Liouvillian operator (Eq. 2.139).

In the following discussion we use Mori's definition of the (normalized) projection operator $\bar{P}$,[59]

$$\bar{P}_A X = (X, A^*)(A, A^*)^{-1} A, \tag{2.183}$$

which projects a general stochastic variable $X(t)$ into nonoverlapping subspaces of a set of vectors $A, f_1, \ldots, f_n$. This set is therefore orthogonal. The time evolution of each $f$ is different from any other. Furthermore, the finer the time graining—the finer the description of $X(t)$—the larger $n$ of the subspace of

$$A, f_1, \ldots, f_{n-1}$$

into which $X(t)$ has been projected. The lowest term in Eq. 2.180 ($n = 0$) is

$$\frac{d}{dt} K_0(t) = - \int_0^t dt' \left( f_1, f_1(t') \right) K_0(t - t'),$$

which equals Eq. 2.142 for $a = b = c$, the "monovariate case" (see Note 14). Hence $K_0(t)$ is to be identified with the correlation function $C(t)$,

$$C(t) = \langle A(0)A(t) \rangle = \langle A(0) | \exp(iLt) | A(0) \rangle,$$

where the second equality may be verified from Eq. 2.113 and Appendix F.

Taking the Laplace transform of Eq. 2.180 gives[60]

$$\tilde{K}_n(s) = K_n(0) \left[ s + \tilde{K}_{n+1}(s) \right]^{-1}.$$

This allows an iteration, the "Mori's continued fraction representation" mentioned before. We get

$$\tilde{C}(s) = \cfrac{C(0)}{s + \cfrac{K_1(0)}{s + \cfrac{K_2(0)}{s + \cdots}}} \qquad \cdots \frac{K_{n-1}(0)}{s + \tilde{K}_n(s)}. \qquad (2.184)$$

The first step in the Mori expansion gives

$$\tilde{C}(s) = \left[ s + \tilde{K}_1(s) \right]^{-1} C(0), \qquad (2.185)$$

or, formally[143e],

$$C(t) = C(0) \exp \left( - \frac{t}{\tau_{\omega'}} \right),$$

where $\tau_{\omega'}$ is the inverse of $\tilde{K}_1(s)$—in other words, $\tau_{\omega'}$ is a *frequency-dependent* correlation time ($s = -i\omega$). But if we demand that the memory function of $C(t)$ decays instantaneously, $K_1(t) = \tau_{\omega'}^{-1} \delta(t)$, we can consider $\tau_{\omega'}$ frequency independent and solve Eq. 2.185 without difficulty (See also Eq. 2.173a). The price we pay is the failure of $C(t)$ for $t < \tau_{\omega'}$—as discussed in Sections 5.6 and 5.7.

The second step in Eq. 2.184 yields

$$\tilde{C}(s) = \left[s + \tilde{K}_2(s)\right]\left[s^2 + s\tilde{K}_2(s) + K_1(0)\right]^{-1}. \qquad (2.186)$$

A priori, only $K_1(0)$ is obtainable from the experiment. To see how, we solve Eq. 2.181 setting $t = 0$. Computing first $f_1(0) \equiv f_1$, we get, with the help of Eqs. 2.139 and 2.183 (see Note 4, Chapter 1),

$$f_1 = (1 - \bar{P})iLA(A, A)^{-1/2}$$

$$= \left\{\dot{A} - \left[(\dot{A}, A)(A, A)^{-1}\right]A\right\}(A, A)^{-1/2}$$

$$= \dot{A}(A, A)^{-1/2}$$

since $(\dot{A}, A) = 0$ (Eq. 2.153). Hence

$$(f_1, f_1) = (\dot{A}, \dot{A})(A, A)^{-1}, \qquad (2.187)$$

which equals $\omega_j^2$ (Eq. 2.163$b$). Again, if we make the assumption that $\tilde{K}_2(s)$ is also frequency independent,

$$\tilde{K}_2(s) \to K_2(0),$$

we can readily solve Eq. 2.186. (We have already shown this; see Eq. 2.170.)
To obtain $K_2(0)$, we proceed as above

$$f_2 = (1 - \Sigma\bar{P})iLf_1(f_1, f_1)^{-1/2} = (1 - \bar{P}_1 - \bar{P}_0)iLf_1(f_1, f_1)^{-1/2}$$

$$= iLf_1(f_1, f_1)^{-1/2} - (iLf_1, f_1)(f_1, f_1)^{-3/2}f_1$$

$$- (iLf_1, A)(A, A)^{-1}(f_1, f_1)^{-1/2}A,$$

using $\bar{P}_1 iLf_1 = (iLf_1, f_1)(f_1, f_1)^{-1}f_1$, $\bar{P}_0 iLf_1 = (iLf_1, f_0)(f_0, f_0)^{-1}f_0$ (see Eq. 2.183) and noting that $f_0 \equiv A$. Here $(iLf_1, f_1) = (\dot{f}_1, f_1) = 0$. Substituting $f_1$ by $\dot{A}(A, A)^{-1/2}$ gives, recalling again Eq. 2.139,

$$f_2(f_1, f_1)^{1/2} = \left\{\ddot{A} - (\ddot{A}, A)(A, A)^{-1}A\right\}(A, A)^{-1/2}.$$

Noting Eq. 2.163$a$, we obtain after a little algebra

$$(f_2, f_2) = K_2(0) = (\ddot{A}, \ddot{A})(\dot{A}, \dot{A})^{-1} - (\dot{A}, \dot{A}). \qquad (2.188)$$

Hence the particular two-variable approximation (Eq. 2.159) corresponds to the second step of the Mori expansion, with

$$K_2(0) = \omega_T^2$$

(refer to Eq. 2.167$b$) under the proviso that the torques decay instantaneously. Again, we pay for the ready solution of Eq. 2.184 with our ignorance about the behavior of $C(t)$ for $t < \tau_\omega$. And so on, to higher and higher approximations.

We recall that $K_1(0) = \omega_j^2$ is equal to the rotational second spectral moment, $M^{(j,0)}(2)$ (Section 4 and Appendix E). With respect to $K_2(0)$, multiplying Eq. 2.188 by $(\dot{A}, \dot{A})$ yields

$$(\ddot{A}, \ddot{A}) - (\dot{A}, \dot{A})^2 = (\ddot{A}, \ddot{A})^c, \tag{2.189}$$

which has the general form of a second-order cumulant—in particular the mean-square fluctuation of $\dot{\omega}$ about $\langle \omega \rangle$ (the dispersion, Eq. H.3, Appendix H). On the other hand, from the first-rank tensor equivalent of Eq. 2.165,

$$\ddot{A} = \omega^2 + A\dot{\omega},$$

we obtain (set $\langle AA \rangle = 1$ and $\langle \dot{\omega} \rangle = 0$)

$$\langle \ddot{A} | \ddot{A} \rangle = \langle \omega^4 \rangle + \langle \dot{\omega}^2 \rangle,$$

which is equal to the infrared rotational fourth spectral moment (see end of Section 4). In fact,[2c] a (even) $2n$th-order moment of $K(t)$ [$K(t)$ is real and even] is given by a cumulant expansion containing the $(2n+2)$th and all lower-order even moments of the associated correlation function of $K(t)$. In other words, the finer we take the time graining of a model of rotational relaxation, the higher is the order of the rotational spectral moments we have tacitly included. In terms of intermolecular potentials, this means that ever higher-order distribution functions come into play;[93] spectral moments above $M(4)$ already depend on higher than pair distribution functions. Clearly this shows the difficulties of accurately modeling, let alone experimentally verifying, orientational correlation functions; they diverge at a certain order of the series development. It also sheds light on the unavoidable failure of "lower-order" models, such as the Debye diffusion, the $J$-, $M$-extended rotational diffusion, the Ivanov orientational jump, and extended Brownian motion (see below) theories. Therefore, we cannot truly say that such and such model of rotational relaxation is "wrong" as long as we restrict our considerations to the inherent time grain of that model.

An accurate measurement of $M(4)$ is rather difficult; a determination of $M(6)$ is near hopeless. It is therefore of little use to show the expression for $K_3(0)$. (However, this does not mean that we are at the end of our rope—as we show later.)

## B.   Nonrandom Collision Processes

### 6.   GENERAL PRINCIPLES: PURE ROTATIONAL SPECTRA AND GENERAL ROTATIONAL MODELING

In the second part of this chapter we first demonstrate truly serious insufficiencies of the models of orientational motion discussed previously. Thereafter we develop more sophisticated theories of rotational relaxation. We emphasize that the newer models do not render previous theoretical concepts null and void but serve as extensions, modifications, and generalizations related to more refined experimental data, better data acquisition and evaluation techniques, and greater expectations.

It appears that absorption and scattering profiles in the far-infrared spectral range offer the best vehicles for a more thorough exploration of condensed-phase orientational motion by fluctuation spectroscopy. The reasons are as follows:

1.   Internal degrees of freedom give constant contributions. Therefore vibration-rotation interaction effects as well as perturbations from overlapping band profiles are absent.

2.   The orientational spectral density is directly obtainable from the absorption or scattering measurements. No numerical deconvolution from an experimental vibration-rotation profile is necessary.

3.   The transition moment belongs to a first-order (far-infrared) or a totally symmetric second-order (Rayleigh) tensor. This dispenses with the conceptual complications of nontotally symmetric second-order tensor correlation functions (Section 3, Chapter 1). On the other hand, we are no longer able to choose several preferred molecular axes directions.

On the negative side, some unavoidable difficulties in the evaluation of orientational dynamics from pure rotation spectra are encountered. The principal difficulties are as follows:

1.   Presence of cross correlation. Although cross correlation can be minimized by employing dilute solutions (Section 2.2, Chapter 1), we would then no longer deal with the original system of the neat liquid. In fact, at high dilution of the active solute molecule, its orientational dynamics predominantly reflect the motional characteristics of the solvent molecules.[146] (Refer also to Section 5.4.)

2.   Presence of collision-induced absorption or scattering from a multiple-molecule transition moment.[12] As we discussed in Section 4.2.1, this additional intensity appears at higher frequencies of a band profile and therefore modifies the short-time dynamics.

**3.** Effects from local electrical fields. This phenomenon, proper to absorption spectra, is difficult to treat.[10c, 147]

It is now necessary to make a choice among theories that are used to explain more sophisticated data; it is impossible to give a full account of all significantly different developments that are applied to obtain a deeper understanding of molecular orientational motion. We note that we had discussed representative examples of two major classes of theories, namely theories based on stochastic modeling of the orientational motion and theories based on solving a generalized Langevin equation. Among the stochastic models we recall (1) the *J*- and *M*-extended rotational diffusion theories by Gordon and by McClung as examples for the collision-interrupted random walk of the angular momentum (Sections 2) and (2) Ivanov's jump model as an example for the collision-interrupted random walk of a molecular axis (Section 5.2). On the other hand, we recall that the Langevin equation, an integro-differential equation connecting correlation and memory function, yielded a useful formulation of a general multiple-variable theory and established the dynamic and orientational pair correlation factors (Sections 5.6–5.8).

We extend these two main avenues of theoretical modeling in the subsequent parts of this chapter but concentrate on stochastic modeling. Although the Langevin memory-function method has recently become popular, we feel that the division of orientational motion into sequences of rotation and perturbation steps permits great flexibility, is elegant and satisfying, and drives for the physical picture above the mathematics. We are at liberty in proposing the nature of these events, we can readily account for all conceivable situations found in condensed environments. Furthermore, we can easily deal with higher-order transition moment tensors. Thus we retain the convenience but enlarge the possibility of mathematically describing orientational motion by detailed microscopic processes.

The Langevin memory-function approach remains for us a method for pointing at the applicability and significance of orientational models; the computational results during different stages of development, the quality of the choice of the memory function, and so on, can be compared with other spectral data.

Among other important theories we mention, briefly, the Brownian motion models[148–150] (solving the rotational analogue of Eq. 2.16). To facilitate the solution, the torque **To** acting on a rotor is split into a sum of an angular retarding force $-(k_B T/\mathbf{D})\omega(t)$, where **D** is the rotational diffusion constant tensor of the molecule, and a rapidly fluctuating component $\mathbf{To}(t)$ arising from the molecular motion of the bath,

$$\mathbf{To} = -\frac{k_B T}{\mathbf{D}}\omega(t) + \mathbf{To}(t); \qquad \langle \mathbf{To}(t) \rangle = 0.$$

Probability distributions of orientational motion of molecular axes are devised in terms of convenient angles between molecule and space-fixed coordinate systems (see, for instance, Eqs. 1.91–1.94, Chapter 1).

The usefulness of applying models of this type to the evaluation of experimental orientational correlation functions $G^{(j, k)}(t)$ seems to have been largely overlooked. The approach is particularly convenient for symmetric top molecules possessing considerable frictional anisotropy[149] and for spherical tops in a strongly anisotropic environment.[148] The models approximate the corresponding Debye anisotropic diffusion results for $D \ll k_B T/I$ and the initial free rotor correlation function for $D \gg k_B T/I$.[151] Furthermore, the models are applicable to parallel- and perpendicular-polarized vibration-rotation bands of interesting molecular systems such as liquid-phase $CH_3I$ and $CH_3CN$. Their rotational diffusion constants $D_\perp$, $D_\parallel$ are known from dielectric and nuclear magnetic relaxation data.[144] We have therefore independent input parameters. Finally, corrections for significant Coriolis coupling (perpendicular bands) are introduced into the theory without mathematical difficulty.

With the increasing improvement in computer technology, Monte Carlo and molecular dynamics calculations are enjoying greater interest.[152–154] As we mentioned previously, their methodology is not within the scope of our discussions.

### 6.1.   Temperature Dependence of the Far-Infrared and Rayleigh Spectra of a Polar Molecule: $CH_3CN$ (Acetonitrile)

Figure 2.36a displays the temperature dependence of the far-infrared spectrum of $CH_3CN$ diluted by n-heptane ($C_7H_{16}$). Figure 2.36b gives the corresponding correlation functions $G^{(1,0)}(t)$.[155] Since the solution is dilute, orientational cross correlation is insignificant (Eq. 1.41, Chapter 1).

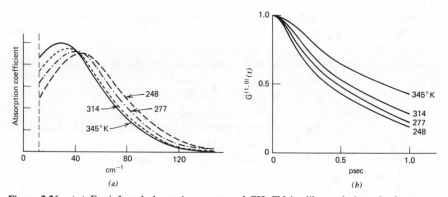

**Figure 2.36.** (a) Far-infrared absorption spectra of $CH_3CN$ in dilute solution of n-heptane at different temperatures. (b) Temperature dependence of the experimental orientational correlation function $G^{(1,0)}(t)$ obtained from the spectra in Fig. 2.36a. [R. M. van Aalst, J. van der Elsken, D. Frenkel, and G. H. Wegdam, *Faraday Disc. Chem. Soc.*, No. 6, 1972, p. 94.]

Figure 2.36$a$ shows the surprising result that the higher the temperature $T$ the lower the root-mean-square angular frequency $\langle \omega^2 \rangle^{1/2}$ of the $CH_3CN$ rotors. Consequently, the higher $T$ the slower the decay of $G^{(1,0)}(t)$ (Fig. 2.36$b$) —an unexpected outcome.

Figure 2.37 gives the temperature dependence of the correlation function $G^{(2,0)}(t)$ of the Rayleigh spectrum of neat $CH_3CN$, corrected for high-frequency collision-induced scattering (see Section 4.2.1).[156] We see that here the temperature dependence of $G(t)$ is according to expectations: The higher the average kinetic energy the faster the decay of the orientational correlation.

It is true that the unknown temperature dependence of the appreciable pair correlation of neat $CH_3CN$ (Table VI, Section 5.6) introduces an ambiguity into the result. However, a third set of solution data[155] also goes contrary to intuitive ideas: $G^{(1,0)}(t)$ of neat $CH_3CN$ relaxes faster than $G^{(1,0)}(t)$ of $n$-heptane-diluted $CH_3CN$ (Fig. 2.38). Why should the highly polar molecule reorient more slowly in a nonpolar environment than in its own structured[141] medium?

A way out of these difficulties is to introduce a non-Poisson distribution of times between collisions.[145a, 155] We no longer maintain that succeeding collisions are independent, and we admit that some memory is retained within a

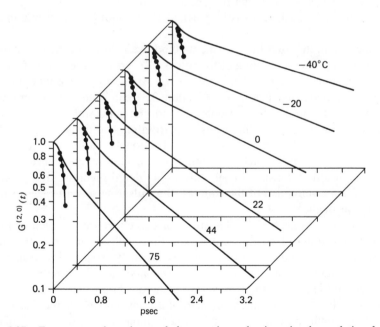

**Figure 2.37.** Temperature dependence of the experimental orientational correlation function $G^{(2,0)}(t)$ of neat $CH_3CN$ from its Rayleigh spectrum, corrected for collision-induced scattering. The dot-curves show the initial decay of the theoretical free rotor correlation functions. [H. Langer and H. Versmold, *Ber. Bunsenges. Phys. Chem.* **83**, 510 (1979).]

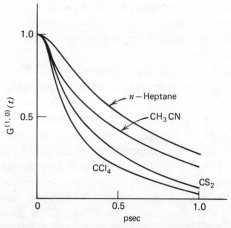

**Figure 2.38.** Solvent dependence of the experimental orientational correlation function $G^{(1,0)}(t)$ of $CH_3CN$ at room temperature. [R. M. van Aalst, J. van der Elsken, D. Frenkel, and G. H. Wegdam, *Faraday Disc. Chem. Soc.*, No. 6, 1972, p. 94.]

sequence of perturbing events. We no longer separate the "allowed" from the "induced" absorption or scattering. Because of the close molecular separations in liquids this approach is realistic, particularly in liquids with strong interactions.

To appreciate the usefulness of the method, we reconsider the orientational model which allows rebounding of the rotors. (Reversal of the sense of rotation upon a collision; see Eq. 2.177, $\cos\theta = -1$.) Assuming that phononlike motions in the condensed medium[141] (Section 3.3) can couple to orientational coordinates of the molecule, a resonance condition is possible between Fourier frequency components of the translatory-rotatory motions of the solvent molecules and of the orientational (or librational) motions of the solute molecule. Consequently, each collision keeps a certain coherence with the previous collision(s). Since we deal with a disordered medium, this coherence dies out. We are reminded of the concept of a pulsating solvent cage (or liquid cage, in neat media).

It is easy to see that the manifestations of this effect are analogous to those from collision-induced intensities. Recall that collision-induced transitions arise by interactions between two (or more) narrowly separated molecules; the transition frequency is modulated by their *common* molecular motion.

It is not necessary to think in terms of actual bouncing of one molecule against another.[100] For instance, the field of a nonspherical charge distribution (permanent quadrupole moment) of a linear molecule polarizes an adjacent solvent molecule.[157] This interaction generates a dipole moment (Eq. 1.2, Chapter 1) *within* the solute–solvent complex. The quadrupole moment is a second-order tensor; the induced effect therefore oscillates at twice the angular fluctuation frequency of the reorienting linear species.[99]

On the other hand, we generate a non-Poisson time distribution of perturbation events by superposing a solidlike phonon mode on the random fluctuations of the medium. For instance, in crystalline NaCl so-called critical points correspond to $1.8 \times 10^{13}$ (transverse-acoustic), $2.9 \times 10^{13}$ (transverse-optic), $3.0 \times 10^{13}$ (longitudinal-acoustic), and $4 \times 10^{13}$ radian sec$^{-1}$ (longitudinal-optic fundamental branch).[158] Clearly, in the liquid with its short-range order, all that remains of the crystal-phase phonon branches is one (or a few) broadened maximum in the phonon number distribution $N(\omega)$. In our example the additional superposed peak occurs at $3 \times 10^{13}$ radian sec$^{-1}$ approximately 160 cm$^{-1}$, approximately $0.2 \times 10^{-12}$ sec in the Poisson distribution (Note 13).

The success of this approach is displayed in Fig. 2.39 on the previous example of CH$_3$CN in $n$-heptane. The dashed curve spectrum is the experimental spectrum at 314°K (taken from Fig. 2.36$a$), the solid curve spectrum is computed using the non-Poisson distribution of times between collisions shown in the inset, the proper moment of inertia $I_\perp$ of the molecule, and the amount of $\Delta J = 7.5 \times 10^{-26}$ g cm$^2$ sec$^{-1}$ transferred per collision. (The stochastic model to accomplish this is discussed later.) The narrower contour is simulated in the same manner, except that the Poisson distribution is used (Eqs. 2.177, 2.179). It is evident that it is the nonzero-time peak in the collision distribution that blue-shifts the simulated spectral peak absorption frequency (from 30 to 40 cm$^{-1}$) to make it coincident with the experiment; a Poisson distribution is not able to do this. (Changing $\tau_\omega$ does not help; we know that it only affects the bandwidth.)

A general non-Poisson distribution can be represented by the relation[159]

$$f_\Gamma(t) = \frac{(\alpha/\tau)(\alpha t/\tau)^{\alpha-1}\exp(-\alpha t/\tau)}{\Gamma(\alpha)}, \qquad 0 \le \alpha \le \infty; t \ge 0,$$

(2.190)

the so-called gamma distribution, where $\tau$ is the mean time between collisions

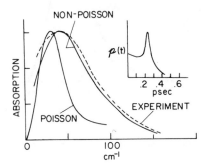

Figure 2.39. Experimental (from Fig. 2.36$a$) and simulated far-infrared absorption spectra of CH$_3$CN in $n$-heptane solution at 314°K. The distribution of times between collisions for the non-Poisson simulation is shown in the inset. [From D. Frenkel, G. H. Wegdam, and J. van der Elsken, *J. Chem. Phys.* **57**, 2691 (1972).]

and $\Gamma(\alpha)$ the gamma function of the parameter $\alpha$. Note that the maximum of $\rlap{/}{\rho}_\Gamma(t)$ lies at

$$t_{max} = \frac{\tau(\alpha-1)}{\alpha}, \tag{2.191}$$

and that the limiting values of $\rlap{/}{\rho}_\Gamma(t)$ are the Poisson distribution for $\alpha=1$ (Eq. 2.10) and the delta function $\delta(\tau)$ for $\alpha \to \infty$.

Equation 2.190 is indeed convenient since its Laplace transform is simply[143f]

$$\rlap{/}{\rho}_\Gamma(s) = \left[1+\frac{s\tau}{\alpha}\right]^{-\alpha}, \tag{2.192}$$

from which we immediately see that

$$\lim_{\alpha \to \infty} \rlap{/}{\rho}_\Gamma(s) = \exp(-s\tau)$$

or

$$\lim_{\alpha \to \infty} \rlap{/}{\rho}_\Gamma(t) = \delta(\tau).$$

### 6.2.   Pure Induced Absorption in Liquids: $CCl_4$ (Carbon Tetrachloride) and $C_6H_6$ (Benzene)

A liquid of spherical molecules, such as $CCl_4$, or of an axial molecule with a center of inversion, such as $C_6H_6$, has no allowed far-infrared spectrum. Hence the observed absorption is solely due to induced phenomena. Figure 2.40 gives an example of the room-temperature spectra observed down to low frequencies

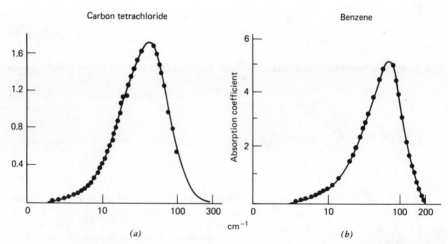

**Figure 2.40.**   Observed (points) and simulated (curves) far-infrared absorption spectra of liquid (a) $CCl_4$ and (b) $C_6H_6$ at 296°K. [G. J. Davis and M. Evans, *J. Chem. Soc. Faraday Trans. II*, **72**, 1194 (1976).]

and of their theoretical reconstruction by the multivariable theory.[160] We now discuss how well experiment and theory agree.

To circumvent the difficulties of accurately measuring higher spectral moments to obtain $K_1(0)$, $K_2(0)$,... (see Section 5.8), we solve Eq. 2.184 by guessing $K(t)$. We then compare the simulated spectral density $\mathrm{Re}\,\tilde{C}(-i\omega)$ with the experimental spectral density $\mathcal{I}(\omega)$ or the absorption coefficient $\sigma(\omega)$ [Eq. 1.3, Chapter 1].

We assume

$$K_2(t) = K_2(0)\exp(-\gamma t), \qquad \gamma \equiv \tau_m^{-1}, \tag{2.193}$$

hence[143e]

$$\tilde{K}_2(s) = K_2(0)(s+\gamma)^{-1}.$$

This choice approximates Eq. 2.184 to

$$\tilde{C}(s) = \left[s^2 + s\gamma + K_2(0)\right]\left\{s^3 + s^2\gamma + s\left[K_2(0) + K_1(0)\right] + \gamma K_1(0)\right\}^{-1} \tag{2.194}$$

and gives the spectral density

$$\mathrm{Re}\,\tilde{C}(-i\omega) = K_1(0)K_2(0)\gamma\left\{\gamma^2\left[K_1(0) - \omega^2\right]^2 + \omega^2\left[\omega^2 - (K_1(0) + K_2(0))\right]^2\right\}^{-1} \tag{2.195}$$

or the absorption coefficient

$$\sigma(\omega) = N\frac{\omega^2}{n(\omega)}\mathrm{Re}\,\tilde{C}(-i\omega), \tag{2.196}$$

where $N$ is a constant and $\hbar\omega \ll k_BT$ (Eq. 1.25, Chapter 1). Note that Eq. 2.194 corresponds to a three-variable approximation. Memory function $K_1(t)$ has now the place equivalent to $C(t)$ of Eqs. 2.171 or 2.177, and $\gamma = K_3(0)$.

The values of $K_1(0)$, $K_2(0)$, and $\gamma$ were obtained by a numerical fit (minimization procedure) to the experimental spectra (Note 15). It is by no means trivial to relate the $K_n(0)$ to the rotational spectral moments; the $K_n(0)$ are theoretically defined only in terms of single-molecule variables (Sections 4 and 5.8). In reduced units of $\gamma^* = \gamma(I_\perp/2k_BT)^{1/2}$, $K_1^* = K_1(I_\perp/2k_BT)$, and $K_2^* = K_2(I_\perp/2k_BT)$, best-fit values are 14.2, 10.9, and 80.6 for $CCl_4$ and 12.8, 20.8, 100.6 for $C_6H_6$ at 296°K. They form the bases for the simulations in Fig. 2.40.

In spite of our ignorance of the theoretical meaning of the $K_n(0)$ for induced phenomena, the $K_n(0)$ afford quantitative parameters for constructing line-shape functions of collision-induced profiles. Such results may serve to correct a profile for collision-induced absorption or scattering. Theoretically, however, we have a long way to go in attempts to explain the data. Angle-dependent collision-induced absorption in spherical $CCl_4$ is caused by the octupolar, in the axial $C_6H_6$ molecule it is caused by the quadrupolar electronic charge distributions.[12] We can only surmise the complicated molecular distribution functions involved here.

In conclusion, we show in Fig. 2.41 the corresponding experimental correlation functions (Eqs. 1.25, Chapter 1) that describe the temporal evolution of the collision-induced coherence in liquid $CCl_4$ and $C_6H_6$, respectively.[160] We see that they reflect the rapidly decaying correlation of multiple-molecule dynamics; their lifetime is indeed significantly shorter than that of reorientational single-molecule dynamics (Section 5.2).

## 7.  GENERAL STOCHASTIC MODEL OF ROTATIONAL RELAXATION

### 7.1.  General Remarks: Difficulty of Extending the Langevin Method

On first sight it appears that the Langevin formalism (Eq. 2.184) is the most extensive and satisfactory approach of constructing model orientational correlation functions. However, scrutiny of the basic formalism as well as its failures

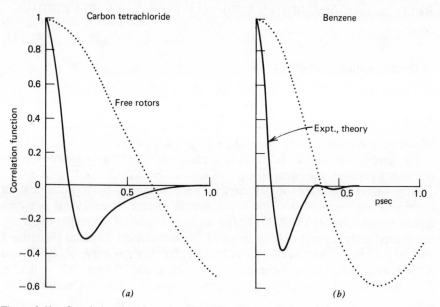

**Figure 2.41.**  Correlation functions describing the time evolution of the coherence of purely collision-induced dynamics of liquid (*a*) $CCl_4$ and (*b*) $C_6H_6$ at 296°K. Solid curves: experiment, from Fourier transformation of the spectra in Fig. 2.40 and from computations (Eqs. 2.195 and 2.196). Experiments and computations coincide. Dot curve: free-rotor decay, Appendix E. [G. J. Davis and M. Evans, *J. Chem. Soc. Faraday Trans. II*, **72**, 1194 (1976).]

at attempts to fit higher-variable approximations to the data show us differently.

Fundamentally, we deal with a mathematical construction (of coupled Volterra equations). It is up to us to give meaning to the members $K_n(t)$. And there lies a possibility of deception since we are tempted to relate them a posteriori to a physical picture. It will be a simplified picture and a simplified function because this is the advantage of using the hierarchy method of associated correlation and memory functions. In "going up the ladder," however, the physical meaning of the resulting $C(t)$ becomes obscured once we are above the two- or three-variable stage. Of course, we may start "higher up" with a sufficiently complicated memory function. But then we do not always grasp its physical context. For instance, inspection of Fig. 2.35 demonstrates that a sophisticated memory function is not necessarily less complicated than its associated correlation function. The contrary may be the case, particularly if $C(t)$ has appreciable oscillatory character. Furthermore, the convolution procedure (Eq. 2.180) erases some of the finer details of the memory function: Hence our labors in deriving the proper $K(t)$ may be too costly (since many ways lead to Rome).

Experimentally, attempts at rigorously (without numerical minimization) fitting far-infrared spectra to the three- or four-variable expansion of the theory (Eq. 2.184) have, at present, met only with limited success[161]; nonpolarizable systems are an exception. Conceivably, a local field exerts dipole-induced dipole coupling[147b] which modifies the rotational spectral fourth moment $M(4)$ and therefore affects $K_2(0)$; the relation between $M(4)$ and the field is[147a]

$$M'(4) \approx M(4)\left\{1 - \frac{36(\varepsilon - 1)\alpha_\parallel}{5a^3(3\varepsilon + 2)} + \cdots \right\},$$

where $M'(4)$ is the true rotational fourth spectral moment, $\varepsilon$ the (optical) dielectric constant of the (nonpolar) solvent, $a$ is the radius of the solvent cage, and $\alpha_\parallel$ is the axial component of the solute (linear or symmetric top) polarizability. For example, HCl placed in a 4-Å diameter cavity of $CCl_4$ gives $M'(4) \sim 0.6M(4)$—a significant decrease.[147a]

We conclude that we might as well start with a general stochastic model of rotational relaxation, a model that describes orientational motion in terms of alternating rotation and collision events of *arbitrary* nature. The actual events are expressed in simple physical concepts; the generality and usefulness of the model are given by the choice and complexity of the event distribution functions.

## 7.2. Theory

The general stochastic model of rotational relaxation in the following discussion is by Lindenberg and Cukier.[159] It comprises novel features and an approach flexible enough to include all previously discussed models. It permits construction of detailed and complicated situations with unexpected predictions and results.

The orientational autocorrelation function is written in Wigner rotation matrices $D_{lk}^{(j)}(\Omega)$ of Eulerian angles $\Omega = \theta, \varphi, \phi$ which determine the orientation of the molecule relative to the laboratory-fixed coordinate system (see also Eq. 2.58a),

$$G(t) = (2j+1)\left\langle D_{lk}^{(j)*}(\Omega(t))D_{lk}^{(j)}(\Omega(0))\right\rangle$$

$$= (2j+1)\int d\Omega \int d\Omega_0 \, P_{eq}(\Omega_0)P(\Omega; t|\Omega_0; t_0)D_{lk}^{(j)*}(\Omega)D_{lk}^{(j)}(\Omega_0).$$

$$(2.197)$$

$P_{eq}(\Omega_0)\,d\Omega_0$ is the equilibrium probability of finding a molecule with orientation between $\Omega_0 \equiv \Omega(t_0)$ and $\Omega_0 + d\Omega_0$. In an isotropic medium (see Section 2.2), the only case we consider here, $P_{eq}(\Omega_0) = (8\pi^2)^{-1}$. $P(\Omega; t|\Omega_0; t_0)\,d\Omega$ is the conditional probability that a molecule has orientation between $\Omega$ and $\Omega + d\Omega$ at time $t$ when it had orientation $\Omega_0$ at $t = t_0$.

The conditional probability $P$ is constructed in terms of an alternating sequence of elementary events $A$ and $B$. They are defined as follows:

1. During event $A$ the angular momentum $\mathbf{J}$ shall remain constant. The transition probability that $\Omega' \rightarrow \Omega$, $\mathbf{J}' \rightarrow \mathbf{J}$, when $t' \rightarrow t$ is then

$$A_1(\Omega; \mathbf{J}; t|\Omega'; \mathbf{J}'; t') \equiv A(\Omega - \Omega'|\mathbf{J}'; t - t')\,\delta(\mathbf{J} - \mathbf{J}'), \qquad (2.198)$$

where the delta function gives the condition of constant angular momentum.

The duration of event $A$ is characterized by an arbitrary (to be chosen) probability distribution

$$p_A(t) = \lim_{\Delta t \rightarrow 0+} \frac{\text{prob}(t < t' \leq t + \Delta t)}{\Delta t}, \qquad (2.199)$$

normalized

$$\int_0^\infty dt\, p_A(t) = 1,$$

and of average duration $\tau_A$,

$$\tau_A = \int_0^\infty dt\, t p_A(t). \qquad (2.200)$$

The conditional probability for event $A$, designated by $P_A$, is then conveniently written

$$P_A(\Omega; t|\Omega'; t') = A(\Omega - \Omega'|\mathbf{J}'; t - t')p_A(t - t'). \qquad (2.201)$$

2.   During event $B$ the angular momentum $\mathbf{J}$ shall be random at any $t$. The transition probability is therefore

$$B_1(\Omega; \mathbf{J}; t | \Omega'; \mathbf{J}'; t') \equiv B(\Omega - \Omega' | t - t') p_{eq}(\mathbf{J}), \qquad (2.202)$$

where $p_{eq}(\mathbf{J})$ is a Maxwell distribution (see, for instance, Eqs. 2.2, 2.38–2.40). The conditional probability for event $B$ is thus

$$P_B = B(\Omega - \Omega' | t - t') p_B(t - t'), \qquad (2.203)$$

where $p_B(t)$ is the corresponding time distribution of average duration $\tau_B = \int dt\, t p_B(t)$.

We assume that the events are independent of each other in the sense that the next event depends only on the present event but has no memory of previous events (Markov process, Eq. 2.13).[162]

It is easy to realize that this alternating sequence of $P_A$ and $P_B$ encompasses previously discussed stochastic rotational relaxation models. For instance, consider first Debye's rotational diffusion model. It demands the following:

1.   Zero-angle free rotation steps. In other words, there is no motion during the randomly occurring $A$-events. Hence we set

$$A(\Omega | \mathbf{J}; t) = \delta(\Omega), \qquad (2.204)$$

and take for its (random) time distribution the Poisson function (Eq. 2.10)

$$p_A(t) = \tau_A^{-1} \exp\left(-\frac{t}{\tau_A}\right). \qquad (2.205)$$

2.   Instantaneously occurring diffusion steps through infinitesimal angles. Hence we characterize the $B$-events by the transition probability

$$B(\Omega | t) = D \nabla^2(\Omega),$$

where $D$ is the diffusion coefficient and $\nabla^2$ the spherical Laplacian operator (Eq. 2.46a), and by the time distribution

$$p_B(t) = \delta(t). \qquad (2.206)$$

Next, we outline that the theory contains the $J$-extended diffusion model (Section 2) as follows:

1.   Probability distribution $A$ is a free rotation step (Eq. 2.37).
2.   Distribution $p_A(t)$ is Poisson (Eq. 2.205).
3.   There are no reorientation steps during $B$-events (see Eq. 2.204)

$$B(\Omega | t) = \delta(\Omega). \qquad (2.207)$$

4.  The time distribution of $B$-events defines instantaneous collisional events (Eq. 2.206).

Finally, let us look at the orientational jump model discussed in Sections 3.3 and 5.2. Events $A$ are identical to those of the Debye model. The distributions $P_A$ and $p_A$ are therefore given by Eqs. 2.204 and 2.205, respectively. On the other hand, $B(\Omega|t)$ is arbitrary (the size of the orientational jump); its time distribution probability $p_B(t)$ is again instantaneous.

To apply the generalized stochastic model to a wide variety of rotational relaxation phenomena, we have to describe the probabilities $A$, $B$, $p_A$, and $p_B$ in greater detail; we consider that an event is in progress at time $t$ (and does not begin or terminate at $t$). We therefore write "intermediate time probabilities" for events $A$ and $B$, respectively. They are designated by

$$F_A(\Omega'; t'|\mathbf{J}; \Omega; t),$$
$$F_B(\Omega'; t'|\Omega; t), \qquad (2.208)$$

and defined as the conditional probabilities that it takes time $t' < t$ to reorient the molecule by $\Omega'$ ("incomplete event") if the molecule reorients by $\Omega$ during $t$ ("complete event"). A priori, $F_A$ and $F_B$ are complicated functions; they can be simplified under the conditions that determine the transition probabilities $A$ and $B$ (Eq. 2.198) for the complete events. Since we assumed that the dynamic variable ($\mathbf{J}$) does not change during event $A$ (or $B$) and is randomly distributed during event $B$ (or $A$), we modify Eq. 2.198 to read

$$A(\Omega - \Omega'; \omega_A; \omega_B | \omega'_A; \omega'_B; t - t')$$
$$= p_{\mathrm{eq}}(\omega_B)\delta(\omega_A - \omega'_A)A(\Omega - \Omega'|\omega'_A; t - t'),$$
$$B(\Omega - \Omega'; \omega_B; \omega_A | \omega'_B; \omega'_A; t - t')$$
$$= p_{\mathrm{eq}}(\omega_A)\delta(\omega_B - \omega'_B)B(\Omega - \Omega'|\omega'_B; t - t'), \qquad (2.209)$$

where a set of dynamic variables $\omega_A(\omega_B)$ has been introduced for the sake of generality ($\omega_A = J$ in Eq. 2.198, for instance). We note that the complicated dynamics of the dynamic variables can be ignored—we hardly know them anyway (see Section 5.4); we need only their equilibrium distribution functions. Hence $\omega_A(\omega_B)$ is not time-dependent and we can set

$$F_A(\Omega''; t''|\omega_A; \Omega; t) = A(\Omega''|\omega_A; t''),$$
$$F_B(\Omega''; t''|\omega_B; \Omega; t) = B(\Omega''|\omega_B; t''). \qquad (2.210)$$

Therefore the transition probabilities for the molecule to reorient by $\Omega''$ during $t''$ when it has reoriented by $\Omega$ during $t$ are equal (as long as the set of dynamical variables $\omega_A$, $\omega_B$ obeys the foregoing conditions).

We now average over all $\tau_0$ of the start of the *first* event ($A$ or $B$) since $\tau_0$ need not coincide with the initial time $t_0$ in the correlation function ( = time of observation). Similarly, we consider that the *last* event may stretch beyond the longest time of observation $t$. Let us then assume that the starting time of the first event (for instance, $A$), $\tau_0$, is likely to be any time before $t_0$. Therefore the probability for the occurrence of the first event is equal to the difference between the probability that the event $A$ definitely occurs once during $t$, $1/\tau_A$, and the probability that it will occur any time after $t = t_0$,

$$p_{1A}(t - t_0) = \tau_A^{-1}\left\{ 1 - \int_{t_0}^{t} dt \, p_A(t)\right\}. \tag{2.211}$$

Replacing the unit term by the normalization condition (Eq. 2.199) we obtain

$$p_{1A}(t - t_0) = \tau_A^{-1}\int_{t - t_0}^{\infty} dt \, p_A(t), \tag{2.212}$$

where $\tau_A^{-1}\int_{t}^{\infty} dt \, p_A(t)$ is the so-called survivor function of $p_A(t)$, with

$$p_{1A}(0) = 1; \qquad p_{1A}(\infty) = 0; \qquad -\frac{d}{dt} p_{1A}(t) = p_A(t). \tag{2.213}$$

Similarly, the time probability distribution of the first $B$-event is

$$p_{1B}(t - t_0) = \tau_B^{-1}\int_{t - t_0}^{\infty} dt \, p_B(t). \tag{2.214}$$

Notice that if $p_A(t)$ is a Poisson distribution (Eq. 2.205), Eq. 2.213 yields

$$p_{1A}(t) = p_A(t). \tag{2.215}$$

We had always set $\tau_0 \equiv t_0$. For most of the orientational models discussed so far this assumption is therefore correct. (We show later the error this assumption introduces for non-Poisson distributions.)

At this point it is useful and informative to mention that this stochastic model of orientational motion is an example of the following alternating renewal process in statistics[159, 163]:

1. A "system component" $A$ is in use ( = event $A$) until it fails after a certain time which is characterized by a time distribution function ( = $p_A(t)$) and its average duration ( = $\tau_A$).
2. After failure, some other component $B$ takes over ( = event $B$) until it fails.
3. This is to be replaced by a system component $A$; and so forth.

It is clear that either system component may have been running for some time before its first observation ("backward recurrence time") and/or may still be running after its last observation ("residual lifetime" or "forward recurrence time"). The distributions of forward and backward recurrence times are equal[163] for an equilibrium process (a process that has started long before its first observation; Eq. 1.4, Chapter 1).

To combine the various $P$ to the overall probability in Eq. 2.197, we write the following possible sequences of alternating (four) and single events (two):

$$ABABAB\ldots AB = A(BA)^{n-1}B \equiv \{AB\}$$

$$ABABAB\ldots ABA = A(BA)^{n-1}BA \equiv \{AA\}$$

$$BABABA\ldots BA = B(AB)^{n-1}A \equiv \{BA\}$$

$$BABABA\ldots BAB = B(AB)^{n-1}AB \equiv \{BB\}$$

$$A \qquad\qquad\qquad \equiv \{A\}$$

$$B \qquad\qquad\qquad \equiv \{B\}. \tag{2.216}$$

A graphic representation of process $\{AB\}$ is given in Fig. 2.42, where $A$-events are denoted by a thin, $B$-events by a thick horizontal.

Let us then take the $\{AB\}$ sequence and express it by the various probabilities established, keeping for the time the expressions for $F_A$, $F_B$ as originally defined (Eq. 2.208); the simplifying relations (Eq. 2.210) are conveniently applied later. We denote time variables between $t = -\infty$ and $t = t_0$ by $t_-$, their associated angles and angular momenta by $\Omega_-$ and $\mathbf{J}_-$. We denote variables between $t$ and $t = \infty$ by $t_+$, $\Omega_+$, and $\mathbf{J}_+$. We obtain $P(\Omega; t|\Omega_0; t_0)$ in the form

$$P_n^{\{AB\}}(\Omega; t|\Omega_0; t_0)$$

$$= \tau_A^{-1}\int_t^\infty dt_+ \int_{t_0}^t dt_n \int_{t_0}^{t_n} dt'_{n-1} \int_{t_0}^{t'_{n-1}} dt_{n-1} \cdots \int_{t_0}^{t_2} dt'_1 \int_{t_0}^{t'_1} dt_1 \int_{-\infty}^{t_0} dt_-$$

$$\times \int d\Omega_+ \int d\Omega_n \int d\Omega'_{n-1} \int d\Omega_{n-1} \cdots \int d\Omega'_1 \int d\Omega_1 \int d\Omega_-$$

$$\times \int d\mathbf{J}_+ \int d\mathbf{J}_n \int d\mathbf{J}'_{n-1} \int d\mathbf{J}_{n-1} \cdots \int d\mathbf{J}'_1 \int d\mathbf{J}_1 \int d\mathbf{J}_- \, p_{\text{eq}}(\mathbf{J}_-)$$

$$\times F_A(\Omega_0 - \Omega_-; t_0 - t_-|\mathbf{J}_-; \Omega_1 - \Omega_-; t_1 - t_-)A(\Omega_1 - \Omega_-|\mathbf{J}_-; t_1 - t_-)$$

$$\times \delta(\mathbf{J}_1 - \mathbf{J}_-)p_A(t_1 - t_-)B(\Omega'_1 - \Omega_1|t'_1 - t_1)p_{\text{eq}}(\mathbf{J}'_1)p_B(t'_1 - t_1)\cdots$$

$$\cdots \times A(\Omega_n - \Omega'_{n-1}|\mathbf{J}'_{n-1}; t_n - t'_{n-1})\delta(\mathbf{J}_n - \mathbf{J}'_{n-1})p_A(t_n - t'_{n-1})$$

$$\times B(\Omega_+ - \Omega_n|t_+ - t_n)p_{\text{eq}}(\mathbf{J}_+)$$

$$\times F_B(\Omega - \Omega_n; t - t_n|\Omega_+ - \Omega_n; t_+ - t_n)p_B(t_+ - t_n). \tag{2.217}$$

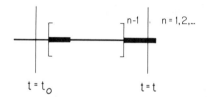

**Figure 2.42.** Schematic of an $ABAB\ldots$ sequence of alternating $A$- and $B$-events. The thin line represents the $A$-, the thick the $B$-event. Note that the first event $A$ is already in progress at $t = t_0$, the zero-time of the correlation function, and that the last event $B$ is still in progress at $t = t$. [From K. Lindenberg and R. I. Cukier, *J. Chem. Phys.* **62**, 3271 (1975).]

The correlation function for the sequence $\{AB\}$ is therefore

$$G_{lk}^{(j)}(t) = (2j+1) \sum_{n=1}^{\infty} \int d\Omega \int d\Omega_0\, p_{eq}(\Omega_0) P_n^{\langle AB \rangle}(\Omega; t|\Omega_0; t_0) D_{lk}^{(j)*}(\Omega) D_{lk}^{(j)}(\Omega_0).$$

(2.218)

Equation 2.217 can be written more concisely. First, recall that we only consider isotropic media; hence only angle differences enter (Eq. 2.56). Second, we restrict the theory to molecules of $C_3$ or higher point-group symmetry. Consequently each transition probability has one Wigner rotation matrix as eigenfunction (Eqs. C.6, C.7; Appendix C). The expectation values of a rotation matrix under the four transition probabilities of $G_{lk}^{(j)}$ are then

$$\int d\Omega \int dJ\, p_{eq}(J) A(\Omega|J; t) p_A(t) D_{lk}^{(j)*}(\Omega) \equiv a_k^{(j)}(t), \qquad (2.219)$$

$$(8\pi^2)^{-1} \int d\Omega \int d\Omega' \int d\Omega'' \int dJ\, p_{eq}(J) \int_t^{\infty} dt'\, A(\Omega - \Omega''|J; t')$$

$$\times F_A(\Omega' - \Omega''; t' - t|J; \Omega - \Omega''; t') p_A(t') D_{kk}^{(j)*}(\Omega - \Omega') \equiv a_k^{i(j)}(t),$$

(2.220)

$$\int d\Omega\, B(\Omega|t) p_B(t) D_{lk}^{(j)*}(\Omega) \equiv b_k^{(j)}(t), \qquad (2.221)$$

$$\int d\Omega \int d\Omega' \int_t^{\infty} dt'\, B(\Omega'|t') F_B(\Omega; t|\Omega'; t') p_B(t') D_{lk}^{(j)*}(\Omega) \equiv b_k^{f(j)}(t),$$

(2.222)

where superscripts $i$ and $f$ denote initial and final event. [For instance, to connect Eq. 2.220 to the corresponding terms in Eq. 2.217, set in the former $t = t_1$, in the latter $t_1 - t_- = t'$, $t_0 = 0$, $dt_- = -d(t' - t_1)$. Furthermore, note that

$$\int d\Omega''\, D_{lk}^{(j)*}(\Omega - \Omega'') D_{lk}^{(j)}(\Omega' - \Omega'') = D_{kk}^{(j)*}(\Omega - \Omega')$$

according to the angle addition theorem (Appendix C) and the normalization condition of the $D_{lk}^{(j)}$.] Insertion of Eqs. 2.217 and 2.219–2.222 into Eq. 2.218 expresses the autocorrelation function for the $\{AB\}$ sequence as the time convolution ($t_0 = 0$)

$$
G_k^{(j)}(t) = \tau_A^{-1} \sum_{n=1}^{\infty} \int_0^t dt_n \int_0^{t_n} dt'_{n-1} \int_0^{t'_{n-1}} dt_{n-1} \cdots \int_0^{t_2} dt'_1 \int_0^{t'_1} dt_1 \, a_k^{i(j)}(t_1)
$$

$$
\times b_k^{(j)}(t'_1 - t_1) a_k^{(j)}(t_2 - t'_1) \cdots b_k^{(j)}(t'_{n-1} - t_{n-1})
$$

$$
\times a_k^{(j)}(t_n - t'_{n-1}) b_k^{f(j)}(t - t_n), \qquad (2.223)
$$

and its Laplace transform[61]

$$
\tilde{G}_k^{(j)}(s) = \tau_A^{-1} \sum_{n=1}^{\infty} \tilde{a}^i(s) \left[ \tilde{b}(s)\tilde{a}(s) \right]^{n-1} \tilde{b}^f(s) \equiv \tilde{G}(AB), \qquad (2.224)
$$

where we omitted the $j$ and $k$ indices.

Inspection of Eq. 2.216 permits us to write the remaining $\tilde{G}$ of the sequences $\{AA\}$, $\{BA\}$, and $\{BB\}$. We obtain

$$
\tilde{G}(AA) = \tau_A^{-1} \tilde{a}^i(s) \sum_{n=1}^{\infty} \left[ \tilde{b}(s)\tilde{a}(s) \right]^{n-1} \tilde{b}(s)\tilde{a}^f(s),
$$

$$
\tilde{G}(BA) = \tau_B^{-1} \tilde{b}^i(s) \sum_{n=1}^{\infty} \left[ \tilde{a}(s)\tilde{b}(s) \right]^{n-1} \tilde{a}^f(s),
$$

$$
\tilde{G}(BB) = \tau_B^{-1} \tilde{b}^i(s) \sum_{n=1}^{\infty} \left[ \tilde{a}(s)\tilde{b}(s) \right]^{n-1} \tilde{a}(s)\tilde{b}^f(s). \qquad (2.225)
$$

Quantities $\tilde{a}^f(s)$ and $\tilde{b}^i(s)$ denote the Laplace transforms of a final $A$- and an initial $B$-event, respectively:

$$
a_k^{f(j)}(t) = \int d\Omega \int d\Omega' \int d\mathbf{J} \, p_{eq}(\mathbf{J}) \int_t^{\infty} dt' \, A(\Omega'|\mathbf{J}; t') F_A(\Omega; t|\mathbf{J}; \Omega'; t')
$$

$$
\times p_A(t') D_{lk}^{(j)*}(\Omega), \qquad (2.226)
$$

$$
b_k^{i(j)}(t) = (8\pi^2)^{-1} \int d\Omega \int d\Omega' \int d\Omega'' \int_t^{\infty} dt'
$$

$$
\times B(\Omega - \Omega''|t') F_B(\Omega' - \Omega''; t' - t|\Omega - \Omega''; t') p_B(t') D_{kk}^{(j)*}(\Omega - \Omega'). \qquad (2.227)
$$

(Note that $a^f$ and $b^i$ differ formally from $b^f$ and $a^i$ only by the subscripts $A$, $B$ and the different $J$-dependence.)

The Laplace transform of the overall autocorrelation function is the sum of the terms of Eqs. 2.224–2.225, each weighted by the probability $\tau_A/(\tau_A + \tau_B)$ that the starting event is $A$, or by $\tau_B/(\tau_A + \tau_B)$ that it is $B$. Furthermore, we include the single events $\langle A \rangle$ and $\langle B \rangle$ (Eq. 2.216), denoting the correlation functions by $G(A)$ and $G(B)$, respectively. This gives

$$\tilde{G}(s) = \frac{\tau_A}{\tau_A + \tau_B} \{ \tilde{G}(A) + [\tilde{G}(AA) + \tilde{G}(AB)] \}$$

$$+ \frac{\tau_B}{\tau_A + \tau_B} \{ \tilde{G}(B) + [\tilde{G}(BB) + \tilde{G}(BA)] \}$$

or, by employing the summation formula for the infinite geometric series,

$$\tilde{G}(s) = (\tau_A + \tau_B)^{-1} \{ \tau_A \tilde{G}(A) + \tau_B \tilde{G}(B)$$

$$+ [\tilde{a}^i(s)\tilde{b}(s)\tilde{a}^f(s) + \tilde{a}^i(s)\tilde{b}^f(s) + \tilde{b}^i(s)\tilde{a}^f(s) + \tilde{b}^i(s)\tilde{a}(s)\tilde{b}^f(s)]$$

$$\times [1 - \tilde{a}(s)\tilde{b}(s)]^{-1} \}. \tag{2.228}$$

Equation 2.228 is difficult to evaluate because of the terms $\tilde{a}^i$, $\tilde{a}^f$, $\tilde{b}^i$, $\tilde{b}^f$, $\tilde{G}(A)$, and $\tilde{G}(B)$. As we discussed before, these quantities describe events that are in progress at $t = 0$ and/or $t = t$; their transition probabilities can be expressed in terms of the transition probabilities $A$ and $B$ (Eq. 2.210). For instance, take $a^i(t)$ of Eq. 2.220. We substitute $F_A(\Omega' - \Omega''; t' - t | \mathbf{J}; \Omega - \Omega''; t')$ by $A(\Omega' - \Omega'' | \mathbf{J}; t' - t)$, obtaining

$$\int d\Omega \int d\Omega' \int d\Omega'' \int d\mathbf{J} \, p_{eq}(\mathbf{J}) \int_t^\infty dt' \, p_A(t') A(\Omega - \Omega'' | \mathbf{J}; t') A(\Omega' - \Omega'' | \mathbf{J}; t' - t)$$

$$= \int d\Omega \int d\Omega' \int d\mathbf{J} \, p_{eq}(\mathbf{J}) \int_t^\infty dt' \, p_A(t') A(\Omega - \Omega' | \mathbf{J}; t)$$

$$= 8\pi^2 \int d\Omega \int d\mathbf{J} \, p_{eq}(\mathbf{J}) A(\Omega | \mathbf{J}; t) \tau_A p_{1A}(t),$$

where we have used the Markovian property[162]

$$\int d\Omega'' \, A(\Omega'' - \Omega' | \mathbf{J}; t - t') A(\Omega - \Omega'' | \mathbf{J}; t') = A(\Omega - \Omega' | \mathbf{J}; t)$$

and Eq. 2.212. It follows that

$$a^i(t) = a^f(t) = \frac{\tau_A a(t) p_{1A}(t)}{p_A(t)},$$

$$b^i(t) = b^f(t) = \frac{\tau_B b(t) p_{1B}(t)}{p_B(t)}. \tag{2.229}$$

It is easy to see that the correlation functions for the single events (see Eqs. 2.217 and 2.228) are then given by

$$G(A) = \frac{a(t) \int_t^\infty dt'\, p_{1A}(t')}{p_A(t)},$$

$$G(B) = \frac{b(t) \int_t^\infty dt'\, p_{1B}(t')}{p_B(t)}. \tag{2.230}$$

For instance, note that $a(t)/p_A(t)$ represents the expectation value for the event $A$ (Eqs. 2.218, 2.219), multiplied by the corresponding survivor probability $\int_t^\infty dt\, p_{1A}(t)$ that this event will continue uninterruptedly. In fact, this result had been derived previously, with the survivor probability of $p_{1A}$ written in the form

$$p_{1A}(t) = \tau_A^{-1} \int_t^\infty dt'\, (t' - t) p_A(t')$$

and its Laplace transform[145a, 164]

$$\tilde{p}_{1A}(s) = s^{-1} - \frac{[1 - \tilde{p}_A(s)]}{\tau_A s^2}.$$

This concludes the derivation of the theory of the general stochastic orientational model as far as it serves our purposes. In the next section we discuss its application.

### 7.3. Applications: Dynamics of Hydrogen Bonding

It makes sense to apply the general rotational relaxation model preferentially to situations where previous, simpler theories proved wanting. For instance, we discussed in Section 3.3 that orientational motion in liquids may be sensitive to remnants of structural order that arises either from strongly angle-dependent components of intermolecular forces or from a short-lived, quasi-crystalline environment. Under such conditions, a model in terms of large free rotation steps interrupted by random instantaneous collisions can no longer be expected to account even approximately for accurate observations. Instead, we now describe such orientational motion in terms of partially cooperative effects, steered by pulsations of the solvent or liquid cage.

We then assume that during a relatively long time the small molecule is orientationally trapped by high local density fluctuations. It no longer performs appreciably orientational motion during this interval ("structure-limited liquids").[18b] To simplify the model, we forgo characterizing in detail transition probability $A$ (we may have defined small-angle, high-frequency librational motion[82b]) and just use Eq. 2.204.[159] The associated non-Poisson (Section 6.1) time probability function $p_A(t)$ is conveniently represented by the gamma distribution (Eq. 2.190).

We define the $B$-event by its expectation value $\lambda_{lk}^{(j)}$ of a single orientation step through arbitrary angle $\Omega$ (Eq. 2.98) with a delta function time distribution (Eq. 2.206). We need not specify event $B$ further (we could have stipulated free rotation steps between preferred potential minima[82b]) since $\tau_A \gg \tau_B$ (relatively long trapping times).

We summarize the conditions of the model:

$$A(\Omega | J; t) = \delta(\Omega)$$

$$p_A(t) = f_\Gamma(t) = \frac{(\alpha/\tau_A)(\alpha t/\tau_A)^{\alpha-1}\exp(-\alpha t/\tau_A)}{\Gamma(\alpha)}$$

$$B(\Omega | t) = \lambda_{lk}^{(j)}$$

$$p_B(t) = \delta(t). \tag{2.231}$$

They yield (Eqs. 2.219, 2.221, 2.229, and 2.230)

$$a(t) = p_A(t)$$

$$b(t) = \int d\Omega \, B(\Omega | t) D_{lk}^{(j)*}(\Omega) \, \delta(t)$$

$$a^i(t) = \tau_A p_{1A}(t) = a^f(t)$$

$$G(A) = \int_t^\infty dt \, p_{1A}(t)$$

$$b^i(t) = 0 = b^f(t). \tag{2.232}$$

(For the last equation, recall that the survivor function of a delta pulse is zero.)

Insertion of the Laplace transform of the terms of Eq. 2.232 into Eq. 2.228 leads (after a little algebra) to[143g]

$$\tilde{G}(s) = s^{-1} + \frac{(\lambda - 1)\left[1 - \tilde{p}_A(s)\right]}{\tau_A s^2 \left[1 - \lambda \tilde{p}_A(s)\right]}, \qquad (2.233)$$

where the $j$, $l$, $k$ dependence of $G$ and $\lambda$ is understood, $\tilde{p}_A(s)$ is given by Eq. 2.192, and $0 \leq \alpha \leq \infty$.

An important and generally interesting phenomenon for which the conditions of this model are tailor-made is furnished by the solvation of small anions in water, for example, $CN^-$.[165] The dynamics of such a system cover a wide time scale. For instance, (1) the lifetime of an aggregate of several water molecules in the solvation sphere of small ions is of the order of $10^{-11}$ sec,[166a] (2) average times of interaction between anion and water are about $10^{-12}$ sec,[166b] and (3) time periods for $CN^-$ to reorient freely by 1 radian amount to $(2k_B T/I)^{-1/2}$ approximately $0.14 \times 10^{-12}$ sec. Hence we find indeed the model conditions of a reorienting species which is blocked by a quasi-rigid environment during intervals (orientational residence time) orders of magnitude longer than intervals of appreciable free orientational motion of the species. Conceivably, the orientational residence time of the anion is equivalent to the lifetime of the ion–water hydrogen bond; an orientational jump of the anion occurs only once a hydrogen bond to a surrounding water molecule is ruptured, whereas orientational motion is rearrested once an anion–$H_2O$ hydrogen bond is reformed.

Relevant data[165] are shown in Fig. 2.43. The solid traces represent the room-temperature Raman VH profile (Eq. 1.50, Chapter 1) and the infrared profile (Eq. 1.35) of the fundamental $CN^-$ stretch of a 4.4 mole/1 aqueous solution of KCN. The (full) bandwidths of the nearly identical contours are 19 and 18 cm$^{-1}$, respectively. The width of the purely vibrational contour $I_{iso}$ (Eq. 1.46) is 9 cm$^{-1}$ (not shown here). This leads to a purely rotational bandwidth

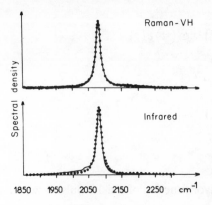

**Figure 2.43.** Room-temperature observed and computed VH-Raman and infrared profile of the $CN^-$ stretch of a 4.4 mol/1 KCN-$H_2O$ solution. The dots represent the computed spectra (Eq. 2.233). [J. Lascombe and M. Perrot, *Faraday Disc. Chem. Soc.* **66**, 216 (1978).]

(Eqs. 1.36, 1.38, 1.39) of 9–10 cm$^{-1}$ for both spectral profiles. Consequently

$$\frac{\tau_R^{(1,k)}}{\tau_R^{(2,k)}} = \frac{\tau_R^{(1)}}{\tau_R^{(2)}} \sim 1,$$

with

$$\tau_R^{(1)} \approx \tau_R^{(2)} \sim 1 \times 10^{-12} \text{ sec.}$$

(We have used Eq. 2.88 although it is not strictly valid unless the profile is Lorentzian.)

To relate this result to the general orientational model, we form (see Eq. 1.68, Chapter 1) $\tau_R^{(j)} = 2\pi\tilde{G}(0)$ by (twice) differentiating numerator and denominator of Eq. 2.233 with respect to $s$ and setting in the result $s = 0$,

$$\tau_R^{(j)} = \pi\tau_A \left[ 1 + \lambda^{(j)} + \frac{1 - \lambda^{(j)}}{\alpha} \right] (1 - \lambda^{(j)})^{-1}.$$

For the linear anion $\lambda_{lk}^{(j)}$ simplifies to $\langle \cos\theta \rangle$ for $j = 1$ and to $(\frac{1}{2})\langle 3\cos^2\theta - 1 \rangle$ for $j = 2$ (see also Eq. 2.98).

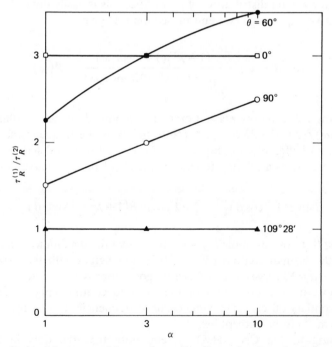

**Figure 2.44.** Ratio of the computed first- and second-order purely orientational correlation times of the orientational jump model (Eq. 2.233) as a function of average angular jump angle $\theta$ and of the deviation from a Poisson time distribution $\alpha$. [Unpublished calculations, J. Lascombe and M. Perrot, Lab. Spectry. Infrarouge, Université de Bordeaux I.]

Figure 2.44 shows the results for $\theta = 60$, 90, and 109 degrees 28 minutes. Note that $\tau_R^{(1)}/\tau_R^{(2)}$ equals or *exceeds* the value of 3, a maximum which is usually ascribed to Debye orientational motion ($\theta = 0$ degrees or $\lambda \to 1$; refer to Eq. 2.97). This already shows one unexpected aspect arising from a non-Poisson distribution of $p_A(t)$ (Note 16).

We conclude from Fig. 2.44 and the experimental $\tau_R^{(1)}/\tau_R^{(2)}$ approximately 1 that solvated $CN^-$ in this system reorients by average jumps of $\theta$ approximately 109 degrees.

It is also easy to demonstrate the modifications that arise in the rotational infrared and Raman profiles from a non-Poisson time distribution of the $A$-events. (The spectra are calculated from the real part of Eq. 2.233, with $s = -i\omega$.) Figure 2.45 shows the ensuing profiles; they are remarkable. We see that the more $p_A(t)$ diverges from random ($\alpha > 1$), the broader the profile. Notice also that the Raman profile broadens to a greater extent than the infrared profile since $\tau_R^{(1)}/\tau_R^{(2)}$ increases with $\alpha$ (Fig. 2.44). Furthermore, with increasing non-Poisson characteristics the Raman contour develops a maximum off the central frequency because $p_A(t)$, which determines the instance $t$ of the orientational event, moves away from $t = 0$ to approach $t = \tau_A$; simultaneously, the distribution of $t$ becomes sharper (see Eq. 2.191).[159]

We can quantify this phenomenon by computing the spectral density from Eq. 2.233 for a delta distribution of $p_A(t)$ since this is easily done: Insertion of $\tilde{p}_A(s, \alpha = \infty) = \exp(-s\tau_A)$ (see end of Section 6.1) gives

$$\mathrm{Re}\,\tilde{G}(-i\omega) = \frac{(1-\lambda^2)(1-\cos \tau_A\omega)}{\tau_A\omega^2(1+\lambda^2 - 2\lambda\cos \tau_A\omega)} \equiv \mathscr{I}(\omega). \qquad (2.234)$$

Equation 2.234 represents a Lorentzian modified by an oscillating term $(1 - \cos \tau_A\omega)/(1 + \lambda^2 - 2\lambda\cos \tau_A\omega)$. (Note that $\omega$ is the frequency shift off the band center.) Differentiating Eq. 2.234 with respect to $\omega$ and setting the result equal to zero yields a relation for the frequency of the side maximum,

$$\beta \sin \beta(1 - \cos \beta)^{-1} = 2 + 2\lambda\beta\sin \beta(1 + \lambda^2 - 2\lambda\cos \beta)^{-1},$$

with $\tau_A\omega \equiv \beta$. In our example ($\tau_A = 1 \times 10^{-12}$ sec) the maximum occurs near 15 $cm^{-1}$ in the Raman contour ($\lambda^{(2)} = -0.5$); this agrees with the maximum in Fig. 2.45c. [$\alpha = 10$ serves as a sufficiently good approximation for $\alpha \to \infty$.]

The most intense maximum of the infrared contour (Fig. 2.45c) lies at $\omega = 0$; secondary off-center maxima occur at frequencies too far into the wings (near 33 $cm^{-1}$) to be perceptible.

Returning to the $CN^- - H_2O$ system, note that the dots in Fig. 2.43 represent Eq. 2.233 with a Poisson time distribution ($\alpha = 1$), using $\tau_A = 1 \times 10^{-12}$ sec. Since experiment and theory agree, any deviation from a random distribution of events of orientational arrest of the anion must be small.

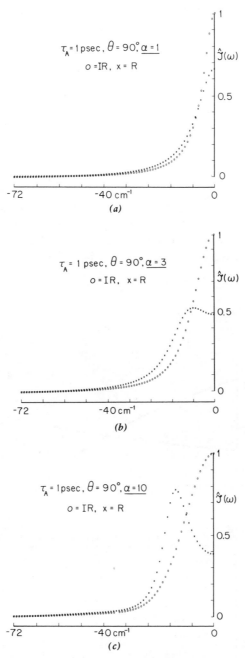

**Figure 2.45.** Simulated spectral densities of the purely orientational component of the infrared and Raman vibration-rotation contours of a reorienting linear molecule obeying the 90-degree jump model of Eq. 2.233. The contours are normalized to unit area; only the low-frequency half is shown. (*a*) Poisson distribution of the duration of orientational arrest; (*b*) non-Poisson distribution, $\alpha = 3$; (*c*) non-Poisson distribution, $\alpha = 10$. Average duration of orientational arrest $= 1 \times 10^{-12}$ sec. [Unpublished calculations, J. Lascombe and M. Perrot, Lab. Spectry. Infrarouge, Université de Bordeaux I.]

**209**

The temperature dependence of the average $CN^-$ reorientation rate in the $CN^-$–$H_2O$ system also leads to interesting conclusions. Figure 2.46 shows a plot of $\log \Delta\nu_R^{(2)}$ versus $1/T$, where

$$\Delta\nu_R^{(2)} \approx \Delta\nu_{VH} - \Delta\nu_{vib}$$

is the purely Raman orientational halfwidth (Eq. 2.96). From it we estimate an apparent activation energy $\Delta E$ of 1.4 kcal/mole near 40°C and 3.5 kcal/mole near −10°C. Closer inspection of Fig. 2.46 shows that the increase of $\Delta E$ with decreasing temperature becomes pronounced below approximately 5°C, paralleling the temperature dependence of the NMR orientational correlation time of $H_2O$ in neat $H_2O$ (3.5 kcal/mole above 40°C and approximately 5 kcal/mole at 0°C).[166c] (Note that the purely vibrational halfwidth decreases slightly with increasing temperature. We have more to say about this in Chapter 3.)

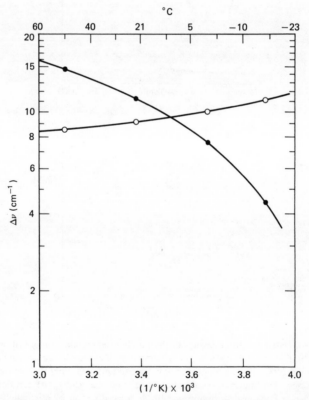

**Figure 2.46.** Activation energies of the Raman halfwidths of the purely vibrational (○) and purely orientational (●) contours of the $CN^-$ fundamental of Fig. 2.43.

However, comparisons of this genre should be considered with caution. First, we deal with anion–water bonds and not with the (stronger) water–water bonds of neat water. Second, anion–cation association in the concentrated medium (4.4 mole/1) must be expected to reduce the effective charge of the anion. In fact, the dynamics of anion–water bonding are poorly understood at present. The potential depth that binds a water molecule to anions such as $I^-$, $Br^-$, and $ClO_4^-$ is of the order of $-1.3$ kcal/mole, compared to $-2.3$ kcal/mole for neat water. We therefore cannot safely conclude that the apparent activation energies of the orientational motion of $CN^-$ in water reflect indeed hydrogen bond *rupture* (and reformation) dynamics between anion and water or among water molecules. Qualitative or method-oriented concepts such as the "structure-braking" effect of the anion are of no help; only a considerably greater effort in fluctuation-spectroscopic studies of aqueous solutions of anions will bring us further. Such work should not only be of general interest but also of fundamental value since pure charge effects (long-range interactions) are predominant.

### 7.4. Special Effects of Nonrandom Time Distributions

Three particularly interesting subjects of the general rotational relaxation model follow, namely:

1. The peculiar characteristics of the correlation function for large angular jumps and large deviations from a Poisson distribution of events of motional arrest.

2. The effects of neglecting to average over the initial times of an event that is governed by a nonrandom time distribution function.

3. The limiting correlation function if the time distribution becomes random (Poisson).

Figure 2.47 shows the infrared and Raman correlation functions $G^{(1)}(t)$ and $G^{(2)}(t)$, calculated from the spectral profiles in Fig. 2.45$b$ and $c$ by numerical Fourier inversion. We see that $G^{(2)}(t)$ becomes negative, even oscillatory if the orientational jump angles are large ($\theta \gtrsim 60$ degrees). In fact, the larger the deviation from Poisson behavior ($\alpha > 1$) and the more negative $\lambda$ ($\theta \gtrsim 60$ degrees), the larger the oscillations. The oscillatory behavior is, of course, a consequence of the pronounced side maximum in the Raman spectral density (see particularly Fig. 2.45$c$). We recall that this reflects a quasi-resonance condition between the mean $\tau_A$ and the effective duration $\tau$ of orientational arrest of the rotors. We realize that a correlation function that expresses large-angle orientational jump motion steered by fluctuating liquid-phase cages (structure-limited liquid[18b]) *can* resemble a correlation function (Fig. 2.1) that reflects collision-interrupted nearly free-rotor kinetic reorientation steps (collision-limited liquid[18b]).

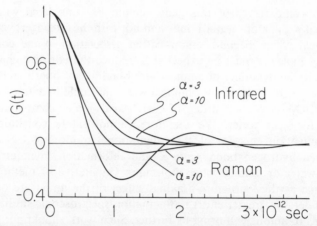

**Figure 2.47.** Correlation functions $G^{(1)}(t)$ and $G^{(2)}(t)$ associated with the infrared and Raman spectral densities for $\alpha = 3$ (Fig. 2.45$b$) and $\alpha = 10$ (Fig. 2.45$c$). The initial curvature is exaggerated by incomplete integrations over $\omega$. Since purely kinetic effects are absent in the model, $G(t)$ does not approach $t = 0$ with zero slope. Frequency range of spectral density: 150 cm$^{-1}$.

We now consider the consequence of neglecting to average over the starting times of the first event if it is governed by a non-Poisson $p_A(t)$. We therefore replace $p_{1A}(t)$ by $p_A(t)$ in $a^i(t)$ and in $G(A)$ [but not in $a^f(t)$]. This gives (compare with Eq. 2.232)

$$a(t) = p_A(t)$$

$$b(t) = \lambda \delta(t)$$

$$a^i(t) = \tau_A p_A(t)$$

$$a^f(t) = \tau_A p_{1A}(t)$$

$$G(A) = \int_t^\infty dt\, p_A(t).\tag{2.235}$$

Insertion of their Laplace transforms into Eq. 2.228 leads to[143g, 159]

$$\tilde{G}(s) = s^{-1}\left[1 - \tilde{p}_A(s)\right]\left[1 - \lambda \tilde{p}_A(s)\right]^{-1}.\tag{2.236}$$

The associated correlation function $G^{(1)}(t)$, obtained by numerical Fourier transformation of the real part of Eq. 2.236, is shown in Fig. 2.48 by the dots. The correct correlation function (Fourier transform of the real part of Eq. 2.233) is represented by crosses. Note that the incorrect $G^{(1)}(t)$—where we have not averaged over initial times although $p_A(t)$ is non-Poisson—decays

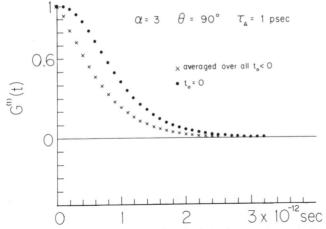

**Figure 2.48.** Infrared correlation function $G^{(1)}(t)$ associated with the spectral density of a non-Poisson orientational jump model. Solid points: time $t_0$ of the initial event is arbitrarily set equal to time zero of $G^{(1)}(t)$. Crosses: time $t_0$ is correctly averaged over all $t_0 < 0$. Frequency range of spectral density: 300 cm$^{-1}$.

more slowly, greatly underestimates the short-time rotational relaxation rate, and even simulates (nonexistent) kinetic behavior.

Finally, we consider point 3, namely, the form of the correlation function if $p_A(t)$ becomes random. Inserting $(1 + \tau_A s)^{-1}$ (Eq. 2.192) into Eq. 2.233 or 2.236 gives

$$\tilde{G}(s) = \left\{ \left(1 - \lambda^{(j)}\right)\tau_A^{-1} + s \right\}^{-1} \qquad (2.237)$$

or[143e]

$$G(t) = \exp\left\{ -\frac{\left(1 - \lambda^{(j)}\right)t}{\tau_A} \right\},$$

which represents Ivanov's orientational jump model (Eq. 2.102).

It should now be clear that nonrandom time distributions of stochastic events introduce unusual, unsuspected, and complicating effects into rotational relaxation phenomena.

## 8.  ROTATIONAL RELAXATION: PRESENT STATUS AND FUTURE DEVELOPMENTS

Interest in rotational relaxation has cooled and progress has considerably slowed. Papers that have appeared during the writing of this book concern themselves predominantly with collision-induced phenomena by far-infrared absorption and Rayleigh scattering using data evaluation following simple empirical approaches, with elaborations and extensions of the Langevin method, and improvements and generalizations of the $J$- and $M$-extended rotational diffusion models. If separation of individual and collective orientational motion is attempted, the results usually end on a pessimistic note since little more than Debye-type orientational motion is (incorrectly, as we know) inferred or assumed a priori from the (often) exponential tail of the orientational decay; such method does not make for progress.

What then has the future in store for further developments of rotational relaxation phenomena in liquids?

In spite of its present hiatus, this field is by no means exhausted or even thoroughly understood. On the contrary, we think that the mere surface of knowledge has been scratched, that almost everything we have discussed in these sections is open to be questioned,[167] to be redone, reexplained, and extended. With respect to information and knowledge that goes beyond the simplest concepts of orientational motion, the discussions in this chapter lead to the conclusion that *individual* molecular orientational motion does not occur. Regarding the orientational motion of *one* probe molecule in the sea of its neighbors as representative of liquid-phase orientational dynamics is too great a simplification for most liquids. Further work must assume or seriously consider that the orientational motion of one molecule entrains that of several of its neighbors in a rather complicated fashion; hints at this are all too obvious.

Consider, for instance, that there are few liquids whose pure rotational second spectral moment has been measured (carefully, of course) to reach the theoretical equipartition value, a value that corresponds (apart from minor corrections) to the trivial average kinetic energy of the rotor. Hence we suspect that the *effective* moment of inertia must be larger than that of the *single* molecule.[104]

Physically, this means that we deal with some "cluster" or "in-cluster" orientational motion. In fact, a glance at the structure of some rather mobile and only weakly associated liquids—benzene is a good example[135]—hardly leaves another possibility but "cogwheel reorientation" of several molecules or a process of "pseudorotation" of the liquid cage.[168]

On the other hand, in network systems such as water, amorphous ice, or high-concentrated aqueous solutions of inorganic salts,[169] the concept of reorientation of individual or small-cluster species loses meaning. Rayleigh scattering[169a] and far-infrared absorption profiles[169b] reflect only the density of vibrational states of the lattice phonons (since the momentum selection rule is lifted). This phononlike collective motion is the extreme example of a cooperative effect in systems of interest to us.

All these are exciting and promising aspects. They should lead naturally to a more satisfying knowledge of cluster formation, short-range liquid-phase order, the meaning of liquid-phase complex formation and solvation with its ramification to solvent dependence of chemical kinetics, and so forth. Fluctuation spectroscopy should attempt to answer thoroughly questions concerning the microscopic structure of condensed phases possessing some local but dynamic order. This indeed would better meet the picture we have of liquids. With some exceptions, we would not longer have to simplify liquid-phase dynamics to resemble those of a strongly compressed gas of individually reorienting molecules or, in the other limit, to reflect an ensemble of solvent cages with individual, jump-reorienting molecules.

Work along such lines would fill the trench between microscopic and macroscopic quantities. Orientational motion of small molecules would then be relatable to viscosity and to free volume in a satisfactory, theoretically sound, and useful manner. In our opinion the connection between microscopic and macroscopic quantities is an important ultimate goal to be strived for if fluctuation spectroscopy of rotational relaxation is not to encrust as an isolated scientific discipline that failed to reach a more general level of interest, acceptance, and application.[170]

# NOTES

1. For a detailed description of "precession" and "nutation," see *Classical Mechanics*, H. Goldstein, Addison-Wesley, 1953. The angular precession frequency $(I_1 - I_3)\omega_z I_1^{-1}$ [Eq. (5.40)] compares to $\xi J_z I_x^{-1}$, the angular nutation frequency $I_3\omega_z/I_1$ [Eq. (5.65)] to $J_z I_x^{-1}$, with $I_3 = I_z$, $I_1 = I_x$.

2. Equation 2.10 describes a process that is "stationary" (its $k$-fold occurrence during finite, nonoverlapping time intervals between $t$ and $t + \tau$ depends only on $k$ and $\tau$ but not on $t$), "without memory" (the probability of its occurrence does not depend on the number of previous occurrences), and "ordinary" (two- or threefold occurrences during $\Delta\tau$ are nil).

3. The Dirac bracket notation is particularly useful. We can automatically "shove" the operators and operands together to arrive at the result. We simultaneously imply with the $\langle \cdots \rangle$ notation that the necessary ensemble averages are to be taken.

4. It is useful to point out here again that the complicated form of the appropriate Cartesian anisotropic Raman correlation function (see Eq. 1.86b, Chapter 1) is inconvenient for model computations because of the presence of the cross-axis term. Although such terms may, by themselves, be of considerable interest, no work along these lines has been published.

5. The correlation function is a linear combination of Legendre polynomials of order $l = 1, 2, \ldots$ (Eq. 1.81a and 1.81b, Chapter 1). This is an inherent complication of the neutron technique. These ensemble-averaged polynomials are frequently designated in the literature of neutron scattering by the symbol $F_l(t)$; it should not be confused with the free rotor correlation function $F(t)$.

6. We should use the symmetrized spectral density $\hat{\mathcal{J}}_s(\omega)$ [Eq. A.19, Appendix A] to include detailed balancing. Neglecting this is not serious unless the profiles are broad.

7. The central portion of a spectral density corresponds to the long-time tail of the correlation functions. Since the tail is almost always exponential (Markovian region), the central portion is near-Lorentzian. Clearly, the larger the region of exponential decay of the correlation function, the larger the range of the central Lorentzian profile.

8. If $\rho(0) = \rho_a(0)\rho_b(0)$, then

$$\rho^{\times}(0) = \left[\rho_a(0)\rho_b(0)\right]^{\times} = \rho_a^{\times}(0)\rho_b(0) + \rho_a(0)\rho_b^{\times}(0),$$

where we have used the commutator relation $[AB, C] = [A, C]B + A[B, C]$ with Eq. 2.112. Hence

$$\mathrm{Tr}\{\rho^{\times}(0)mU^{\times}(t)m\} = \mathrm{Tr}\{\left[\rho_a^{\times}(0)\rho_b(0) + \rho_a(0)\rho_b^{\times}(0)\right]mU^{\times}(t)m\}$$

$$= \mathrm{Tr}_a\{\rho_a^{\times}(0)m\,\mathrm{Tr}_b\rho_b(0)U^{\times}(t)m\}$$

$$+ \mathrm{Tr}_a\{\rho_a(0)\,\mathrm{Tr}_b\rho_b^{\times}(0)mU^{\times}(t)m\}.$$

But the second term vanishes since $\rho_b^{\times}(0)m = [\rho_b(0), m] = 0$.

9. To keep matrix element notation and ensemble averaging apart, we do not write the matrix elements in Dirac vector notation here.

10. Note that this relation appears to be in error by a factor of $1/\sqrt{2}$ : For a (spherical or linear) molecule the classical average root-mean-square angular frequency of axis reorientation of the free rotors is $\langle\omega^2\rangle^{1/2} = (2k_BT/I)^{1/2}$—see Eq. 2.2. Hence orientation by $(41/360)2\pi \sim (2\pi/9)$ requires $t \sim (2\pi/9)(I/2k_BT)^{1/2}$. Furthermore, a correlation value of $1/e$ for an ensemble of free rotors is devoid of theoretical significance.

11. A differential equation of the type $\dot{\gamma}(t) - \beta(t)\gamma(t) - \alpha(t) = 0$, where $\gamma(t) = (1 - \bar{P})|A(t)\rangle$, $\alpha(t) = (1 - \bar{P})iL|A\rangle C(t)$, and $\beta = (1 - \bar{P})iL$, must be solved. The "variation of the constant" leads to

$$\gamma(t) = \left(\exp\int\beta(t)\,dt\right)\left(C + \int_0^t\alpha(t)\exp\left\{-\int_0^t\beta(t)\,dt\right\}\,dt\right),$$

where $C$ is an integration constant. Since $\gamma(0) = 0$ for $t = 0$, we can set $C = 0$. Since $\beta$ is constant,

$$\gamma(t) = \int_0^t dt'\exp\{\beta(t - t')\}\alpha(t') = \int_0^t dt'\exp\{\beta t'\}\alpha(t - t').$$

12. For the monovariable case (Section 5.6),

$$|\dot{D}^+\rangle \equiv (1 - \bar{P})|\dot{D}\rangle = (1 - |\hat{D}\rangle\langle\hat{D}|)|\dot{D}\rangle = |\dot{D}\rangle.$$

13. The peak of the Poisson distribution (Eq. 2.10 or 2.205) lies at $t = 0$. A useful discussion about this process begins on p. 29 of Ref. 163.

14. The designations "multi*variate*" or "multi*variable*", and so on, mean the same thing in our context. It should not be confused with cross correlation between different molecules.

15. Absorption measurements to very low frequencies are difficult since we require $\mathscr{I}(\omega) \approx \sigma(\omega)/\omega^2$ (see, for instance, Fig. 3 of Ref. 99).

16. It is readily shown that $\tau_R^{(1)}/\tau_R^{(2)} \to 3$ for $\theta = 0$, irrespective of the value of $\alpha$, by letting $\lambda \to 1$ in $\tau_R^{(j)}$. Hence develop $\lambda = \cos\theta$ and $(\frac{1}{2})(3\cos^2\theta - 1)$, respectively, into their series expansions for small $\theta$.

# CHAPTER 3

# VIBRATIONAL
# RELAXATION

## A.  Two-Level Systems

### 1.  GENERAL CONTEXT AND OUTLOOK

In this chapter we discuss broadening of an infrared or Raman vibrational band due to coupling between the atomic displacement coordinates of the mode and the static and dynamic forces in the surrounding medium. Numerical Fourier transformation of the band profiles leads to autocorrelation functions that describe the time evolution of the coherence decay of fluctuations of the rapid periodic motions of an oscillator-in-a-molecule. This phenomenon, called vibrational relaxation, is therefore of a character and variety different from rotational relaxation (Section 4.1, Chapter 1).

In the earlier heyday of modern infrared fluctuation spectroscopy of liquids, vibrational relaxation was, at worst, shrugged off as unimportant and, at best, considered a nuisance. All efforts were directed toward learning about orientational motion, for instance, to see how molecules orient in a viscous phase,[133a] to examine whether a concept of free rotation steps interrupted by sudden collisions truthfully described molecular motion in liquids,[117] to quantitatively explore the anisotropy of orientational motion of low-symmetry condensed-phase systems,[34] and so forth. It was not until 1970–1971—six years after Gordon's first papers—when extended theories[171] emphasized that a certain amount of concomitant vibrational relaxation was, after all, complicating the simple picture. The epithet "residual width"[18b, 45] evokes the spirit and the attitude with which purely vibrational band broadening in condensed phases was then considered.

This has now drastically changed. Hardly a month passes without publication of a theory on vibrational relaxation or a modification of an existing theory. This abundance of approaches to an experimentally limited phenomenon requires selection of subjects to be discussed as well as a choice of viewpoints to be taken. We concentrate from the outset on three questions.

1. How does the vibrational relaxation process fit within the larger frame of liquid-phase phenomena of general interest?
2. What does and can vibrational relaxation teach us about molecular motion of the liquid phase?
3. Are there any relations or fundamental analogies between a vibrational relaxation process and rotational relaxation?

With respect to the three points raised, we feel that published vibrational relaxation studies have frequently "set their sight too low." Most work merely has persued a quest about the characteristics of the time evolution of the vibrational amplitude correlation of a particular mode of a particular molecule, without addressing itself to the larger problems of molecular dynamics. To a large extent this has been a consequence of the ubiquitous and all-pervading vibrational dephasing process (Section 4.1, Chapter 1). In our opinion the physical preponderance of this process is not matched by an accruing wealth of useful and *generally* interesting information. We discuss later that vibrational dephasing teaches us little about atomic and molecular motion beyond simple, rudimentary collision dynamics in ordinary liquids. However, vibrational dephasing is suited to describe the dynamics of strong interactions (hydrogen bonding, cage effects in strongly associated liquids) and, in particular, *static* or *quasi-static* perturbations from local distributions of condensed-phase inhomogeneities on the observed oscillator energy (and was, indeed, so originally understood and characterized[171]).

In contrast, the vibrational energy relaxation or phonon population relaxation process (Section 4.1, Chapter 1) is sensitive to dynamic phenomena since its relaxation rate is determined by the very nature and speed of molecular motions in the system. Further, the process is of great general interest since it leads naturally to a better understanding of important subjects, such as transition state theory and reaction kinetics in the condensed phase. Of course, this entails the study of highly excited vibrational levels, requiring special instrumental techniques such as picosecond pulsing (Section 4.3, Chapter 1) or spontaneous relaxation in the near-infrared. None of this is experimentally easy or generally practiced at present. Also, the theory of a relaxing two-level system, that is to say, a system of ground and upper vibrational level, would have to be extended to include vibrational depopulation as well as population to and from other levels below and above (multilevel systems). We leave this for the second part of this chapter.

The vibrational resonance transfer process, which describes how an excited phonon diffuses through a disordered condensed-phase medium ("vibrational excitons"), is also of considerable general interest. Despite the simplicity of the required experimental technique (isotopic substitution spectra) and of its interpretation (Section 4.1, Chapter 1), vibrational resonance relaxation in liquid-phase systems has received little attention (in contrast to its wide interest for molecular crystals[172]). A great deal remains to be done here.

In the following sections we discuss first the vibrational dephasing mechanism. We do not go into all the many theories presently known about this process.[173] We prefer to deal more extensively with the physical ideas, to introduce theoretical considerations in accordance with the quality of the experimental evidence, and to draw useful comparisons to related model concepts. In particular, we elaborate on simple collision dynamics in liquids.

## 2.  VIBRATIONAL DEPHASING

### 2.1.  Basic Theory

Imagine a two-level vibrational system of an *individual* oscillator. The lifetime of its upper vibrational level is considered "very long" (much longer than times of interest here). In a disordered condensed phase (neat liquid, solution, plastic crystal, heterogeneous surface), the sharp transition frequency $\omega_0$ of this two-level oscillator system is "smeared out" into a *continuous, nonresolvable* distribution of vibrational transition frequencies; the interactions between the oscillator and the surrounding molecules in their various positions and orientations have caused perturbative "instantaneous" shifts of the lower and upper oscillator energy levels. (Consider positioning $n$ identical and unperturbed oscillators of eigenfrequency $\omega_0$ at various locations $i, j, \ldots$ within a microscopic volume of a locally disordered medium. This leads to a "splitting" of the "$n$-fold degenerate" frequency $\omega_0$ into a certain distribution of frequencies around $\omega_0$, often referred to as "inhomogeneous broadening.")

A simple theoretical model for the amplitude correlation function (Eqs. 1.32, 1.34, and 1.37, Chapter 1) of this phenomenon is easily written. Using the superoperator formalism (Eq. F.1 and F.11, Appendix F), the equation of motion for the vibrational amplitude $m(t)$ (Eq. 1.73, Chapter 1) reads

$$\frac{d}{dt}m(t) = \frac{i}{\hbar}H^{\times}m(t).$$

(3.1a)

The total Hamiltonian $H$ is the sum of the Hamiltonian of the unperturbed vibrational system $H_0$ and a time-dependent perturbation $H_1(t)$. The formal solution of Eq. 3.1a is (Eq. G.5, Appendix G; Eq. F.5)

$$m(t) = \exp\left\{\frac{i}{\hbar}\int_0^t dt'\, H_1^{I\times}(t')\right\}\exp\left(iH_0^{\times}\frac{t}{\hbar}\right)m = \exp\left\{i\int_0^t dt'\, \omega_1^I(t')\right\}m^I,$$

(3.1b)

where

$$\omega_1(t) = \frac{H_1(t)}{\hbar} = \omega(t) - \omega_0 \tag{3.1c}$$

is the instantaneous oscillator transition frequency shift caused by $H_1(t)$. Splitting off the uninteresting constant modulation factor $\exp(i\omega_0 t)$ about the unperturbed band center $\omega_0$, we get after multiplying each side by $m(0)$,

$$\langle \hat{m}(t)\hat{m}(0) \rangle = \langle m(t)m(0) \rangle \langle |m(0)|^2 \rangle^{-1} = \left\langle \exp\left\{ i\int_0^t dt'\,\omega_1(t') \right\} \right\rangle \tag{3.1d}$$

(See also Eq. 1.102, Chapter 1). Equation 3.1d indeed describes an amplitude correlation function as function of a random phase angle $\eta(t) = \int dt\,\omega_1(t)$; for the rotational analogue, see Eq. 2.6 in Chapter 2.

The average of the exponential in Eq. 3.1d is now expressed as the exponential of an average (Appendix H). Using Eq. H.6, setting $\langle \omega_1(t) \rangle^c = 0$, and assuming the validity of the central limit theorem (Eq. H.8), we obtain

$$\langle \hat{m}(t)\hat{m}(0) \rangle = \exp\left\{ -\frac{1}{2}\int_0^t dt'' \int_0^t dt'\,\langle \omega_1(t')\omega_1(t'') \rangle \right\}. \tag{3.2}$$

Translational time invariance of the correlation function (Section 1.3, Chapter 1) permits us to write

$$\langle \omega_1(t')\omega_1(t'') \rangle = \langle \omega_1(t'-t'')\omega_1(0) \rangle \equiv \hat{\psi}(\bar{t})\langle |\omega_1(0)|^2 \rangle$$

where $\bar{t} = t' - t''$. This changes Eq. 3.2 to (see Note 1 at end of chapter)

$$\langle \hat{m}(t)\hat{m}(0) \rangle = \exp\left\{ -\langle |\omega_1(0)|^2 \rangle \int_0^t d\bar{t}\,(t - \bar{t})\hat{\psi}(\bar{t}) \right\} \equiv \hat{\phi}(t). \tag{3.3}$$

Notice that we have now shifted our attention from the (normalized) vibrational amplitude correlation function $\langle \hat{m}(t)\hat{m}(0) \rangle$ to the (normalized) autocorrelation function $\hat{\psi}(t)$ of the oscillator frequency shift $\omega_1(t)$ or of the perturbation Hamiltonian $H_1(t)$ [Eq. 3.1c].

Let us then assume that at present we know nothing else about

$$\langle H_1(t)H_1(0) \rangle = \hbar^2 \langle \omega_1(t)\omega_1(0) \rangle$$

but (1) its correlation time

$$\tau_c = \int_0^\infty dt\,\langle \hat{H}_1(t)\hat{H}_1(0) \rangle = \int_0^\infty dt\,\langle \hat{\omega}_1(t)\hat{\omega}_1(0) \rangle = \int_0^\infty dt\,\hat{\psi}(t), \tag{3.4}$$

and (2) that $\langle \hat{H}_1(t)\hat{H}_1(0) \rangle$ "behaves"

$$\langle \hat{H}_1(0)\hat{H}_1(0) \rangle = 1,$$

$$\lim_{t \to \infty} \langle \hat{H}_1(t)\hat{H}_1(0) \rangle = 0.$$

Two limiting, interesting situations can then be written down, namely, that the perturbation coherence (correlation time $\tau_c$) decays much slower or much faster, respectively, than the amplitude coherence (correlation or lifetime $\tau_a$)

$$\tau_c \gg \tau_a, \tag{3.5a}$$

and

$$\tau_c \ll \tau_a, \tag{3.5b}$$

respectively, where

$$\tau_a = \int_0^\infty dt \langle \hat{m}(t)\hat{m}(0) \rangle \equiv \int_0^\infty dt\, \hat{\phi}(t). \tag{3.5c}$$

We first discuss condition $\tau_c \gg \tau_a$. It permits us to set $\hat{\psi}(t)$ approximately 1 during the lifetime $\tau_a$ of $\hat{\phi}(t)$. Equation 3.3 then simplifies to (omitting the normalization sign)

$$\phi(t) \approx \exp\left\{ -\langle |\omega_1(0)|^2 \rangle \frac{t^2}{2} \right\}, \tag{3.6a}$$

with an associated Gaussian profile (Eq. A.20, Appendix A)

$$\mathscr{I}(\omega) \propto \exp\left\{ -\frac{\omega^2}{2\langle |\omega_1(0)|^2 \rangle} \right\}. \tag{3.6b}$$

Equation 3.5a describes a static situation, that is to say, a time-independent distribution $\langle |\omega_1(0)|^2 \rangle^{1/2}$ of line shifts around the unperturbed oscillator transition frequency. Early work[174] modeled such distribution in terms of transition-dipole–transition-dipole interactions between the centrally placed oscillator and a rigid cage of surrounding octahedrally arranged molecules (point charges). Broadening was predicted to be proportional to the integrated band intensity $A = \int d\omega\, \mathscr{I}(\omega)$ since $A \propto \langle |m(0)|^2 \rangle$ (Eq. 1.25, Chapter 1). (We go into more detail about transition-dipole–transition-dipole interactions further along.) Later studies introduced interaction between the cubic term of the oscillator normal coordinate $Q$ and a general intermolecular potential $V_s$ as principal cause of the frequency shift $\omega_1(t)$ ("anharmonicity coupling").[171] Neglecting effects of molecular motion, a theoretical expression for $\omega_1(t = 0) \equiv$

$\omega_1(0)$ is established by developing $V_s$ as power series in $Q$ ($Q$ is proportional to $m$; see Eqs. 1.32, 1.37, and 1.39, Chapter 1) and defining a static perturbation term $H_1(0) = \hbar\omega_1(0)$,

$$H_1(0) = \frac{1}{6}fQ^3 + V_s^0 + \left(\frac{\partial V_s}{\partial Q}\right)_{Q=0} Q + \frac{1}{2}\left(\frac{\partial^2 V_s}{\partial Q^2}\right)_{Q=0} Q^2 + \frac{1}{6}\left(\frac{\partial^3 V_s}{\partial Q^3}\right)_{Q=0} Q^3$$

up to third order in $Q$. $f$ is the constant of the cubic (anharmonic) terms of the intraoscillator potential. To obtain the frequency shift of the perturbed oscillator transition frequency, $(E_v - E_v^0)/\hbar = \omega_1(0)$, we apply [171] stationary perturbation theory. By well-known principles, this gives the vibrational energy level of the perturbed oscillator in quantum state $v$,

$$E_v = E_v^0 + \lambda(H_1)_{vv} + \lambda^2 \sum_{u \neq v} \frac{(H_1)_{uv}(H_1)_{vu}}{E_v^0 - E_u^0} + \cdots$$

Keeping only the lowest-order product interaction terms in $f$ and in the $\partial^n V_s/\partial Q^n$, the second-order contributions to $E_v$ are

$$\frac{2}{36}f\left(\frac{\partial^3 V_s}{\partial Q^3}\right)_{Q=0} \sum \frac{Q_{uv}^3 Q_{vu}^3}{E_u^0 - E_v^0}$$

from term $\qquad \dfrac{1}{6}\left\{fQ^3 + \left(\dfrac{\partial^3 V_s}{\partial Q^3}\right)_{Q=0} Q^3\right\}$

and

$$\frac{1}{3}f\left(\frac{\partial V_s}{\partial Q}\right)_{Q=0} \sum \frac{Q_{uv}Q_{vu}^3}{E_u^0 - E_v^0}$$

from term $\qquad \dfrac{1}{6}fQ^3 + \left(\dfrac{\partial V_s}{\partial Q}\right)_{Q=0} Q.$

(Terms mixing products of even and odd powers of $Q$ vanish since they have no common subscripts $u, v$.[175]) With the known matrix elements of $Q^n$ and using $E^0 = \hbar\omega_0(v + \frac{1}{2})$, we get, after a little algebra, the static frequency shift[171, 175]

$$\omega_1 = \frac{E_v - E_v^0}{\hbar} = v(2\mu_M\omega_0)^{-1}\left\{-\left(\frac{f}{\mu_M\omega_0^2}\right)\left(\frac{\partial V_s}{\partial Q}\right) + \frac{\partial^2 V_s}{\partial Q^2}\right.$$

$$\left. -\left(\frac{5\hbar f}{12\mu_M^2\omega_0^3}\right)\left(\frac{\partial^3 V_s}{\partial Q^3}\right)\right\} + v^2\left(\frac{5\hbar f}{24\mu_M^3\omega_0^4}\right)\left(\frac{\partial^3 V_s}{\partial Q^3}\right), \qquad (3.7)$$

(where we omitted to specifically write the boundary condition $Q = 0$ on the differential quotients). Here, $\mu_M$ denotes the reduced mass of the oscillator, $v$ its quantum number of the excited level. Notice that Eq. 3.7 contains a harmonic term arising from $(H_1)_{vv}$. Hence dephasing is not contingent on anharmonicity.

Squaring Eq. 3.7 and taking the average over all coordinates of the potential derivatives then gives $\langle |\omega_1(0)|^2 \rangle$ (Note 2).

We now discuss the other limiting situation of Eq. 3.5,

$$\tau_c \ll \tau_a.$$

It permits us to approximate $\hat{\psi}(t)$ in Eq. 3.3 by a delta function. Hence we can substitute $t$ in the upper integration limit by $\infty$. Multiplying out shows that the first integral gives $t\tau_c$ (see Eq. 3.4) and that the second integral vanishes. The result is

$$\phi(t) \approx \exp\{-\langle |\omega_1(0)|^2 \rangle \tau_c t\}. \tag{3.8}$$

This predicts a Lorentzian band profile of halfwidth (full width at half-peak height; see Appendix J)

$$\Delta \nu = 2\{\tau_c \langle |\omega_1(0)|^2 \rangle\}. \tag{3.9}$$

Experimentally, few pure-Gaussian or pure-Lorentzian profiles have been encountered. It is physically unrealistic to expect either limiting condition, $\tau_c \gtrless \tau_a$, to hold *generally* since a local environment is neither completely rigid nor extremely mobile. We therefore define a more sophisticated form for $\psi(t)$,[126, 176, 177]

$$\psi(t) = \exp\left(-\frac{t}{\tau_c}\right), \tag{3.10}$$

to interpolate between the extreme regimes $\tau_c \gtrless \tau_a$. Insertion of Eq. 3.10 into Eq. 3.3 yields

$$\phi(t) = \exp\left\{-\langle |\omega_1(0)|^2 \rangle \left(\tau_c^2 \left[\exp\left(-\frac{t}{\tau_c}\right) - 1\right] + \tau_c t\right)\right\}, \tag{3.11}$$

a formulation with several interesting properties.

1. For $t \ll \tau_c$, Eq. 3.11 approaches Eq. 3.6a. (To see this, develop the inner exponential into a power series up to and including $t^2$.)

2. For $t \gg \tau_c$, Eq. 3.11 approaches Eq. 3.8. (This is obvious.) Hence the associated band profile of Eq. 3.11—which cannot be written in closed form for a general $\tau_c$—is Lorentzian in its central section (long $t \leftrightarrow$ small $\omega$) and Gaussian in its wings (short $t \leftrightarrow$ high $\omega$).

3.  Defining (see Eq. 3.9)

$$\left\{\tau_c \langle |\omega_1(0)|^2 \rangle \right\}^{-1} \equiv \tau_a,$$

hence

$$\langle |\omega_1(0)|^2 \rangle = (\tau_c \tau_a)^{-1}, \tag{3.12}$$

we rewrite Eq. 3.11 accordingly to[118b]

$$\phi(t) = \exp\left[ -\frac{\tau_c}{\tau_a} \exp\left( -\frac{t}{\tau_c} \right) \right] \exp\left( \frac{\tau_c}{\tau_a} \right) \exp\left( -\frac{t}{\tau_a} \right). \tag{3.13}$$

Now we set the condition

$$\tau_a > \tau_c, \tag{3.14}$$

which permits us to simplify Eq. 3.13. Using the expansion

$$\exp\left[ -\left( \frac{\tau_c}{\tau_a} \right) \exp\left( -\frac{t}{\tau_c} \right) \right] \approx 1 - \frac{\tau_c}{\tau_a} \exp\left( -\frac{t}{\tau_c} \right),$$

we obtain

$$\phi(t) = \tau_c \exp\left( \frac{\tau_c}{\tau_a} \right) \left\{ \tau_c^{-1} \exp\left( -\frac{t}{\tau_a} \right) - \tau_a^{-1} \exp\left( -\frac{t}{\tau_c} \right) \right\}$$

or [normalized to $\phi(0) = 1$]

$$\hat{\phi}(t) = (\tau_c \tau_a)(\tau_a - \tau_c)^{-1} \left\{ \tau_c^{-1} \exp\left( -\frac{t}{\tau_a} \right) - \tau_a^{-1} \exp\left( -\frac{t}{\tau_c} \right) \right\}. \tag{3.15}$$

Inserting Eq. 3.15 into the Langevin equation (Eq. 2.32, Chapter 2), taking the Laplace transforms,[143c] and solving for the memory function $\tilde{K}(s)$ gives

$$\tilde{K}(s) = (\tau_c \tau_a)^{-1} \left( s + \tau_c^{-1} + \tau_a^{-1} \right)^{-1}$$

or, upon inversion,[143e]

$$K(t) = \langle |\omega_1(0)|^2 \rangle \exp\left( -\left[ \tau_c^{-1} + \tau_a^{-1} \right] t \right). \tag{3.16}$$

We conclude: The assumed exponential autocorrelation function of the intermolecular perturbation Hamiltonian,

$$\psi(t) = \exp\left( -\frac{t}{\tau_c} \right),$$

(Eq. 3.10) is a good approximation of the memory function associated with the amplitude autocorrelation function of the vibrational dephasing process in the regime of "fairly rapid modulation" ($\tau_a > \tau_c$).[177] In turn, we realize that the associated memory function of $K(t)$ of Eq. 3.16, $K_1(t)$, is a delta pulse (see Eq. 2.173, Chapter 2). Also, we recall that $K(t)$ is itself to be equated with a stochastic intermolecular force (see Eq. 2.30, Chapter 2). Consequently, in the range of the "very rapid" to the "fairly rapid" modulation regimes as defined by Eqs. 3.5b and 3.14, we can attach an approximate but meaningful physical concept to the modulation time $\tau_c$: We identify $\tau_c$ with the average time *between* near-instantaneous perturbation events; we may think of "collisions."

The effect of the perturbing collisions results in "motional (band) narrowing"; the vibrational band profile would be broader (more Gaussian) if no motion existed in the medium. To see this, consider first a system in the static condition $\tau_c \gg \tau_a$ (Eq. 3.5a), called "extremely slow" modulation regime. During time $\tau_a$ (the lifetime of amplitude correlation) the perturbation exerted by the environment on the observed oscillator remains fully effective and would lead to a pure Gaussian band shape (Eq. 3.6b) with full width at half-peak height,

$$\Delta\nu = 2(2\ln 2)^{1/2}\langle|\omega_1(0)|^2\rangle^{1/2}. \tag{3.17}$$

Now let us allow for motion in the environment on a time scale $\tau_c \ll \tau_a$ (Eq. 3.5b; "extremely rapid modulation"). During $\tau_a$ each oscillator has now the time to "wander" (or to "sample") the whole breadth of the distribution $\rlap{/}{p}(H_1)$ of $H_1$. This results in an *effective* homogenizing, averaging, or smoothing process of $\rlap{/}{p}(H_1)$. The influence of the intermolecular perturbation $H_1(t)$ is diminished, the band contour therefore narrows [recall that it would be a delta function, $\mathscr{I}(\omega) = \delta(\omega_0)$, for $\rlap{/}{p}(H_1) = 0$] and assumes a more Lorentzian shape (Eq. 3.8). Its halfwidth approximates Eq. 3.9.

Our model correlation function of Eq. 3.11 reflects these different time domains and smoothly interpolates between them. (1) At—and shortly after—establishment of amplitude coherence in the system ($0 \le t \le \tau_c$), collisions have not yet had time to make themselves felt. In this initial interval, $\phi(t)$ decreases parabolically according to Eq. 3.6a—the quasi-static regime. (2) At later times, $t$ approximately $\tau_c$, perturbative collisions are beginning to intervene. The rate of decay of $\phi(t)$ is slowed down (the bandwidth decreases) since the full effect of $\langle|\omega_1(0)|^2\rangle$ is curtailed, memory loss is less rapid. (3) Finally, at $t \ge \tau_c$, the decay of $\phi(t)$ is purely exponential (Eq. 3.8), Markovian, memory function $K(t)$ approximates 0, all memory of the original state is lost.

Of course, this picture is a "Gedankenexperiment." The bandwidth does not decrease with time; it stays the same. We do not start to look at a given instant at a frozen or motionless environment which, thereafter, becomes mobile; at any instant, we observe whatever motion is present (equilibrium process). But the correlation function $\phi(t)$ exactly represents this Gedankenexperiment. At short times $\phi(t)$ singles out the rigid system. At longer times it lets us see how the effects of collisions gradually drive the system toward randomness, thereby introducing irreversibility. (Refer also to end of Section 1.3, Chapter 1.)

The theoretical method applied here is an example of simplifying a complicated time-dependent phenomenon to manageable proportions. Clearly the perturbation Hamiltonian $H_1(t)$ is an intricate quantity. But by decomposing its effects into the quasi-static factor $\langle|H_1(0)|^2\rangle$—which we can measure and model—and into the time-dependent modulation factor $\psi(t)$—which we can devise, at least empirically—we have arrived at a useful solution without difficulty. Incidentally, notice that the derivations of Eqs. 3.14–3.16 are analogous to a two-variable approximation (Chapter 2, Section 5.7, Eq. 2.171). Furthermore, we encountered a collision-induced slowing-down of memory loss in a model of rotational relaxation, namely the $J$- and $M$-extended rotational diffusion model (Sections 2, Chapter 2).

Finally, some remarks on nomenclature. We note that the initial coherence of the amplitude autocorrelation function of *other* vibrational relaxation processes dephases with increasing time. In this respect the restriction of the notation "vibrational dephasing" to the particular process discussed in this section is not precise—but it enjoys historical preference. It would indeed be clearer to name this vibrational dephasing mechanism "inhomogeneous broadening," reserving the notation "homogeneous broadening" to vibrational energy relaxation phenomena. These terms are indeed found in the literature but their use is not adhered to consistently. In fact, such an inhomogeneous process is sometimes termed "homogeneous" if its rate of modulation is very rapid. This has some justification since, under this condition, the effects of the (inhomogeneous) environment are diminished.

## 2.2. Vibrational Dephasing in the Fast Modulation Regime

### 2.2.1. Dipole-Dipole Interaction in Liquid Methyl Iodide ($CH_3I$).

We now direct attention to the manifestations of dephasing processes, and we deal in greater detail with the type of information they furnish. At first, we assume that the examples describe vibrational dephasing. We show later how to ascertain whether such a process is in fact a major contributor to the width and shape of an experimental spectral profile.

Once an experimental purely vibrational correlation function $\langle m(0)m(t)\rangle = \phi_{\exp}(t)$ is obtained by numerical Fourier transformation of the spectral profile, application of the dephasing model of Eq. 3.11 is straightforward as demonstrated in the following paragraphs.

1. The mean squared frequency shift $\langle|\omega_1(0)|^2\rangle$ is an observable spectral quantity. It is equal to the second spectral moment of the (normalized) spectral density,

$$\langle|\omega_1(0)|^2\rangle = \int d\omega\, \hat{\mathscr{I}}_{\text{vib}}(\omega)(\omega - \omega_s)^2, \tag{3.18}$$

where $\omega_s$ is the solvent-shifted band center. (Usually, $\omega_s \sim \omega_0$. Hence we no longer notationally distinguish between these quantities but keep their difference in mind—see Eq. 2.75, Chapter 2.) Equation 3.18 is readily understood by remembering that we can develop any $\phi(t)$ into an even Taylor series (we assume classical correlation functions; see Appendix A). Hence

$$\hat{\phi}_{exp}(t) = 1 + \frac{1}{2}\frac{d^2}{dt^2}\langle\hat{m}(0)\hat{m}(t)\rangle_{t=0}t^2 + O(t^4)$$

$$= \int d\omega\,\hat{\mathscr{I}}_{vib}(\omega)\cos\left[(\omega - \omega_0)t\right]$$

$$= \int d\omega\,\hat{\mathscr{I}}_{vib}(\omega)\left\{1 - \frac{(\omega - \omega_0)t^2}{2} + O(t^4)\right\}.$$

Comparing terms of equal order yields

$$\frac{d^2}{dt^2}\langle\hat{m}(0)\hat{m}(t)\rangle_{t=0} = -\int d\omega\,\hat{\mathscr{I}}_{vib}(\omega)(\omega - \omega_0)^2.$$

On the other hand, differentiating Eq. 3.3 twice gives (Note 3)

$$\frac{d^2}{dt^2}\langle\hat{m}(0)\hat{m}(t)\rangle = -\hat{\phi}(t)\langle\omega_1(0)\omega_1(t)\rangle$$

$$-\frac{d}{dt}\hat{\phi}(t)\int_0^t dt\langle\omega_1(0)\omega_1(t)\rangle,$$

hence

$$-\frac{d^2}{dt^2}\langle\hat{m}(0)\hat{m}(t)\rangle_{t=0} = \langle|\omega_1(0)|^2\rangle \equiv M_{ph}(2), \qquad (3.19)$$

where $M_{ph}(2)$ is the vibrational second spectral moment for a vibrational dephasing process.

2.  Once $\phi_{exp}(t)$ and $M_{ph}(2)$ are determined, we let the computer adjust modulation time $\tau_c$ until $\phi_{exp}(t)$ and Eq. 3.11 agree.

We give an example. The totally symmetric C–I stretch of liquid $CH_3I$ (the $\nu_3$ fundamental at 526 $cm^{-1}$) was the first system whose vibrational relaxation was studied by a VV-VH polarization Raman experiment.[118b] The results were surprising, and their importance was not then widely appreciated. First, the system was found to be in the fast modulation regime (Eq. 3.5b). Second, the width of the purely vibrational contour $I_{iso}(\omega)$ decreased with increasing

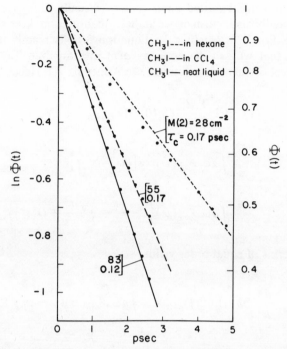

**Figure 3.1.** Correlation functions describing vibrational dephasing of the carbon–iodine stretching fundamental of neat $CH_3I$ and its 20 mole % solutions in $CCl_4$ and hexane, respectively, at 30°C. The points give the experimental data, the curve the theory with the computer-fitted modulation times $\tau_c$ as indicated. The experimental values of $M_{ph}(2)$ are also given. [M. Constant, Thèse, Université de Lille, 1972; M. Constant and R. Fauquembergue, *J. Chem. Phys.* **58**, 4030 (1975).]

temperature (at constant pressure)[178]. Both experimental observations were at odds with then generally accepted notions. It was hard to see what type of dynamic processes could bring a vibrational system into a fast modulation regime;[171] it was not understood how a bandwidth could decrease with increasing molecular kinetic energy.

Figure 3.1 shows a comparison of the experimental vibrational correlation functions (Eq. 1.47, Chapter 1) from this study[118b] with those computed from the vibrational dephasing model (Eq. 3.11), using the measured values of $M_{ph}(2) = M_{iso}(2)$ (Eq. 3.18) and the best-fit $\tau_c$. We notice the following:

1.  The parabolic part of $\phi(t)$ is relatively short-lived ($0.2 \times 10^{-12}$ sec).
2.  Correlation decay is essentially exponential in all environments.
3.  The modulation times are near-equal, regardless of the nature of the medium.
4.  But $M_{ph}(2) = \langle |\omega_1(0)|^2 \rangle$ is a strong function of the nature of the solvent.

Clearly, the systems are in the rapid modulation regime; amplitude correlation time $\tau_a$, which we can approximate here by the time where $\phi(t)$ approximates $e^{-1}$ (Eqs. 3.8 and 3.12), exceeds modulation time $\tau_c$ by one order of magnitude. Using Eq. 3.12, we rewrite the rapid modulation condition (Eq. 3.5b) as

$$\{M_{ph}(2)\tau_c\}^{-1} \gg \tau_c$$

or

$$\langle |\omega_1(0)|^2\rangle^{1/2}\tau_c \ll 1, \tag{3.20}$$

an expression familiar from nuclear magnetic relaxation ("extreme narrowing condition").

The range of the experimental modulation times $\tau_c$ in Fig. 3.1 is $0.1 \times 10^{-12}$ to $0.2 \times 10^{-12}$ sec. This is one order of magnitude shorter than intervals of typical orientational or translational correlation times (see Section 5.2, Chapter 2). This was the surprising and hard-to-believe aspect because it had been the widespread consensus that dipole-dipole interactions (long-range intermolecular forces due to permanent or transition-induced dipole vectors) are responsible for vibrational relaxation processes in liquids.

On the other hand, we had previously discussed that long-range interactions are not always effective in bringing about rotational relaxation (Sections 3.3 and 5.5, Chapter 2). We now show that long-range forces also do not generally cause fast modulation in vibrational dephasing: The reason is that effective modulation by long-range forces requires fluctuations that persist for several picoseconds.

Take, for instance, the lowest term of the well-known perturbation expansion[179a] of the interaction between two nonoverlapping dipoles $\mu_i = \mu(\mathbf{R}_i, \Omega^i)$ on molecules $i, j$, where $\mathbf{R}$ and $\Omega$ denote sets of radial and orientational coordinates with respect to a laboratory-fixed reference system,

$$H_1 = \sum_{i<j} R_{ij}^{-3}\{\mu_i \cdot \mu_j - 3(\mu_i \cdot \hat{\mathbf{R}}_{ij})(\mu_j \cdot \hat{\mathbf{R}}_{ij})\}, \tag{3.21a}$$

with $\hat{\mathbf{R}} = \mathbf{R}/R$; $R_{ij} \equiv |\mathbf{R}_i - \mathbf{R}_j|$. Casting this into a spherical harmonic expansion,[138b, 179b, c] we get

$$H_1(t) = \sum_{i<j} \mu_i\mu_j \sum_{mm'} a^{mm'}R_{ij}^{-3}(t)Y_1^m(\Omega^i(t))Y_1^{m'}(\Omega^j(t))$$

$$\times Y_2^{-m-m'}(\Omega^{ij}(t)) \tag{3.21b}$$

with $\Omega \equiv (\theta, \varphi)$. (The $a$-coefficients are simple numerical factors.[179d])

We now form the autocorrelation of $H_1(t)$ by individually averaging over the translational (radial coordinates of the intermolecular axis $R_{ij}$) and the orientational (coordinates $\Omega^i, \Omega^j$) motions. We assume that translational and orientational motions are statistically uncorrelated. We then introduce the appropriate, known dynamic parameters; thereby, we get an idea on the time scales of long-range potential fluctuations.

1. *Translational Motion.* We write the autocorrelation of the position of the intermolecular dipole-dipole Hamiltonian,

$$\left\langle R_{ij}^{-3}(0) R_{ij}^{-3}(t) Y_2^{-m-m'}\left(\Omega^{ij}(0)\right) Y_2^{-n-n'}\left(\Omega^{ij}(t)\right)\right\rangle \equiv \left\langle H_1(0) H_1(t)\right\rangle_{\text{transl}}$$

$$= \int p_0(R') P(\mathbf{R}, \mathbf{R}', t) R_{ij}'^{-3} R_{ij}^{-3} d^3\mathbf{R}' d^3\mathbf{R}$$

$$= (8\pi Dt)^{-3/2} \rho \int_\sigma^\infty d^3 R_{ij} R_{ij}^{-6} \exp\left(\frac{-R_{ij}^2}{8Dt}\right) Y_2^{-m-m'}(\Omega^{ij}) Y_2^{-n-n'}(\Omega^{ij}).$$

$$(3.22)$$

$P$ is the conditional probability that at time $t$ the two molecules have diffused by a distance $\mathbf{R} - \mathbf{R}'$ relative to each other, with zero distance at $t = 0$,[55, 180a]

$$P(\mathbf{R}, \mathbf{R}', t) = (8\pi Dt)^{-3/2} \exp\left(-\frac{[\mathbf{R} - \mathbf{R}']^2}{8Dt}\right). \qquad (3.23a)$$

$D$ is the translational self-diffusion coefficient,[180b]

$$D = \frac{1}{3}\int_0^\infty dt \left\langle \left(\frac{d\mathbf{R}_{ij}(t)}{dt}\right) \cdot \left(\frac{d\mathbf{R}_{ij}(0)}{dt}\right)\right\rangle. \qquad (3.23b)$$

$p_0$ is the probability of finding a molecule in a volume element $d^3\mathbf{R}$ at position $\mathbf{R}$,

$$p_0 = 4\pi\rho R^2 \, dR. \qquad (3.23c)$$

$\sigma$ is the distance of closest approach, and $\rho$ is the number density. (Note that there is no time dependence in the angle coordinates $\Omega^{ij}$ since the relative orientation of $R_{ij}$ with respect to the external coordinate system is immaterial for an isotropic liquid. Also, we assume that the translational motion of molecule $i$ is not statistically correlated with that of molecule $j$.)

Equation 3.22 is readily integrated over $R_{ij}$ from $\sigma$ to $\infty$ and over all $\Omega^{ij}$,[180a, b]

$$\left\langle H_1(0) H_1(t)\right\rangle_{\text{transl}} = \left(\frac{\rho}{\sigma^3}\right)\int_0^\infty du \, J_{3/2}^2(u)\exp\left(-\frac{2Du^2 t}{\sigma^2}\right) u^{-1}, \quad (3.24)$$

where $J_{3/2}$ is a Bessel function,

$$J_{3/2}(u) = \left(\frac{2}{\pi u}\right)^{1/2}\left[\frac{\sin(u)}{u} - \cos(u)\right].$$

To give an example: For $CH_3I$ at room temperature, $D = 3.5 \times 10^{-5}$ $cm^2$ $sec^{-1}$,[181] $\sigma = 4.75$ Å,[118a] giving the translational correlation time $\sigma^2/2D = 3.2 \times 10^{-11}$ sec. Inserting this value into Eq. 3.24, we find

$$\langle H_1(0)H_1(t)\rangle_{transl} \text{ approximately constant}$$

for $t \lesssim 10^{-12}$ sec.[182]

2. *Orientational Motion.* We write the correlation function

$$\langle Y_1^m(\Omega^i(t))Y_1^{m'}(\Omega^j(t))Y_1^n(\Omega^i(0))Y_1^{n'}(\Omega^j(0))\rangle$$

$$= \langle Y_1^m(\Omega^i(t))Y_1^n(\Omega^i(0))\rangle\langle Y_1^{m'}(\Omega^j(t))Y_1^{n'}(\Omega^j(0))\rangle$$

$$= (4\pi)^{-2}\exp\left(-\frac{2t}{\tau_R}\right), \qquad (3.25)$$

assuming that the orientational motion of the two dipoles is not statistically correlated and that each follows a simple, noninertial orientational model (Eq. C.11, Appendix C) with correlation time $\tau_R$. To give an example: For $CH_3I$ at room temperature, $\tau_R^{(1,0)}$ approximates $5 \times 10^{-12}$ sec.[144] Consequently, forming the product of Eqs. 3.24 and 3.25, we find

$$\langle H_1(t)H_1(0)\rangle_{dip\text{-}dip} \text{ approximately constant}$$

for $t \lesssim 1 \times 10^{-12}$ sec.

We conclude that only particularly rapid orientational or translational component motions of long-range dipole-dipole interactions can contribute effectively to motional narrowing in vibrational dephasing. We then see that the usual time scale of reorientational or translatory-diffusional intermolecular dynamics from long-range interactions is too long to account for the observed modulation times of (most) vibrational dephasing processes in common liquids. Consequently it appears that the short-range forces of the repulsive regions of the intermolecular potential are responsible for the short modulation times observed in vibrational dephasing. This, naturally, leads to the concept of a collision-induced modulation process; molecules are very close to each other. The question then arises whether we can obtain some ideas of prevalent collision mechanisms in liquids from a study of vibrational dephasing. We discuss this in the following section.

## 2.2.2. Dynamically Correlated Collision Frequencies in Liquids.

We get detailed information about collision frequencies and processes in liquids from the dynamics of a simulated liquid, the so-called hard-sphere liquid. As the name indicates, it consists of impenetrable spheres ("molecules"). This idealized system is nevertheless quite useful since it is amenable to theoretical modeling and represents well many important phenomena and features of a real liquid.[183]

We now use this concept to outline theories on translational diffusion. Translational diffusion naturally addresses our present concern about the character and magnitude of collision frequencies in liquids and about the effects that are introduced by a relatively high density (closeness of adjacent molecules). It turns out that high density introduces non-Poisson distributions of times between collisions, a phenomenon met already in rotational relaxation of strongly anisotropic liquids (Chapter 2, B).

From simple ideal gas theory, we write the velocity-dependent head-on collision frequency between two hard spheres,[184a]

$$\tau^{-1}(v) = \pi\rho\sigma^2\langle v\rangle, \qquad \langle v\rangle = \left(\frac{8k_B T}{\pi\mu_M}\right)^{1/2}. \qquad (3.26)$$

Here $\langle v\rangle$ is the average (Boltzmann) velocity, $\mu_M$ the reduced mass, $\sigma$ the molecular diameter, and $\rho$ the number density. We first merely modify this relation for the shielding effect of nearest neighbors at higher $\rho$. Figure 3.2 explains this in detail.

1.  Each molecule is in the center of a spherical boundary shell $S$ of radius $\sigma$ (that is, molecular diameter). $S$ is inaccessible to the *center* of all other molecules. This eliminates volume element $(\frac{4}{3})\pi\rho\sigma^3$ per unit volume. The collision probability is thus increased by $[1 - (\frac{4}{3})\pi\rho\sigma^3]^{-1}$.

2.  We define a critical shell volume, delimitated by two concentric shells of radii $\sigma$ and $2\sigma$. The probability that the center of a molecule lies between $x$ and $x + dx$ ($\sigma \leq x \leq 2\sigma$) within this shell is $4\pi\rho x^2\,dx$. Since the boundary shell of a molecule (say molecule 3) cuts a cap of height $h = \sigma - (\frac{1}{2})x$ and area $2\pi\sigma h$ from

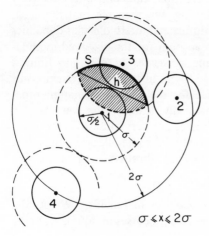

$\sigma \leqslant x \leqslant 2\sigma$

**Figure 3.2.** Screening by third molecules in a binary hard-sphere collision. During collisions of molecules 1 and 2, the center of 2 cannot lie within the upper half of the shaded area because of shielding by molecule 3. Molecule 4 is located at the minimum no-shielding distance; the associated spheres of diameter $\sigma$ just touch.

the boundary shell of the reference molecule (molecule 1), the probable overall area cut off the boundary shell of molecule 1 by the shielding of all other molecules is[184b]

$$\int_{\sigma}^{2\sigma} 2\pi\sigma(\sigma - \tfrac{1}{2}x)4\pi\rho x^2 \, dx = (\tfrac{11}{4})\pi^2\rho\sigma^5,$$

or $(\tfrac{11}{12})\pi\rho\sigma^3$ relative to the total surface area $3\pi\sigma^2$. Consequently $1 - (\tfrac{11}{12})\pi\rho\sigma^3$ is the fraction where centers of molecules can lie at collision (fractional reduction of the collision probability).

3.  Therefore

$$\frac{1 - \tfrac{11}{12}\pi\rho\sigma^3}{1 - \tfrac{4}{3}\pi\rho\sigma^3} \approx 1 + \frac{5}{12}\pi\rho\sigma^3 \equiv \chi(\rho)$$

gives the correction factor for $\tau^{-1}(v)$ (to first order in $\rho\sigma^3$) owing to shielding. The collision frequency is

$$4\pi\rho\sigma^2\left(\frac{k_B T}{\pi M}\right)^{1/2}\chi(\rho) = 4\pi\rho\sigma^2 g(\sigma)\left(\frac{k_B T}{\pi M}\right)^{1/2} \equiv \tau_E^{-1}. \qquad (3.27)$$

Its inverse, $\tau_E$ (Enskog time), is the time between collisions in terms of $M = 2\mu_M$ (molecular mass) and $g(\sigma)$ (equilibrium radial distribution function of the hard-sphere liquid).[183, 184b] Since $g(\sigma) > 1$, the collision frequency is higher in the liquid than in the dilute gas at equal temperature. The numerical estimate $\tau_E$ approximately equal to $0.1 \times 10^{-12}$ sec for liquid $CH_3I$ at 30°C agrees with the data in Fig. 3.1.[185]

If succeeding collisions are correlated, Eq. 3.27 fails. It does not account for cage effects (the bouncing of the molecule between a small number of its neighbors) we expect in a more realistic liquid. In the following discussion we outline a theory that considers such cage effects, specifically on collision frequencies, and thereby clarifies the meaning of the modulation time $\tau_c$ in the rapid modulation regime of vibrational dephasing.

The theory[186] is the translational analogue of the general stochastic model of rotational relaxation[159] (Section 7, Chapter 2). It describes the translational dynamics of a hard-sphere molecule, during a collision and between successive collisions, by an alternating sequence of stochastic events (renewal process). It is convenient to employ the van Hove self-space-time correlation function[87] (autocorrelation function of the particle density)

$$G_s(\mathbf{R} - \mathbf{R}(0), t) = \int d\mathbf{R}' \delta[\mathbf{R} + \mathbf{R}(0) - \mathbf{R}']\delta[\mathbf{R}' - \mathbf{R}(t)]$$

$$= \int d\mathbf{R}' \langle \rho(\mathbf{R}' - \mathbf{R}, 0)\rho(\mathbf{R}', t)\rangle.$$

We consider here $G_s(\mathbf{R} - \mathbf{R}(0), t)$ as the probability distribution for finding the probe molecule at a position within $dR$ at time within $dt$ when it was initially at $\mathbf{R}(0) = \mathbf{R}_0$; all other molecules are in equilibrium at $t = 0$.

Experimental methods probing $G_s(\mathbf{R}, t)$, such as neutron and Brillouin scattering[43] (Section 3.2, Chapter 1), are more closely related to the Fourier space-transform of $G_s$, the so-called intermediate scattering function

$$I_s(\mathbf{k}, t) = \langle \exp[i\mathbf{k} \cdot \mathbf{R}(t)] \exp[-i\mathbf{k} \cdot \mathbf{R}(0)] \rangle$$

$$= \int d\mathbf{R}_0 \int d\mathbf{v}_0 \int d\mathbf{R} \int d\mathbf{v} \, P(\mathbf{R}\mathbf{v}t | \mathbf{R}_0\mathbf{v}_0 0) p_{eq}(\mathbf{R}_0\mathbf{v}_0) \exp[ik \cdot (\mathbf{R} - \mathbf{R}_0)].$$

Here $\mathbf{k}$ is the scattering vector (the momentum$/\hbar$ transferred to the molecule in the scattering process—see Note 4), $\mathbf{v}$ the velocity vector of the molecule, and $P(\mathbf{R}\mathbf{v}t | \mathbf{R}_0\mathbf{v}_0 0)$ the conditional probability of finding the probe molecule within $d\mathbf{R} \, d\mathbf{v}$ of phase point $\mathbf{R}\mathbf{v}$ at a time within $dt$ of $t$ when it was initially at phase point $\mathbf{R}_0\mathbf{v}_0$. The equilibrium probability is $p_{eq}(\mathbf{R}_0\mathbf{v}_0) = V^{-1}\not{p}_B(\mathbf{v}_0)$, $V$ is the volume, and $\not{p}_B$ stands for the Maxwellian distribution of the velocities. Hence

$$I_s(\mathbf{k}, t) = \left\langle \int d\mathbf{R} \exp(i\mathbf{k} \cdot \mathbf{R}) P(\mathbf{R}\mathbf{v}t | 0\mathbf{v}_0 0) \right\rangle, \tag{3.28}$$

where $\mathbf{R}$ now signifies the displacement of the probe molecule from its original position. The angular brackets denote the average over all $\mathbf{v}_0$. (Equation 3.23$a$ gives an example for the special case of Brownian diffusion between two molecules.)

Each event is described by the product of an event duration probability function

$$p_i(t - t' | \mathbf{R}'\mathbf{v}'), \tag{3.29a}$$

and an event transition probability

$$\bar{p}_i(\mathbf{R}\mathbf{v} | \mathbf{R}'\mathbf{v}'; t - t'). \tag{3.29b}$$

Index $i$ takes the value 1 for the collisional event and the value 2 for the free translational motion ("streaming" or "drifting") between collisional events.

Introducing delta function events as required by the hard-sphere model, the single event probabilities are, for $i = 1$,

$$P_1 = \bar{p}_1(\mathbf{R}\mathbf{v} | \mathbf{R}'\mathbf{v}'; t - t') p_1(t - t' | \mathbf{R}'\mathbf{v}') = \delta(\mathbf{R} - \mathbf{R}') \mathscr{M}(\mathbf{v}|\mathbf{v}') \delta(t - t'), \tag{3.30}$$

since the collision is instantaneous and the position of the probe molecule does not change during such collision. (Recall the $J$- and $M$-extended rotational diffusion models.) The quantity $\mathscr{M}(\mathbf{v}|\mathbf{v}')$ denotes the probability that the molecule changes velocity from $\mathbf{v}'$ to $\mathbf{v}$ upon collision with equilibrium (Boltzmann distribution) collision partners.

For $i = 2$, the free translational motion event, we get

$$P_2 = \delta[\mathbf{R} - \mathbf{R}' - \mathbf{v}(t - t')]\delta(\mathbf{v} - \mathbf{v}')p_2(t - t'|\mathbf{v}') \tag{3.31}$$

since, by hypothesis, the velocity of the probe molecule stays constant and its motion is free during $t - t'$.

Notice that Eqs. 3.30 and 3.31 are the translational analogues of Eqs. 2.203 and 2.206 (collisional event) and 2.201 (free rotational motion) of Chapter 2. Hence we can at once write down the general $n$-collision term, that is to say, the probability that the probe molecule is within $d\mathbf{R}\,d\mathbf{v}$ at a time within $dt$ of $t$ after having experienced $n \geq 1$ collisions (Eqs. 2.217 and 2.210, Chapter 2). We get, dropping the subscript on $p(t|\mathbf{v})$,

$$P_n(\mathbf{R}\mathbf{v}t|0\mathbf{v}_0 0) = \tau^{-1}(v_0)\int_t^\infty dt_+ \int_0^t dt_n \int_0^{t_n} dt_{n-1} \cdots \int_0^{t_2} dt_1 \int_{-\infty}^0 dt_- \cdots$$

$$\times \int d\mathbf{R}_n \cdots \int d\mathbf{R}_1 \int d\mathbf{v}_{n-1} \cdots \int d\mathbf{v}_1\, p(t_+ - t_n|\mathbf{v})\delta[\mathbf{R} - \mathbf{R}_n - \mathbf{v}(t - t_n)]$$

$$\times \mathscr{M}(\mathbf{v}|\mathbf{v}_{n-1})p(t_n - t_{n-1}|\mathbf{v}_{n-1})\delta[\mathbf{R}_n - \mathbf{R}_{n-1} - \mathbf{v}_{n-1}(t_n - t_{n-1})]$$

$$\times \cdots \mathscr{M}(\mathbf{v}_1|\mathbf{v}_0)p(t_1 - t_-|\mathbf{v}_0)\delta(\mathbf{R}_1 - \mathbf{v}_0 t_1). \tag{3.32a}$$

For no collisions during $t_-$ and $t_+$, we obtain

$$P_0(\mathbf{R}\mathbf{v}t|0\mathbf{v}_0 0) = \tau^{-1}(v_0)\int_t^\infty dt_+ \int_{-\infty}^0 dt_- \,\delta(\mathbf{R} - \mathbf{v}_0 t)p(t_+ - t_-|\mathbf{v}_0). \tag{3.32b}$$

The mean time between collisions $\tau(v)$ is velocity dependent (Eq. 3.26):

$$\tau(v) = \int_0^\infty dt\, tp(t|\mathbf{v}). \tag{3.32c}$$

$\tau(v_0)$ in Eq. 3.32$a$ and $b$ is the mean time between collisions of the molecule emerging with velocity $\mathbf{v}_0$ after its collision at $t_- < t = 0$, having had velocity $\mathbf{v}_-$ at $t_-$. (See also Eqs. 2.211 and 2.212, Chapter 2.)

The intermediate scattering functions are obtained by inserting Eqs. 3.32 into Eq. 3.28 and integrating over $\mathbf{R}$ with the help of the delta function:

$$I_{s,n}(\mathbf{k}, t) = \left\langle \tau^{-1}(v_0)\int d\mathbf{v}_{n-1} \cdots \int d\mathbf{v}_2 \int d\mathbf{v}_1 \int_t^\infty dt_+ \int_0^t dt_n \int_0^{t_n} dt_{n-1} \right.$$

$$\times \cdots \int_0^{t_2} dt_1 \int_{-\infty}^0 dt_- \exp[i\mathbf{k}\cdot\mathbf{v}(t - t_n)]\, p(t_+ - t_n|\mathbf{v})\mathscr{M}(\mathbf{v}|\mathbf{v}_{n-1})$$

$$\times \exp[i\mathbf{k}\cdot\mathbf{v}_{n-1}(t_n - t_{n-1})]\, p(t_n - t_{n-1}|\mathbf{v}_{n-1}) \cdots \mathscr{M}(\mathbf{v}_2|\mathbf{v}_1)$$

$$\times \left. \exp[i\mathbf{k}\cdot\mathbf{v}_1(t_2 - t_1)]\, p(t_2 - t_1|\mathbf{v}_1)\mathscr{M}(\mathbf{v}_1|\mathbf{v}_0)\exp[i\mathbf{k}\cdot\mathbf{v}_0 t_1]\, p(t_1 - t_-|\mathbf{v}_0) \right\rangle$$

$$\tag{3.33a}$$

for $n \geq 1$, and

$$I_{s,0}(\mathbf{k}, t) = \left\langle \tau^{-1}(v_0) \int_t^\infty dt_+ \int_{-\infty}^0 dt_- \exp[i\mathbf{k} \cdot \mathbf{v}_0 t] \, p(t_+ - t_- |\mathbf{v}_0) \delta(\mathbf{v} - \mathbf{v}_0) \right\rangle$$

(3.33b)

for $n = 0$.

Taking the Laplace transforms yields[143g, i]

$$\tilde{I}_{s,n}(\mathbf{k}, s) = \left\langle \tau^{-1}(v_0) \int d\mathbf{v}_{n-1} \cdots \int d\mathbf{v}_1 z^{-1} [1 - \tilde{p}(z|\mathbf{v})] \mathcal{M}(\mathbf{v}|\mathbf{v}_{n-1}) \right.$$

$$\times \tilde{p}(z_{n-1}|\mathbf{v}_{n-1}) \mathcal{M}(\mathbf{v}_{n-1}|\mathbf{v}_{n-2}) \tilde{p}(z_{n-2}|\mathbf{v}_{n-2}) \cdots$$

$$\left. \times \mathcal{M}(\mathbf{v}_2|\mathbf{v}_1) \tilde{p}(z_1|\mathbf{v}_1) \mathcal{M}(\mathbf{v}_1|\mathbf{v}_0) z_0^{-1} [1 - \tilde{p}(z_0|\mathbf{v}_0)] \right\rangle \quad (3.34a)$$

and

$$\tilde{I}_{s,0}(\mathbf{k}, s) = \left\langle \{z_0^{-1} - \tau^{-1}(v_0) z_0^{-2} [1 - \tilde{p}(z_0|\mathbf{v}_0)]\} \delta(\mathbf{v} - \mathbf{v}_0) \right\rangle \quad (3.34b)$$

where

$$z_j = s - i\mathbf{k} \cdot \mathbf{v}_j$$

and

$$\tilde{p}(z_j|\mathbf{v}_j) = \int_0^\infty dt \exp\left[-(s - i\mathbf{k} \cdot \mathbf{v}_j)t\right] p(t|\mathbf{v}_j).$$

To obtain Eq. 3.34b it is convenient to define new coordinates, $x = t_+ - t_-$, $y = t_+ + t_-$. The Jacobian is $\frac{1}{2}$, the integration limits $2t - x \leq y \leq x$, $t \leq x \leq \infty$. Then split the integral from $t$ to $\infty$ into the domains $t$ to $0$ and $0$ to $\infty$ and apply the Laplace transforms.[143i]

Note the resemblance of Eqs. 3.33 and Eq. 2.217 (Chapter 2) and recall that the integrals from $t$ to $\infty$ and $-\infty$ to $0$ involve the averaging over the backward and forward recurrence times.[163] (Refer also to Eqs. 2.211 and 2.212 in Chapter 2.)

Writing the quantities in Eq. 3.34 as operators in a velocity basis,

$$(\mathbf{v}_m| \mathcal{M} |\mathbf{v}_n) = \mathcal{M}(\mathbf{v}_m|\mathbf{v}_n)$$

$$(\mathbf{v}_m| \tilde{\mathbf{p}} |\mathbf{v}_n) = \tilde{p}(\mathbf{z}_m|\mathbf{v}_m) \delta(\mathbf{v}_m - \mathbf{v}_n)$$

$$(\mathbf{v}_m| \tau |\mathbf{v}_n) = \tau(\mathbf{v}_m) \delta(\mathbf{v}_m - \mathbf{v}_n)$$

$$(\mathbf{v}_m| \mathbf{z} |\mathbf{v}_n) = (s - i\mathbf{k} \cdot \mathbf{v}_m) \delta(\mathbf{v}_m - \mathbf{v}_n), \quad (3.35)$$

where $\delta(\mathbf{v}_m - \mathbf{v}_n) = (\mathbf{v}_m | \mathbf{1} | \mathbf{v}_n)$, we get the more concise expressions

$$\tilde{I}_{s,0}(\mathbf{k}, s) = \left\langle (\mathbf{v} | z^{-1} - (1 - \tilde{\mathbf{p}})(z^2 \tau)^{-1} | \mathbf{v}_0) \right\rangle,$$

$$\tilde{I}_{s,n}(\mathbf{k}, s) = \left\langle (\mathbf{v} | z^{-1} (1 - \tilde{\mathbf{p}})(\mathcal{M}\tilde{\mathbf{p}})^{n-1} \mathcal{M}(1 - \tilde{\mathbf{p}})(z\tau)^{-1} | \mathbf{v}_0) \right\rangle.$$

Summing over $n = 0$ to $\infty$,

$$\tilde{I}_s(\mathbf{k}, s) = \sum_{n=0}^{\infty} \tilde{I}_{s,n}(\mathbf{k}, s),$$

and noting that the infinite sum is a geometric series,

$$\sum_{n}^{\infty} (\mathcal{M}\tilde{\mathbf{p}})^n = (1 - \mathcal{M}\tilde{\mathbf{p}})^{-1},$$

we obtain

$$\tilde{I}_s(\mathbf{k}, s)$$

$$= \left\langle \left( \mathbf{v} | z^{-1} - (1 - \tilde{\mathbf{p}})(z^2 \tau)^{-1} + z^{-1}(1 - \tilde{\mathbf{p}})(1 - \mathcal{M}\tilde{\mathbf{p}})^{-1} \mathcal{M}(1 - \tilde{\mathbf{p}})(\tau z)^{-1} | \mathbf{v}_0 \right) \right\rangle.$$

It will be more instructive to write this in terms of its memory function $K(t)$ (Eq. 2.21, Chapter 2; note the change of sign),

$$\tilde{I}_s(k, s) = \left\langle (\mathbf{v} | (z - \tilde{\mathbf{K}})^{-1} | \mathbf{v}_0) \right\rangle,$$

where

$$\tilde{\mathbf{K}} = (1 - \mathbf{E}\tilde{\mathbf{g}})^{-1} \mathbf{E},$$

$$\tilde{\mathbf{g}} = \tau \tilde{\mathbf{p}}(1 - \tilde{\mathbf{p}})^{-1} - z^{-1},$$

$$\mathbf{E} = (\mathcal{M} - 1)\tau^{-1}. \tag{3.36}$$

The quantity $\tilde{g}$ is called "event duration cumulant function" or "cumulant free-flight distribution";[163] we show the reason for this in the next paragraph. $\mathbf{E} = (\mathcal{M} - 1)/\tau$ is the Boltzmann-Enskog collision operator (for hard spheres); for molecules of velocity $\mathbf{v}$, $1/\tau(v)$ is the "loss," $\mathcal{M}/\tau(v)$ is the "gain" rate.[186, 187]

First, let us look at the significance of $\tilde{g}$; it is most pertinent to our discussion. We replace the general probability distribution for the free-streaming event $\tilde{p}(z | \mathbf{v})$ by the Poisson distribution $[1 + z\tau(v)]^{-1}$ (Eq. 2.192, Chapter

2, with $\alpha = 1$). Inserting this into the matrix element of $\tilde{g}(z|\mathbf{v})$ of Eq. 3.36 gives

$$\tilde{g}(z|\mathbf{v}) = 0. \tag{3.37}$$

Hence, for a random distribution of event durations (uncorrelated collisions of the hard-sphere molecules), $\tilde{g}(z|\mathbf{v})$ vanishes and $\tilde{\mathbf{K}} = \mathbf{E}$. We are, indeed, reminded here of a cumulant property: In the absence of statistical correlation between (two) variables, their cumulant vanishes (Appendix H).

Further discussion is facilitated by taking the limit $k \to 0$ valid for optical spectra (infinitely small momentum transfer). This leads to the vector velocity autocorrelation function[186, 188]

$$C_v(t) = \left(\tfrac{1}{3}\right)\langle \mathbf{v}(t)\cdot\mathbf{v}(0)\rangle, \tag{3.38}$$

which gives us a conceptually simpler, yet sufficiently accurate, picture of collision times and dynamics than $I_s(\mathbf{k}, t)$.

To outline this, we differentiate $I_s(\mathbf{k}, t)$ twice with respect to $t$,

$$\frac{d^2}{dt^2}\left\langle \exp[i\mathbf{k}\cdot\mathbf{R}(t)]\exp[-i\mathbf{k}\cdot\mathbf{R}_0]\right\rangle$$

$$= -\mathbf{k}\cdot\left\langle \frac{d}{dt}\mathbf{R}(t)\{\exp[i\mathbf{k}\cdot\mathbf{R}(t)]\}\exp[-i\mathbf{k}\cdot\mathbf{R}_0]\frac{d}{dt}\mathbf{R}_0\right\rangle\cdot\mathbf{k} \tag{3.39}$$

(refer to Eq. B.2, Appendix B). Putting $(d/dt)\mathbf{R}(t) = \mathbf{v}(t)$, we get

$$\frac{d^2}{dt^2}I_s(\mathbf{k}, t) = -\mathbf{k}\cdot\left\langle \exp(iLt)\mathbf{v}|\exp[i\mathbf{k}\cdot\mathbf{R}(t)]\exp[-i\mathbf{k}\cdot\mathbf{R}_0]|\mathbf{v}\right\rangle\cdot\mathbf{k}, \tag{3.40}$$

where we have set (omitting subscripts of $v$)

$$\mathbf{v}(t) = \exp(iLt)\mathbf{v} \tag{3.41}$$

and used Eq. 2.139 (Chapter 2).

Now we form the Laplace transforms, divide by $\mathbf{k}\cdot\mathbf{k} = k^2$, and take the limit $k \to 0$. This gives[143a, h]

$$\lim_{k \to 0}\frac{1}{k^2}\{s^2\tilde{I}_s(\mathbf{k}, s) - sI_s(\mathbf{k}, t=0)\} = -\langle\mathbf{v}|(s - iL)^{-1}|\mathbf{v}\rangle. \tag{3.42}$$

The inverse of Eq. 3.42 is indeed the velocity autocorrelation function,[143e]

$$\mathscr{L}\{\langle\mathbf{v}|(s - iL)^{-1}|\mathbf{v}\rangle\} = \langle\mathbf{v}(t)\mathbf{v}(0)\rangle \equiv C_v(t), \tag{3.43}$$

with $v(t)$ expressed by Eq. 3.41. Writing this in terms of the components $v_i$ ($i = 1, 2, 3$) of $v$, we realize that $\langle v(0)v(t) \rangle$ is diagonal[188] (see also Appendix B). Hence

$$\langle v_i(0)v_i(t) \rangle = \left(\tfrac{1}{3}\right)\langle v(0) \cdot v(t) \rangle \tag{3.44}$$

or

$$\langle v(0)v(t) \rangle = \left(\tfrac{1}{3}\right)\langle v(0) \cdot v(t) \rangle \mathbf{1}, \tag{3.45}$$

where $\mathbf{1}$ is the unit tensor.

Now we take the limit $s \to 0$. We get

$$\lim_{s \to 0} \tilde{C}_v(s, \mathbf{k} = 0) = \lim_{s \to 0} \langle v | (s - iL)^{-1} | v \rangle$$

$$= \lim_{s \to 0} \frac{1}{3} \int_0^\infty dt \exp(-st) \langle v(0) \cdot v(t) \rangle$$

$$= D, \tag{3.46}$$

where the last equality follows from Eq. 3.23$b$.

We then continue our discussion on the collision times in liquids, using $\tilde{C}_v(s, \mathbf{k} = 0)$ or coefficient $D$ instead of $\tilde{I}_s(\mathbf{k}, s)$. We write, again in terms of operator matrix elements in the v-basis,

$$\tilde{C}_v(s, \mathbf{k} = 0) = \left(\tfrac{1}{3}\right)\langle v \cdot (v | (s - \tilde{K})^{-1} | v_0) v_0 \rangle. \tag{3.47}$$

As a first approximation, we consider a Poisson process ($\tilde{K} = E$, see Eqs. 3.36 and 3.37) and omit nondiagonal matrix elements of collision operator $E$. This gives[143h]

$$C_v(t) = \exp(Et) = \exp\left(-\frac{t}{t_v}\right), \tag{3.48}$$

where (Eq. 3.35)

$$\langle v | E | v' \rangle = (\mathscr{M} - 1)\tau^{-1} = -\tau_v^{-1}\delta(v - v')$$

(since $\mathscr{M}$ is nondiagonal). The velocity relaxation constant $\tau_v^{-1}$ is related to the Enskog collision frequency $\tau_E^{-1}$ by[183c, 189]

$$\tau_v^{-1} = \tau_E^{-1}, \tag{3.49}$$

smooth or slip sphere, and

$$\tau_v^{-1} = \frac{\tau_E^{-1}(1 + 2\kappa)}{1 + \kappa}, \tag{3.50}$$

completely rough or nonslip sphere. (See also Section 5.5, Chapter 2.) The parameter $\kappa = 4I/M^2$ ($I$ = moment of inertia, $M$ = mass) is of the order of 0.2 for molecules of interest.

Next, we consider the more interesting and realistic case of correlated collisional events. We again make use of the gamma distribution for $\tilde{p}(s|v)$ (Eq. 2.192, Chapter 2) and therefore write for the cumulant event function of the free-drifting event (Eq. 3.36), with $\tau(v)$ = mean time between collisions,

$$\tilde{g}(s|v) = \tau(v) \left\{ \left[ 1 + \frac{s}{\alpha} \tau(v) \right]^{\alpha} - 1 \right\}^{-1} - s^{-1}. \tag{3.51a}$$

Inversion gives[186]

$$g(t|v) \approx 2 \exp\left[ -\frac{2\pi^2 t}{\alpha \tau(v)} \right] \cos\left[ \frac{2\pi t}{\tau(v)} \right] \tag{3.51b}$$

for $\alpha \geq 2\pi$ and long times. The cumulant event function decays exponentially; the faster the decay, the more $p(t|v)$ is peaked at $\tau(v)$ (the larger $\alpha$). Furthermore, $g(t|v)$ oscillates with maxima at time intervals approximating $\tau(v)$.

The resulting expression for $C_v(t)$ is too cumbersome to be useful to us. We consider instead the simpler but sufficiently instructive frequency-independent translational diffusion coefficient $D$. Keeping only diagonal terms, we have (Eqs. 3.36 and 3.47)

$$\tilde{C}_v(s) = \tfrac{1}{3} \langle \mathbf{v} \cdot | \left[ s - \mathbf{E}(1 - \tilde{g}\mathbf{E})^{-1} \right]^{-1} | \mathbf{v}_0 \rangle$$

$$= \tfrac{1}{3} \langle \mathbf{v} \cdot | \left[ s + \{ \tau_v + \tilde{g}(s|v) \}^{-1} \right]^{-1} | \mathbf{v}_0 \rangle \tag{3.52}$$

with

$$\langle \mathbf{v} | \mathbf{E} | \mathbf{v}' \rangle = -\tau_v^{-1} \delta(\mathbf{v} - \mathbf{v}').$$

Now letting $s \to 0$ gives

$$D = \lim_{s \to 0} \tilde{C}_v(s, \mathbf{k} = 0) = \tfrac{1}{3} \{ \langle \tau_v v^2 \rangle + \langle v^2 \tilde{g}(0|v) \rangle \}. \tag{3.53}$$

The first term,

$$\tfrac{1}{3} \langle \tau_v v^2 \rangle = \frac{k_B T}{M} \tau_v \equiv D_E,$$

is the diffusion coefficient from the Boltzmann-Enskog theory (in the diagonal approximation). The second term reads, after substituting $\tilde{g}(s|v)$ by Eq. 3.51a and performing the limit $s \to 0$ (with the help of L'Hospital's rule),

$$\frac{1}{3} \langle v^2 \tilde{g}(0|v) \rangle = -\frac{1}{3} \frac{\alpha - 1}{2\alpha} \langle v^2 \tau(v) \rangle \equiv D_{\text{corr}}. \tag{3.54}$$

$D_{corr}$, which arises from dynamic correlations of the collisional events, is *negative*. The self-diffusion in the liquid is therefore diminished relative to the Boltzmann-Enskog value. Clearly, we can consider this the result of frequent molecular encounters of the probe molecule with the same neighbors, within the solvent cage, at periodic intervals.[186]

We have now obtained a better understanding of collisional processes in liquids. We have seen that collision events where the probe molecule bounces or rattles between the same neighbors have event cumulant functions with a rapid exponential decay and oscillations with maxima at $\tau(v)$. The time evolution of the velocity correlation function (Eq. 3.52) is correspondingly complicated. At shorter times, the oscillations of $g(t|v)$ are washed out by the ensemble average over **v** with the Maxwellian distribution function $f_B(\mathbf{v})$,

$$\langle f(\mathbf{v}\mathbf{v}_0)\rangle = \int d\mathbf{v} \int d\mathbf{v}_0 f_B(\mathbf{v}) f(\mathbf{v}\mathbf{v}_0).$$

At long times and for large $\alpha$, $C_v(t)$ is exponential,

$$C_v(t) \approx \exp\left(-\frac{t}{\tau_E}\right),$$

(each collision randomizes $v$). It is obvious that there is no way in which a modulation time of a vibrational dephasing process could be used to distinguish the nature of the collision dynamics. On the other hand, it now turns out that an exponential modulation function (Eq. 3.10) is a reasonable choice for longer times, even in systems with dynamically correlated collisions. However, this does not tell us anything new: At long times we (almost) always encounter Markovian behavior, regardless of the dynamics of the system.

### 2.2.3. Collision Frequencies in Symmetric Top and Linear Molecules.

We first address ourselves to an important question: How do we know that this or that vibrational band profile or correlation function represents a vibrational dephasing process and not any other of the possible avenues of vibrational relaxation? The point here is not to ascertain the presence of a vibrational dephasing process—local inhomogeneities of the environment always cause vibrational dephasing—but to determine its relative importance among vibrational energy relaxation, vibrational resonance transfer, and intramolecular vibrational relaxation. We concern ourselves first with spontaneous relaxation, leaving stimulated relaxation (Section 4.3, Chapter 1) for later.

Figure 3.3a–d demonstrates how solvent and temperature effects allow us to tag the importance of the various concomitant vibrational relaxation mechanisms. As example, we take two totally symmetric modes of methyl iodide,[190] $\nu_1$ (2950 cm$^{-1}$), the carbon–hydrogen stretching, and $\nu_2$ (1250 cm$^{-1}$), the carbon–hydrogen deformation fundamental.

First we look at solvent effects. From Fig. 3.3a and c we notice that $\nu_1$ is little but $\nu_2$ is greatly affected by dilution. In particular, dilution by $CD_3I$

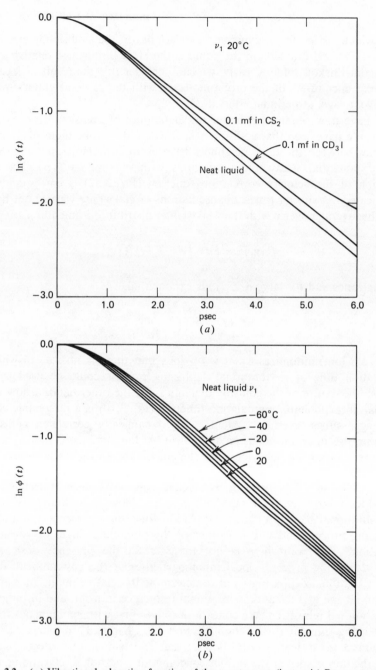

**Figure 3.3.** (*a*) Vibrational relaxation function of the spontaneous (isotropic) Raman contour of the $\nu_1$ mode of methyl iodide (2950 cm$^{-1}$) at 20°C, neat liquid and dissolved in CD$_3$I and CS$_2$. The abbreviation mf represents mole fraction. (*b*) Vibrational relaxation function of the spontaneous (isotropic) Raman contour of the $\nu_1$ mode of neat methyl iodide at several temperatures. (*c*) Vibrational relaxation function of the spontaneous (isotropic) Raman contour of the $\nu_2$ mode of methyl iodide (1250 cm$^{-1}$) at 20°C, neat liquid and dissolved in CD$_3$I and CS$_2$. (*d*) Vibrational relaxation function of the spontaneous (isotropic) Raman contour of the $\nu_2$ mode of neat methyl iodide at several temperatures. [G. Döge, R. Arndt, and A. Khuen, *Chem. Phys.* **21**, 53 (1977).]

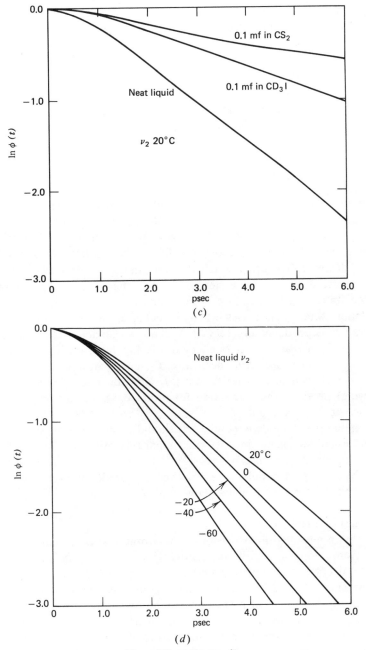

Figure 3.3. (*Continued*)

shows that $\nu_1$ is slightly and $\nu_2$ is strongly influenced by vibrational resonance energy transfer since the correlation functions decrease more slowly in the isotopically diluted medium.

Next we consider the temperature dependence. It is slightly positive for $\nu_1$ (Fig. 3.3b); its relaxation rate increases roughly with the square root of the absolute temperature. In contrast, the temperature dependence of the relaxation rate of $\nu_2$ (Fig. 3.3d) is large and negative; its decrease is faster than the increase of $(k_B T)^{1/2}$.

From the strong solvent dependence and the negative temperature dependence of $\nu_2$ we conclude that this mode is mainly influenced by vibrational dephasing. Why?

Clearly, changing a solvent changes the molecular environment of the oscillator $[\langle | \omega_1(0)|^2 \rangle$, Eq. 3.18] as well as the collision rate $(\tau_c^{-1}$, Eq. 3.4). The direction and magnitude of the changes, and hence the characteristics of the resulting correlation function, are difficult to predict: *Usually*, nonpolar solvents decrease $\langle \omega |_1(0)|^2 \rangle$, smaller solvent molecules decrease $\tau_c$ (Eq. 3.26).

Temperature effects are easier to foretell than are solvent effects. For instance, increasing temperature $T$ must raise the rate of collisions; therefore, modulation time $\tau_c$ decreases when $T$ increases. But the direction of the temperature dependence of $\langle | \omega_1(0)|^2 \rangle$ is not so obvious. The quantity may increase with increasing $T$ (less energetically favorable environments become populated) or decrease with increasing $T$ (the environment becomes more homogeneous, its distribution narrower). *Usually*, the effect due to $\tau_c$ (motional narrowing) prevails; the correlation function decays more slowly (the bandwidth is narrower) at higher $T$.

By these criteria mode $\nu_1$ (Fig. 3.3b) should possess a relatively low degree of vibrational dephasing since its vibrational relaxation rate increases with $T$ and its solvent dependence is weak.[190, 191] In fact, mode $\nu_1$ relaxes predominantly by phonon depopulation of its excited vibrational level—as surprising as this appears for such a high-energy mode (see Section 4.1, Chapter 1). We give the experimental proof and its explanation (furnished by induced Raman scattering) in a later section.

To obtain meaningful conclusions from vibrational dephasing data without prior assumption of a collision model,[192] it is necessary to measure the vibrational correlation function under conditions where the second spectral moment (Eqs. 3.18, 3.19) has *absolutely* converged. Otherwise, conclusions based on Eq. 3.11, give, at best, (upper) limiting values of modulation time $\tau_c$ (assuming that, as usual, $M(2)$ is experimentally underestimated). Unfortunately, not many experimental situations lend themselves to such thorough analysis. For instance, data of the totally symmetric C–N fundamental stretching mode of $CH_3CN$ (acetonitrile), $\nu_2$ (near 2250 cm$^{-1}$), fall into this category.[193] It turns out that $M(2)$ has only relatively converged since the baseline of the Raman profile is not white-noise Rayleigh scattering but the wing of the extremely broad degenerate $\nu_5$ C–H stretching fundamental near 3000 cm$^{-1}$.[20, 194] Similarly, we are concerned about the totally symmetric C–H stretching fundamental $\nu_1$ of $CH_3CN$ (near 2940 cm$^{-1}$).[20] Its experimental

$M(2)$ is suspect since the baseline of its contour is also influenced by the wing of $\nu_5$.[193]

We discuss in the following paragraphs the vibrational dephasing phenomena of the totally symmetric C–D stretching fundamental $\nu_1$ (2255 cm$^{-1}$) of deuterochloroform, CDCl$_3$. Its profile is not perturbed by adjacent modes. The baseline is pure Rayleigh scattering. Fluorescence of the sample can be readily eliminated by purification of the compound. Chlorine isotope effects are insignificant. The contribution of rotational relaxation to the overall bandwidth is small;[10b] we need not worry about the success of the VV-VH polarization experiment (vibration-rotation coupling should be minimal).[195] Crystalline-phase $\nu_1$ shows no significant factor or site group splittings; hence we deal only with a single oscillator (Section 3.3 and 4.2.1, Chapter 2). Strong dependence of the bandwidth on the nature of the sovent[113b] but insignificant changes upon isotopic dilution (in CHCl$_3$)[10b] indeed indicate the prevalence of vibrational dephasing. The only contradictory evidence is a slightly positive temperature dependence of the vibrational relaxation rate[112, 196a]. Also, the contour of the mode is slightly asymmetric to higher frequencies. We ignore this asymmetry which, curiously, is much stronger in the infrared (and particularly in the protonated compound) than in the Raman profile.[113b]

Figure 3.4 shows the experimental (dots) and the theoretical vibrational correlation function $\phi(t)$ (Eq. 3.11, solid curve) with the best-fit value of $\tau_c = 0.13 \times 10^{-12}$ sec.[177] The dash curve is the exponential memory function $K(t)$ (Eq. 3.16). The open circles represent the experimental memory function obtained numerically from the experimental correlation function by way of the Langevin equation (Eq. 2.32, Chapter 2).[145b] Notice that the dip of $K(t)$ into the negative region is not unusual (Fig. 2.35), but it is difficult to say here how much of it is caused by the inherent inaccuracy of $\phi(t)$ at short times. For instance, we have no knowledge of $\phi(t)$ for times shorter than[14]

$$t_{min} \sim \frac{\pi}{\Delta\omega_{\pm}}, \tag{3.55}$$

where $\Delta\omega_{\pm}$ denotes the angular frequency range (counted from the band center) of the band profile. We compute[177]

$$t_{min} \sim 0.17 \times 10^{-12} \text{ sec,}$$

close to the time where the experimental $K(t)$ begins to reach negative values (see Fig. 3.4).

From Eq. 3.20 we compute that the system is in the rapid modulation regime

$$M(2)^{1/2}\tau_c = 0.26.$$

Vibrational correlation time $\tau_a$ (Eq. 3.5c) amounts to $2.0 \times 10^{-12}$ sec; hence even Eq. 3.5b is satisfied. The assumed exponential modulation function (Eq.

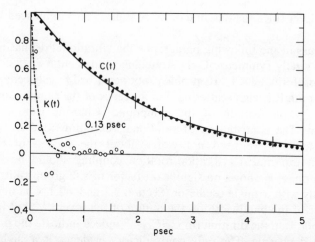

**Figure 3.4.** Vibrational dephasing of the C–D stretch (2255 cm$^{-1}$) of liquid deuterochloroform at about 40°C. (Note that the temperature in the original reference (10*b*) is in error.) Solid points: experimental correlation function of the isotropic Raman contour. Solid curve: computed correlation function (Eq. 3.11) using the best-fit value $\tau_c = 0.13 \times 10^{-12}$ sec. Open points: experimental memory function, from (●) with Eq. 2.32, Chapter 2. The error bars show the computed range with $\tau_c = 0.11 \times 10^{-12}$ sec (upper limit) and $0.15 \times 10^{-12}$ sec (lower). [W. G. Rothschild, *J. Chem. Phys.* **65**, 455 (1976).]

3.10) is indeed a good approximation of the effective memory function (Eq. 3.16) of $\phi(t)$.

Spectral investigations of the isotropic $\nu_1$ Raman contour of CDCl$_3$ as a function of density permit us to characterize modulation time $\tau_c$ more closely.[196] In particular, we are interested in relating $\tau_c$ (Eq. 3.10) to collision times. We have plotted in Fig. 3.5 the density dependence of $\tau_c$ and of $\tau_E$ (Eq. 3.27).[196a] Disagreement between the absolute values of $\tau_c$ and $\tau_E$ is not upsetting; because of the sensitivity of $\tau_E$ to the value of the pair correlation $g(\sigma)$, we do not expect otherwise. But it is curious to note that $\tau_c$ is density independent

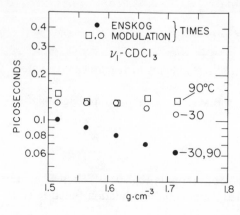

**Figure 3.5.** Density dependence of the Enskog and the modulation time of the isotropic Raman C–D stretching fundamental of deuterochloroform (2255 cm$^{-1}$) at 30 and 90°C. [Data from Ref. 196*a*.]

whereas $\tau_E$ drops (as expected) with decreasing molecular separation in the liquid.

In contrast to optimism expressed in the literature,[177, 192, 196a] this observation does not seem to permit us to relate $\tau_c$ with the simple model of random collision events. In fact, it demonstrates the limitations of applying vibrational dephasing studies (in the fast modulation regime) to liquid-phase collision dynamics. Although the order of magnitude of the $\tau_c$ values agrees with collision times in liquids, and although the principle of relating $\tau_c$ with $\tau_E$ or even with $\tau_J$ (the angular momentum correlation time, see Eq. 2.11 of Chapter 2)[118a] is sound, the information remains too meager and too coarse to permit *detailed* predictions of collision effects.

Let us now discuss the dynamics of nonassociated, nonpolar liquids. The relatively simple structure of liquid nitrogen and oxygen puts these fluids high on the list of interesting candidates for vibrational dephasing studies. Intermolecular potential, radial distribution function, and anharmonicity parameter of their (only) vibrational fundamental are known.

In particular, liquid nitrogen near its boiling point has been thoroughly studied experimentally and theoretically. To obtain estimates of $\tau_c$, calculations of $\langle |\omega_1(0)|^2 \rangle$ (Eq. 3.7), using the known parameters of liquid nitrogen, were performed. The result was inserted into Eq. 3.11, the halfwidths (parametrized in $\tau_c$) were computed, and their values compared with that from a spectral measurement.[197, 198] The comparison yielded $\tau_c = 0.06 \times 10^{-12}$ sec. A value of $\tau_c = 0.15 \times 10^{-12}$ sec was obtained from molecular dynamics calculations (simulations of molecular motions by computer).[199]

The spontaneous (Raman) contour of the liquid nitrogen fundamental is Lorentzian over its range of observation; its halfwidth, $0.067$ cm$^{-1}$,[198] corresponds to amplitude correlation time $\tau_a$,

$$\tau_a^{(\text{spont})} = \frac{1}{\pi c 0.067} = 158 \times 10^{-12} \text{ sec.}$$

In addition, the decay time $\tau_a^{(\text{ind})}$ of the induced amplitude phase coherence (picosecond pulsing, see Section 4.3, Chapter 1) of the liquid-phase nitrogen fundamental was measured,[200]

$$\tau_a^{(\text{ind})} = 75 \times 10^{-12} \text{ sec.}$$

That this result is equivalent to $\tau_a^{(\text{spont})} = 158 \times 10^{-12}$ sec is explained in Appendix J.

It later turned out that the computations of $\langle |\omega_1(0)|^2 \rangle$[197] contain several errors. [The derivatives of the intermolecular potential (Eq. 3.7) of binary interactions between two $N_2$ molecules (with a four-center atom-atom 6-12 potential)[138a, 201a] had been formed with respect to the center-of-mass separation of the two molecules, instead with respect to the atom-atom distances. Furthermore, the anharmonicity constant $f$ had been underestimated by 40%.] The apparent agreement of the values of $\tau_c$ from these calculations with the molecular dynamics simulations is therefore fortuitous. Nevertheless, the essence of the calculation of $\langle|\omega_1(0)|^2\rangle$ by way of Eq. 3.7 remains valid. However, we should realize that the results depend critically on the radial distribution function, particularly at short distances.

It is interesting that modulation time $\tau_c = 0.15 \times 10^{-12}$ is close to the velocity correlation time $\tau_v = 0.11 \times 10^{-12}$ sec.[201b] (Refer also to Section 2.2.2.) It appears that translational motion is responsible for the fast modulation regime in liquid nitrogen. Incidentally,

$$\langle|\omega_1(0)|^2\rangle^{1/2}\tau_c = 0.035.$$

Hence, for all practical purposes, an exponential memory function (Eq. 3.16) is valid for the dephasing in the liquid nitrogen system.

In situations where modulation is less rapid, an exponential memory function is no longer useful (certainly not at short times). Attempts have been made to improve the description of the dynamics of dephasing by modeling the memory function and to solve the Langevin equation (Section 5.6–5.8, Chapter 2). We now outline such an attempt.[202]

We define the function

$$K_2(t) = N_2 K_2^{RL}(t)\exp\left(-\frac{t}{\tau}\right) \tag{3.56}$$

as a second-rank memory function of correlation function $\phi(t)$ in the Mori scheme (Eq. 2.184, Chapter 2). $N_2 = [\kappa(4)/2M(2)^2]+1$ is the normalization constant in terms of a reduced moment[202] $\kappa(4)$ and the second moment $M(2)$ of the vibrational contour (see also Eq. 2.188, Chapter 2). Specifically, $K_2^{RL}(t)$ is considered the "rigid lattice" second-rank memory function of a reference system that is characterized by a vibrational correlation function in the quasi-static modulation regime (Eq. 3.6a), $K_0^{RL}(t)$. Parameter $\tau$ is an adjustable correlation time (see also Eq. 3.10).

$K_2^{RL}(t)$ is generated from Eq. 3.6a by twice applying Eq. 2.180 (Chapter 2). Starting with $\exp\{-M(2)t^2/2\} \equiv K_0^{RL}(t)$ gives $K_1^{RL}(t)$; repeating with $K_1^{RL}(t)$ gives $K_2^{RL}(t)$. Then, the reversed procedure is performed to construct the final correlation function $K_0(t)$: From $K_2^{RL}(t)$ we get $K_2(t)$ (by way of Eq. 3.56); we insert $K_2(t)$ into Eq. 2.180 and solve it successively for $K_1(t)$ and $K_0(t) = \phi(t)$. Schematically,

$$\text{Eq. } 3.6a \equiv K_0^{RL}(t) \xrightarrow{\text{Eq. } 2.180} K_1^{RL}(t) \xrightarrow{\text{Eq. } 2.180} K_2^{RL}(t)$$

Static system        Eq. 3.56

$$K_0(t) \equiv \phi(t) \xleftarrow{\text{Eq. } 2.180} K_1(t) \xleftarrow{\text{Eq. } 2.180} K_2(t)$$

Dynamic system

(Note that the $K_i^{RL}(t)$ and $K_i(t)$, save $K_0^{RL}(t)$, are not analytical.)

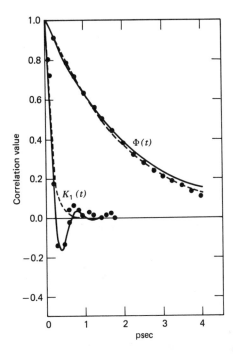

**Figure 3.6.** Comparison of experimental and computed vibrational correlation $\phi(t)$ and first-rank memory function $K_1(t)$ of the C–D stretch of CDCl$_3$. Dots: data from Fig. 3.4. Solid curves: computations based on the rigid lattice memory function approach. Broken curves: dephasing model, from Fig. 3.4 (Eq. 3.11). [S. S. Cohen and R. E. Wilde, *J. Chem. Phys.* **68**, 1138 (1978).]

Figure 3.6 shows the results of this memory function approach on the example of the C–D $\nu_1$ mode of CDCl$_3$.[202] The data fit is improved over that predicted by Eq. 3.11.

For us the interesting aspect of this method is the choice of $K_2^{RL}(t)$. It points again to analogies between the simple vibrational dephasing models and the *J*- and *M*-diffusion models of rotational relaxation. In both rotational and vibrational relaxation models, (1) the population and depopulation of a particular energy level is not considered; (2) effects of collisions are pictured by purely exponential (Poisson) collision distributions with an adjustable autocorrelation time of momentum (angular and linear, respectively); (3) a "bimodal" concept of the time evolution is assumed, namely, a short-time pure dephasing process due to a Maxwellian distribution of angular rotational velocities or a Gaussian distribution of vibrational frequency shifts, respectively, and its collision-induced motional narrowing at longer times. Correlation decay of the simple model (Eq. 3.11) is initially parabolic and later purely exponential (Section 2.1). Correlation decay of the modified model (Eq. 3.56) is closer in spirit to the *M*-model since memory is partly regained.

The essential difference between the extended rotational diffusion models and the fast-modulation vibrational dephasing models is that dephasing in the rotation models reflects the ensemble-averaged loss of the orientational coherence of *free* rotor energies, whereas dephasing in the vibrational model reflects an ensemble-averaged loss of the amplitude coherence due to a *perturbation-induced* vibrational energy distribution. In the vibrational dephasing model we need first a perturbation to split the single unperturbed "degenerate" (see Section 2.1) oscillator energy into a distribution about its unperturbed value. As we mentioned before, the distribution is (usually) Gaussian (Eq. H.9, Appendix H); the Maxwell distribution of free rotor energies (Eq. 2.2, Chapter 2) has a "tail."

As an obvious corollary, we recall that the rotational second spectral moment (Appendix E) gives no information on intermolecular forces, whereas the second spectral vibrational moment $\langle|\omega_1(0)|^2\rangle = \hbar^{-2}\langle|H_1(0)|^2\rangle$ does.

Of course, we recall that the physical meaning of the respective correlation times $\tau_J$ for the rotational and $\tau_c$ for the vibrational model differ: $\tau_J$ causes energy population-depopulation ("hard collisions") for the $J$-model (and the $M$-model for symmetric tops), whereas $\tau_c$ describes "soft collisions" (elastic collision, no energy exchange between oscillator and bath or among vibrational levels).

In the following section we direct our attention to a discussion of vibrational dephasing in the slow or intermediate modulation regimes, with particular reference to hydrogen-bonding effects and dynamics of adsorbed molecules.

## 2.3.  Vibrational Dephasing in the Intermediate and Slow Modulation Regimes

Studies of vibrational dephasing appear more worthwhile in the slower than in the rapid modulation regimes. In particular, effects with pronounced directional interaction characteristics, such as hydrogen bonding in aqueous solutions and interaction between a surface and an adsorbed species, can be profitably examined. The dephasing characteristics of an oscillator connected with or near the "exalted" bond—in the foregoing examples it is the hydrogen or the adsorbate–adsorbent bond—are different from those of an oscillator located elsewhere in the molecule. We see that in such a situation the selective response of each oscillator of the molecule to the environment is of particular advantage.

An interesting system whose dynamic properties are influenced by hydrogen bonding is the pyridine $(C_6H_5N)$–water system. Of particular concern are the vibrational dephasing phenomena of the totally symmetric fundamental at 998 $cm^{-1}$, involving an in-plane-radial-to-the-ring nitrogen atom displacement, and the 1030-$cm^{-1}$ fundamental, representing the standing-wave-in-plane carbon–nitrogen angle deformation (with the N atom at its node).[203] The vibrational dephasing of the 998-$cm^{-1}$ mode should be sensitive to hydrogen bonding between the nitrogen atom of $C_6H_5N$ and the hydrogen atom of water. Vibrational dephasing of the 1030-$cm^{-1}$ band should be insensitive to hydrogen bonding.

Figure 3.7 shows the experimental results. We consider first the isotropic Raman contour (Eq. 1.46, Chapter 1) of the 998-$cm^{-1}$ mode. $M(2) = 14$ $cm^{-2}$ (Eq. 3.18) is constant over the whole concentration range. $\tau_c$ is about $1.2 \times 10^{-12}$

sec. Hence, $M(2)^{1/2}\tau_c$ is approximately 0.8; the system is in the intermediate modulation regime. Two aspects of the data are striking.

First, we notice that $\tau_c$ agrees with typical lifetimes of hydrogen bonds in liquid systems involving solvent water (Section 7.3, Chapter 2). Second, we speculate that the constancy of $M(2)$ with increasing dilution by water shows an insensitivity of $M(2)$ to increasing coordination by surrounding water molecules once a N $\cdots$ H hydrogen bond between pyridine and water has been formed.[203]

Next, we discuss the vibrational relaxation behavior of the isotropic Raman contour of the 1030-cm$^{-1}$ mode. Figure 3.7 shows that $M(2)$ stays constant at mole fraction $C_6H_5N = 0.1$–$0.5$ and increases for mf $> 0.5$. Modulation time $\tau_c$ amounts to $0.8 \times 10^{-12}$ sec at mf $= 0.1$, reaches a maximum of $1 \times 10^{-12}$ sec at mf about 0.3, and drops to $0.4 \times 10^{-12}$ sec for mf $> 0.4$. The concentration

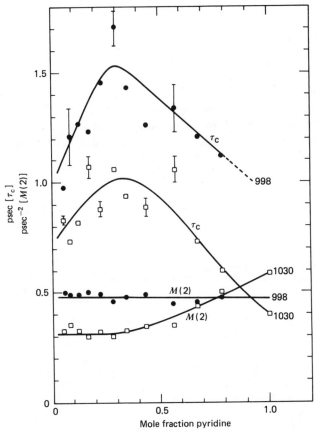

**Figure 3.7.** Concentration dependence of the modulation times and of the second vibrational spectral moments of the 998-cm$^{-1}$ and 1030-cm$^{-1}$ pyridine isotropic Raman modes of pyridine dissolved in water at 25°C. [From W. Schindler and H. A. Posch, *Chem. Phys.* **43**, 9 (1979).]

dependence of $M(2)$ corresponds to the vibrational dephasing behavior of an oscillator undergoing nonspecific interactions with its environment (see Fig. 3.3c). The behavior of $\tau_c$ is interesting: Its maximum occurs at a pyridine concentration where the shear viscosity has a maximum.[203]

Several questions are raised by this work. Does the proximity between the two modes (30 cm$^{-1}$) cause perturbations by interoscillator coupling by way of nonresonant energy transfer? Are the $M(2)$ data accurate, considering that the contours overlap? What is the effect of concentration fluctuations, particularly for near-equimolar mixtures? Indeed, band broadening by concentration fluctuations seems to have been largely overlooked in solution studies of vibrational dephasing; we discuss this in Section 5.

A more sophisticated model for dephasing in hydrogen bonded systems has been applied to phenol–dioxane solutions,[204]

$$C_6H_5OH \cdots O \underset{(CH_2)_2}{\overset{(CH_2)_2}{\diagup \diagdown}} O.$$

The H atom of the OH group of phenol is the donor, the oxygen atom of dioxane is the acceptor site. In this study, the O–H stretching oscillator of the O–H $\cdots$ O hydrogen bond, the so-called $\nu_s$ fundamental (near 3420 cm$^{-1}$),[205] was analyzed. Instead of a simple exponential modulation function (Eq. 3.10), a damped-oscillatory modulation function,

$$\hat{\psi}(t) = \exp(-\gamma t)\left\{\cos(\beta t) + \left(\frac{\gamma}{\beta}\right)\sin(\beta t)\right\}, \qquad \beta^2 \equiv \omega_\sigma^2 - \gamma^2, \quad (3.57)$$

$$\frac{2\gamma}{\beta^2 + \gamma^2} = \frac{2\gamma}{\omega_\sigma^2} = \tau_c$$

was used. (For $\tau_c$ see Eq. 3.4; for $\omega_\sigma$ see below.) This function is familiar to us; we derived it previously (Eq. 2.177, Chapter 2) to describe the autocorrelation of orientational motion in the two-variable approximation. The rationale for using this complicated modulation function is as follows: The observed $\nu_s$ oscillator (O–H) can no longer be considered "isolated"; it couples to the O $\cdots$ O stretching fundamental of the O–H $\cdots$ O hydrogen bond, the so-called $\nu_\sigma$ mode.[205] As the frequency $\omega_\sigma$ of $\nu_\sigma$ lies in the far-infrared, it interacts strongly with the random bath (or lattice) motions of the liquid. Thereby, $\nu_\sigma$ transmits environment-induced stochastic motion to the near-infrared $\nu_s$ mode.[204, 206]

Insertion of Eq. 3.57 into Eq. 3.3 leads (after some ordering) to the vibrational dephasing function $\phi(t)$ of mode $\nu_s$,

$$\phi(t) = \exp\left\{ -\langle|\omega_1(0)|^2\rangle\omega_\sigma^{-2}\left[\exp(-\gamma t)\left(\frac{\gamma}{\beta}\right)\omega_\sigma^{-2}(4\gamma^2 - 3\omega_\sigma^2)\sin(\beta t)\right.\right.$$

$$\left.\left. + 2\gamma t + \omega_\sigma^{-2}(4\gamma^2 - \omega_\sigma^2)(\exp(-\gamma t)\cos(\beta t)-1)\right]\right\}. \tag{3.58}$$

The relation corresponds to a three-variable approximation (Sections 5.8, and 6.2 of Chapter 2).

Note that the lattice-coupled $\nu_\sigma$ oscillator, which modulates mode $\nu_s$, most likely moves under a non-Poisson distribution of perturbing ("collisional") events. In fact, although modulation function $\psi(t)$ (Eq. 3.57) is not the memory function of correlation function $\phi(t)$ (Eq. 3.58), $\psi(t)$ has a similar form as the cumulant event function $g(t)$ (Eq. 3.51b) of non-Poisson translational diffusion. Hence we have simulated (at least in spirit) the cage effect of real liquids. We propose that modulation function $\psi(t)$ should be applicable to other situations of strong coupling between near-, medium-, or infrared modes with far-infrared modes during dephasing processes: An obvious example is interaction between a fundamental internal vibration with its own librational (external) motions. In particular, we expect this to be observable in associated liquids.

Let us now show an application of Eq. 3.58 to the phenol–dioxane spectral data. The smooth curve in Fig. 3.8 shows the spectrum of $\nu_s$ of the O–D oscillator of $C_6H_5OD$ (0.026 molar) complexed with dioxane (1.3 molar) in $CCl_4$-solution. The crosses represent the computed spectrum, $\int dt\, \phi(t)$

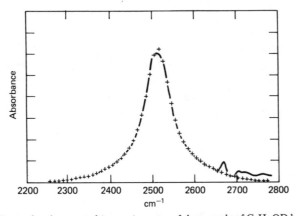

**Figure 3.8.** Observed and computed (crosses) spectra of the $\nu_s$ mode of $C_6H_5OD$ hydrogen-bonded to $O(CH_2)_4O$ in $CCl_4$ at 40°C. [J. Yarwood, R. Ackroyd, and G. N. Robertson, *Chem. Phys.* **32**, 283 (1978).]

$\times \exp(-i\omega t)$, with the parameters

$$\nu_s = 2512 \text{ cm}^{-1},$$

$$\nu_\sigma = \frac{\omega_\sigma}{2\pi c} = 148 \text{ cm}^{-1},$$

$$\langle |\omega_1(0)|^2 \rangle^{1/2} = 14.7 \times 10^{12} \text{ radian sec}^{-1} = 78 \text{ cm}^{-1},$$

$$\gamma = 24.2 \times 10^{12} \text{ radian sec}^{-1} = 128 \text{ cm}^{-1}.$$

From Eq. 3.57,

$$\tau_c = 0.031 \times 10^{-12} \text{ sec},$$

hence

$$\langle |\omega_1(0)|^2 \rangle^{1/2} \tau_c = 0.46.$$

The phenol–dioxane system is in an intermediate modulation regime. Because of the oscillatory behavior of $\phi(t)$, the meaning of modulation time $\tau_c$ is obscured.[204] Nevertheless, taking this time at face value, we see that it corresponds to 170 cm$^{-1}$, a frequency at the upper limit of the far-infrared range.

Next, we discuss the vibrational relaxation behavior of the carbon–silicon stretching (633 cm$^{-1}$) and the silicon–chlorine stretching (464 cm$^{-1}$) fundamentals of $(CH_3)_3SiCl$, the silicon analogue of tertiary butyl chloride, $(CH_3)_3CCl$ (see also Section 5.2, Chapter 2). We present this work since it shows the diverse vibrational relaxation behavior among modes of the same symmetry species ($A_1$) of the molecular point group $C_{3v}$.[207]

Figure 3.9 shows the vibrational correlation functions at 298°K. Circles and triangles represent experimental values (Eq. 1.47, Chapter 1), the curves show the computed values (Eq. 3.11) with $\tau_c = 0.40 \times 10^{-12}$ (C–Si) and $0.75 \times 10^{-12}$ sec (Si–Cl). The experimental $M_{iso}(2)$ (Eqs. 3.18 and 3.19) are 45 and 68 cm$^{-2}$, respectively. Amplitude correlation times $\tau_a$ (Eq. 3.5c) are $1.9 \times 10^{-12}$ (C–Si) and $1.1 \times 10^{-12}$ sec (Si–Cl). Both oscillator-environment systems are in the intermediate modulation regime,

$$M_{iso}(2)^{1/2} \tau_c = 0.50 \text{ (C–Si)},$$

$$= 1.2 \text{ (Si–Cl)}.$$

Modulation times $\tau_c$ are too long to represent binary collision times in a nearest-neighbor regime (Section 2.2.2); the modulation mechanism involves slower dynamic effects. Possibly, rotational motion of the molecules (Eq.

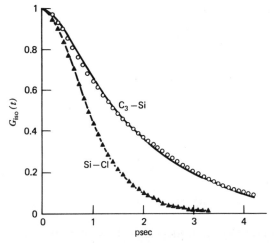

**Figure 3.9.** Vibrational correlation functions for the $C_3$–Si and Si–Cl stretching fundamentals of $(CH_3)_3SiCl$ at 298°K. Circles and closed triangles represent the experimental values, curves the theoretical values. [P. Reich, A. Reklat, G. Siefert, and Th. Steiger, *Acta Phys. Polon.* **A58**, 665 (1980).]

3.21$b$) is the modulating agent. From the difference of the measured infrared and isotropic Raman widths of the Si–Cl fundamental,[207] $16.9 - 14.0 = 2.9$ cm$^{-1}$, we estimate (Eq. J.2, Appendix J) an orientational correlation time of $\tau_R^{(1,0)}$ approximately $3 \times 10^{-12}$ sec. We recall that $\tau_R^{(1,0)}$ describes rotational motion around the perpendicular axes of the molecule. This motion tilts the permanent dipole moment; yet we now question whether long-range interactions cause motional narrowing in this system since $\tau_R \gg \tau_c$ (see also Section 2.2.1).

The observed difference in the dephasing rates of the two modes is an interesting aspect. We might have expected that the rate $\tau_c^{-1}$ of a collisional modulation mechanism should affect all vibrational modes of the molecule to the same extent. However, we introduced $\tau_c$ as an effective correlation time that describes the time-evolution of the coherence decay of the (many-body) perturbation Hamiltonian, a complicated quantity. This Hamiltonian couples an oscillator coordinate to dynamic coordinates of the surrounding molecules, for instance, by way of the oscillator anharmonicity. It would indeed be fortuitous if the quasi-static as well as dynamic aspects of this coupling were identical for different modes of the molecule, even if they belonged to the same symmetry species of the molecular point group.

Finally, a few caveats concerning these particular results must be mentioned. First, there is no proof here that the isotropic Raman band shapes of the two modes represent solely vibrational dephasing—as has been assumed.[207] Although direct vibrational energy transfer to the lattice, even at the low oscillator frequency of 464 cm$^{-1}$, is still a relatively slow process (we present some relevant data in Section 7), the molecule possesses three skeletal far-infrared fundamentals (186–242 cm$^{-1}$). Contribution from bath-modulated intraoscillator coupling therefore cannot be ruled out. Second, the natural Cl isotope distribution, which leads to a $1:3$ mixture of $(CH_3)_3Si^{37}Cl$ and $(CH_3)_3Si^{35}Cl$ molecules, contributes to band broadening of the Si–Cl stretch (Section 4.2.1, Chapter 2). We estimate an isotopic frequency displacement of about 2 cm$^{-1}$, at worst.

As the last example of this section we discuss vibrational dephasing data of CO (carbon monoxide) chemisorbed on rhodium (Rh) dispersed on $\gamma$-Al$_2$O$_3$. Not only does this system demonstrate that the simple dephasing model of Eq. 3.11 can be extended to describe adsorbate-adsorbent interactions of a translationally immobilized molecule, but it also shows the extremes of modulational behavior (very fast and very slow librations) within the same macroscopic system.[208a]

It is known that CO chemisorbs on $\gamma$-Al$_2$O$_3$-supported Rh as three species, a "bridged" $(a)$, a "single" $(b)$, and a "double" configuration $(c)$:

We are concerned only with configurations $b$ and $c$, the so-called linear and doubly adsorbed species. In both cases we analyze the absorption profile of the C=O stretching fundamental near 2000 cm$^{-1}$. (The more interesting metal–carbon bond is unfortunately obscured by a reststrahlen band of Al$_2$O$_3$, starting near 1900 cm$^{-1}$.)

Spectrum C in Fig. 3.10a represents that of species $c$. It forms a doublet, representing the symmetric and antisymmetric CO oscillation.[208b] It is obvious that the twin profiles are essentially Gaussian (notice the short wings). (The near-equal intensity of the two band components indicates that the angle between the two adsorbed CO molecules is $\sim 90°$.) Consequently, Eq. 3.6a is a good approximation of Eq. 3.11; the fit of the experimental $\phi(t)$ with the experimental $M(2) = 4.0 \times 10^{24}$ radian$^2$ sec$^{-2}$ leads only to a *range* of modulation times $\tau_c$ (Fig. 3.10b); $\tau_c$ approximates $1.5 \times 10^{-12}$–$5 \times 10^{-12}$ sec. The system is in the slow modulation regime,

$$\langle |\omega_1(0)|^2 \rangle^{1/2} \tau_c \sim 6.$$

The CO molecules adsorbed as species $c$ (two CO molecules on one Rh atom) are thus rigidly fixed during amplitude correlation time $\tau_a$ (Eq. 3.5c) approximately $1.6 \times 10^{-12}$ sec.

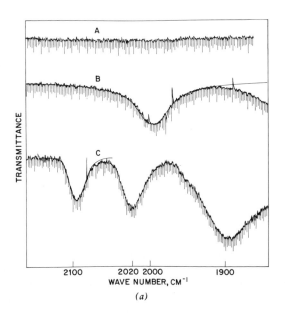

(a)

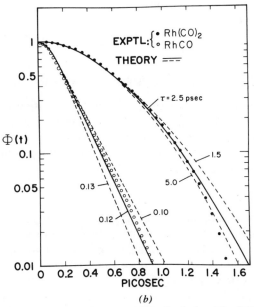

(b)

**Figure 3.10.** (a) Infrared transmittance spectra of CO adsorbed on Rh of Rh/γ-Al₂O₃ catalysts. Spectrum A: transmittance spectrum of a Rh/Al₂O₃ catalyst. Spectrum B: linear species, RhCO. Rh is in the form of crystallites. Spectrum C: double species, $Rh(CO)_2$; Rh is in the form of atomic clusters and two-dimensional patches. Temperature, 27°C. Coverage: less than a monolayer. The broad profile centered near 1900 cm⁻¹ represents the bridge structure, $Rh_2CO$. (b) Vibrational correlation functions of CO adsorbed as species RhCO and $Rh(CO)_2$. Dots: numerical Fourier transform of experimental absorption coefficient data obtained from Fig. 3.10a. Curves: model computations (Eq. 3.11) with the best-fit modulation times indicated in the body of the figure. [H. C. Yao and W. G. Rothschild, *J. Chem. Phys.* **68**, 4774 (1978).]

Figure 3.10$b$ shows that the modulation regime of $\phi(t)$ belonging to species $b$, RhCO, is faster; $\langle|\omega_1(0)|^2\rangle^{1/2}\tau_c$ amounts to 0.9. The singly adsorbed CO molecules perform rapid (probably librational) motions on their adsorption sites ($\tau_c = 0.12\times10^{-12}$ sec, $\tau_a \sim 0.9\times10^{-12}$ sec). We expect that a faster modulation regime generates a band profile with more Lorentzian character; this is clearly exhibited by the spectrum B of species $b$ (see Fig. 3.10$a$).

At present, application of the vibrational dephasing process to the static and dynamic characteristics of adsorbed molecules is in its infancy. Clearly, this is an interesting subject; a great deal of work needs to be done. Selectivity of catalytic sites to chemical reactions, adsorption-desorption rates, effect of coverage, conversion of adsorbed precursors to stable species—to name just a few—are topics of general interest that could be studied by vibrational fluctuation spectroscopy of molecules adsorbed on dispersed metal states. In fact, the nature of the dispersed surface state can be explored by this method. The weaker the interactions between the metal and its dispersing agent, the more metallic the state of the catalytically active metal. We therefore expect that the vibrational relaxation process of an adsorbate oscillator mainly shows aspects of pure dephasing for metals dispersed on a semi or nonconductor support (inhomogeneous broadening due to a wide distribution of adsorption sites), but relaxes principally by phonon depopulation for neat metal surfaces (homogeneous broadening due to oscillator-electron coupling).[208a, 209]

## 2.4. Vibrational Dephasing in Molecular Subenvironments

To perform a meaningful analysis of a vibration-rotation or a purely vibrational band profile by the methods of fluctuation spectroscopy, the band profile must represent an *individual, single* (not overlapping) oscillator mode (Sections 3.3 and 4.2.1 of Chapter 2). If this condition is not satisfied—for instance, in the presence of band broadening due to isotope splittings, from factor or site group splittings that persist in a partially ordered phase, or through overlap by a hot or summation band—computational corrections must be applied to decompose the profile or to decouple the correlation functions. The importance of necessary corrections in such a situation cannot be overemphasized. We believe that their omission may have rendered published results suspect, particularly since these effects are not always obvious. Let us discuss this now in detail.

Overlap of isotopic components is the most easily correctable case. We assume that the overlapping profiles have the same shape and differ in intensity according to the natural abundance of the isotopic species. We write[10a]

$$\hat{C}_a(t) = \int d\omega\, \hat{\mathcal{I}}_a(\omega)\exp\{i(\omega - \omega_c)t\} = \sum_k \int d\omega\, x_k \hat{\mathcal{I}}_k(\omega)\exp\{i(\omega - \omega_k^0)t\},$$

where $\hat{C}_a(t)$ and $\hat{\mathcal{I}}_a(\omega)$ are the apparent (experimental) normalized correlation function and spectral density. $\hat{\mathscr{I}}_k(\omega)$ is the normalized spectral density of component $k$ and $x_k$, $\omega_k^0$ is its relative intensity and band center. Since $\hat{\mathscr{I}}_k(\omega) = \hat{\mathscr{I}}_j(\omega) = \cdots \hat{\mathscr{I}}(\omega)$, we get

$$\hat{C}_a(t) = \int d\omega\, \hat{\mathscr{I}}(\omega) \exp\{i(\omega - \omega_c)t\} \sum_k x_k \exp\{i(\omega_c - \omega_k^0)t\},$$

where $\omega_c$ is a convenient band center.

We assume that $\hat{\mathscr{I}}_k(\omega)$ is symmetric and identify the band center $\omega_k^0$ with the respective peak frequencies. Therefore,

$$\hat{C}_a(t) = \hat{C}(t) \sum_k x_k \cos\{(\omega_k^0 - \omega_c)t\} \tag{3.59}$$

with

$$\hat{C}(t) = \int d\omega\, \hat{\mathscr{I}}(\omega) \cos\{(\omega - \omega_c)t\},$$

$$\omega_c = \frac{\int d\omega\, \omega \mathscr{I}_a(\omega)}{\int d\omega\, \mathscr{I}_a(\omega)}.$$

Of course, for Eq. 3.59 to be useful, we must know the $\omega_k^0$. Since we (tacitly) assume that solvent shifts affect all $\omega_k^0$ equally, a good approximation of the frequency displacements of the $\omega_k^0$ is always obtainable from vapor-phase values.

Equation 3.59 has been applied to the symmetric C–Cl Raman deformation fundamental of naturally abundant $CDCl_3$ at 367.1 cm$^{-1}$ (Cl–35), 364.2 cm$^{-1}$ ($2\times35, 1\times37$), 361.3 cm$^{-1}$ ($1\times35, 2\times37$), and 358.5 cm$^{-1}$ (Cl–37). Success was verified with isotopically pure $CH^{35}Cl_3$.[10b]

Equation 3.59 has also been used to disentangle overlapping hot bands from the polarized C–I Raman stretching fundamental of $CH_3I$.[10a] However, the profile of a hot band usually differs from that of its fundamental; for instance, their anharmonicities need not be identical. Hence, it is questionable whether Eq. 3.59 can lead to more than a rough approximation of $C(t)$.

In such a situation it is preferable to use the real and imaginary parts of $\hat{C}_a(t)$; this gives two numerical values for each $t$ and can thus be used to disentangle overlapping profiles. To show this,[11] we write the normalized correlation function of the (asymmetric) global profile,

$$\hat{C}_a(t) = \int d\omega\, \hat{\mathscr{I}}_a(\omega) \exp\{i(\omega - \omega_c)t\}, \tag{3.60a}$$

and express each of its components $\hat{C}_k(t)$ by

$$\hat{C}_k(t) = \frac{\int_k d\omega \,\mathscr{I}_k(\omega)\exp\{i(\omega - \omega_k^0)t\}}{M_k(0)}, \qquad (3.60b)$$

where

$$M_k(0) = \int_k d\omega \,\mathscr{I}_k(\omega)$$

is the intensity (zero moment) of profile $\mathscr{I}_k(\omega)$. We wish to write $\hat{C}_a(t)$ in terms of the $\hat{C}_k(t)$. We define the frequency shift

$$\Delta\omega_k = \omega_c - \omega_k^0,$$

which permits us to transform Eq. 3.60b to

$$M_k(0)\hat{C}_k(t) = \exp(i\Delta\omega_k t)\int_k d\omega \,\mathscr{I}_k(\omega)\exp\{i(\omega - \omega_c)t\}. \qquad (3.60c)$$

Obviously,

$$\hat{\mathscr{I}}_a(\omega) = \frac{\sum_k \mathscr{I}_k(\omega)}{\sum_k M_k(0)}.$$

Introducing this relation into Eq. 3.60c lets us express Eq. 3.60a as

$$\hat{C}_a(t) = \frac{\sum_k M_k(0)\hat{C}_k(t)\exp(-i\Delta\omega_k t)}{\sum_k M_k(0)}.$$

We normalize the global intensity,

$$\sum_k M_k(0) = 1,$$

and obtain the desired relation,

$$\hat{C}_a(t) = \sum_k M_k(0)\hat{C}_k(t)\exp(-i\Delta\omega_k t). \qquad (3.60d)$$

Splitting Eq. 3.60d into its real and imaginary parts, we obtain the individual correlation function $\hat{C}_k(t)$ of two sets $(k = p, q)$ of overlapping symmetric profiles of *different* shape (within each set, the shape must be the same),

$$\operatorname{Re}\hat{C}_a(t) = \hat{C}_p(t)\sum_p M_p(0)\cos(\Delta\omega_p t) + \hat{C}_q(t)\sum_q M_q(0)\cos(\Delta\omega_q t),$$

$$-\operatorname{Im}\hat{C}_a(t) = \hat{C}_p(t)\sum_p M_p(0)\sin(\Delta\omega_p t) + \hat{C}_q(t)\sum_q M_q(0)\sin(\Delta\omega_q t),$$

$$(3.60e)$$

if the relative intensities are known. Usually the nodes of the fundamental determinant $\sum M_p(0)M_q(0)\sin\{(\omega_p - \omega_q)t\}$ cause little trouble.

As successful applications,[11] we mention strongly overlapping vibrations in acetylene (two modes), in ethylene dissolved in $CCl_4$ (three modes), and in $N_2O$ dissolved in $CCl_4$ (two modes). The last example is particularly useful and opportune since it represents overlap of a fundamental $\nu_p$ with the difference band $\nu_p + \nu_q - \nu_q$, a configuration frequently encountered in the infrared.

Notice that the imaginary component of $\hat{C}_a(t)$ is a consequence of overlap between different oscillators; it has no deeper physical meaning (see Appendix A and Section 4 that follows). Furthermore, for $k$ overlapping symmetric profiles of equal shape, Eq. 3.60$d$ obviously gives Eq. 3.59.

Initial hopes that Raman picosecond pulsing techniques (under coherent wave vector matching) would also lead to a clear-cut identification of overlapping band profile regions in liquids have not been sustained.[210]

Finally, we discuss briefly an example of the errors that arise if we fail to reduce factor group splittings in evaluating dynamic phenomena from band broadening.[211a]

Figure 3.11$a$ shows the experimental infrared bandwidths of the degenerate $\nu_6$ mode of $CH_3I$ and of $CD_3I$ in their solid and liquid phases between $-110°C$ and $+40°C$. Assuming that end-over-end rotation has ceased in the solid, we assign the bandwidth below the melting point to purely vibrational relaxation processes (Eq. 1.36, Chapter 1);[45] their nature is of no concern here. Hence, on the basis of the data, we would predict for both molecules a purely orientational width of about 2 cm$^{-1}$ at the melting point (and about 10 cm$^{-1}$ at $+40°C$).

As it turns out, these results are wrong since the profile of the solid consists of three overlapping transitions due to factor and site group splittings.[211] Upon melting, this multiplicity collapses into the single oscillator of the disordered liquid; taking the apparent solid-state halfwidths of 18 ($CH_3I$) or 14 cm$^{-1}$ ($CD_3I$) for the true vibrational widths therefore greatly underestimates the contribution of orientational broadening in the liquid phase.

Figure 3.11$b$ shows the solid-state spectrum of $\nu_6$ of $CH_3I$ at two temperatures; the splittings become apparent only at the lowest temperatures and after annealing.

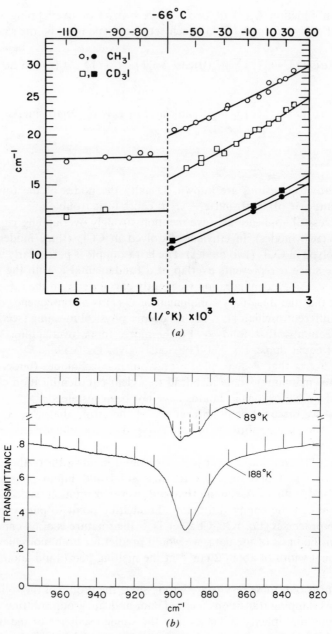

**Figure 3.11.** (*a*) Experimental infrared bandwidth (full width at half-peak height, open points) and pure rotational relaxation simulation (model of Ref. 149, solid points) of the $\nu_6$ modes of neat CH$_3$I and CD$_3$I in their amorphous-solid and liquid phases. The vertical line denotes the melting point temperature. (*b*) Infrared solid-phase spectra of the $\nu_6$ modes of neat CH$_3$I. Upper spectrum: annealed sample; the frequencies of the crystalline-phase splittings (Ref. 211*b*) are indicated by the vertical dashes. Lower spectrum: amorphous sample. [W. G. Rothschild, J. Soussen-Jacob, J. Bessière, and J. Vincent-Geisse, *J. Chem. Phys.* **79**, 3002 (1983).]

**264**

## 2.5.   Vibrational Dephasing in a Glassy System: Quinoline ($C_9H_7N$)

The usefulness of employing vibrational dephasing data to characterize changes in the local order upon phase transitions is demonstrated by the 1030-cm$^{-1}$ mode of crystalline, glassy, and liquid quinoline, $C_9H_7N$.[47] Figure 3.12$a$ shows that the infrared spectral profile changes from a typical, relatively symmetric contour to a broad, asymmetric shape if the compound is quenched from the liquid into a glassy phase. Upon annealing to a crystalline configuration, the profile narrows and weak sidebands appear. Simultaneously, the peak frequency undergoes a blue shift, but static frequency shifts are not of our concern here.

The points in Fig. 3.12$b$ display the correlation functions from the glassy and crystalline spectra of Fig. 3.12$a$. The curves are the predictions of Eq. 3.11, using the measured spectral moments $M(2)$ and the best-fit modulation times $\tau_c$. [The correlation functions of the room-temperature liquid are displayed in Fig. 1.6; $\tau_c(\text{liquid}) = 0.3 \times 10^{-12}$ sec.][177] Although no care had been taken to anneal the glass to the maximum degree of crystallinity, the different aspects of the correlation functions of the two solid phases are interesting: The modulation of the 1030-cm$^{-1}$ oscillator changes from a slow $M(2)^{1/2}\tau_c = 2.3$ to a more rapid regime ($= 0.6$) when the compound passes from a glassy to a crystalline conformation. These changes are caused by a narrowing of the width of the quasi-static environment by about 20% and, more significantly, by an increase in the modulation rate $\tau_c^{-1}$ by a factor of about 3. Apparently, the modulation mechanism works through the longer-range (crystal) phonons.

The strong asymmetry of the band profile in the amorphous, low-temperature solid quinoline requires comment. Equation 3.11, which predicts symmetric contours, evidently fails here. Mathematically, an appearance of odd-order terms in the cumulant development of the stochastic phase angle (Eq. 3.1) means that the series (Eq. H.6, Appendix H) has not converged with the second-order term.[127a] Physically, this signifies that the effective perturbation Hamiltonian $H_1(t)$ is the composite of less than four to six independent contributions (Note 5),

$$H_1(t) = \int_0^t dt \cdots \int_0^{t_2} dt_1\, h_1(t - t_{n-1}) h_2(t_{n-1} - t_{n-2}) \cdots h_n(t_1), \qquad n \leq 6,$$

where $h_n(t)$ is the interaction term of the oscillator with each environmental state at instant $t$.

Choosing $n = 4$, picking a reasonable function (not necessarily the same) for each of the perturbation "pulses" $h_n(t)$—such as an exponential, a triangular, or rectangular step function—and constructing the spectrum of $H_1(t)$,

$$\text{Re}\,\mathscr{L}\{H_1(t)\} = \text{Re}\,\tilde{H}_1(s) = \text{Re}[\tilde{h}_1(s)\tilde{h}_2(s)\cdots\tilde{h}_n(s)],$$

shows that $\text{Re}\,\tilde{H}_1(s)$ is barely Gaussian.[127b] (It is Gaussian with $n > 6$–8.)

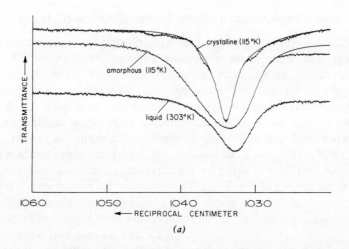

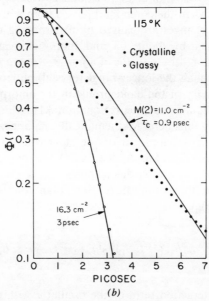

**Figure 3.12.** (*a*) Infrared band contour of the 1033-cm$^{-1}$ fundamental of quinoline in its crystalline state (upper trace), glassy state (intermediate), and liquid phase (lowest). The spectra are displaced vertically for better viewing. (*b*) Correlation functions describing vibrational dephasing of the 1033-cm$^{-1}$ fundamental in crystalline and glassy phases of quinoline at 115°K. The points are the experimental values by Fourier inversion of the spectra of Fig. 3.12*a*. [Part *b* is from W. G. Rothschild, *Molecular Motions in Liquids*, J. Lascombe, Ed. Copyright 1974 by D. Reidel Publishing Co., Dordrecht, p. 247.]

Consequently the correlation of the intermolecular coupling Hamiltonian $H_1(t)$ stretches over more than two successive time points. Evidently this requires more detailed knowledge of the dynamics of the system than for a Gaussian $H_1(t)$ (see Section 2.1). We go into more detail on the required modifications of the theory for asymmetric band profiles in Section 4.

On the other hand, it is conceivable that the asymmetry in the profile of the 1030-cm$^{-1}$ mode of amorphous, low-temperature quinoline represents the superposition of a few strongly overlapping, relatively homogeneous subdistributions of molecular environments. In this situation the foregoing analysis would be suspect; numerical and experimental methods discussed in Section 2.4 must be applied. At any rate, the data presented here corroborate the particular utility of vibrational dephasing studies in the slow-to-intermediate modulation regimes as probes of the nature of local environments. Other examples are molecular morphology of crystalline and amorphous regions in polymers, local changes caused by phase transitions, solvent effects, solvent cage structure, and so forth. A related phenomenon, "concentration fluctuations," is discussed in Section 5.

Incidentally, the 1030-cm$^{-1}$ mode of quinoline corresponds to the 1030-cm$^{-1}$ mode of pyridine (see Section 2.3). It is surprising that the experimentally deduced modulation times $\tau_c$ for both neat liquids at room temperature are nearly equal, $0.3 \times 10^{-12}$ sec for $C_9H_7N$ and $0.4 \times 10^{-12}$ for $C_6H_5N$.[177, 203] Notice that the mass of the molecules varies by a factor of about 2, their room-temperature viscosities by a factor of about 4. Geometrical (steric) effects also differ owing to the diverse shape of the two molecules. Orientational correlation times (from Rayleigh scattering) differ by more than one order of magnitude, $\tau_R^{(2)} = 2.5 \times 10^{-12}$ for pyridine, $4 \times 10^{-11}$ sec for quinoline. Clearly, we do not understand the nature of the dynamic intermolecular perturbation process that leads to the observed $\tau_c$ in liquid quinoline.

## 2.6.  Effect of Vibrational Dephasing on Rotational Relaxation

In situations where vibrational and rotational relaxation processes cannot be separated, we can estimate the orientational correlation function (or spectral density) on the basis of a reasonable model that is parametrized with extraneous data, such as dielectric and nuclear magnetic correlation times. The purely vibrational correlation function is then obtained by deconvolution. Although the results from this approach are not precise, they indicate the magnitude of the contribution of vibrational relaxation. In view of frequently considerable band broadening of infrared and Raman profiles by vibrational dephasing, this is helpful information.

Let us discuss two examples briefly. The first is the infrared perpendicular (degenerate) band of $CH_3I$ at 884 cm$^{-1}$; we recall that combined molecular rotation about the $C_3$ symmetry axis (spinning) and about any one of the two perpendicular axes of inertia (tumbling) is observed (Section 2.1, Chapter 1). The spinning motion involves a relatively small moment of inertia and does not tilt the molecular permanent dipole moment vector (Fig. 1.3$b$). It is faster than tumbling around the perpendicular axes.

Let us then estimate the contribution of rotational relaxation to the correlation function (displayed in Fig. 1.5). We use a rotational Brownian motion model for symmetric top molecules,[149] properly modified for the Coriolis interaction ($\zeta = 0.208$) of the mode and parametrized with rotational diffusion coefficient $D_\perp$ and $D_\parallel$ (Table VI, Section 5.6, Chapter 2) from nuclear magnetic relaxation data.[144] Details of the straightforward computations are not of interest here. The resulting orientational model correlation function $G^{(1,1)}(t)$ is entered into Fig. 1.5. We now realize that vibrational relaxation in the infrared 884-cm$^{-1}$ mode of $CH_3I$ is as important as rotational relaxation since the total rate of decay is about twice that of $G^{(1,1)}(t)$.

Our second example is the 923-cm$^{-1}$ Raman band of a neopentane monocrystal in its plastic phase.[84b] The contribution of vibrational relaxation was determined by deconvolution of the experimental spectral density with that of an orientational correlation function based on the general stochastic model of rotational relaxation (Section 7.2, Chapter 2). The model assumed free rotational motion for event $A$ (Eq. 2.198), librational motion for event $B$ (Eq. 2.202), and random (Poisson) event distribution probabilities (Eq. 2.205) for both $A$ and $B$. The experimental parameters of the orientational model function were taken from neutron scattering data of plastic neopentane at the same temperature.[82b]

The resulting vibrational broadening of the 923-cm$^{-1}$ Raman band contour of plastic neopentane (at 173°K) is about 5 cm$^{-1}$. This compares to a halfwidth of about 8 cm$^{-1}$ of the infrared vibration-rotation profile (Fig. 2.11). It therefore signifies a considerable contribution of vibrational relaxation (Note 6).

### 2.7. Vibrational Dephasing: Concluding Remarks

The mathematical formalism of the spontaneous vibrational dephasing process, as discussed in this and the previous sections, is simple, but the physical aspects that the theory attempts to explain are complicated and subtle. This has led to misunderstanding and to ambiguous statements in the literature. We therefore raise here some of the important problems and questions.

1. Is there a hard-and-fast rule that the quasi-static distribution of environments $M(2)$ must increase with density or pressure? The answer is no. $\langle |\omega_1(0)|^2 \rangle$ depends not only on intermolecular distances but also on the functional behavior of the intermolecular potential derivatives with the normal coordinate (Eq. 3.7).

2. Can the temperature dependence of $M(2)$ be predicted? Again, the answer is no. Increasing temperatures may increase $M(2)$ since energetically less favorable states of the environment become populated. But increasing temperatures may also decrease $M(2)$ if higher local order ensues.

3. Can the temperature dependence of modulation time $\tau_c$ be predicted? Once more, the answer is no. Steric or kinetic effects may predominate.

4. Can we predict the *particular* type of motion from values of $\tau_c$? For instance, if $\tau_c$ does not undergo a discontinuity when the system passes through a phase transition (liquid–solid, ordered-solid–plastic-solid), does this mean the modulation by translational motion is ineffective? If translational diffusion over molecular distances is required to bring about modulation, the answer is yes (Section 2.2.1). But if the modulation arises from a short-distance, high-frequency translatory rattling motion (for instance in a cage), the change in the modulation regime upon passage of the system through the phase transition may not lead to a detectable difference in the observed $\tau_c$.

We note that the bandwidth of a profile that is broadened mainly by vibrational dephasing usually decreases ($\tau_a$ increases) with increasing temperatures (at constant pressure). This signifies that a temperature-induced increase of the modulation rate $\tau_c^{-1}$ overshadows any temperature-induced increase of the width $\langle |\omega_1(0)^2|\rangle^{1/2}$ of the environmental states that interact with the oscillator. In other words, increasing the temperature usually (but not always) brings the system into a faster modulation regime: $\langle |\omega_1(0)^2|\rangle^{1/2}\tau_c$ decreases.

## 3. RESONANCE VIBRATIONAL ENERGY TRANSFER

### 3.1. Basic Theory and General Aspects

We consider identical oscillators on separate, identical molecules. An oscillator is excited to an upper level by the incident radiation field. The oscillator returns subsequently from its excited to its ground level by giving (by way of some interoscillator coupling potential) its vibrational energy to an adjacent oscillator in its ground vibrational level, thereby inducing this oscillator to perform the upward transition ("flip-flop"). Clearly, the greater the relative mobility of the adjacent molecules bearing the oscillators, the less efficient the resonance transfer process. For instance, imagine a translational jump. It cuts off the interoscillator coupling and thereby keeps the original oscillator in its excited level for a longer time (assuming it finds no other effective means to relax). According to the energy uncertainty relation (Eq. 1.101, Chapter 1), the bandwidth is thereby decreased (motional narrowing).

For a basic mathematical formalism of this process, we can take that derived for vibrational dephasing. We employ Eq. 3.3 (Section 2.1) and make use of its predictions (Eqs. 3.3–3.6).[212] But we should realize that the instantaneous stochastic frequency $\omega(t) = \omega_0 + \omega_1(t)$ is now determined by the decrease in the *lifetime* of vibrational states owing to environment-induced resonance vibrational depopulation-population steps and not (as in the dephasing process) by the environment-induced *shifts* of the long-lived oscillator energy levels. Although the end result for both processes is a change in the stochastic phase angle (Eq. 3.1) and both processes may thus be called "dephasing processes" (which we will not do), this fundamental physical difference should be kept in mind.

We then realize that the vibrational dephasing and the resonant vibrational energy transfer processes are sensitive to certain aspects of the environment of the oscillator. An active vibrational dephasing process requires a wide distribution of interaction sites, each site coupling to the oscillator. An active resonance process requires effective oscillator-oscillator coupling over molecular distances. Whether this is furthered or hindered by a wide distribution of local environmental states is an interesting point that merits study. Nevertheless, we must be prepared to find that the resonance vibrational energy transfer and the vibrational dephasing processes are coupled. In favorable cases, this shows up in the experiment. Let us now explore this.

We formally include resonance vibrational energy transfer within our dephasing model by adding a cross term to the intermolecular solvent potential (see Section 2.1),[199]

$$\sum_{i<j} \left( \frac{\partial^2 V_s}{\partial Q_i \partial Q_j} \right)_{Q=0} Q_i Q_j,$$

where indices $i, j$ refer to the same oscillator (normal mode) on different molecules $i, j$. The intermolecular coupling operator $H_1(t)$ (Eq. 3.7) is then

$$\hbar^{-1} H_1^i(t) = \omega_1^i(t) = \left( -f/2\mu_M^2 \omega_0^3 \right) \frac{\partial V_s(t)}{\partial Q_i}$$

$$+ \left( 2\mu_M \omega_0 \right)^{-1} \left\{ \frac{\partial^2 V_s(t)}{\partial Q_i^2} + \sum_{i \neq j} \frac{\partial^2 V_s(t)}{\partial Q_i \partial Q_j} \right\} \qquad (3.61)$$

for the $0 \rightarrow 1$ vibrational transition of the observed oscillator on molecule $i$. (We omit the third-order term in $Q$.)

To get an idea on the influence of the additional term, again we take recourse to molecular dynamics calculations on the example of the Raman fundamental of liquid nitrogen near its boiling point.[199] They predict an amplitude correlation time of

$$\tau_a = 125 \times 10^{-12} \text{ sec}$$

for a coupled dephasing-resonance vibrational energy transfer process (Eq. 3.61), and

$$\tau_a = 115 \times 10^{-12} \text{ sec}$$

for a pure dephasing process (refer to Section 2.2.3). In other words, suppressing the concomitant resonance vibrational energy transfer [$\partial^2 V_s / \partial Q_i \partial Q_j$ = 0] yields a faster vibrational dephasing rate. Usually, two simultaneous relaxation processes give a reciprocal, combined correlation time which is the sum (or the squared sum[213]) of the individual reciprocal correlation times; the multiple process is faster than each component process. Here, we find that the relaxation rate of the combined relaxation process is slower than the sum of the individual effects; the interference term is *negative*. Consequently, the bandwidth of a vibrational profile broadened by resonance energy transfer may *increase* upon dilution [$\partial^2 V_s / \partial Q_i \partial Q_j = 0$] in an isotopically substituted medium. Although this prediction has not yet been demonstrated by a spectroscopic experiment, it may rationalize the curious fact that resonance vibrational energy transfer is rarely observed to be a predominant or even active vibrational relaxation mechanism in situations where it is expected.[10b] We have more to say about this in Section 3.3.

Of course, we must not lose sight of the fact that the interference term between resonance and dephasing is of relatively small magnitude; a difference of 10% between correlation times is on the border of detectibility. Nevertheless, the interesting aspect is its negative sign.

### 3.2. Resonance Coupling by Long-Range Forces with Vibration-Rotation Interaction: Methylene Chloride ($CH_2Cl_2$)

There are two main ideas for calculating the quasi-static bandwidth (absence of motional narrowing) $\langle | \omega_1(0)|^2 \rangle^{1/2}$ or the correlation function $\phi(t)$ of a resonance vibrational energy transfer process. One is based on the assumption that the oscillator coupling takes place by way of a short-range, the other assumes that it takes place by way of a long-range intermolecular potential.

For the short-range potential, the model calculations use an exponential (purely repulsive) potential,

$$V_s = V_0 \exp(-\alpha R_{ij}). \tag{3.62}$$

The theory predicts that the rate $\hbar^{-1} \langle |H_1|^2 \rangle^{1/2}$ of the resonance transfer mechanism is proportional to the sum of the squared derivatives of the actual average atomic displacement motions $q_1, q_2, \ldots q_n$ with respect to the normal coordinate $Q_k$ of molecule $k$,[214]

$$\hbar^{-1} \langle |H_1|^2 \rangle^{1/2} = \left(\frac{1}{6}\right) V_0 \left(\frac{\rho}{\pi}\right)^{1/2} \exp(-\alpha\sigma) \left\{ \frac{\sigma^2 \alpha^3}{2} + \frac{\sigma \alpha^2}{2} \right\}^{1/2} \sum_n \left(\frac{\partial q_n}{\partial Q_k}\right)^2 \omega_k^{-1},$$

where $\rho, \sigma$ denote the number density and contact distance of the radial distribution function (Eq. 3.23c). At present, potential constants $V_0, \alpha$ are known only for the simplest fluids (liquefied rare gases).

For long-range intermolecular forces, dipole-dipole interactions (Eq. 3.21$a$) are invoked. In this case, the rate of the resonance transfer is proportional to $(\partial\mu/\partial Q_k)^2$ (Eq. 1.32, Chapter 1), a quantity obtainable from the integrated intensity of the band. Let us then discuss coupling by dipole forces in the example of resonance infrared vibrational energy transfer[179d] in the $\nu_8$ ($B_2$ of $C_{2v}$) mode of $CH_2Cl_2$, the $CH_2$-rocking fundamental of methylene chloride at 1265 cm$^{-1}$.[34]

Our approach is as follows. We construct a theoretical correlation function $\phi(t)$ for resonance vibrational energy transfer by way of binary coupling between the (identical) vibrationally induced dipoles of a vibration-rotation mode $\alpha$, $(\partial\mu/\partial Q)_\alpha Q^\alpha$, on molecules $i$ and $j$. We relate this theoretical correlation function to the numerical Fourier transforms of the observed band profiles of mode $\alpha$ in the neat system and in the isotopically diluted system. The experimental correlation function of the neat liquid $\phi^\alpha(t)$ describes the totality of occurring (vibrational-rotational) relaxation processes of the system. We assume that isotopic dilution decouples the vibrational resonance transfer process (we neglect interference terms—see Section 3.1). Hence the correlation function for the isotopically diluted system $\phi_0^\alpha(t)$ describes all relaxation processes *except* resonance vibrational energy transfer; the quotient $\phi^\alpha(t)/\phi_0^\alpha(t)$ is the experimental correlation function of the pure resonance process.

We write the vibrational-rotational correlation function in symmetrized form (Eq. A.4, Appendix A),

$$\phi(t) = \left\langle\left[m(0), m(t)\right]_+\right\rangle = \mathrm{Tr}\rho\left[m(0), m(t)\right]_+$$

$$= \left\langle\left[m(0), \exp_0\left\{i\hbar^{-1}\int_0^t dt\,\overline{\overline{H}}_1^\times(t)\right\}\overline{\overline{m}}(t)\right]_+\right\rangle, \qquad (3.63)$$

where $H_1$ is the (yet to be specified) perturbation Hamiltonian inducing the resonance vibrational energy transfer process, and $m(t)$ is the macroscopic infrared vibrational-rotational transition moment. The double bar over a quantity indicates that it moves in the interaction representation (Eq. G.5, Appendix G) under operator

$$H_0 + B,$$

where $H_0$ is the Hamiltonian of the unperturbed oscillator and $B$ that of the bath or lattice.

Notice that Eq. 3.63 is an analogue of Eqs. 2.115 and 2.119 in Chapter 2. However, we introduce here a finer characterization of the dynamic variables by assuming that the orientational (external) motion of the vibrational transition moment vector is statistically correlated with the orientational motion of the lattice-modulated interaction Hamiltonian $H_1$. (We neglect translational motion.) In other words, we no longer ensemble-average separately over the orientational motion of the oscillator and over the orientational motion of the interoscillator coupling Hamiltonian. This introduces a certain measure of vibration-rotation coupling (Section 3.2, Chapter 1); we still average separately over the (internal) vibrational $H_0$ and (external) lattice $B$ coordinates.

To facilitate ensemble averaging we rewrite $m(t)$ in Eq. 3.63 in the interaction representation under lattice coordinates $B$, indicating this by a single upper bar. Since

$$\overline{\overline{m}}(t) = \exp\{i\hbar^{-1}(H_0 + B)^\times t\}m(t) \qquad (3.64a)$$

(Eq. G.7, Appendix G), it follows that

$$\overline{\overline{m}}(t) = \exp\{i\hbar^{-1}H_0^\times t\}\overline{m}(t), \qquad (3.64b)$$

where

$$\overline{m}(t) = \exp\{i\hbar^{-1}B^\times t\}m(t). \qquad (3.64c)$$

Recalling that the solution of the unperturbed oscillator is known, $\overline{\overline{m}}(t)$ simplifies to (Eqs. F.5 and F.11, Appendix F)

$$\overline{\overline{m}}(t) = \sum_\alpha \exp(i\alpha t)\overline{m}(t); \qquad \alpha \equiv \omega_\alpha = \frac{E' - E''}{\hbar},$$

where we have abbreviated the vibrational transition frequency $\omega_\alpha$ between a set of unperturbed vibrational levels $E'$, $E''$ by $\alpha$.

We insert Eq. 3.64b into Eq. 3.63, divide by the zero-order ($H_1^\times(t) = 0$) correlation function

$$\langle [m(0), \overline{m}(t)]_+ \rangle \exp(i\alpha t) \qquad (3.65)$$

(to obtain a dimensionless result), and perform the cumulant expansion (Appendix H). This yields the infrared correlation function for resonance vibrational energy transfer (Eq. H.8)

$$\phi(t) = \exp\left\{ -\hbar^{-2}\int_0^t dt' \int_0^{t'} dt'' \left\langle \left[ m^{-\alpha}(0), \overline{\overline{H}}_1^\times(t'')\overline{\overline{H}}_1^\times(t')\overline{m}^\alpha(t) \right]_+ \right\rangle \right.$$

$$\left. \times \left\langle [m^{-\alpha}(0), \overline{m}^\alpha(t)]_+ \right\rangle^{-1} \right\} \qquad (3.66)$$

with

$$\left\langle \overline{\overline{H}}_1^\times(t) \right\rangle = 0.$$

[Superscript $-\alpha$ indicates negative frequencies.[215]]

According to our intention, we use a transition-dipole–transition-dipole interaction mechanism. We define (Note 7)

$$m = M(a^\dagger + a), \qquad Q^\alpha = a^\dagger + a, \qquad \gamma = \frac{\hbar}{2\alpha}, \qquad M = \frac{\partial \mu}{\partial Q},$$

$$H_1^{ij}(t) = \gamma c_{ij}(t) Q_i^\alpha Q_j^{-\alpha}$$

$$= \gamma c_{ij}(t)\{a_i^\dagger a_j + a_i a_j^\dagger\}, \tag{3.67}$$

in terms of the Boson creation and destruction operators.[216] [Note that $c_{ij}(t)$ is the part of the coupling parameter $H_1$ which moves under $B$, whereas normal coordinate $Q$ is the part that moves under $H_0$.] Indices $i$, $j$ remind us that $H_1$ couples molecules $i$ and $j$; $c_{ii}(t) = 0$.

We decompose $m(t)$ into its microscopic components and their direction cosines with respect to a space-fixed axis,

$$\bar{m}^\alpha(t) = \sum_l \cos(\theta_l(t)) \frac{\partial \mu}{\partial Q} Q^\alpha \equiv \sum_l \cos((\theta_l(t)) M Q^\alpha. \tag{3.68}$$

Notice that $\cos(\theta_l(t))$ moves under bath Hamiltonian $B$. Insertion of Eqs. 3.67 and 3.68 into Eq. 3.66 and recalling the definition of the Kubo operator (Appendix F) gives

$$\phi(t) = \exp\left\{ -\frac{\hbar^{-2}\gamma^2}{N(t)} \int_0^t dt' \int_0^{t'} dt'' \sum_{\substack{i<j \\ l}} \left\langle \cos(\theta_l(0))\cos(\theta_l(t))c_{ij}(t'')c_{ij}(t') \right.\right.$$

$$\left.\left. \times \left[a_i + a_i^\dagger, \left[a_i^\dagger a_j + a_i a_j^\dagger, \left[a_j a_i^\dagger + a_j^\dagger a_i, a_i^\dagger + a_i\right]\right]\right]_+ \right\rangle \right\} \tag{3.69}$$

with (see Eq. 3.65)

$$N(t) = \sum_l \left\langle \cos(\theta_l(0))\cos(\theta_l(t))\left[a_i + a_i^\dagger, a_i^\dagger + a_i\right]_+ \right\rangle.$$

Note that the multiple commutator is written to give a vanishing frequency sum

$$\left[Q_i^{-\alpha}, \left[Q_i^\alpha Q_j^{-\alpha}, \left[Q_j^\alpha Q_i^{-\alpha}, Q_i^\alpha\right]\right]\right]_+$$

because we require resonance.

We solve the inner and outer commutator in Eq. 3.69 using $[a_i, a_j^\dagger] = \delta_{ij}$.[216] This gives $a_i^\dagger + a_i$. Next, we take the ensemble average over the internal coordinate by performing a trace operation over $H_0$. This yields (dropping subscript $i$ on the oscillator)

$$\left\langle [a + a^\dagger, a^\dagger + a]_+ \right\rangle = \mathrm{Tr}\rho_{H_0}\left(\tfrac{1}{2}\right)\{(a + a^\dagger)(a^\dagger + a) + (a^\dagger + a)(a + a^\dagger)\}$$

$$\approx \mathrm{Tr}\rho_{H_0}(aa^\dagger + a^\dagger a)$$

$$= 2\mathrm{Tr}\rho_{H_0}\left(\frac{H_0}{\hbar\alpha}\right) = \coth\left(\frac{\hbar\alpha}{2k_B T}\right). \tag{3.70}$$

[For the last equality in Eq. 3.70, see Appendix K.] We insert Eq. 3.70 into Eq. 3.69, set $\coth(\hbar\alpha/2k_B T)$ approximately 1 $(\hbar\alpha/2k_B T \sim 3)$, and express the cosines by their spherical harmonics. This gives the correlation function

$$\phi(t) = \exp\left\{ -\left[\alpha^2 N'(t)\right]^{-1} \int_0^t dt' \int_0^{t'} dt'' \sum_{\substack{i < j \\ l}} \mathrm{Tr}\rho_B Y_1^0(\Omega^l(0)) Y_1^0(\Omega^l(t)) \right.$$

$$\left. \times c_{ij}(t'')c_{ij}(t') \right\}, \tag{3.71}$$

$$N'(t) = \sum_l \mathrm{Tr}\rho_B Y_1^0(\Omega^l(0)) Y_1^0(\Omega^l(t)).$$

We identify $c_{ij}(t)$ with $H_1$ of Eq. 3.21. Of course, we must replace permanent dipole moments $\mu_i, \mu_j$ by their vibration-induced moment derivatives $M_i, M_j = (\partial\mu/\partial Q)_i, (\partial\mu/\partial Q)_j$. Inserting these terms into Eq. 3.71 requires the trace operations

$$\mathrm{Tr}\rho_B \sum a^{mm'}a^{nn'}M_l^2 M_k^2 Y_1^0(\Omega^l(0)) Y_1^0(\Omega^l(t))$$

$$\times R_{lk}^{-3}(t'') Y_1^m(\Omega^l(t'')) Y_1^{m'}(\Omega^k(t'')) Y_2^{-(m+m')}(\Omega^{lk}(t''))$$

$$\times R_{lk}^{-3}(t') Y_1^n(\Omega^l(t')) Y_1^{n'}(\Omega^k(t')) Y_2^{-(n+n')}(\Omega^{lk}(t'))$$

and

$$\mathrm{Tr}\rho_B Y_1^0(\Omega^l(0)) Y_1^0(\Omega^l(t)).$$

Summation is over all $m, m', n, n', k$. Averaging first over an isotropic distribution of the (statistically independent) center-of-mass separations yields

$$\left\langle H_1(t'') H_1(t') \right\rangle_{\text{transl}} = -(1)^{n+n'} \frac{\rho}{3\sigma^3} \delta_{-m-m', n+n'} \tag{3.72}$$

(see Eqs. 3.22 and 3.24; set $u = 0$). Averaging $N'(t)$ with an exponential orientational probability gives (Eq. 3.25)

$$N'(t) = (4\pi)^{-1} \exp\left(-\frac{t}{\tau_R^{(1)}}\right). \tag{3.73}$$

Similarly, averaging over all orientational coordinates of oscillator $k$ yields (we assume statistical independence between the orientational motion of adjacent molecules)

$$\left\langle Y_1^{m'}\left(\Omega^k(t'')\right) Y_1^{n'}\left(\Omega^k(t')\right)\right\rangle = (-1)^{n'}(4\pi)^{-1}\exp\left\{-\frac{(t'-t'')}{\tau_R^{(1)}}\right\}\delta_{m',-n'}. \tag{3.74}$$

We are then left with the ensemble average over the orientational motion of oscillator $l$. The correlation function (we drop superscript $l$) is complicated,

$$\left\langle Y_1^0(\Omega(0)) Y_1^m(\Omega(t'')) Y_1^n(\Omega(t')) Y_1^0(\Omega(t))\right\rangle, \tag{3.75}$$

since the orientational correlation of the oscillator stretches over four succeeding time points, $0 \le t'' \le t' \le t$; we do not a priori assume that the $H_1(t)$ commute at different times. To solve Eq. 3.75, we define a conditional probability $P$ that the orientation of the oscillator at time $t$ is $\Omega(t)$ when it was $\Omega(0), \Omega(t''), \Omega(t')$ at preceding times $0, \le t'', \le t'$, respectively.[179d] Choosing again a simple exponential for the individual probabilities (Eq. 3.25), we write $[\Omega(0) = \Omega^0]$

$$P(t|t';t'';0) = (4\pi)^{-1} \sum_{p''q''} Y_{p''}^{q''*}(\Omega^0) Y_{p''}^{q''}(\Omega'')\exp\left(-\frac{t''}{\tau_R^{(p'')}}\right)$$

$$\times \sum_{p'q'} Y_{p'}^{q'*}(\Omega'') Y_{p'}^{q'}(\Omega')\exp\left\{-\frac{(t'-t'')}{\tau_R^{(p')}}\right\}$$

$$\times \sum_{pq} Y_p^{q*}(\Omega') Y_p^{q}(\Omega)\exp\left\{-\frac{(t-t')}{\tau_R^{(p)}}\right\}. \tag{3.76}$$

Inserting this into Eq. 3.75 and reordering gives

$$(4\pi)^{-1} \sum_{p''q''p'q'pq} \exp\left(-\frac{t''}{\tau_R^{(p'')}}\right)\exp\left\{-\frac{(t'-t'')}{\tau_R^{(p')}}\right\}$$

$$\times\exp\left\{-\frac{(t-t')}{\tau_R^{(p)}}\right\} \int\int\int\int d\Omega^0\, d\Omega''\, d\Omega'\, d\Omega\, Y_{p''}^{q''*}(\Omega^0) Y_1^0(\Omega^0)$$

$$\times Y_{p''}^{q''}(\Omega'') Y_p^{q*}(\Omega'') Y_1^m(\Omega'') Y_{p'}^{q'}(\Omega') Y_p^{q*}(\Omega') Y_1^n(\Omega') Y_p^{q}(\Omega) Y_1^0(\Omega). \tag{3.77}$$

This somewhat formidable expression is nevertheless readily averaged over $\Omega^0$ and $\Omega$ by noting the orthogonality conditions of the spherical harmonics (with $Y_p^{q*} = (-1)^q Y_p^{-q}$).[217a] They require $p'' = 1$, $q'' = 0$ (for $\Omega^0$) and $p = 1$, $q = 0$ for $\Omega$. This gives

$$(4\pi)^{-1} \exp\left(-\frac{|t''|}{\tau_R^{(1)}}\right) \exp\left(-\frac{|t-t'|}{\tau_R^{(1)}}\right) \sum_{p'q'} \int d\Omega'' \, Y_p^{q*}(\Omega'') Y_1^m(\Omega'') Y_1^0(\Omega'')$$

$$\times \int d\Omega' \, Y_1^0(\Omega') Y_1^n(\Omega') Y_{p'}^{q'}(\Omega') \exp\left(-\frac{|t'-t''|}{\tau_R^{(p')}}\right).$$

For the terms in $\Omega''$ and $\Omega'$ we use twice the relation between three spherical harmonics of the same argument. This leads to simple products of normalization constants and Clebsch-Gordon coefficients of the form $C(p'11; q'n0)$, $C(p'11; 000)$, $C(11p'; 0mq')$, and $C(11p'; 000)$.[217b] The algebra is straightforward but uninteresting; symmetry lets only terms with $p' = 0, 2$ survive.

Inserting the result and Eqs. 3.72–3.74 into Eq. 3.71 and looking up the nonvanishing product coefficients $a^{mm'}a^{nn'}$ ($m' = -n'$; $m = -n$; see Eqs. 3.72 and 3.74) finally gives[179d]

$$\ln\phi(t) = \sum_p \left[ -f\frac{\pi\rho(\partial\mu/\partial Q)^4}{27\sigma^3\alpha^2} \int_0^t dt' \int_0^{t'} dt'' \exp\left\{-\frac{(t'-t'')}{\tau_R^{(p)}}\right\} \right] \quad (3.78)$$

with

$$\begin{cases} p = 0; f = 2 \\ p = 2; f = 4. \end{cases}$$

For $p = 0$, $\tau_R^{(0)}$ is a zero-rank tensor correlation time; we set $\tau_R^{(0)} \to \infty$. Integration over the intermediate times leads at once to the rigid lattice approximation (see also Eq. 3.6$a$).

For $p = 2$, we set $\tau_R^{(2)} = \tau_2$, a second-rank tensor correlation time. Integration over intermediate times yields

$$\phi(t) = \exp\left\{ -B_0^2\left(\frac{1}{2}t^2 + 2\left[\tau_2^2\left(\exp\left(-\frac{t}{\tau_2}\right) - 1\right) + \tau_2 t\right]\right) \right\},$$

with

$$B_0^2 = \langle |c_{ij}(t)|^2 \rangle = \frac{2\pi\rho(\partial\mu/\partial Q)^4}{27\sigma^3\alpha^2}. \quad (3.79)$$

(Notice that Eq. 3.79 does not become exponential for $t \geq \tau_2$.)

It is instructive to perform the ensemble average of Eq. 3.75 *separately* over the orientational coordinates of $m(t)$ and $H_1(t)$ and to disregard time-ordering. This means (1) we neglect statistical correlation between the orientational motions of vibrational coordinate $Q$ and of perturbation operator $H_1(t)$, (2) we assume that the $Y(t)$ commute. Time integration in Eq. 3.75 is then written

$$\left\langle Y_1^0(\Omega(0))Y_1^0(\Omega(t))\right\rangle \int_0^t dt' \int_0^{t'} dt'' \left\langle Y_1^m(\Omega(t''))Y_1^n(\Omega(t'))\right\rangle.$$

Introducing isotropic exponentials for the individual orientational correlations, Eq. 3.71 now leads to Eq. 3.11 except for the different definitions of

$$B_0^2 = \left\langle |c_{ij}(t)|^2 \right\rangle = \left\langle |c_{ij}(0)|^2 \right\rangle$$

and of $\tau_R^{(j)}$ (Note 8).

Figure 3.13 shows the comparison between experiment and theory on the 1265-cm$^{-1}$ mode of $CH_2Cl_2$.[179d] The points represent the quotient $\phi^\alpha(t)/\phi_0^\alpha(t)$ of the experimental correlation values for the neat, $\phi^\alpha(t)$, and isotopically diluted liquid. The curves are computed with Eq. 3.79; correlation time $\tau_2$ is taken from nuclear magnetic relaxation data, $\sigma$ is 4 Å, and $B_0^2$ is adjusted by a best fit. Notice the motional narrowing effect with decreasing $\tau_2$.

We have discussed this theory in detail to demonstrate the success of modeling vibrational relaxation of complicated interaction regimes, including

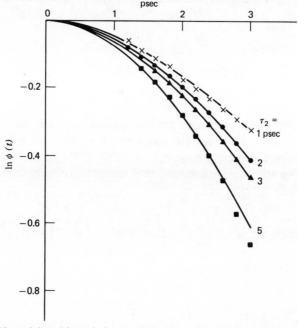

**Figure 3.13.** Natural logarithm of the correlation function for resonance vibrational energy transfer in the $B_2$ mode of $CH_2Cl_2$ at 1265 cm$^{-1}$. Points: experimental results. Curves: computed from Eq. 3.79. [P. C. M. van Woerkom, J. de Bleyser, M. de Zwart, and J. C. Leyte, *Chem. Phys.* **4**, 236 (1974).]

statistical correlation between vibrational and orientational coordinates, on the basis of a few principles. We now show how significant the vibration-rotation interaction is.

Suppose we write a limiting relation between $\tau_2$ ($j = 2$) and $\tau_1$ ($j = 1$), namely, $\tau_1/\tau_2 = 3$ (Eq. 2.97, Chapter 2). Now we can readily compare Eq. 3.79 and Eq. 3.11, assuming the same $\tau_1$ and $B_0^2$ (Note 8). Developing the respective inner exponentials into a series, we obtain, from Eq. 3.79,

$$\phi^+(t) = \exp\left\{-B_0^2\left[\frac{3t^2}{2} + 2\sum_{n=3}^{\infty}(-1)^n\frac{1}{n!}\left(\frac{3}{\tau_1}\right)^{n-2}t^n\right]\right\},$$

and from Eq. 3.11,

$$\phi^-(t) = \exp\left\{-B_0^2\left[\frac{3t^2}{2} + 2\sum_{n=3}^{\infty}(-1)^n\left(\frac{2}{3}\right)^{n-3}\frac{1}{n!}\left(\frac{3}{\tau_1}\right)^{n-2}t^n\right]\right\}.$$

Notice that Eq. 3.79 and Eq. 3.11 begin to diverge with term $t^4$; resonance vibrational energy transfer with coupling $\phi^+(t)$ decays faster. Hence neglect of strong coupling between the bath-modulated motions of the vibrational transition moment and the interoscillator coupling Hamiltonian leads to an *underestimate* of the rate of resonance vibrational energy transfer. As physical reason of this phenomenon, we propose that the rate of coupled orientational motions is in concordance with the rate of the slowest-moving lattice coordinate. Consequently, motional narrowing is lessened relative to uncorrelated orientational motions; the bandwidth increases. A numerical example is instructive: Using $B_0^2 = 0.010 \times 10^{24}$ radian$^2$ sec$^{-2}$ and $\tau_1 = 4 \times 10^{-12}$ sec, we compute [$t$ in $10^{-12}$ sec, $\phi^+(t), \phi^-(t)$] 1, 0.961, 0.961; 4, 0.631, 0.664; 8, 0.255, 0.336.

Finally, we remark that correlation time $\tau_2$ must describe the orientational motion of an axis direction that coincides with that of $\mathbf{m}(t)$ of the observed mode. This requirement is not always easily met. If we deal with degenerate modes or low-symmetry molecules (Section 3.4, Chapter 1), $\tau_2$ may have to be estimated from nuclear magnetic relaxation data (possibly different axes directions) or depolarized Rayleigh scattering results (complicated by orientational pair correlations).

## 3.3.  Phonon Diffusion in Liquid Benzene

The most intense vibrational transition in the infrared spectrum of liquid benzene is the out-of-plane carbon–hydrogen deformation fundamental of symmetry species $A_{2u}$ (point group $D_{6h}$) at 674 cm$^{-1}$. Its integrated intensity $\int d\omega\, I(\omega)$ is $2.17 \times 10^4$ cm$^2$ mole$^{-1}$ or $1.46 \times 10^7$ cm mole$^{-1}$.[218] Hence we can no longer omit the effect of reflectivity changes and must use the optical constants $n(\omega)$, $\kappa(\omega)$ instead of transmission data.[9] Equating $\sigma$, the distance of closest approach, with the Lennard-Jones parameter $\sigma = 5.35$ Å, we compute (Eq. 3.79)

$$B_0 = \left\langle |c_{ij}(0)|^2 \right\rangle^{1/2} = 1.3 \times 10^{12} \text{ radian sec}^{-1}$$

from[174]

$$\int d\omega\, I(\omega) = \left( \frac{4\pi^2 g}{3ch} \right) \rho \omega_0 \left( \frac{\partial \mu}{\partial Q} \right)^2 |(v|Q|v')|^2 \ [\text{cm}^{-1} \text{ sec}^{-1}]$$

$$= \left( \frac{2\pi^2 g}{3c\omega_0} \right) N_M v' \left( \frac{\partial \mu}{\partial Q} \right)^2 \ [\text{cm}^2 \text{ mole}^{-1}]. \tag{3.80}$$

$g$ is the degeneracy of the vibrational lower level $v$ and $N_M$ Avogadro's number. This is equivalent to a vibrational resonance transfer width (Eq. 3.17) of 16 cm$^{-1}$.

The isotopic dilution data, displayed in Fig. 3.14, tell another story;[219] band broadening from vibrational resonance energy transfer is a mere 1.5 cm$^{-1}$, one order of magnitude off the estimated value of 16 cm$^{-1}$. (A resonance bandwidth of 15 cm$^{-1}$, reported in a similar study,[220] is suspect. The result is based

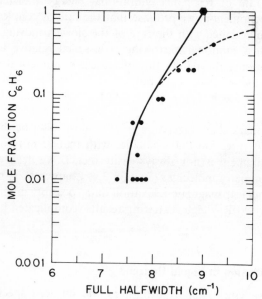

**Figure 3.14.** Infrared band width of the $A_{2u}$ mode of liquid benzene at room temperature as a function of dilution by perdeuterobenzene. The dash curve shows the effect of neglecting the frequency variation of the refractive index (transmission data). The width for neat $C_6H_6$ (hexagonal point) is taken from optical constants data.[9, 218] [W. G. Rothschild and T. Tokuhiro, unpublished work, 1975.]

on infrared transmission data and undoubtedly overestimated the bandwidth for the neat liquid and its more concentrated $C_6H_6$–$C_6D_6$ mixtures because of reflection changes over the range of the profile.[9,218])

The large discrepancy between experiment and theory may arise from several causes. A fast modulation regime could narrow the bandwidth. Strong coupling between resonance vibrational energy transfer and dephasing may yield an interference term that counterbalances the pure resonance width (Section 3.1). At any rate, many questions about the dynamic and static characteristics of liquid-phase benzene are raised by these results. The system possesses intriguing optical properties;[220] it would appear worthwhile to re-study its vibrational dynamics thoroughly. Fortunately, a great deal is known about the liquid-phase structure of benzene;[135] such data may thus be profitably incorporated and utilized.

Taking the experimental resonance energy transfer width of 1.5 cm$^{-1}$ (Fig. 3.14) and turning it into the average time $\tau$ of a vibrational resonance flip-flop step (Appendix J), gives

$$\tau = (2\pi 1.5c)^{-1} = 3.5 \times 10^{-12} \text{ sec.}$$

From the resonance transfer rate $\tau^{-1}$ and the average distance $\sigma$ between adjacent molecules $i$ and $j$, we can estimate the diffusion coefficient $D_{\text{res}}$ of the excited phonon as it travels in any direction through the liquid,[221]

$$D_{\text{res}} = \left(\tfrac{1}{6}\right)\tau_{i \to j}^{-1}\sigma^2.$$

We compute $D_{\text{res}} = 1.4 \times 10^{-4}$ cm$^2$ sec$^{-1}$, a value that is two orders of magnitude larger than the translational diffusion coefficient of the room-temperature $C_6H_6$.[181]

A more sophisticated (and more realistic) view of excitation diffusion involves (partial) return to the original site and excitations arriving from other loci in the liquid. We discuss this in the latter half of the Chapter.

## 4. ASYMMETRIC BAND PROFILES: BREAKDOWN OF THE CENTRAL LIMIT THEOREM

In the majority of cases, the observed profiles of isolated, individual vibrational modes of condensed-phase molecules are symmetric—apart from a small asymmetry due to the factor of detailed balance (see Appendix A).

However, cases of strongly, inherently asymmetric band profiles have been reported that require special consideration. For instance, the points in Fig. 3.15a display the experimental isotropic Raman profile of the $0 \to 1$ fundamental transition of HCl in dilute solution (0.1 mf) of $CCl_4$.[222a] The points in Fig. 3.15b display the experimental infrared profile of the $0 \to 1$ fundamental transition of CO chemisorbed on platinum dispersed on $\gamma$-$Al_2O_3$.[223] In both examples the profiles are significantly skewed to lower frequencies. The experimental correlation functions are complex (Appendix A), their physical meaning is obscured.

Since the profiles are no longer symmetric, the distribution of the interaction Hamiltonian $H_1(t)$ (Eq. 3.1) is no longer Gaussian (normal), the central limit theorem is no longer satisfied (Appendix H), and the perturbation series of $\langle \exp i \int dt\, \omega_1(t) \rangle$ has not converged with the second-order term. As discussed in Section 2.5, the active oscillator senses relatively few (less than about four) interaction sites in its environment. Such physical situations are conceivably realized in our examples: A HCl molecule in an "incomplete" $CCl_4$-solvent cage[171, 224], an adsorbed CO molecule on an "exposed" adsorption site on the Pt metal crystallite.[223]

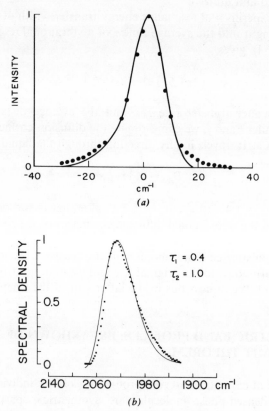

**Figure 3.15.** (*a*) Room-temperature experimental isotropic Raman profile (dots) of HCl in $CCl_4$ (0.1 mf) and theoretical profile (curve) in the rigid lattice approximation using $\langle \omega_1(0)^2 \rangle^c = 52$ $cm^{-2}$, $\langle \omega_1(0)^3 \rangle^c = -300$ $cm^{-3}$, and $\langle \omega_1(0)^4 \rangle^c = 1200$ $cm^{-4}$. [Y. Guissani and J. C. Leicknam, *Can. J. Phys.* **51**, 938 (1973). Reproduced by permission of the National Research Council of Canada.] (*b*) Simulated (curve) and experimental infrared spectra (dots) of CO chemisorbed at 30°C on Pt/$\gamma$-$Al_2O_3$. Best-fit values of modulation times $\tau_1$ and $\tau_2$ are in picoseconds. Experimental values of $\langle \omega_1(0)^2 \rangle = 756$ $cm^{-2}$, $\langle \omega_1(0)^3 \rangle = -15170$ $cm^{-3}$. [W. G. Rothschild and H. C. Yao, *J. Chem. Phys.* **74**, 4186 (1981).]

The smooth curves in Fig. 3.15$a$ and 3.15$b$ show theoretical predictions of asymmetric band profiles. For the HCl–CCl$_4$ system (Fig. 3.15$a$), the quasi-static modulation regime is assumed and the cumulants developed and optimized to the data up to the term $t^4$ (Eq. H.6, Appendix H),[222b]

$$\mathscr{I}(\omega) = \int_{-\infty}^{\infty} dt \exp(i\omega t) \exp\left\{ i \int_0^t dt \left\langle \omega_1(0) \right\rangle^c + \frac{i^2}{2} \int_0^t dt' \int_0^t dt \left\langle \omega_1(0)^2 \right\rangle^c \right.$$

$$\left. + \frac{i^3}{6} \int_0^t dt'' \int_0^t dt' \int_0^t dt \left\langle \omega_1(0)^3 \right\rangle^c + \frac{i^4}{24} \int_0^t dt''' \int_0^t dt'' \int_0^t dt' \int_0^t dt \left\langle \omega_1(0)^4 \right\rangle^c \right\}.$$

$$(3.81a)$$

[Note that the cumulants are taken with respect to the first moment $\left\langle \omega_1(0) \right\rangle^c$.] For the CO–Pt/Al$_2$O$_3$-system (Fig. 3.15$b$), the cumulant series is broken off after the $t^3$ term, and modulation is admitted by proposing the modulation functions

$$\left\langle \omega_1(t)\omega_1(t') \right\rangle \left\langle |\omega_1(0)|^2 \right\rangle^{-1} = \exp\left[ -\frac{(t-t')}{\tau_c} \right],$$

$$\left\langle \omega_1(t)\omega_1(t')\omega_1(t'') \right\rangle \left\langle |\omega_1(0)|^3 \right\rangle^{-1} = \exp\left[ -\frac{(t'-t'')}{\tau_2} \right] \exp\left[ -\frac{(t-t')}{\tau_1} \right],$$

$$(3.81b)$$

with $\left\langle \omega_1(t) \right\rangle^c = 0$. The normalization factors $\left\langle |\omega_1(0)|^2 \right\rangle$ and $\left\langle |\omega_1(0)|^3 \right\rangle$ are obtained experimentally from the vibrational second and third spectral moments (see Section 2.2.1). Insertion of Eq. 3.81$b$ and Eq. 3.10 into the cumulant expansion (Eq. H.6, Appendix H), integrating over intermediate times, and taking the Fourier transform, gives the spectral density (up to third order) shown in Fig. 3.15$b$. The fit to the experimental spectral density necessitates adjusting three independent modulation times $\tau_c$, $\tau_1$ ($\sim \tau_c$), and $\tau_2$ ($\gg \tau_1$).

Both these attempts of explaining inherent strongly asymmetric band profiles are promising but need further development. Although numerical criteria can be offered to estimate how well the (a priori infinite) cumulant series has converged,[223] the simple assumption of a quasi-rigid lattice for the system HCl–CCl$_4$[222b] is suspect from what we have learned about dephasing. On the other hand, for intermediate modulation regimes, the meaning of the higher-order modulation times $\tau_2$ is not obvious; they involve complicated correlations of interaction phenomena too detailed to be of ready use to us (Section 2.5). But for fast modulation $\tau_1$, $\tau_2 \ll t$, integration of Eq. 3.81$b$ over $t'$, $t''$, $t$ gives $\left\langle |\omega_1(0)|^3 \right\rangle \tau_1 \tau_2 t$.[223] Together with the fast-modulation (even) main term

$\langle|\omega_1(0)|^2\rangle\tau_c t$ (Eq. 3.8), this leads to the Lorentzian spectral density

$$\mathscr{I}(\omega_1) = \int dt \exp(-i\omega_1 t)\exp\left\{-\langle|\omega_1(0)|^2\rangle\tau_c t - i\langle|\omega_1(0)|^3\rangle\tau_1\tau_2 t\right\}$$

$$= 2M(2)\tau_c\left\{\left[M(2)\tau_c\right]^2 + \left[\omega_1 + M(3)\tau_1\tau_2\right]^2\right\}^{-1}$$

with (see also Eq. 3.18)

$$\langle|\omega_1(0)|^3\rangle = M(3) = \int d\omega\, \hat{\mathscr{I}}(\omega)(\omega - \omega_s)^3.$$

We see that the higher-order modulation terms contribute only a frequency shift in a fast modulation regime. The physical reason is obvious: The rapid sampling of the environment by the oscillator effectively *increases* the number of interaction sites.

## 5.  BAND BROADENING BY CONCENTRATION FLUCTUATIONS

In Section 2.2.3 we discussed the effect of dilution by solvents on the rate of vibrational relaxation processes of an active solute molecule. On the examples of several fundamental vibrational modes of methyl iodide, we demonstrated how dilution by $CS_2$ led to a considerable decrease in the vibrational dephasing rate of the $\nu_2$ mode (refer to Fig. 3.3c)—with the obvious explanation that the different static and dynamic local environments of the active oscillator were responsible for this effect.

Of course, we recall that there is no a priori reason that the dephasing rate or the corresponding bandwidth of the observed oscillator must decrease upon dilution by solvents; it may also increase. This depends entirely on the particular system perturbation Hamiltonian $H_1(t)$. (Refer to Section 2.7.) Furthermore, the curious aspect here is that a solvent-induced rise in the vibrational dephasing rate is not necessarily a monotonous or asymptotic function of increasing dilution, but may reach a *maximum* at mole fraction of approximately 0.5. This interesting effect, which has nothing to do with the formation of liquid-phase complexes of equimolar (or similar) composition (Section 5.1, Chapter 2), was first reported in the early 1970s[225a] but has not received the attention it deserves until recently.

In brief, the phenomenon is caused by quasi-static local concentration fluctuations in solutions. Taking a typical molecular translational correlation time about $10^{-11}$ sec (Eq. 3.24), it is easy to see that concentration fluctuations may generate contributions to vibrational dephasing processes that are in the slow modulation regime. Clearly, if the modulation is rapid, concentration fluctuations are averaged out (Section 2.1).

The quantitative relations describing the ensuing band broadening are very simple.[224,225a,b] Restricting our considerations to binary systems, we express by

$$x_1 = x_1^0 + \Delta x_1 \quad [\text{mf}]$$

$$\nu_1 = \nu_1^0 + \Delta \nu_1 \quad [\text{cm}^{-1}]$$

the instantaneous molecular mole fraction and oscillator frequency, respectively, of component 1 (the observed species). The zero superscript indicates the macroscopic (analytical) mole fraction and the unperturbed transition frequency of the oscillator, respectively. $\Delta x_1, \Delta \nu_1$ denote random excursions from these values owing to concentration fluctuations. Putting

$$\Delta \nu_1 = \frac{\partial \nu_1}{\partial x_1} \Delta x_1$$

and assuming 8–10 nearest neighbors around the active oscillator, we predict a Gaussian distribution of the $\Delta \nu_1$.[127] Thus we expect a Gaussian band profile $\mathcal{I}(\Delta \nu_1)$ (Eq. 3.6b) with a full width at half-peak height $\Delta \nu_{1/2}$ given by Eq. 3.17:

$$\Delta \nu_{1/2} = 2\left\{ (2\ln 2)\left\langle (\Delta \nu_1)^2 \right\rangle^c \right\}^{1/2}$$

$$= 2(2\ln 2)^{1/2} \left| \frac{\partial \nu_1}{\partial x_1} \right| \left\{ \left\langle (\Delta x_1)^2 \right\rangle^c \right\}^{1/2}.$$

We recognize $\langle (\Delta x_1)^2 \rangle^c$ as the dispersion of the instantaneous mole fraction of the active oscillator (Eq. H.3, Appendix H). Designating by $N_1, N_2$ the average number of molecules in the "fluctuation volume," we write

$$x_1 = \frac{N_1}{(N_1 + N_2)}, \qquad \Delta x_1 = \frac{N_2 \Delta N_1}{(N_1 + N_2)^2}.$$

Employing the well-known thermodynamic relation[225c]

$$\left\langle (\Delta N_1)^2 \right\rangle^c = \frac{k_B T}{(\partial \Upsilon_1 / \partial N_1)},$$

$$\Upsilon_1 = k_B T \ln x_1,$$

where $\Upsilon_1$ is the chemical potential at constant temperature and pressure, we get

$$\left\langle (\Delta N_1)^2 \right\rangle^c = \frac{N_1 N}{N_2}; \qquad \left\langle (\Delta x_1)^2 \right\rangle^c = \frac{x_1(1 - x_1)}{N},$$

where $N = N_1 + N_2$. Therefore the bandwidth increment caused by concentration fluctuations in the slow modulation regime is

$$\Delta \nu_{1/2} = 2(2\ln 2)^{1/2} \left| \frac{\partial \nu_1}{\partial x_1} \right| \left\{ \frac{x_1(1 - x_1)}{N} \right\}^{1/2}. \tag{3.82}$$

Notice that we have assumed ideal solutions (otherwise mole fraction $x_1$ needs to be replaced by the activity).

Equation 3.82 then tells us that we may observe concentration fluctuations in situations where the frequency shift with increasing dilution $(\partial \nu_1 / \partial x_1) \Delta x_1$ is large. We see that the maximum of the bandwidth is predicted at $x_1 = 0.5$.

Figure 3.16$a$ displays the concentration dependence of the bandwidth of the infrared $\nu_2$ mode of $CH_3I$ dissolved in $CH_3CN$ and in $CS_2$ as well as of the $\nu_1$ mode of $CHCl_3$ dissolved in $CS_2$ and $CCl_4$. Figure 3.16$b$ shows the corresponding peak frequencies. Notice that large frequency shifts and a pronounced maximum in the bandwidth go together indeed. Notice also that the maximum of the bandwidth of $CH_3I$ dissolved in $CH_3CN$ is not at $x_1 = 0.5$ as predicted by Eq. 3.82 but is shifted to approximately 0.6 mole fraction $CH_3I$. Apart from perturbations due to other participating concentration-dependent band-broadening effects, this deviation may be a consequence of the nonideality of the $CH_3I$–$CH_3CN$ solutions. As suggested before, in such situations we should replace the mole fraction by the activity $a = \gamma x$, where $\gamma \lesssim 1$ is the activity coefficient. This indeed shifts the maximum of $x_1$ above 0.5 mole fraction $CH_3I$ if $\gamma < 1$.

Band broadening by concentration fluctuations has also been observed in purely vibrational Raman profiles (Eq. 1.46, Chapter 1).[224] For instance, the profile of the isotropic $\nu_1$ mode of $CH_3I$ in $CDCl_3$ solution is broadened and its peak frequency shifted as predicted by Eq. 3.82. Interestingly, the Gaussian character of the profile augments with decreasing temperatures, pointing to a diminished speed of site changes of solute and solvent molecules surrounding the active $CH_3I$ oscillator.

It is not difficult to see that the effect of concentration fluctuations affords interesting insights and answers to questions on the solvation-dependent range of intermolecular forces (the size of the fluctuation volume) and effective number of molecules surrounding that carrying the active oscillator.[224] In fact, we expect that closer investigation shows variations depending on the spectroscopies (infrared absorption, spontaneous and stimulated Raman scattering).

There is also a detrimental aspect to concentration fluctuations; their presence may obscure solution studies of vibrational relaxation processes. For instance, a study of solvation of the molecule carrying the active oscillator—performed, say, to probe static and dynamic coupling between molecules of diverse shape, polarity, and so on—may well be perturbed by fluctuation concentrations that cause band broadening irrelevant to the

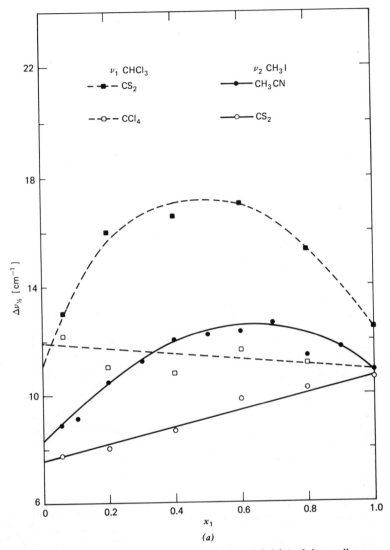

**Figure 3.16.** (*a*) Infrared bandwidth (full width at half-peak height) of the totally symmetric $\nu_1$ C–H stretching mode of $CHCl_3$ dissolved in $CS_2$ and $CCl_4$ and the totally symmetric $\nu_2$ C–H deformation mode of $CH_3I$ dissolved in $CH_3CN$ and $CS_2$ at room temperature. The abscissa gives the concentration (mole fraction) of solutes $CHCl_3$ and $CH_3I$, respectively. (*b*) Infrared frequencies of the $\nu_1$ $CHCl_3$ mode (left ordinate) and the $\nu_2$ $CH_3I$ mode (right ordinate) as a function of the mole fraction of solute. [From T. Fujiyama, M. Kakimoto, and T. Suzuki, *Bull. Chem. Soc. Jap.* **49**, 606 (1976).]

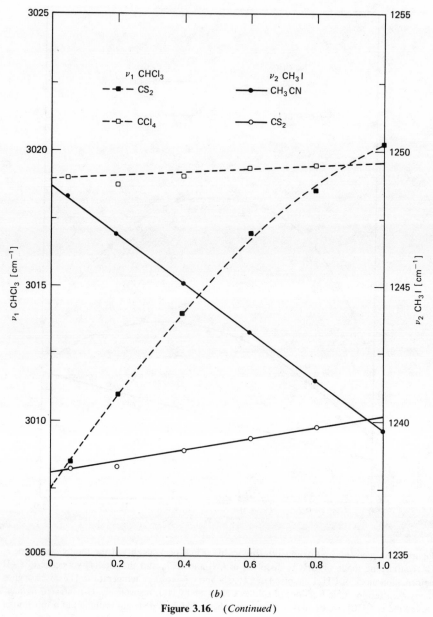

*(b)*

**Figure 3.16.** (*Continued*)

sought-after effect. (Obvious candidates are the complex systems pyridine–water and phenol–dioxane discussed in Section 2.3.)

## B.    Multilevel Systems

## 6.    VIBRATIONAL ENERGY RELAXATION (DISSIPATION)

### 6.1.    Carbon–Hydrogen Stretching Fundamentals of Methyl Iodide

We mentioned on previous occasions that vibrational energy relaxation is a relatively slow process since there is a large mismatch between the Fourier frequency components of a near- or mid-infrared excited oscillator phonon and the far-infrared bath motions or lattice phonons (Section 4.1, Chapter 1).

Consequently it is not surprising that the stimulated relaxation methods (picosecond pulsing), which permit distinction between vibrational dephasing and population relaxation processes, are here particularly useful and frequently irreplaceable. Complicated vibrational energy relaxation pathways of several energy levels in the same or between different molecules can be examined. Such processes would be difficult to follow with the spontaneous relaxation techniques we have discussed in this chapter.

Nevertheless, it would be erroneous to assume that spontaneous vibrational energy relaxation is never observable. We recall that the $\nu_1$ mode of $CH_3I$ at 2950 cm$^{-1}$ has a spontaneous vibrational correlation time of approximately $2 \times 10^{-12}$ sec (20°C) which is rather independent of solvents and increases with decreasing temperature (Fig. 3.3$a$ and $b$). As we discussed (Section 2.2.3), all this strongly indicates that $\nu_1$ relaxes by vibrational energy dissipation.

Of course, such a fast relaxation rate cannot signify that the $\nu_1$ quantum is directly transformed into lattice or bath motion of average energy $k_B T$ approximately 200 cm$^{-1}$ (Note 9). Which then is the relaxation pathway?

Ultrashort infrared excitation experiments (the Raman pumping pulse is converted into a tunable infrared pumping pulse) have furnished the answer.[49] They show that the phonon population of the first excited level of $\nu_1$ relaxes rapidly to the nearby (doubly degenerate) $\nu_4$ C–H stretching fundamental at 3050 cm$^{-1}$. This pathway needs 100 cm$^{-1}$ additional energy—which is readily available from the bath. In turn, the population of the $\nu_4$ mode decays equally rapidly to adjacent modes within the molecule, coupling to summation and overtone bands ($v = 2$) involving the C–H deformation fundamental near 1400 cm$^{-1}$.[226] Subsequently, the excitation of the overtone drifts, at a considerably slower rate, to thermal equilibrium.

The ultrashort infrared excitation experiment gives the true energy relaxation time $\tau_{ED}(\nu_1)$ approximately $1 \times 10^{-12}$ sec.[49] Comparing this with the correlation time of the observed spontaneous contour $\tau(\nu_1)$ approximately

$2 \times 10^{-12}$ sec [where $\phi(t) \sim e^{-1}$; Fig. 3.3$a$],[190] shows that $\tau$ approximates $2\tau_{ED}$, as expected (Appendix J).

Clearly this result should put us on our guard: It is wrong to assume that vibrational energy dissipation is always a relatively slow process and that spontaneous contours are never observably broadened by this relaxation route. Strictly, however, the process just described is a (bath-assisted) vibrational nonresonant energy transfer between two internal modes and not pure and simple depopulation to the bath motions.

## 6.2.   Vibrational Energy Dissipation and Stochastic Intermolecular Forces

To relate the measured rate of a vibrational energy relaxation (dissipation) process to the intermolecular forces of the medium, we use the golden rule (Eqs. 1.63 and 1.66$b$, Chapter 1): It gives the probability per unit time (angular transition rate) that system oscillator $k$, under the influence of the stochastic perturbation Hamiltonian $A(t)$, has transferred a vibrational quantum $\hbar\omega_k$ to the bath regardless of initial and final states $s$,

$$
\begin{aligned}
W_{ss'}(\omega_k) &= \int_{-\infty}^{\infty} dt \exp\left[i(\omega_{ss'} - \omega_k)t\right]\langle A(t)A(0)\rangle \\
&= \frac{2\pi}{\hbar} \sum_{ss'} \rho_{s'} |(s|A|s')|^2 \, \delta\{E_s - E_{s'} - \hbar\omega_k\} \\
&= \frac{2\pi}{\hbar^2} \sum_{ss'} \rho_{s'} |(s|A|s')|^2 \, \delta\{\omega_{ss'} - \omega_k\}, \\
\omega_{ss'} &= \omega_s - \omega_{s'}.
\end{aligned}
\tag{3.83a}
$$

We have multiplied this expression by 4 to remove the normalization factor of $H_1(t)$ (Eq. 1.57); we also omit the superscript on $E$.

We now set

$$
\hbar^{-2} W_{ss'}(\omega_k) = K_s = \frac{1}{\tau_{ED}},
\tag{3.83b}
$$

where $\tau_{ED}^{-1}$ is the average rate of the vibrational energy dissipation process (Eqs. J.8 and J.9, Appendix J).

We proceed by modeling $\langle A(t)A(0)\rangle$, inserting the expression into Eq. 3.83$a$ with known parameters and constants, and comparing the result with the experiments.

The matrix elements of $A(t)$ are written over a product basis of oscillator $|v)$ and bath (or lattice) coordinates $|b)$, $|s) = |v)|b)$,

$$
(s|A|s') = (v|b|Q_k B_{kb}|b'|v') = (v|Q_k|v')(b|B_{kb}|b').
\tag{3.83c}
$$

In other words, we neglect statistical correlations between $|v\rangle$ and $|b\rangle$ (see also Eq. 1.33 of Chapter 1, Section 5.4 of Chapter 2, and Section 3.2 of this chapter). The double index of $B$ is convenient; it signifies that $B_{kb}$ contains constant factors belonging to the oscillator $k$ (see below). Evaluation of Eq. 3.83 depends then solely on the definitions or choice of the autocorrelation function of the intermolecular forces.

First, we discuss the effects of long-range forces. We invoke interactions of the vibrational transition moment of mode $k$, $(\partial\mu/\partial Q_k)Q_k$ (Eq. 1.32, Chapter 1), with the permanent dipole of molecule $j$, $\mu_j$. We get

$$(s|A|s') = \left(\frac{\hbar v'}{2\omega_k}\right)^{1/2} \frac{\partial\mu}{\partial Q_k} \sum_j \mu_j B_b(t), \qquad (3.84)$$

where

$$\left(\frac{\hbar v'}{2\omega_k}\right)^{1/2} = (v|Q_k|v') = (v'|Q_k|v)$$

is the expectation value of normal coordinate $Q_k$ between vibrational levels $v'$ and $v = v' - 1$ (Eq. 3.67). $B(t)$, the time-dependent factor of $A(t)$, moves under bath (lattice) coordinates: $B_{kb}(t)$ is the quantity of greatest interest to us since it contains the information we seek.

We insert Eqs. 3.83c and 3.84 into Eq. 3.83a, denoting the ensemble average over the bath coordinates by the angular brackets. This yields

$$W_{ss'}(\omega_k) = \frac{\hbar}{2\omega_k} \left(\frac{\partial\mu}{\partial Q_k}\right)^2 v' \sum_j \mu_j^2$$

$$\times \int_{-\infty}^{\infty} dt \exp[i(\omega_{bb'} - \omega_k)t]\langle B_b(t)B_b(0)\rangle \qquad (3.85)$$

with

$$B_{kb}(t) = \left(\frac{\hbar}{2\omega_k}\right)^{1/2} \mu_j \frac{\partial\mu}{\partial Q_k} B_b(t).$$

Notice that the integral in Eq. 3.85 is the spectral density $\mathscr{I}_b(\omega)$ of the translational-orientational autocorrelation function of the bath molecules. Hence the rate of vibrational energy relaxation is proportional to this quantity.

We assume, as previously (Section 2.2.1), that $\langle B(t)B(0)\rangle$ is the product of the pure translational and pure orientational autocorrelation functions of the bath motions. Using Eq. 3.25 and Eq. J.1 (Appendix J), this leads to the orientational spectral density

$$\mathscr{I}_R(\omega) = (\tau_R)^{-1}\{(\omega - \omega_{bb'})^2 + (2\tau_R)^{-2}\}^{-1}. \qquad (3.86a)$$

For the translational contribution we obtain (Eq. 3.24 and Appendix L)

$$\mathscr{I}_T(\omega) = \mathrm{Re}\left\{ (3s)^{-1} - s^{-1}\left(\left[2(s\tau_T)^{1/2}\right]^{-1} - \left[2(s\tau_T)^{3/2}\right]^{-1}\right.\right.$$

$$\left.\left. + \left[2(s\tau_T)^{1/2}\right]^{-1}\exp\left[-2(s\tau_T)^{1/2}\right]\left[1+(s\tau_T)^{-1/2}\right]^2\right)\right\} \qquad (3.86b)$$

where $s = -i\omega$; $\tau_T = \sigma^2/2D$.

For mid- and near-infrared $\omega_k$,

$$(\omega_k - \omega_{bb'})\tau \gg 1. \qquad (3.87)$$

Hence the system is in a slow modulation regime; the exact nature of the bath motion is of little concern. In fact, on the basis of condition 3.87, we approximate $\mathscr{I}_R(\omega)$ and $\mathscr{I}_T(\omega)$ by

$$\mathscr{I}_R(\omega_k) = 4\left(\omega_k^2\tau_R\right)^{-1}, \qquad (3.88a)$$

$$\mathscr{I}_T(\omega_k) = \left(2\sqrt{2}\,\omega_k^{3/2}\tau_T^{1/2}\right)^{-1} + O\left[\left(\omega_k^{5/2}\tau_T^{3/2}\right)^{-1}\right]. \qquad (3.88b)$$

Substituting the integral expression of Eq. 3.85 by the product of Eqs. 3.88a and 3.88b and noting the correct constant factors (Eqs. 3.21b, 3.72, and 3.74) gives (all frequencies are angular frequencies)

$$\hbar^{-2}W_{ss'}(\omega_k) = K_s \; (\mathrm{radian\;sec}^{-1})$$

$$= \frac{4\pi}{9\hbar\omega_k}\left(\frac{\rho v'}{\sigma^3}\right)\left(\frac{\partial\mu}{\partial Q}\right)^2\mu_j^2\left[4\left(\tau_R\omega_k^2\right)^{-1} + \frac{1}{2\sqrt{2}}\left(\tau_T^{1/2}\omega_k^{3/2}\right)^{-1}\right]$$

$$(3.89)$$

where $v'$ is the vibrational quantum number of the excited level and $g$ its degeneracy. Equation 3.89 is in essential agreement with a corresponding relation of the earlier literature.[179c] Expressing $(\partial\mu/\partial Q)^2$ by the integrated intensity $\Gamma_k$ of (fundamental) oscillator mode $k$ in the medium of molecules $j$ (Eq. 3.80), we obtain

$$K_s = \frac{4c}{3hg}\frac{\mu_j^2}{\sigma^3}\Gamma_{kj}V_M^{-1}\left[4\left(\tau_R\omega_k^2\right)^{-1} + \frac{1}{2\sqrt{2}}\left(\tau_T^{1/2}\omega_k^{3/2}\right)^{-1}\right], \qquad (3.90)$$

where $V_M$ is the molar volume.

As example we estimate the energy-dissipation-induced spontaneous band-width increment of the C–I stretching fundamental of neat, liquid $CH_3I$, using the known parameters

$$\mu_j = 1 \times 10^{-18} \text{ g}^{1/2} \text{ cm}^{5/2} \text{ sec}^{-1} \ ( = \text{esu cm})$$

$$\Gamma_{kj} = 730 \text{ cm}^2 \text{ mole}^{-1}$$

$$\sigma = 4.8 \times 10^{-8} \text{ cm}$$

$$c = 3 \times 10^{10} \text{ cm sec}^{-1} \tag{3.91}$$

$$\tau_R^{(1,0)} = 5 \times 10^{-12} \text{ sec}$$

$$\tau_T = 3.2 \times 10^{-11} \text{ sec}$$

$$\omega_k = 0.99 \times 10^{14} \text{ radian sec}^{-1} = 525 \text{ cm}^{-1}$$

$$g = 1; \quad T = 300°\text{K}; \quad V_M = 62.3 \text{ cm}^3 \text{ mole}^{-1}.$$

This gives $K = 0.9 \times 10^8$ radian sec$^{-1}$ or $\tau_{\text{ED}} = 1.1 \times 10^{-8}$ sec (see Eq. 3.83$b$). The corresponding spontaneous bandwidth increment amounts[4] to $(2\pi c \tau_{\text{ED}})^{-1}$ $= 0.5 \times 10^{-3}$ cm$^{-1}$—a value too small to be observable. (Incidentally, translational and orientational motion contribute nearly equally to the result.)

Next we consider short-range, repulsive (steric) perturbation forces. We take a purely exponential potential function (Eq. 3.62). We write the distance between the centers of interaction separating oscillator $j$ and medium site $i$, $R_{ij}$, as vector sum of the distance between the centers of molecules $i$ and $j$, $\mathbf{R}_{ij}^0$, and the atomic displacements of the oscillator, $\mathbf{s}_j$,[227]

$$\mathbf{R}_{ij} = \mathbf{R}_{ij}^0 + \mathbf{s}_j.$$

The length of $\mathbf{R}_{ij}$, $|\mathbf{R}_{ij}^2|^{1/2} = R_{ij}$, is

$$R_{ij} \approx R_{ij}^0 \left[ 1 + \frac{(\tfrac{1}{2})s_j^2}{R_{ij}^{0\,2}} + \frac{\mathbf{R}_{ij}^0 \cdot \mathbf{s}_j}{R_{ij}^{0\,2}} \right],$$

since $s \ll R$. Only term $\mathbf{R}_{ij} \cdot \mathbf{s}_j$ is relevant for vibrational energy relaxation to the lattice. [A term $(\mathbf{R}_{ij} \cdot \mathbf{s}_j)(\mathbf{R}_{ij} \cdot \mathbf{s}_i)$ is pertinent for resonance transfer; see Section 3.2.] The effective short-range perturbation potential is therefore (Eq. 3.62)

$$V \approx V_0 \exp\left( -\alpha R_{ij}^0 \right) \exp\left[ -\alpha \left( \frac{\mathbf{R}_{ij}^0}{R_{ij}^0} \right) \cdot \mathbf{s}_j \right].$$

Developing the second exponential into a Taylor expansion, truncating it after the first-order term, and setting $\mathbf{R}^0_{ij}/R^0_{ij} = \hat{\mathbf{R}}^0_{ij}$, we obtain the short-range perturbation Hamiltonian

$$A = -\alpha V_0 \hat{\mathbf{R}}^0_{ij} \cdot \mathbf{s}_j \exp\left(-\alpha R^0_{ij}\right). \tag{3.92}$$

We identify the magnitude of vector $\mathbf{s}$ with the sum of the atomic displacements that form the normal coordinate $Q$,

$$s_j = \sum_n \frac{\partial s_{j,n}}{\partial Q_j} Q_j, \tag{3.93}$$

where index $n$ runs over all symmetry coordinates that span $Q_j$. In many cases, $\partial s/\partial Q$ is known from tabulations of the $L$ matrix of normal coordinate analyses,[228]

$$S = LQ.$$

Insertion of Eqs. 3.92 and 3.93 into Eq. 3.83$a$ gives

$$W_{ss'}(\omega_j) = \alpha^2 V_0^2 \left(\sum_n \frac{\partial s_{j,n}}{\partial Q_j}\right)^2 \langle |(v|Q_j|v')|^2 \rangle_v \langle \exp\left(-2\alpha R^0_{ij}\right) \rangle_R$$

$$\times \left\langle \left(\hat{\mathbf{R}}^0_{ij} \cdot \hat{\mathbf{s}}_j\right)^2 \right\rangle_\Omega \int_{-\infty}^{\infty} dt \langle B(t)B(0) \rangle_{R,\Omega} \exp(i\omega_j t). \tag{3.94}$$

The subscripts at the angular brackets remind us over which coordinates we must do the ensemble average. We average the internal coordinate $Q$ over $v$ according to Eq. 3.84. To average over the external coordinates $(R, \Omega)$, we introduce the isotropic distribution functions (Section 2.2.1)

$$p_0(R) = 4\pi\rho \int_\sigma^\infty dR\, R^2,$$

$$p_0(\Omega) = (4\pi)^{-1} d\Omega.$$

Next, we define

$$\langle B(t)B(0) \rangle = \exp\left(\frac{-t}{\tau_a^*}\right) \tag{3.95}$$

to represent the time evolution of the autocorrelation of some activated process of average time $\tau_a^*$ between relaxation events,[227]

$$\tau_a^* = \tau_0 \exp\left(\frac{\Delta E}{k_B T}\right).$$

(With increasing temperature the events occur more frequently; refer to Fig. 3.3$b$ for an example.) Integrating over $R_{ij}^0$ is elementary. To integrate $\hat{\mathbf{R}}_{ij}^0 \cdot \hat{\mathbf{s}}_j$ over $\Omega$, we consider that $\hat{\mathbf{R}}_{ij}^0$ (direction of separation between molecules) and $\hat{\mathbf{s}}_j$ (direction of atomic displacement vector of mode $j$) are independent; $\hat{\mathbf{R}}_{ij}^0 \cdot \hat{\mathbf{s}}_j$ is the direction cosine of the angle $\Omega'(\theta, \varphi)$ between them. Hence

$$\left\langle \left( \hat{\mathbf{R}}_{ij}^0 \cdot \hat{\mathbf{s}}_j \right)^2 \right\rangle = (4\pi)^{-1} \int_0^{2\pi} d\varphi \int_0^\pi d\theta \cos^2\theta \sin\theta = \frac{1}{3}.$$

To solve the integral in Eq. 3.94, we recall Eq. J.1 (Appendix J) with the slow modulation condition $\omega_j^2 \tau^2 \gg 1$. Equation 3.94 then reads

$$W_{ss'}(\omega_j) = \alpha^2 V_0^2 \left( \sum_n \frac{\partial s_{j,n}}{\partial Q_j} \right)^2 \left( \frac{\hbar v'}{2\omega_j} \right) 4\pi\rho \int_\sigma^\infty dR\, R^2 \exp(-2\alpha R)$$

$$\times (4\pi)^{-1} \int d\Omega \cos^2\Omega \int dt \exp\left( \frac{-t}{\tau_a^*} \right) \exp(i\omega_j t),$$

or

$$K_s = \left( 4\pi\rho V_0^2 \right) \frac{v'}{3\hbar \mu_M \tau_a^* \omega_j^3} \left( \sum_n \frac{\partial s_{j,n}}{\partial Q_j} \right)^2$$

$$\times \exp(-2\alpha\sigma) \left\{ \frac{1}{4\alpha} + \frac{\sigma}{2} + \frac{\alpha\sigma^2}{2} \right\}. \tag{3.96}$$

$\mu_M$ is the reduced mass of the oscillator; the other symbols have their previous meanings. (Notice that the dimension of $s_j$ is cm, that of $Q_j$ is $g^{1/2}$ cm.)

Using the input parameter $v' = 1$, $V_0 = 500$ eV $= 8 \times 10^{-10}$ erg, $\omega = 1.9 \times 10^{14}$ radian sec$^{-1}$ (1000 cm$^{-1}$), $\rho = 3 \times 10^{22}$ cm$^{-3}$, $\alpha = 2.4$ Å$^{-1}$, $\tau_a^* = 5 \times 10^{-12}$ sec, $\sigma = 3$ Å, $\mu_M = 2 \times 10^{-24}$ g, and $(\Sigma \partial s/\partial Q)^2 = 1.0$ g$^{-1}$, we compute[229]

$$K_s = 8 \times 10^{10} \text{ radian sec}^{-1}$$

or a spontaneous bandwidth increment of

$$\frac{K_s}{2\pi c} = 0.4 \text{ cm}^{-1},$$

an amount near the observation limit of standard-precision measurements.

The difficulty here (as with all short-range potential functions) is the strong dependence of $K_s$ on potential parameter $\alpha$ and $\sigma$; the factor $\exp(-2\alpha\sigma)$ varies by one order of magnitude when its argument changes by 20%. We recall similar problems in calculating the dephasing rate of the fundamental stretch of liquid nitrogen (Section 2.2.3).

The calculations presented here indicate that direct vibrational energy dissipation to the lattice motions in the short-range potential range, particularly for lower-frequency modes and under elevated pressures, may be an important pathway of vibrational relaxation and should be observable in spontaneous band profiles under favorable conditions. For instance, if at constant density and at high external pressures a temperature rise causes an appreciable increase in the observed vibrational bandwidth, it is likely that the prevalent vibrational relaxation process is energy dissipation to the lattice, in particular since any contributions from concomitant vibrational dephasing may cause increased motional narrowing ($\tau_c$ decreases).

At any rate, we cannot ascertain that the vibrational relaxation behavior of even thoroughly studied modes, such as the carbon–iodine stretching fundamental ($525 \text{ cm}^{-1}$) of methyl iodide, is at present completely understood. It appears unlikely that pure energy dissipation by dipole forces is a rapid relaxation mechanism (Eq. 3.91), but energy dissipation may become significant if repulsive forces are invoked.[118a]

Concluding this section, we reconsider the formal resemblance between vibrational and nuclear magnetic resonance relaxation (Section 4.1, Chapter 1). We can now sharpen this comparison by the following arguments.

The inequality of Eq. 3.87, $(\omega_k - \omega_{bb'})\tau \gg 1$, can be considered a slow modulation regime; the average frequencies of the bath motions $\omega_{bb'}$ are considerably lower than vibrational frequencies $\omega_k$ of oscillators $k$. On the other hand, in ordinary liquids a $T_1$-nuclear magnetic resonance (NMR) relaxation process transfers spin energy to the lattice motions in the very rapid modulation regime $\omega\tau \ll 1$ ($\omega$ = Larmor frequency $\sim 3 \times 10^8$ radian $\sec^{-1}$).[180a] The $T_1$-process of vibrational energy relaxation by direct transfer to the bath motions should thus be compared to a $T_1$-spin energy relaxation process that has been stepped down into a slow modulation regime, for instance, in near-glassy or viscous states or in ordinary liquids at sufficiently low temperatures ($\tau \sim 10^{-8}$ sec). In such situations the rate of the $T_2$-NMR process (dephasing) exceeds that of the $T_1$-NMR process;[230] we recall that in ordinary liquids $T_1(\text{NMR}) = T_2(\text{NMR})$. In other words, the formal resemblance between the $T_1$, $T_2$-processes of nuclear magnetic resonance and of optical spectra is based on different "effective temperatures" of the respective systems.

## 6.3. Vibrational Energy Dissipation in Liquid Hydrocarbons

In this section we discuss energy dissipation of C–H stretching fundamentals of liquid hydrocarbons obtained from picosecond-pulsing measurements with a 90-degree geometry arrangement (Section 4.3, Chapter 1). We wish to show the complexity of possible dissipative relaxation pathways and to pinpoint the time domains of the vibrational energy dissipation processes.

Figure 3.17 shows the incoherent anti-Stokes intensity from the time evolution of the phonon population of the first excited level of the closely spaced set of carbon–hydrogen stretching fundamentals of $n$-heptane ($C_7H_{16}$), 1-heptene

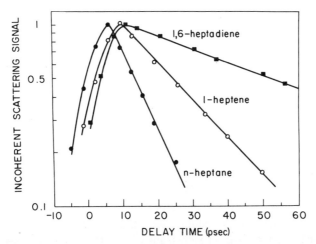

**Figure 3.17.** Stimulated vibrational dissipative relaxation of the carbon–hydrogen stretching region ( ~ 3000 cm$^{-1}$) of three liquid $C_7$ hydrocarbons at 20°C. [P. R. Monson, S. Patumtevapibal, K. J. Kaufmann, and G. W. Robinson, *Chem. Phys. Lett.* **28**, 312 (1974).]

($C_7H_{14}$, double bond between carbon atoms 1 and 2), and 1,6 heptadiene ($C_7H_{12}$, double bond between each $C_1$ and $C_2$, $C_6$ and $C_7$; no methyl groups).[231] The relaxation time $\tau_{ED}$ (delay time of the probe pulse for which the signal has dropped to $1/e$ of its peak value) is $16 - 5 = 11$ psec for heptane, $30 - 9 = 21$ for heptene, and about 60 psec for heptadiene.

We notice the inverse trend between $\tau_{ED}$ and number of $CH_3$ groups; $\tau_{ED}$ is the longest for heptadiene (no methyl groups). However, whether the rate of vibrational energy dissipation of the set of C–H stretches is indeed related to the presence of methyl groups (dissipation by way of internal $CH_3$-rotation)[232] is an interesting aspect but needs corroboration. The obvious experiment to do is to look at a series of multiple-branched hydrocarbons. Indeed, at 20°C, $\tau_{ED}$ of isodecane (three methyl groups) amounts to $11 \times 10^{-12}$ sec versus $16 \times 10^{-12}$ sec in *n*-decane (two methyl groups).[231]

Interestingly, $\tau_{ED}$ of C–H stretching fundamentals in saturated, straight-chain hydrocarbons increases with chainlength. For instance, from $C_7$ to $C_{15}$ $\tau_{ED}$ rises from $11 \times 10^{-12}$ to $25 \times 10^{-12}$ sec (at 20°C).[231]

It is difficult to explain this observation. Increasing the number of $CH_2$ groups does not alter the eigenfrequencies of the C–H stretching and deformation fundamentals. On the other hand, it affects the nature of the translatory-rotatory lattice modes since these depend on the flexibility of the molecular skeleton and on molecule-molecule interactions ("entanglements" of longer chains). Hence we may speculate that such interactions are reflected in some way or other by the measured relaxation times. Temperature studies of $\tau_{ED}$ are in agreement with such ideas. For instance, lowering the temperature from 20 to $-25$°C increases $\tau_{ED}$ in *n*-decane ($C_{10}H_{22}$) from 16 to $21 \times 10^{-12}$ sec,[231] with an apparent activation energy of $\Delta E = 870$ cal mole$^{-1}$ (Eq. 3.95). For *n*-heptane, $\Delta E$ is approximately 630 cal mole$^{-1}$. It would be presumptuous to attach much significance to the absolute value of this difference, but it lies in the right direction: The longer the carbon skeleton, the higher the energy to decouple chain-chain interaction ( = rigidity), the less effective the motion of carbon chain segments to cause relaxation. Apparently, a great deal of more effort and work is needed here. Precise relaxation time measurements as a function of temperature, pressure, solvents, length and branching of the carbon chain, atomic substitution (H by Cl, for instance), and degree of crystallinity for solid phases, can be envisaged.

The suggested pathway of vibrational energy relaxation in the hydrocarbons is initial transfer of the C–H stretch excitation to an overtone of the C–H bending fundamentals and its subsequent decay to the lattice.[49, 226, 231] Indeed, we demonstrate in the next section some time-resolved data that indicate a rapid, initial energy relaxation step from the C–H stretch to the overtone of the C–H bend, followed by considerably slower energy relaxation of the excited overtone to the lattice.

### 6.4.  Nonequivalence of Spontaneous and Stimulated Relaxation Times

We have discussed in the preceding section that population lifetimes obtained by stimulated techniques point to a variety of vibrational relaxation pathways. We have also demonstrated that evaluation of a spontaneous band profile with respect to a vibrational energy dissipation process is generally not successful because the relative contribution of the dissipation process is small.

Yet, we have seen that the rates of certain spontaneous dissipation steps can be rapid (Sections 2.2.1 and 6.1). In addition, we have predicted that in a repulsive potential domain vibrational population decay may significantly contribute to a spontaneous width (end of Section 6.2). Hence it is useful to inquire whether there is a fundamental difference between the $\tau_{ED}$ obtained from spontaneous and stimulated techniques. (We are not concerned here about the difference "of the factor of 2," discussed in Appendix J.) The problem is not trivial; the two measuring techniques are different, and the observed phenomena are complicated.

Indeed, in certain situations the two techniques yield different relaxation times. Fig. 3.18 tells the story[233] for the incoherent anti-Stokes signal of the carbon–hydrogen stretching mode (2928 cm$^{-1}$) of 0.04 molar ethyl alcohol, $CH_3CH_2OH$, in $CCl_4$ (22°C). Notice that the anti-Stokes signal decays initially with a relaxation time of $2 \times 10^{-12}$ sec but then slows to a rate corresponding to $40 \times 10^{-12}$ sec. The fast decay reflects the energy relaxation of the C–H stretch to the adjacent overtone of the C–H bending mode; the long-time, slower decay describes the proper relaxation of the C–H bending fundamental (see also the previous section). Clearly, the corresponding *spontaneous* width (from which we obtain $\tau$) would encompass some weighted average of the two different relaxation rates.

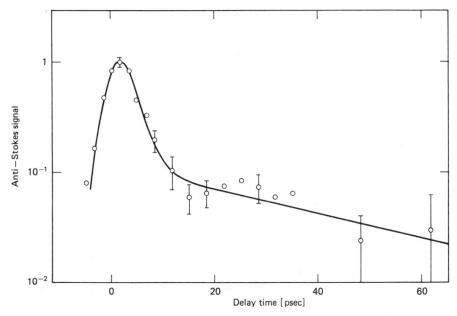

**Figure 3.18.**  Incoherent anti-Stokes signal from the 2928-cm$^{-1}$ infrared C–H stretching mode of ethanol diluted in CCl$_4$ (0.04 molar). The faster initial decay corresponds to a relaxation time of about 2 psec, the subsequent slower decay to about 40 psec. [K. Spanner, A. Laubereau, and W. Kaiser, *Chem. Phys. Lett.* **44**, 88 (1976).]

Therefore the induced relaxation techniques allow here a more flexible approach than spontaneous relaxation methods. [In fact, with a modified version of the picosecond laser technique, probing the *coherent* scattering signal leads to a direct observation of modulation time $\tau_c$ (Eq. 3.11).][234]

Incidentally, the CH$_3$CH$_2$OH/CCl$_4$ system represents another well-documented case where energy dissipation of a high-frequency mode occurs on a relatively short time scale (refer also to Sections 2.2.3 and 6.1). Yet, let us not forget that the depopulation step is actually an energy transfer (between adjacent modes) step; we discuss this further in following sections. At any rate, we should now be concerned about the ramifications this may have on the indiscriminate assignment of band broadening to vibrational *dephasing*.

### 6.5.  The Master Equation of Vibrational Energy Dissipation

**6.5.1.  Basic Principles.**  The quality of understanding we obtain from studying vibrational energy dissipation depends on the level of quantitative knowledge of the rates of the various pathways of the vibrational decay. As we have seen, the complexity of such processes involves direct energy relaxation to the lattice modes and competitive vibrational energy transfer between adjacent vibrational levels within the molecule or between such levels of different molecules. As a further complication, it appears that we cannot consider the nonexcited vibrational modes of the molecule as bath (lattice) motions without introducing an error.[235] Whereas a bath or lattice contains, by definition, many closely spaced rotational-translational levels (quasi-continuous energy distribution), the density of vibrational states in the near- and medium-infrared spectral ranges is considerably thinner.

Consequently our theoretical scheme should allow us to describe the relaxation of the chosen vibrational mode within a wide and general framework of coupling processes which include energy transfer to a few adjacent modes as well as to the quasi-continuous lattice motions. We have discussed in detail the rate expression for vibrational relaxation between two levels. Now we need a relation that weaves the most important vibrational relaxation steps into a global scheme consisting of *sets* of such two-level transitions (multilevel system).

The expression we are going to use, a so-called master or rate-balancing equation, describes in its simplest version the rate of change $(d/dt)\rho(n,t)$ of the probability $\rho(n,t)$ of finding the oscillator at time $t$ in vibrational level $n$,

$$\frac{d}{dt}\rho(n,t) = \sum_m B(n,m)\rho(m,t) - \rho(n,t)\sum_m B(m,n), \qquad (3.97)$$

where $B(n,m)$ is the time-independent transition probability per unit time for the transition $m$ to $n$.[236, 237]

Equation 3.97 looks simple—as if we could have guessed it. It balances the rate of change of the population density probability of level $n$ between the flux into $n$ from all other levels $m$ ($m \to n$) and the flux out of $n$ into all other levels $m$ ($n \to m$). (See also Eq. 2.52, Chapter 2.) The range of validity of Eq. 3.97 is, however, coupled to several restrictive conditions as follows.[238]

We first write the equation of motion of the density matrix $\rho(t)$ [see Eq. 1.9, Chapter 1, or Eq. 2.139, Chapter 2 (and note change of sign)]. We split the Liouvillian operator $L$ and the total Hamiltonian $H$ into the diagonal (subscript zero) and the small (real) off-diagonal parts (subscript 1)

$$L = L_0 + \lambda L_1, \qquad (3.98a)$$

$$H = H_0 + \lambda H_1, \qquad (3.98b)$$

that represent the unperturbed oscillator system and the perturbation effects, respectively. We keep only the diagonal matrix elements of $\rho(t)$. This means we consider averages of operators that commute with $H_0$ (Eq. 1.6, Chapter 1). Furthermore, we assume that initial phases are random (Eq. 1.10). Introducing projection operators $\bar{P}$ and $1 - \bar{P}$ which select the diagonal $\rho_d$ and off-diagonal $\rho_{od}$ matrix elements of $\rho(t)$,

$$\rho = \rho_d + \rho_{od}; \qquad \rho_d = \bar{P}\rho; \qquad \rho_{od} = (1 - \bar{P})\rho,$$

we get

$$\bar{P} i \frac{d}{dt} \rho = i \frac{d}{dt} \rho_d = \bar{P} L \rho_d + \bar{P} L \rho_{od}. \tag{3.99a}$$

Equation 3.99a is solved for $\rho_{od}(t) = (1 - \bar{P})\rho(t)$; we know the result (Eq. 2.141b, Chapter 2). Inserting it into Eq. 3.99a yields

$$i\left(\frac{d}{dt}\right)\rho_d(t) = \bar{P} L \rho_d(t) - i \int_0^t dt' \bar{P} L \{\exp[-i(1-\bar{P})Lt']\}(1-\bar{P})L\rho_d(t-t').$$

$$\tag{3.99b}$$

The first term in Eq. 3.99b vanishes since $\rho_d$ is diagonal,

$$\bar{P} L \rho_d = (L\rho_d)_{mm} = \hbar^{-1}[H, \rho_d]_{mm} = 0.$$

(Recall that $H_1$ is nondiagonal.) Consequently, the memory function term

$$(\bar{P} L \exp[-i(1-\bar{P})Lt](1-\bar{P})L\rho_d)_{mm}$$

$$= \bar{P} H^\times \exp\left[-i(1-\bar{P})H^\times \frac{t}{\hbar}\right](1-\bar{P})H^\times \rho_d \hbar^{-2}$$

(see Appendix F) simplifies to

$$\frac{\lambda}{\hbar^2}\left(H_1^\times \exp\left[-i(1-\bar{P})H^\times \frac{t}{\hbar}\right](1-\bar{P})H^\times \rho_d\right)_{mm}. \tag{3.100}$$

We replace $H^\times$ in the exponential by $H_0^\times$ (since $H^\times$ gives rise to terms of order $\lambda^3$ and higher) and simplify operator $(1-\bar{P})H^\times \rho_d$ to $\lambda H_1^\times \rho_d$ (Eq. 3.98),[238]

$$(1-\bar{P})H^\times \rho_d = (1-\bar{P})[H, \rho_d]$$

$$= [H, \rho_d] - [H_0, \rho_d]$$

$$= \lambda H_1^\times \rho_d.$$

Equation 3.99b then reads

$$\frac{d}{dt}\rho_{mm}(t) = -\lambda^2 \int_0^t dt' \left( L_1\{\exp(-iL_0 t')\}L_1\rho_d(t-t')\right)_{mm}$$

$$= -\left(\frac{\lambda}{\hbar}\right)^2 \int_0^t dt' \left( H_1^\times \left\{\exp\left(\frac{-iH_0^\times t'}{\hbar}\right)\right\}H_1^\times \rho_d(t-t')\right)_{mm}.$$

$$\tag{3.101}$$

Notice that

$$H_1^{\times} \rho_{\rm d} = (H_1)_{mn}(\rho_{nn} - \rho_{mm})$$

(from Eq. F.4, Appendix F). We write the triple commutator of Eq. 3.101 as (Eq. F.2, Appendix F)

$$\left[ H_1, \exp\!\left( \frac{-iH_0 t'}{\hbar} \right) H_1 \rho_{\rm d}(t - t') \exp\!\left( \frac{iH_0 t'}{\hbar} \right) \right]$$

$$- \left[ H_1, \exp\!\left( \frac{-iH_0 t'}{\hbar} \right) \rho_{\rm d}(t - t') H_1 \exp\!\left( \frac{iH_0 t'}{\hbar} \right) \right]. \qquad (3.102)$$

Operator $\exp(-iH_0 t/\hbar)$ acts on the row, $\exp(iH_0 t/\hbar)$ operates on the column vector of the representation (Eq. F.5). Hence the matrix element of the second commutator of Eq. 3.102 is

$$(m| \left\{ \exp\!\left( \frac{-iH_0 t}{\hbar} \right) \rho_{\rm d} H_1 \exp\!\left( \frac{iH_0 t}{\hbar} \right) \right\}^{\times} H_1 |m)$$

$$= \{ \exp(-i\omega_{mn} t)(m|\rho|m) - \exp(i\omega_{mn} t)(n|\rho|n) \} |(m|H_1|n)|^2$$

where

$$H_0|m) = E_m|m), \qquad \frac{E_m - E_n}{\hbar} = \omega_{mn}.$$

The matrix element of the first commutator is

$$\{ \exp(i\omega_{mn} t)(m|\rho|m) - \exp(-i\omega_{mn} t)(n|\rho|n) \} |(m|H_1|n)|^2.$$

We therefore obtain

$$\frac{d}{dt} \rho_{mm}(t) = -2\left( \frac{\lambda}{\hbar} \right)^2 \int_0^t dt' \, |(H_1)_{mn}|^2 \cos(\omega_{mn} t')$$

$$\times \{ \rho_{mm}(t - t') - \rho_{nn}(t - t') \}.$$

Summing over all $n$ leads to

$$\frac{d}{dt} \rho_{mm}(t) = \lambda^2 \sum_{n \neq m} \int_0^t dt' \, K_{mn}(t') \{ \rho_{nn}(t - t') - \rho_{mm}(t - t') \}$$

with

$$K_{mn}(t) = \frac{2}{\hbar^2} |(H_1)_{mn}|^2 \cos(\omega_{mn} t). \qquad (3.103)$$

The presence of the memory kernel makes numerical evaluation of Eq. 3.103 difficult. Since vibrational energy relaxation times are relatively long, we consider only situations where $(d/dt)\rho(t)$ varies slowly with respect to $K(t)$. (See also Sections 5.6–5.8, Chapter 2.) We therefore drop the $t'$ dependence of $\rho$ ($\lambda$ is small, the "weak coupling limit"),[238]

$$\rho(t - t') \approx \rho(t).$$

Consequently the integral in Eq. 3.103 amounts to

$$\int_0^t dt' K(t') = \frac{2}{\hbar^2} |(H_1)_{mn}|^2 \frac{\sin(\omega_{mn}t)}{\omega_{mn}}. \tag{3.104}$$

Taking the long-time limit, we find[27]

$$\lim_{t \to \infty} \frac{\sin(\omega_{mn}t)}{\omega_{mn}} = \pi\delta(\omega_{mn}),$$

and

$$K_{mn}(\infty) = \frac{2\pi}{\hbar} |(H_1)_{mn}|^2 \delta(E_m - E_n). \tag{3.105}$$

$K_{mn}(\infty)$ is the rate constant expressed as golden rule.[237] (See Section 6.2, Eq. 3.83a, and Section 3.3 of Chapter 1.) Insertion of Eq. 3.105 into Eq. 3.103 leads to (see also Eq. 3.97)

$$\frac{d}{dt}\rho_{mm}(t) = \sum_n K_{mn}(\infty)\{\rho_{nn}(t) - \rho_{mm}(t)\}, \tag{3.106}$$

the conventional Pauli master equation.[236, 237]

We recall the restrictive assumptions we had to make to obtain this simple result: $\rho(0)$ is diagonal, $\lambda$ is small, $t$ is long (order of $1/\lambda^2$). In addition, the degree of freedom of the system must be large (many closely spaced energy levels).[73, 238] In the following sections we discuss applications of Eq. 3.106.

**6.5.2.  Intramolecular Vibrational Relaxation.**  Intramolecular vibrational relaxation is a multistage process that can be modeled by Eq. 3.97 or 3.106. We consider a large, isolated molecule with many vibrational degrees of freedom of about equal energy spacing.[239, 240] Such an idealized example is furnished by a near-harmonic sequence of highly excited vibrational levels. We would like to know the time development of the change of the vibrational population of a chosen level of this set as it loses and gains excitation from the nearest level below and above. This situation is of interest because it has bearing on chemical kinetics under nonequilibrium conditions.

We assume that all nonexcited vibrational oscillators $j$ of eigenfrequency $\omega_j$ are in equilibrium (Boltzmann distribution); their fraction in state $v_j$ (relative to the zero-point vibration energy) is

$$f_{v_j} = \frac{\exp(-\hbar v_j \omega_j / k_B T)}{\sum_{v_j} \exp(-\hbar v_j \omega_j / k_B T)} = \exp\left(\frac{-\hbar v_j \omega_j}{k_B T}\right)\left[1 - \exp\left(\frac{-\hbar \omega_j}{k_B T}\right)\right]$$

$$\equiv Z_j^{-1} \exp\left(\frac{-\hbar v_j \omega_j}{k_B T}\right). \tag{3.107}$$

$\rho_{nn}(t)$ is therefore the probability of finding the system with $v$ vibrational quanta in mode $n$. We write Eq. 3.97[240] (dropping the no longer needed double index of $\rho(t)$) as

$$\frac{d}{dt}\rho_{v_n} = -K_n\left\{\rho_{v_n}\left[\sum_{v_j}^{\infty} f_{v_j} A(v_n, v_n+1; v_j, v_j-1)\right.\right.$$

$$\left. + \sum_{v_j}^{\infty} f_{v_j} A(v_n, v_n-1; v_j, v_j+1)\right]$$

$$- \rho_{v_n+1} \sum_{v_j}^{\infty} f_{v_j} A(v_n+1, v_n; v_j, v_j+1)$$

$$\left. - \rho_{v_n-1} \sum_{v_j}^{\infty} f_{v_j} A(v_n-1, v_n; v_j, v_j-1)\right\}, \tag{3.108}$$

where $A(v_n, v_n'; v_j, v_j')$ is the joint transition kernel for vibrational energy exchange; the excited oscillator makes a transition $v_n \to v_n'$ when a bath oscillator undergoes the transition $v_j \to v_j'$. Notice that Eq. 3.108 represents the balance of the rate of quanta reaching level $v_n$ from $v_n+1$ and $v_n-1$ (third and fourth terms, respectively) and quanta leaving level $v_n$ for $v_n+1$ and $v_n-1$ (first and second terms), with the bath oscillators undergoing the reverse transition (conservation of energy).

We now write the probability $A(v_n, v_n'; v_j, v_j')$ as product of the squared expectation values $|(v_n|Q_n|v_n')|^2$, $|(v_j|Q_j|v_j')|^2$ of normal coordinates $Q_n$ and $Q_j$.[239, 241] This gives

$$A(v_n, v_n+1; v_j, v_j-1) = (v_n+1)v_j \gamma_n \gamma_j$$

$$A(v_n, v_n-1; v_j, v_j+1) = v_n(v_j+1)\gamma_n \gamma_j$$

$$A(v_n+1, v_n; v_j, v_j+1) = (v_n+1)(v_j+1)\gamma_n \gamma_j$$

$$A(v_n-1, v_n; v_j, v_j-1) = v_n v_j \gamma_n \gamma_j, \tag{3.109}$$

where $\gamma_i = (\hbar/2\omega_i)$ (Eqs. 3.67 and 3.84). Insertion of Eqs. 3.109 and 3.107 into 3.108, performing the elementary summations over the equilibrium states $v_j$ (see the summation in Eq. K.2, Appendix K), and setting $\omega_j = \omega_n$, yields

$$\frac{d}{dt}\rho_{v_n} = K_n Z_n^{-1}\left\{(v_n+1)\rho_{v_n+1} - \left[v_n + (v_n+1)\exp\left(\frac{-\hbar\omega_n}{k_B T}\right)\right]\rho_{v_n}\right.$$

$$\left. + v_n\exp\left(\frac{-\hbar\omega_n}{k_B T}\right)\rho_{v_n-1}\right\}. \tag{3.110}$$

Equation 3.110 describes the rate of vibrational energy exchange between mode $n$ in vibrational state $v_n$ and the (many) bath oscillators in their Boltzmann distribution, with oscillator $n$ making transitions between states $v_n$ and $v_n \pm 1$ (three-level scheme). Note that for $t \to \infty$, that is to say, $\rho_{v_n} = \exp(-\hbar v_n\omega_n/k_B T)/Z_n$, $(d/dt)\rho_{v_n} = 0$ (equilibrium).

Rate constant $K_n$ in Eq. 3.110 depends on the choice of the interaction Hamiltonian $H_1$ (Eq. 3.105). Since harmonic oscillators of the same molecule are decoupled, we use as the lowest-order perturbation the three-phonon approximation[240]

$$H_1 = \sum_{nml} \frac{1}{3!}\left(\frac{\partial^3 U}{\partial Q_n \partial Q_m \partial Q_l}\right)\bigg|_{Q=0} Q_n Q_m Q_l. \tag{3.111}$$

It usually represents the largest anharmonicity term in the nuclear potential energy $U$ (see also Section 2.1). $K_n$ then assumes the form

$$K_n = \frac{2\pi}{3!\hbar}\sum_{ml}\gamma_n\gamma_m\gamma_l f_m f_l\left|\left(\frac{\partial^3 U}{\partial Q_n \partial Q_m \partial Q_l}\right)_{Q=0}\right|^2 \delta(E_n - E_m - E_l), \tag{3.112}$$

where we left out terms with two equal indices (although they are not necessarily small). Occupation value $f_l$ is given by Eq. 3.107.

Equation 3.110 can be integrated for several initial distributions of occupations of oscillators $v_n$. For example, if the system is at $t = 0$ in state $u_n$ (the active oscillator of mode $n$ is excited with $u_n$ quanta),[239,240]

$$\rho_{v_n}(t) = \left\{\frac{u_n!}{v_n!(u_n-v_n)!}\right\}\{1-\exp(-K_n t)\}^{u_n-v_n}\exp(-v_n K_n t),$$

where

$$\rho_{v_n}(0) = \delta_{u_n v_n}; \qquad T \to 0°K \ (\hbar\omega_n \gg k_B T). \tag{3.113}$$

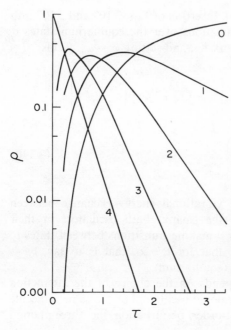

**Figure 3.19.** Vibrational population occupation of the first five levels of a mode initially in quantum level $v = 4$ as a function of the (reduced) time in a low-temperature regime. The numerals at the curves denote the vibrational quantum numbers. [G. R. Fleming, O. L. J. Gijzeman, and S. H. Lin, *J. Chem. Soc. Faraday Trans. II* **70**, 37 (1974).]

This example simulates relaxation of a high-frequency mode (a C–H stretch, $\hbar\omega/k_B T \sim 15$, or some overtone). $T = 0$ insures that there is no population from levels below. With the boundary conditions of Eq. 3.113, we interpret $\rho_{v_n}(t)$ as the conditional probability of finding oscillator $n$ at $t$ in level $v_n$ when it was in level $u_n$ at $t = 0$; all other oscillators have their Boltzmann occupation number. (The number of possible ways of distributing $v_n$ levels among $u_n$ levels is given by the factorial quotient.)

Figure 3.19 gives an example with $u_n = 4$, $v_n = 4, 3, 2, 1, 0$. Notice the population-depopulation cascade with increasing time $\tau \equiv K_n t$. Each of the four equilibrium levels $v_n = 3, 2, 1, 0$ is excited from above and decays, in its own turn, to the level below until the initial Boltzmann distribution is regained. The exponential time dependence of level $v_n = 4$ is, of course, a consequence of the assumed Markovian nature of the process as well as of the condition $T = 0$.

As a further example consider intramolecular coupling of a C–H stretching fundamental $\nu_s$ with the overtone, $2\nu_b$, of a C–H bending mode—the fast step in the relaxation chain of $\nu_s$ (Section 6.3). Figure 3.20 shows this as three-phonon coupling between a quantum of $\nu_s$ and two quanta of $\nu_b$.[242] The rate expression is obtained by setting $m = l$ in Eq. 3.112,[240]

$$K_s = \sum_b \gamma_s \gamma_b^2 f_b^2 \left| \left( \frac{\partial^3 U}{\partial Q_s \partial Q_b^2} \right)_{Q=0} \right|^2 \delta(E_s - 2E_b). \qquad (3.114)$$

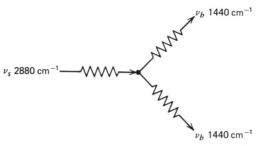

**Figure 3.20.** Energy balance of an intramolecular relaxation process, shown schematically in terms of a three-phonon anharmonicity-induced coupling between a carbon–hydrogen stretching fundamental $\nu_s$ and the overtone $2\nu_b$ of a carbon–hydrogen deformation fundamental $\nu_b$.

**6.5.3. Lattice-Modulated Multistage Vibrational Energy Relaxation and Transfer.** We return to systems where the excited oscillator is in contact with a heat bath of translatory-rotatory molecular motions. We can see that Eq. 3.110 and similar formulations can be readily adapted to participation of bath motions in the vibrational relaxation process (see Note 9).

Inclusion of translatory-rotatory motions requires reinterpretation of transition kernel $A$. Instead of specifying $A$ in terms of transitions between the excited oscillator and the set of (internal) Boltzmann bath oscillators (Eq. 3.109), we now define coupling of the active oscillator to the lattice (or bath) in terms of collision effects and their probabilities. We include direct transfer of vibrational quanta to and from the bath as well as bath-modulated vibrational energy exchange between oscillators. Eventually, we allow these processes to occur simultaneously.

The collision events are specified by a relaxation rate constant $K$ which depends on the appropriate intermolecular coupling forces (Eq. 3.105). Transition kernel $A$ (Eq. 3.109) is modified as follows:[236, 239] We replace the squared matrix element of the transition expectation value of a bath oscillator $j$ by "upward" ($P_b^+$) and "downward" ($P_b^-$) translational-rotational transition probabilities per collision of the lattice (bath) molecules. Their specified nature is of no interest to us since it leads to a more detailed description of lattice motions than we are able to handle at present. (At best, we can write a correlation function of the dynamic bath variable.) $P_b^+$, $P_b^-$ is, of course, the squared matrix element of the expectation value of the interaction Hamiltonian over bath coordinates,

$$P_b^+ = P_b^- = \langle |(b|H_1|b')|^2 \rangle. \tag{3.115}$$

(See, for instance, Eq. 3.83c.) $P_b^\pm$ obeys the principle of microscopic reversibility: The transition probability per collision from $b$ to $b'$ equals that from $b'$ to $b$

(Section 3.3, Chapter 1). Hence

$$A(v_n, v_n + 1; b) = (v_n + 1)\gamma_n P_b^-$$

$$A(v_n, v_n - 1; b) = v_n \gamma_n P_b^+$$

$$A(v_n + 1, v_n; b) = (v_n + 1)\gamma_n P_b^+$$

$$A(v_n - 1, v_n; b) = v_n \gamma_n P_b^-. \tag{3.116}$$

Inserting Eqs. 3.116 into Eq. 3.108 and replacing the occupation number $f_{v_j}$ of the internal bath oscillators by the concentration of bath molecules $N_b$ gives

$$\frac{d}{dt}\rho_{v_n} = -K_n'\left\{(v_n + 1)\left[\rho_{v_n}\sum_b N_b P_b^- - \rho_{v_n+1}\sum_b N_b P_b^+\right]\right.$$

$$\left. + v_n\left[\rho_{v_n}\sum_b N_b P_b^+ - \rho_{v_n-1}\sum_b N_b P_b^-\right]\right\}, \tag{3.117}$$

where the time dependence of $\rho(t)$ is understood. At equilibrium, $(d/dt)\rho(t \to \infty) = 0$. Therefore we require (see also Eq. 3.110)

$$\frac{\rho_{v_n+1}(\infty)}{\rho_{v_n}(\infty)} = \frac{\rho_{v_n}(\infty)}{\rho_{v_n-1}(\infty)} = \exp\left(\frac{-\hbar\omega_n}{k_B T}\right). \tag{3.118a}$$

Hence, for $t = \infty$, Eq. 3.117 reads

$$(v_n + 1)\rho_{v_n}(\infty)\left[\sum_b N_b P_b^- - \exp\left(\frac{-\hbar\omega_n}{k_B T}\right)\sum_b N_b P_b^+\right]$$

$$- v_n\rho_{v_n-1}(\infty)\left[\sum_b N_b P_b^- - \exp\left(\frac{-\hbar\omega_n}{k_B T}\right)\sum_b N_b P_b^+\right] = 0. \tag{3.118b}$$

Notice that the first bracketed term represents all vibrational transitions $v_n \rightleftharpoons v_n + 1$, the second all $v_n \rightleftharpoons v_n - 1$. At equilibrium each term must vanish; the transition probabilities per unit time obey the detailed balance condition (see also Eq. 2.54, Chapter 2)

$$\sum_b N_b P_b^- = \exp\left(\frac{-\hbar\omega_n}{k_B T}\right)\sum_b N_b P_b^+. \tag{3.118c}$$

Substitution of Eq. 3.118c into Eq. 3.117 gives

$$\frac{d}{dt}\rho_{v_n}(t) = K'_n \sum_b N_b P_b^+ \left\{ (v_n + 1)\rho_{v_n+1}(t) \right.$$

$$- \left[ v_n + (v_n + 1)\exp\left( \frac{-\hbar\omega_n}{k_BT} \right) \right]\rho_{v_n}(t)$$

$$\left. + v_n \exp\left( \frac{-\hbar\omega_n}{k_BT} \right)\rho_{v_n-1}(t) \right\}, \tag{3.119}$$

with

$$K'_n \sum_b N_b P_b^+ \equiv K_n = \frac{2\pi}{\hbar}\gamma_n \sum_{bb'} \rho_b |(b|B_{nb}|b')|^2 \delta\{E_{b'} - E_b - \hbar\omega_n\},$$

$$H_1 = B_{nb}Q_n. \tag{3.120}$$

Equation 3.119 is readily extended for a set of oscillators that interacts linearly with the bath. Assuming that each oscillator relaxes independently, the rate of change of $\rho(t)$ of a set of $n$ oscillators is simply the sum over all $n$,

$$\frac{d}{dt}\rho_{(v)}(t) = \sum_n \frac{d}{dt}\rho_{v_n}, \qquad H_1 = \sum_n B_{nb}Q_n. \tag{3.121}$$

In certain cases it is useful to treat the heat bath as ensemble of lattice oscillators (Note 9), for instance, when we wish to consider the interaction of the vibrational oscillator $Q_n$ with long-wavelength phonon motions of the condensed medium $Q_l$ about which we have some special knowledge or ideas (see, for instance, Section 6.1, Chapter 2). We express this by rewriting $H_1$ (Eq. 3.121) as

$$H_1 = \sum_n B_{nb}Q_n = \sum_{nl} F_{nl}Q_nQ_l, \tag{3.122}$$

where $F_{nl}$ is the force[237] acting between each vibrational oscillator $n$ and lattice oscillator $l$. Inserting Eq. 3.122 into Eq. 3.115 and performing the trace operation over the lattice oscillators yields (Eqs. K.6–K.8, Appendix K)

$$\left\langle |(b|F_{nl}Q_l|b')|^2 \right\rangle = \gamma_l |F_{nl}|^2 \left\{ 1 - \exp\left( -\frac{\hbar\omega_l}{k_BT} \right) \right\} \sum_l \exp\left( -\frac{l\hbar\omega_l}{k_BT} \right)(l|l^\dagger)$$

$$= \gamma_l |F_{nl}|^2 (\overline{N}_l + 1). \tag{3.123}$$

Notice that this formulation, if inserted into the expression for $K'_n$ of Eq. 3.120,

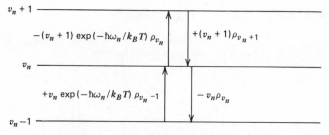

**Figure 3.21.** Schematic of the population flux per unit time in a transition scheme between adjacent vibrational levels in contact with a heat bath. A Boltzmann factor appears for all upward transitions.

shows a formal equivalence of the rate relations for relaxation of independent vibrational oscillators into a lattice of Boltzmann oscillators (intramolecular relaxation) and into a lattice of narrowly spaced equilibrium phonon modes (intermolecular relaxation).

We extend our discussion to joint probabilities of coupled oscillators. We can see that this introduces interesting and realistic effects. Figure 3.21 shows as an example the vibrational levels and the indexing of $\rho(t)$ in a three-level scheme of two coupled oscillators. Notice the appearance of the detailed balance factor $\exp(-\hbar\omega_n/k_BT)$ whenever a transition proceeds upward. It is not difficult to write down the master equation for relaxation schemes in which vibrational transitions involve adjacent energy levels only. (Index $m$ in Eq. 3.97 is bounded; the transition kernel $A$ is said to be "local.") Let us then give the rate-balancing equation for two coupled oscillators of eigenfrequencies $\omega_n > \omega_m$ that transfer energy between each other, giving or receiving the balance to or from the lattice (bath-assisted energy transfer). Defining perturbation operator $H_1$ for the two vibrational oscillators $n, m$ by (Eq. 3.122)

$$H_1 = \sum_{nml} F_{nml} Q_n Q_m Q_l \tag{3.124}$$

($Q_l$ represents the equilibrium phonons of the bath), we write

$$\frac{d}{dt}\rho_{(v)} = \sum_{n<m} K_{nm}\left\{(v_n+1)v_m\left(\rho_{v_n+1,\,v_m-1} - \rho_{v_n,\,v_m}\exp\left[-\frac{\hbar(\omega_n-\omega_m)}{k_BT}\right]\right)\right.$$

$$\left. + v_n(v_m+1)\left(\rho_{v_n-1,\,v_m+1}\exp\left[-\frac{\hbar(\omega_n-\omega_m)}{k_BT}\right] - \rho_{v_n,\,v_m}\right)\right\},$$

with

$$K_{nm} = \frac{2\pi}{\hbar^2}\gamma_n\gamma_m\sum_l \gamma_l |F_{nml}|^2(\overline{N}_l+1)\,\delta(\omega_n-\omega_m-\omega_l). \tag{3.125}$$

We define $\rho_{v_n+1,\,v_m-1}$ as the joint probability of finding the system in state $v_1, v_2, \ldots v_n+1, v_m-1, \ldots, v_k$ and $\rho_{(v)}$ as the joint probability of finding the system in state $v_1, v_2, \ldots, v_n, v_m, \ldots, v_k$, with the lattice oscillators $Q_l$ in thermal equilibrium. Notice that the number of quanta is constant at $v_n + v_m$ (energy exchange).

Clearly the sum of Eq. 3.125 and Eq. 3.119 (noting Eq. 3.121) describes a relaxation model of simultaneous redistribution of vibrational energy between two oscillators $n, m$ (with excess energy $|E_n - E_m|$ given or taken from the equilibrium lattice) and independent, direct energy relaxation of $E_n$, $E_m$ to the lattice. Setting[237] (with $\omega_n > \omega_m$)

$$W_{v_n, v_n + 1} = (v_n + 1)K_n$$

$$W_{v_n, v_n - 1} = v_n \exp\left(-\frac{\hbar \omega_n}{k_B T}\right)K_n$$

$$W_{v_n, v_m; v_n + 1, v_m - 1} = (v_n + 1)v_m K_{nm}$$

$$W_{v_n, v_m; v_n - 1, v_m + 1} = v_n(v_m + 1)\exp\left[-\frac{\hbar(\omega_n - \omega_m)}{k_B T}\right]K_{nm} \qquad (3.126)$$

we get, abbreviating $\hbar \omega_n / k_B T \equiv \beta_n$,

$$\frac{d}{dt}\rho_{v_n, v_m} = W_{v_n, v_m; v_n + 1, v_m - 1}\left[\rho_{v_n + 1, v_m - 1} - \rho_{v_n, v_m}\exp(\beta_m - \beta_n)\right]$$

$$+ W_{v_n, v_m; v_n - 1, v_m + 1}\left[\rho_{v_n - 1, v_m + 1} - \rho_{v_n, v_m}\exp(\beta_n - \beta_m)\right]$$

$$+ W_{v_n, v_n + 1}\left[\rho_{v_n + 1, v_m} - \rho_{v_n, v_m}\exp(-\beta_n)\right]$$

$$+ W_{v_n, v_n - 1}\left[\rho_{v_n - 1, v_m} - \rho_{v_n, v_m}\exp(\beta_n)\right] + \{m\}. \qquad (3.127)$$

Symbol $\{m\}$ represents the repeat of the last four terms but $n$ ($m$) replaced by $m$ ($n$). [Summation over all indices (Eq. 3.125) and the time dependence of $\rho$ is understood.] For $H_1$ we employ

$$H_1 = \sum_l F_{nl}Q_n Q_l + \sum_l F_{ml}Q_m Q_l + \sum_l F_{nml}Q_n Q_m Q_l. \qquad (3.128)$$

Although Eq. 3.127 looks complicated, we can give a simplified version[237] by considering low effective temperatures, $\hbar \omega \gg k_B T$. As we know, this cuts off upward transitions. Equation 3.127 reduces then to

$$\frac{d}{dt}\rho_{v_n, v_m}(t) = \sum_n K_n\left[(v_n + 1)\rho_{v_n + 1, v_m}(t) - v_n\rho_{v_n, v_m}(t)\right] + \{m\}$$

$$+ \sum_{n < m} K_{nm}\left[(v_n + 1)v_m\rho_{v_n + 1, v_m - 1}(t) - v_n(v_m + 1)\rho_{v_n, v_m}(t)\right],$$

$$(3.129)$$

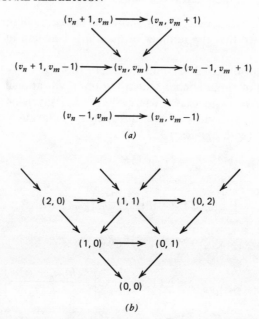

**Figure 3.22.** (*a*) Section of a transition scheme for simultaneous vibrational relaxation to the lattice (down arrows) and vibrational energy exchange (horizontal arrows) of (and between) two independent oscillators *n* and *m* at $\hbar\omega \gg k_B T$ (no upward transitions). (*b*) Transition scheme with three levels initially populated.

where $\{m\}$ designates a repeat of the first two sums with index $n$ ($m$) replaced by $m$ ($n$). Equation 3.129 describes the level scheme shown by Fig. 3.22*a*. Notice that whenever two indices change (their sum stays invariant), the process is energy transfer between the oscillators (with the balance going to the lattice). If only one index changes, the vibrational energy is given directly to the lattice. (Recall that we do not allow upward transitions here.)

A more detailed example, which we will then evaluate numerically, is presented by Fig. 3.22*b*. We consider here two oscillators, indexed $n$ and $m$, which are placed at $t = 0$ in vibrational levels $n = 1, m = 0$; $n = 0, m = 1$; and $n = 0, m = 0$. We separate Eq. 3.129 into its three components, getting ($\omega_n > \omega_m$)

$$\frac{d}{dt}\rho_{1_n,0_m} = - K_n \rho_{1_n,0_m} - K_{nm}\rho_{1_n,0_m},$$

$$\frac{d}{dt}\rho_{0_n,1_m} = - K_m \rho_{0_n,1_m} + K_{nm}\rho_{1_n,0_m},$$

$$\frac{d}{dt}\rho_{0_n,0_m} = K_n \rho_{1_n,0_m} + K_m \rho_{0_n,1_m}. \qquad (3.130)$$

Simplifying notation by attributing the first quantum number to oscillator $n$, the second to $m$, integration of Eqs. 3.130 yields[237]

$$\rho_{10}(t) = \rho_{10}(0)\exp\left[-(K_n + K_{nm})t\right]$$

with $\rho(t=0) \equiv \rho(0)$,

$$\rho_{01}(t) = -\left\{\frac{K_{nm}}{K_n + K_{nm} - K_m}\right\}\rho_{10}(0)\exp\left[-(K_n + K_{nm})t\right]$$

$$+ \left[\rho_{01}(0) + \left\{\frac{K_{nm}}{K_n + K_{nm} - K_m}\right\}\rho_{10}(0)\right]\exp(-K_m t),$$

$$\rho_{00}(t) = 1 - \rho_{10}(t) - \rho_{01}(t). \tag{3.131}$$

With this relation we now calculate the vibrational energy exchange rate between the fundamental of liquid nitrogen (2331 cm$^{-1}$) and two fundamental modes (at 1526 and 1306 cm$^{-1}$) of methane, CH$_4$, added as dopant at low concentration. It will come as no surprise that these steps serve as detours around the exceedingly slow oscillator-to-bath dissipation rate (energy gap $\sim 50\ k_B T$); its time constant, $\tau_{ED}$ approximately 1 sec in 99.9995% liquid N$_2$, is considered a *lower* limit.[243a, 244] Figure 3.23 shows the relevant data, the incoherent anti-Stokes intensity from the excess-populated vibrational $v=1$ level of liquid N$_2$ at 77°K as function of the delay time of the interrogating laser pulse.[243b] The square points represent the experimental data for liquid N$_2$ containing 8 parts per million (ppm), the dots for 83 ppm CH$_4$. The solid curves are computed from $\rho_{10}(t)$ of Eq. 3.131.

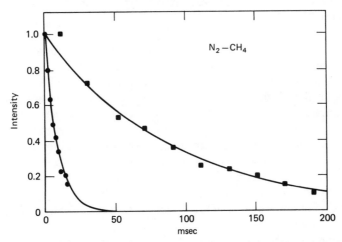

**Figure 3.23.** Vibrational relaxation of the first level of the nitrogen fundamental in liquid N$_2$ at 77°K containing 8 ppm (square points) and 83 ppm (dots) of CH$_4$. The curves are computed on the basis of the level scheme of Fig. 3.22$b$. [W. F. Calaway and G. E. Ewing, *J. Chem. Phys.* **63**, 2842 (1975).]

Considering the uncertainty in the relaxation rate of pure $N_2$, we obtain no more than a rough indication of the transfer rate $K_{nm}$. From the observed rate constants (the inverse time for which the signal intensity has dropped to $1/e$), we get from Fig. 3.23

$$k_{obs}(8 \text{ ppm}) = 12 \text{ sec}^{-1},$$

$$k_{obs}(83 \text{ ppm}) = 115 \text{ sec}^{-1}.$$

Using (see above)

$$k_{obs}(0 \text{ ppm}) = K_n = 1 \text{ sec}^{-1},$$

we obtain

$$115 = 1 + 83 \times 10^{-6} K_{nm},$$

$$12 = 1 + 8 \times 10^{-6} K_{nm}.$$

This yields

$$K_{nm} \sim 1.4 \times 10^6 \text{ sec}^{-1} \text{ mf}^{-1}.$$

# 7.  COUPLING OF LOW-FREQUENCY MODES WITH TRANSLATIONAL LATTICE MOTIONS: ULTRASONIC RELAXATION

At the end of the pathway of the decay of a vibrational excitation there *has* to be a step where a vibrational quantum is transferred to the random motions of the heat bath. This step represents a bottleneck since it proceeds at the slowest rate owing to the energy gap between a (internal) vibrational and a (external) lattice phonon or translatory-rotatory fluctuation (see Note 9).

Relevant theoretical expressions for this rate (Eqs. 3.89 and 3.90) predict that the *translational* lattice or bath motions accept a considerable fraction of the internal energy. Measurements of this effect by spontaneous or induced incoherent Raman scattering are difficult to perform, as we have extensively discussed: Spontaneous spectra are perturbed by vibrational dephasing, stimulated spectra are not always useful because of the weaker Raman polarizability of lower-frequency modes.

The technique uniquely adapted to examine gross effects of transfer of internal vibrational energy to translational degrees of freedom is ultrasonics, the measurement of the velocity of ultrasound as a function of frequency.[183a, 245] Evaluation of the observed dispersion rests on the principle of energy exchange

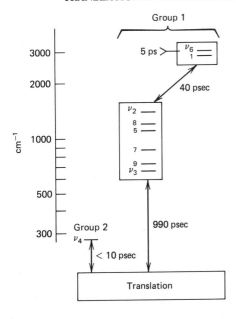

**Figure 3.24.** Vibrational energy transfer and relaxation in liquid methylene chloride. [K. Takagi, P.-K. Choi, and K. Negishi, *J. Chem. Phys.* **74**, 1424 (1981).]

between the internal degrees of freedom and the impressed translational motion. It yields a correlation time of *cooperative* relaxation of sets of several adjacent vibrational modes into the translational coordinates of the medium.

Figure 3.24 shows this process within a scheme of vibrational energy relaxation of liquid methylene chloride, $CH_2Cl_2$ ($C_{2v}$).[246] Let us assume that the highest-frequency fundamental, the $\nu_6$ C–H stretch near 3070 cm$^{-1}$, is initially populated.[247] Its excitation rapidly ($\sim 5 \times 10^{-12}$ sec) exchanges energy with the nearby $\nu_1$ mode at about 2990 cm$^{-1}$. This set of energy levels thereupon relaxes at a considerably slower rate (time constant $40 \times 10^{-12}$ sec), by coupling to an overtone of a C–H deformation mode (Fig. 3.20) or to nearby combination band(s), into a collective of six fundamental levels. These levels comprise frequencies between the $\nu_3$ fundamental at about 700 cm$^{-1}$ and the $\nu_2$ fundamental at 1425 cm$^{-1}$. In turn, this set relaxes cooperatively into the translational lattice motions with the long energy dissipation time $\tau_{ED} = 990 \times 10^{-12}$ sec. This value is one magnitude shorter than that estimated for the $\nu_3$ mode of $CH_3I$ (Eq. 3.91).

The lowest-frequency fundamental of $CH_2Cl_2$, the $\nu_4$ mode at 283 cm$^{-1}$, is a special case. Independently of the other oscillators (Eq. 3.127) of the molecule, it decays into the lattice at the relatively short time constant of $\tau_{ED} < 10 \times 10^{-12}$ sec.[248]

Ultrasonic relaxation measurements do not give a sufficiently detailed picture of vibrational energy dissipation to be of more than passing interest to us. Nevertheless, the data discussed in this section corroborate that energy transfer from a vibrational mode to a fluctuation of lattice motions is indeed slow.

## 8.   EXTENDED GLOBAL MODEL OF VIBRATIONAL RELAXATION IN LIQUIDS

### 8.1.   General Aspects: The Noise-Averaged System Operator

In the concluding sections of this chapter we outline a global model of vibrational relaxation. The theory includes vibrational dephasing, resonance energy transfer, and non-Markovian behavior.

Although the formalism is difficult and experimental applications, as yet, sparse, a global model of vibrational relaxation is needed if we want to go beyond the fundamental principles outlined in the previous sections. Significant interactions between the various vibrational relaxation mechanisms are nearly always present; the experimental techniques to pick out the process of interest have their own built-in limitations and disadvantages.

The theory[235] outlined in the following discussion proceeds by solving the equation of motion of convenient dynamic oscillator variables $A$ in the superoperator formalism (Eq. F.4, Appendix F),

$$\frac{d}{dt}A_{a_N a_N}(t) = \sum U_{\alpha\beta a'_N a_N} A_{a'_N a_N} X_{\alpha\beta}$$

$$+ V_{\alpha\beta a'_N a_N a_M b_M} A_{a'_N a_N} A_{a_M b_M} X_{\alpha\beta} + \cdots,$$

where $A_{a_N a_N}(t)$ is a density matrix of the dynamic oscillator coordinate over excitation states $a_N$, $X$ is a bath operator, and $U, V, \ldots$, are coupling coefficients between oscillator and bath coordinates.

Solution of the equation of motion is accomplished by writing and formally solving the equations of motion of the operator products $A_{a'_N a_N} X_{\alpha\beta}, \ldots$, and inserting the results into the foregoing expression. This leads to a complicated integro-differential equation which is evaluated by making physically realistic, mathematically simplifying assumptions about the interaction coefficient $U_{\alpha\beta a_N b_M}, \ldots$. This allows ready averaging over the bath coordinates. Following this, Boson operators are introduced to reexpress the (internal) oscillator coordinates in terms of pairwise interoscillator interactions (see Fig. 3.21) in the harmonic approximation. We neglect anharmonicity-induced multiphonon effects.[249]

The result is a general equation of motion of "noise-averaged" (bath-averaged) system operators (density operators) from which all other interesting quantities (correlation function, phase relaxation, rate of change of occupation, internal energy) can be derived.

This unified approach proves to be of considerable scope and complexity. It includes, as special cases, previously discussed formalisms of individual and simplified relaxation processes (for instance, Eqs. 3.110 and 3.127). It gives a description of the dynamics of the bath conforming to the knowledge we have about it. It avoids perturbation (cumulant) expansions with their possible lack of rapid convergence (Section 4). Furthermore, the model shows the physical basis for necessary or desired simplifications in the description of vibrational relaxation formalisms that are to the point and applicable. Therefore it forms a natural basis for investigations of cooperative as well as interference effects between various vibrational relaxation processes.

It will be convenient to introduce a detailed subscript nomenclature. We assign Greek subscripts to bath coordinates and their energy levels, lower-case Roman subscripts to system (oscillator) coordinates and their energy levels ("excitation indices"). We identify system coordinates of a particular oscillator by an upper-case Roman subscript ("oscillator index").

We write the total Hamiltonian[235a]

$$H = \sum_N H_N + H_B + H_1, \tag{3.132}$$

where $H_N$ represents the dynamic ( = internal, quasi-harmonic oscillator) subsystem of oscillator index $N$ and of excitation (vibrational quantum number) $a_N$, $H_B$ the Hamiltonian of the dissipative system ( = bath or lattice) of excitation $\alpha$, and $H_1$ the coupling Hamiltonian for direct as well as bath-modulated oscillator-oscillator and oscillator-bath interactions. We define sets of orthonormal eigenvectors (Dirac notation) $|a_N\rangle, \dots, |\alpha\rangle, \dots$ corresponding to the excitation numbers $a_N, \dots, \alpha, \dots$ by

$$H_N|a_N\rangle = \hbar\omega_{a_N}|a_N\rangle,$$

$$H_B|\alpha\rangle = \hbar\omega_\alpha|\alpha\rangle, \tag{3.133}$$

and introduce the operators in Liouville space (see Appendix F)

$$A_{a_N b_N} = |a_N\rangle\langle b_N|,$$

$$X_{\alpha\beta} = |\alpha\rangle\langle\beta|. \tag{3.134}$$

$A_{a_N b_N}$ and $X_{\alpha\beta}$ are matrices. Their elements are in terms of the components of state vectors $|a_N\rangle$ and $|\alpha\rangle$ of the oscillator system and the dissipative system, respectively. Their commutator properties (Eq. F.3) are

$$\left[A_{a_N b_N}, A_{a'_M b'_M}\right] = \delta_{NM}\left(A_{a_N b'_N}\delta_{b_N a'_N} - A_{a'_N b_N}\delta_{b'_N a_N}\right),$$

$$\left[X_{\alpha\beta}, X_{\alpha'\beta'}\right] = X_{\alpha\beta'}\delta_{\beta\alpha'} - X_{\alpha'\beta}\delta_{\beta'\alpha},$$

$$\left[X_{\alpha\beta}, A_{a_N b_N}\right] = 0. \tag{3.135}$$

The operators $H_N$ and $H_B$ are then of the form (Eqs. F.6, F.13)

$$H_N = \hbar \sum_{a_N} \omega_{a_N} A_{a_N a_N},$$

$$H_B = \hbar \sum_{\alpha} \omega_\alpha X_{\alpha\alpha}. \tag{3.136}$$

We can see that these definitions not only permit great flexibility in the formalism but also conveniently lead to density matrices (Eq. F.13).

Since we restrict the formalism to binary interactions between dynamic (internal) variables, the perturbation operator $H_1$ is defined as ($\hbar = 1$)

$$H_1 = U_{\alpha\beta a_N b_N} X_{\alpha\beta} A_{a_N b_N} + \left(\tfrac{1}{2}\right) V_{\alpha\beta a_N b_N a_M b_M} X_{\alpha\beta} A_{a_N b_N} A_{a_M b_M}$$

$$+ \left(\tfrac{1}{2}\right) W_{a_N b_N a_M b_M} A_{a_N b_N} A_{a_M b_M}, \tag{3.137}$$

where the time dependence of $A$, $X$ as well as summation over repeated indices is understood, with $M \neq N$. Notice that the first, second, and third term describes the oscillator-bath, bath-assisted oscillator-oscillator and direct (no participation of the bath) oscillator-oscillator coupling, respectively.

Using Eq. 3.135, the equation of motion for the dynamic subsystem $N$ is

$$\frac{d}{dt} A_{a_N b_N} \equiv \dot{A}_{a_N b_N} = i \left[ H_{N'} + H_B + H_1, A_{a_N b_N} \right].$$

Its nonvanishing commutator terms are

$$i\left\{ \left[ \omega_{a'_N} A_{a'_N a'_N}, A_{a_N b_N} \right] + \left[ U_{\alpha\beta a'_N b'_N} X_{\alpha\beta} A_{a'_N b'_N}, A_{a_N b_N} \right] \right.$$

$$+ \left(\tfrac{1}{2}\right) \left[ V_{\alpha\beta a'_N b'_N a'_M b'_M} X_{\alpha\beta} A_{a'_N b'_N} A_{a'_M b'_M}, A_{a_N b_N} \right]$$

$$+ \left(\tfrac{1}{2}\right) \left[ W_{a'_N b'_N a'_M b'_M} A_{a'_N b'_N} A_{a'_M b'_M}, A_{a_N b_N} \right] \Bigg\}$$

$$= i\omega_{a_N b_N} A_{a_N b_N} + i\left\{ \left( U_{\alpha\beta a'_N a_N} A_{a'_N b_N} - U_{\alpha\beta b_N a'_N} A_{a_N a'_N} \right) X_{\alpha\beta} \right.$$

$$+ \left( V_{\alpha\beta a'_N a_N a_M b_M} A_{a'_N b_N} - V_{\alpha\beta b_N a'_N a_M b_M} A_{a_N a'_N} \right) A_{a_M b_M} X_{\alpha\beta}$$

$$+ \left( W_{a'_N a_N a_M b_M} A_{a'_N b_N} - W_{b_N a'_N a_M b_M} A_{a_N a'_N} \right) A_{a_M b_M} \Bigg\}$$

$$\equiv i\omega_{a_N b_N} A_{a_N b_N} + R_{a_N b_N}, \tag{3.138}$$

where we have used the well-known product commutator relations and have set

$$\omega_{a_N b_N} = \omega_{a_N} - \omega_{b_N}.$$

It is clear from this relation that the algebra will get involved. We shall keep the description to the minimum necessary for understanding and checking.

As we mentioned before, we develop Eq. 3.138 into a useful expression[235a] by (1) writing equations of motion of its operator products such as

$$A_{a_N a'_N} X_{\alpha\beta} \equiv O^{(1)}_{\alpha\beta a_N a'_N}, \tag{3.139a}$$

$$A_{a_N a'_N} A_{a_M b_M} X_{\alpha\beta} \equiv O^{(2)}_{\alpha\beta a_N a'_N a_M b_M}; \tag{3.139b}$$

(2) constructing their formal solution; (3) reinserting it into Eq. 3.138; and (4) discarding unimportant terms. For instance, using Eqs. 3.139a, 3.135, and 3.136, we first find

$$\dot{O}^{(1)}_{\alpha\beta a_N a'_N} = i\left[ H_{N'}, O^{(1)}_{\alpha\beta a_N a'_N} \right]$$

$$= i(\omega_{a_N a'_N} + \omega_{\alpha\beta}) O^{(1)}_{\alpha\beta a_N a'_N} + i\{ U_{\gamma\alpha a_M b_M} X_{\gamma\beta} A_{a_M b_M} A_{a_N a'_N}$$

$$- U_{\beta\gamma a_M b_M} X_{\alpha\gamma} A_{a_N a'_N} A_{a_M b_M} \} + \text{terms in } VXAAA, WAAX.$$

$$\tag{3.140}$$

The formal solution of Eq. 3.140 (see Note 11, Chapter 2) is

$$O^{(1)}_{\alpha\beta a_N a'_N}(t) = O^{(1)}_{\alpha\beta a_N a'_N}(t_0) \exp\left[ i(\omega_{a_N a'_N} + \omega_{\alpha\beta})(t - t_0) \right]$$

$$+ \int_{t_0}^{t} dt' \exp\left[ i(\omega_{a_N a'_N} + \omega_{\alpha\beta})(t - t') \right] i\{ U_{\gamma\alpha a_M b_M} X_{\gamma\beta} A_{a_M b_M}(t') A_{a_N a'_N}(t')$$

$$- U_{\beta\gamma a_M b_M} X_{\alpha\gamma} A_{a_N a'_N}(t') A_{a_M b_M}(t') \} \pm \cdots. \tag{3.141}$$

Inserting this result as well as the corresponding solution for $O^{(2)}$ (see Eq. 3.139b) back into Eq. 3.138 changes the latter into an integro-differential equation. The integrals are simplified by physically realistic assumptions that "remove" the lattice operators $X_{\alpha\beta}$ from the relations. We demonstrate this on the example of the term with coefficient $U_{\alpha\beta a'_N a_N}$.

Replacing $A_{a'_N b_N} X_{\alpha\beta}$ in Eq. 3.138 by $O^{(1)}_{\alpha\beta a'_N b_N}$ of Eq. 3.141 gives

$$U_{\alpha\beta a'_N a_N} A_{a'_N b_N} X_{\alpha\beta} = U_{\alpha\beta a'_N a_N} O^{(1)}_{\alpha\beta a'_N b_N}(t)$$

$$= U_{\alpha\beta a'_N a_N} O^{(1)}_{\alpha\beta a'_N b_N}(t_0) \exp\left[ i(\omega_{a'_N b_N} + \omega_{\alpha\beta})(t - t_0) \right]$$

$$+ U_{\alpha\beta a'_N a_N} \int_{t_0}^{t} dt' \exp\left[ i(\omega_{a'_N b_N} + \omega_{\alpha\beta})(t - t') \right]$$

$$\times i\{ U_{\gamma\alpha a_M b_M} X_{\gamma\beta} A_{a_M b_M}(t') A_{a'_N b_N}(t')$$

$$- U_{\beta\gamma a_M b_M} X_{\alpha\gamma} A_{a'_N b_N}(t') A_{a_M b_M}(t') \pm \cdots \}. \tag{3.142}$$

Recall that summation over repeated indices ($\alpha, \beta, \gamma, a'_N, a_M, b_M$) is implied. Let us evaluate the first term of the integral. Multiplying out yields

$$i U_{\alpha\beta a'_N a_N} U_{\gamma\alpha b'_N a'_N} \int_{t_0}^{t} dt' \exp\left[ i(\omega_{a'_N b_N} + \omega_{\alpha\beta})(t - t') \right]$$

$$\times A_{b'_N b_N}(t') X_{\gamma\beta}(t') \equiv S_{a_N b_N}(t). \tag{3.143}$$

(We have changed dummy index $a_M$ to $b'_N$.) We transfer the rapid time dependence of $A(t)$ to a phase factor,

$$A_{b'_N b_N}(t) = \bar{A}_{b'_N b_N}(t) \exp\left[ i\omega_{b'_N b_N}(t - t_0) \right], \tag{3.144}$$

and consider amplitude $\bar{A}(t)$ a slowly varying function of $t$. Writing $\bar{A}(t)$ as Fourier transform, we get

$$A_{b'_N b_N}(t) = \int d\Omega_N \, \bar{A}_{b'_N b_N}(\Omega_N) \exp\left[ i(\omega_{b'_N b_N} + \Omega_N)(t - t_0) \right]. \tag{3.145}$$

Since on the average the heat bath does not change, we set

$$X_{\alpha\beta}(t) \approx X_{\alpha\beta}(t_0) \exp\left[ i\omega_{\alpha\beta}(t - t_0) \right]. \tag{3.146}$$

Inserting Eqs. 3.145 and 3.146 into Eq. 3.143 and integrating over $t'$ yields (note that $\omega_{\gamma\beta} + \omega_{\alpha\gamma} = \omega_{\alpha\beta}$, etc.)

$$S_{a_N b_N}(t) = i \sum_{a'_N b'_N} \sum_{\alpha\beta\gamma} U_{\alpha\beta a'_N a_N} U_{\gamma\alpha b'_N a'_N} X_{\gamma\beta}(t_0) \int d\Omega_N \, \bar{A}_{b'_N b_N}(\Omega_N)$$

$$\times \exp\left[ i(\omega_{a'_N b_N} + \omega_{\alpha\beta})(t - t_0) \right]$$

$$\times \left\{ 1 - \exp\left[ -i(\omega_{a'_N b'_N} + \omega_{\alpha\gamma} - \Omega_N)(t - t_0) \right] \right\}$$

$$\times \frac{1}{i} \left( \omega_{a'_N b'_N} + \omega_{\alpha\gamma} - \Omega_N \right)^{-1}. \tag{3.147a}$$

To undo the Fourier transforms, we pull out

$$\exp\left[i\left(\omega_{b'_N b_N} + \Omega_N + \omega_{\gamma\beta}\right)(t - t_0)\right].$$

(See Eqs. 3.145 and 3.146.) This gives

$$S_{a_N b_N}(t) = i \sum_{a'_N b'_N} \sum_{\alpha\beta\gamma} U_{\alpha\beta a'_N a_N} U_{\gamma\alpha b'_N a'_N} X_{\gamma\beta}(t) \int d\Omega_N \bar{A}_{b'_N b_N}(\Omega_N)$$

$$\times \exp\left[i\left(\omega_{b'_N b_N} + \Omega_N\right)(t - t_0)\right]$$

$$\times \left\{\exp\left[i\left(\omega_{a'_N b'_N} + \omega_{\alpha\gamma} - \Omega_N\right)(t - t_0)\right] - 1\right\}$$

$$\times \left[i\left(\omega_{a'_N b'_N} + \omega_{\alpha\gamma} - \Omega_N\right)\right]^{-1}. \tag{3.147b}$$

We apply several, physically realistic simplifications to get this relation into manageable form. First, amplitude $\bar{A}_{b'_N b_N}(\Omega_N)$ is only appreciable within a certain frequency interval $\Delta\Omega_N$. Second, since bath energy levels are quasi-continuous, we assign the frequency

$$\omega_{\alpha'} = \omega_\alpha - \Omega_N \tag{3.148}$$

to a bath frequency and set

$$U_{\gamma\alpha b'_N a'_N} \approx U_{\gamma\alpha' b'_N a'_N}. \tag{3.149}$$

In other words, within the frequency interval $\Delta\Omega_N$, operator $U_{\gamma\alpha b'_N a'_N}$ is only weakly dependent on $\alpha$.

These assumptions permit us to avoid integration over $\Omega_N$ in Eq. 3.147b and to do independent ensemble averaging over the bath coordinates. We consider $F_{\gamma\beta}(t)$ a bath fluctuation operator of zero average with respect to its bath average $\langle X_{\gamma\beta}(t)\rangle$,

$$F_{\gamma\beta}(t) = X_{\gamma\beta}(t) - \langle X_{\gamma\beta}(t)\rangle,$$

with

$$\langle X_{\gamma\beta}(t)\rangle = \delta_{\gamma\beta}\rho_\beta, \tag{3.150a}$$

where

$$\rho_\beta = \frac{\exp(-\hbar\omega_\beta / k_B T)}{\sum_\beta \exp(-\hbar\omega_\beta / k_B T)} \tag{3.150b}$$

is the thermal bath occupation (see Eq. K.1, Appendix K). Abbreviating

$$\frac{i}{\Omega}\{1-\exp[i\Omega(t-t_0)]\} \equiv \Delta(\Omega) \tag{3.151}$$

and summing over $\gamma$, transforms Eq. 3.147$b$ into

$$\frac{1}{i}S_{a_N b_N} = \sum_{a'_N b'_N} \sum_{\alpha\beta} U_{\alpha\beta a'_N a_N} U_{\beta a b'_N a'_N}\rho_\beta \Delta(\omega_{a'_N b'_N} + \omega_{\alpha\beta})A_{b'_N b_N}(t)$$

$$+ \sum_{a'_N b'_N} \sum_{\alpha\beta\gamma} U_{\alpha\beta a'_N a_N} U_{\gamma a b'_N a'_N} F_{\gamma\beta}(t)\Delta(\omega_{a'_N b'_N} + \omega_{\alpha\gamma})A_{b'_N b_N}(t). \tag{3.152}$$

The first term of this relation is the sum over all $\gamma = \beta$, the second over $\gamma \neq \beta$. (Equations 3.145 and 3.148–3.151 have been used.)

If we consider times that are sufficiently long for non-Markovian behavior of the bath modes to have died out,

$$\omega_{\alpha\beta}(t-t_0) \gg 1,$$

function $\Delta(\Omega)$ of Eq. 3.151 becomes

$$\Delta(\Omega) = iP\frac{1}{\Omega} + \pi\delta(\Omega), \tag{3.153}$$

where $P$ is the principal value and $\delta$ the delta function. (See also Eqs. 2.135 to 2.137, Section 5.4 of Chapter 2.)

As a practical result, we see that to obtain useful equations of motions for the oscillator system operator, we must formulate the global system operators in terms of their averages over the bath dissipative system,

$$\left\langle S_{a_N b_N}(t) \right\rangle = \mathrm{Tr}\,\rho_\beta S_{a_N b_N}(t)$$

$$= i \sum_{a'_N b'_N} \sum_{\alpha\beta} U_{\alpha\beta a'_N a_N} U_{\beta a b'_N a'_N}\rho_\beta \Delta(\omega_{a'_N b'_N} + \omega_{\alpha\beta})A_{b'_N b_N}(t).$$

$$\tag{3.154}$$

Contributions from fluctuation operators of the bath (Eq. 3.150) are neglected (random phase approximation).[250] The averages retain, however, their operator characteristics with respect to the dynamic (oscillator) system (noise-averaged system operators).

In the following section we discuss the further development of the theory and its application to multilevel vibrational relaxation phenomena of general interest. We are mainly concerned with developing expressions for the rate of change of density matrices of the oscillator system (excitation probabilities) that describe the time evolution of the various vibrational relaxation processes. We are not interested in discussing correlation functions; they are less useful here and, at any rate, easily constructed once the density matrix of the system is known (see Eq. 1.6, Chapter 1).

## 8.2.  Energy Transfer between Coupled Vibrations

On the basis of the results of the preceding section, we can establish a general rate-balancing equation for the diagonal elements of the noise-averaged oscillator system operator $\langle A_{a_N b_N}(t) \rangle$. The restriction to diagonal elements $\langle A_{a_N a_N}(t) \rangle$ is easily understood since they define the occupation probability of the $a$th vibrational energy level of oscillator $N$.

The computations are rather extensive[235a,b] but do not introduce concepts beyond those we have already discussed. For instance, the method we used to remove the operators $X_{\alpha\beta}$ (Eqs. 3.140–3.154) is reemployed to write useful expressions for matrix elements whose time dependence we need not know precisely or that have nondiagonal elements. We therefore outline only key computations and results.

First, we rewrite the interaction Hamiltonian $H_1$ (Eq. 3.137) in terms of Boson operators $B_N$ of pairwise interactions between any (not necessarily adjacent) system oscillator levels. We employ the rotating wave or resonance approximation[250]: Terms in $B_N B_M$ or $B_N^\dagger B_M^\dagger$, which involve rapidly fluctuating far off-resonance contributions

$$B_N^\dagger B_M^\dagger \propto \exp[i(\omega_N + \omega_M)t],$$

are omitted. (We have tacitly employed this approximation in Eqs. 3.67 and Eq. K.5, Appendix K.) Introducing coupling coefficients $D_N^{(1)}$, $D_N^{(2)}$, $D_{NM}$, and $C_{NM}$, we write

$$H_1 = D_N^{(1)}\big(B_N + B_N^\dagger\big) + \big(\tfrac{1}{2}\big) D_N^{(2)}\big(B_N^\dagger B_N + B_N B_N^\dagger\big)$$

$$+ \big(\tfrac{1}{2}\big) D_{NM}\big(B_M^\dagger B_N + B_M B_N^\dagger\big) + \big(\tfrac{1}{2}\big) C_{NM}\big(B_M^\dagger B_N + B_M B_N^\dagger\big), \tag{3.155}$$

where again summation over equal indices is implied, with $M \neq N$.

Comparison with Eq. 3.137 shows that the coupling coefficients $U$, $V$, and $W$ are related to the $D_N$, $D_{NM}$, and $C_{NM}$ by

$$U_{\alpha\beta a_N b_N} = \langle \alpha| D_N^{(1)}|\beta\rangle\langle a_N| B_N + B_N^\dagger|b_N\rangle$$

$$+ \big(\tfrac{1}{2}\big)\langle \alpha| D_N^{(2)}|\beta\rangle\langle a_N| B_N B_N^\dagger + B_N^\dagger B_N|b_N\rangle, \tag{3.156a}$$

$$V_{\alpha\beta a_N b_N a_M b_M} = \langle \alpha| D_{NM}|\beta\rangle\big[\langle a_N| B_N|b_N\rangle\langle a_M| B_M^\dagger|b_M\rangle$$

$$+ \langle a_N| B_N^\dagger|b_N\rangle\langle a_M| B_M|b_M\rangle\big], \tag{3.156b}$$

$$W_{a_N b_N a_M b_M} = C_{NM}\big[\langle a_N| B_N|b_N\rangle\langle a_M| B_M^\dagger|b_M\rangle$$

$$+ \langle a_N| B_N^\dagger|b_N\rangle\langle a_M| B_M|b_M\rangle\big]. \tag{3.156c}$$

Note that $D_N^{(1)}$, $D_N^{(2)}$, $D_{NM}$ still contain operators of the dissipative system since they describe bath-modulated vibrational energy transfer. On the other hand, $C_{NM}$ is a number (direct energy transfer). We assume that $U$ and $V$ are nondiagonal with respect to the dissipative system. We have also set $\hbar = 1$.

To obtain $\langle \dot{A}_{a_N a_N}(t) \rangle = (d/dt) \langle A_{a_N a_N}(t) \rangle$ we first substitute in Eq. 3.138 all operator products $A_{a'_N a_N} X_{\alpha\beta} = O^{(1)}_{\alpha\beta a'_N a_N}$, $O^{(2)}_{\alpha\beta a_N a'_N a_M a'_M}$, and so on (Eq. 3.139), by their noise-averaged expressions. For instance, in Eq. 3.138 the first nonvanishing diagonal element is (see Eqs. 3.142–3.154)

$$i\sum \{ U_{\alpha\beta a'_N a_N} A_{a'_N a_N} X_{\alpha\beta} \} = - \sum_{a'_N b'_N} \sum_{\alpha\beta} U_{\alpha\beta a'_N a_N} U_{\alpha\beta b'_N a'_N}$$

$$\times \rho_\beta \Delta(\omega_{\alpha\beta} + \omega_{a'_N b'_N}) \langle A_{b'_N a_N}(t) \rangle + \cdots .$$

Subsequently, we replace the coupling coefficients $U$, $V$, and $W$ by Eqs. 3.156a–3.156c. After a great deal of algebra this gives for the foregoing element the expression

$$i\sum \{ U_{\alpha\beta a'_N a_N} A_{a'_N a_N} X_{\alpha\beta} \} = - \sum_{a'_N b'_N} \sum_{\alpha\beta} \Big\{ |\langle \alpha | D_N^{(1)} | \beta \rangle|^2 \langle a'_N | B_N + B_N^\dagger | a_N \rangle$$

$$\times \langle b'_N | B_N + B_N^\dagger | a'_N \rangle \rho_\beta \Delta(\omega_{\alpha\beta} + \omega_{a'_N b'_N})$$

$$- \left( \tfrac{1}{4} \right) |\langle \alpha | D_N^{(2)} | \beta \rangle|^2 \langle a'_N | B_N B_N^\dagger + B_N^\dagger B_N | a_N \rangle$$

$$\times \langle b'_N | B_N B_N^\dagger + B_N^\dagger B_N | a'_N \rangle \rho_\beta \Delta(\omega_{\alpha\beta} + \omega_{a'_N b'_N}) \Big\} \langle | b'_N \rangle \langle a_N | \rangle$$

$$+ \cdots , \tag{3.157a}$$

omitting cross terms between $D_N^{(1)}$ and $D_N^{(2)}$. Writing out the matrix elements of Eq. 3.157a with the help of the Boson operator relations (Appendix K) $B_N |a_N\rangle = a_N^{1/2} |a_N - 1\rangle$, $B_N^\dagger |a_N\rangle = (a_N + 1)^{1/2} |a_N + 1\rangle$, and performing the summation over the intermediate states $a'_N$, $b'_N$, leads to terms such as

$$- \sum_{\alpha\beta} |\langle \alpha | D_N^{(1)} | \beta \rangle|^2 a_N \rho_\beta \delta(\omega_{\alpha\beta} - \omega_N) \langle A_{a_N a_N} \rangle. \tag{3.157b}$$

We have set

$$\omega_{a_N} = a_N \omega_N,$$

where $\omega_N$ is the fundamental frequency of oscillator $N$. The total expression is[235a]

$$\langle \dot{A}_{a_N a_N}\rangle = \left[\Gamma_N^{(1)} + \sum_M V_{NM}^{(1)}\right]\left[(a_N+1)\langle A_{a_N+1 a_N+1}\rangle - a_N\langle A_{a_N a_N}\rangle\right]$$

$$+ \Gamma_N^{(2)}\left[a_N\langle A_{a_N-1 a_N-1}\rangle - (a_N+1)\langle A_{a_N a_N}\rangle\right]$$

$$+ \sum_{a_M M} a_M\left\{V_{NM}^{(1)}\left[(a_N+1)\langle A_{a_M a_M}A_{a_N+1 a_N+1}\rangle - a_N\langle A_{a_M a_M}A_{a_N a_N}\rangle\right]\right.$$

$$+ V_{NM}^{(2)}\left[a_N\langle A_{a_M a_M}A_{a_N-1 a_N-1}\rangle - (a_N+1)\langle A_{a_M a_M}A_{a_N a_N}\rangle\right]\right\}$$

$$+ \langle A'_{a_N a_N}\rangle, \tag{3.158}$$

with (we set $\pi = 1$)

$$\Gamma_N^{(1)} = 2\sum_{\alpha\beta}\rho_\beta\delta(\omega_{\alpha\beta}-\omega_N)|\langle\alpha|D_N^{(1)}|\beta\rangle|^2,$$

$$\Gamma_N^{(2)} = 2\sum_{\alpha\beta}\rho_\alpha\delta(\omega_{\alpha\beta}-\omega_N)|\langle\alpha|D_N^{(1)}|\beta\rangle|^2, \tag{3.159}$$

$$V_{NM}^{(1)} = V_{MN}^{(2)} = 2\sum_{\alpha\beta}\rho_\beta\delta(\omega_{\alpha\beta}-\omega_N+\omega_M)|\langle\alpha|D_{NM}|\beta\rangle|^2, \quad (M\neq N).$$

(Note that Eq. 3.157*b* contributes the third term in Eq. 3.158. Also recall that primed symbols *never* imply differentiation.)

The rate coefficients in Eq. 3.159 are in the golden rule form (Eq. 1.63, Chapter 1). $\Gamma_N^{(1)}$ describes a "down," $\Gamma_N^{(2)}$ the "up" oscillator-bath coupling process; $V_{NM}$ describes bath-assisted interoscillator coupling.

The term $\langle A'_{a_N a_N}\rangle$ in Eq. 3.158 introduces mainly products between those elements of $N$ and $M$ that couple without participation of the bath (direct interoscillator coupling, see also Eq. 3.156*c*) as well as those that are caused by pure dephasing. In other words, all processes that do not involve oscillator-bath energy exchange. $\langle A'_{a_N a_N}\rangle$ reads[235a]

$$\langle A'_{a_N a_N}\rangle = i\sum_M \Lambda_{NM}\langle Q_{NM}\rangle + \text{complex conjugate} \tag{3.160a}$$

with

$$Q_{NM} = \left[(a_N+1)^{1/2}A_{a_N+1 a_N} - a_N^{1/2}A_{a_N a_N-1}\right]\sum_{a_M}(a_M+1)^{1/2}A_{a_M a_M+1},$$

$$\Lambda_{NM} = C_{NM} + i\sum_{\alpha\beta}\left\{\langle\alpha|D_N^{(1)}|\beta\rangle\langle\beta|D_M^{(1)}|\alpha\rangle(\rho_\beta-\rho_\alpha)\Delta(\omega_{\alpha\beta}+\omega_M)\right.$$

$$+ \langle\alpha|D_{NM}|\beta\rangle\left[\langle\beta|D_N^{(2)}|\alpha\rangle - \langle\beta|D_M^{(2)}|\alpha\rangle\right]\rho_\beta\Delta(\omega_{\alpha\beta})\right\}. \tag{3.160b}$$

To obtain $\langle Q_{NM} \rangle$, we need the terms

$$\langle A_{a_N+1a_N} A_{a_M a_M+1} \rangle, \qquad \langle A_{a_N a_N-1} A_{a_M a_M+1} \rangle.$$

They are special cases of term

$$\langle A_{a_N b_N} A_{a_M b_M} \rangle.$$

To obtain the latter, we write its equation of motion[235b]

$$\frac{d}{dt} \langle A_{a_N b_N} A_{a_M b_M} \rangle = i\left[ (a_N - b_N)\omega_N + (a_M - b_M)\omega_M \right] \langle A_{a_N b_N} A_{a_M b_M} \rangle$$

$$+ \langle R_{a_N b_N} A_{a_M b_M} \rangle + \langle A_{a_N b_N} R_{a_M b_M} \rangle + \langle R'_{a_N b_N a_M b_M} \rangle.$$

$$(3.161a)$$

[This is obtained in analogy with the steps leading from Eq. 3.138 to Eq. 3.154, with operator products such as $O^{(3)}_{\alpha\beta a_N b_N a_M b_M} = A_{a_N b_N} A_{a_M b_M} X_{\alpha\beta}$, $O^{(4)}_{\alpha\beta a_N b_N a_M b_M a_L b_L} = A_{a_N b_N} A_{a_M b_M} A_{a_L b_L} X_{\alpha\beta}$.] In turn, Eq. 3.161a gives us the equation of motion for $Q_{NM}$ (see Eq. 3.160b),

$$\frac{d}{dt} Q_{NM} = i(\omega_N - \omega_M) Q_{NM}$$

$$+ \left[ (a_N+1)(a_M+1) \right]^{1/2} \left( R_{a_N+1a_N} A_{a_M a_M+1} + A_{a_N+1a_N} R_{a_M a_M+1} \right)$$

$$- \left[ a_N(a_M+1) \right]^{1/2} \left( R_{a_N a_N-1} A_{a_M a_M+1} + A_{a_N a_N-1} R_{a_M a_M+1} \right)$$

$$+ \left[ (a_N+1)(a_M+1) \right]^{1/2} R'_{a_N+1a_N a_M a_M+1}$$

$$- \left[ a_N(a_M+1) \right]^{1/2} R'_{a_N a_N-1a_M a_M+1}.$$

$$(3.161b)$$

What we shall do with Eq. 3.161b is to find all the "diagonal" elements $A_{a_N a_N} A_{b_M b_M}$, and so on, and neglect the "nondiagonal" terms. Since oscillator-bath energy relaxation is usually relatively slow, we drop the corresponding product terms. For instance, if we keep in $R_{a_N b_N}$ only contributions from direct oscillator exchange, we get (Eq. 3.138)

$$R_{a_N b_N} A_{a_M b_M} = i \sum_{b'_N} \sum_{a'_M b'_M} \sum_M \{ W_{b'_N a_N a'_M b'_M} A_{b'_N b_N} A_{a'_M b'_M}$$

$$- W_{b_N b'_N a'_M b'_M} A_{a_N b'_N} A_{a'_M b'_M} \} A_{a_M b_M},$$

where $W_{a_N b_N a_M b_M}$ is defined by Eq. 3.156c. We then operate with the Boson operators $B_N^\dagger, B_N, B_M^\dagger, B_M$ on the basis vectors $|a_N\rangle$, $|b'_M\rangle$ and $|b'_N\rangle, |b'_M\rangle$, sum over all $b'_N, a'_M, b'_M$, and set $a_N, b_N, a_M, b_M$ to their values required by Eq. 3.160$b$. This yields, after some algebra, two diagonal terms each for $R_{a_N+1a_N}A_{a_Ma_M+1}$ and $R_{a_Na_N-1}A_{a_Ma_M+1}$ as well as for $A_{a_N+1a_N}R_{a_Ma_M+1}$, $A_{a_Na_N-1}R_{a_Ma_M+1}$. For instance:

$$R_{a_N+1a_N}A_{a_Ma_M+1} = i\sum_{a'_M}\sum_M C_{NM}\Big\{\big[(a_N+1)a'_M\big]^{1/2}A_{a_Na_N}A_{a'_Ma'_M-1}A_{a_Ma_M+1}$$

$$-\big[(a_N+1)a'_M\big]^{1/2}A_{a_N+1a_N+1}A_{a'_Ma'_M-1}A_{a_Ma_M+1}\Big\}$$

$$= i\sum_{\substack{M\\M\neq N}} C_{NM}\Big\{\big[(a_N+1)(a_M+1)\big]^{1/2}$$

$$\times\Big(A_{a_Na_N}A_{a_M+1a_M+1}-A_{a_N+1a_N+1}A_{a_M+1a_M+1}\Big)\Big\}. \quad (3.162)$$

Furthermore, if we add terms from pure dephasing, we deal with operator products

$$U_{\alpha\beta a_N b_N} = \left(\tfrac{1}{2}\right)\langle\alpha|D_N^{(2)}|\beta\rangle\langle a_N|B_N B_N^\dagger + B_N^\dagger B_N|b_N\rangle$$

(see Eq. 3.156$a$). This leads to (omitting four complementary terms)

$$R_{a_Nb_N}A_{a_Mb_M} + A_{a_Nb_N}R_{a_Mb_M}$$

$$= -\sum_{b'_N a'_N}\sum_{\alpha\beta} U_{\alpha\beta b'_N a_N}U_{\beta a a'_N b'_N}\rho_\beta\Delta\Big[\omega_{\alpha\beta}+(b'_N-a'_N)\omega_N\Big]A_{a'_N b_N}A_{a_M b_M}$$

$$+\sum U_{\alpha\beta b'_N a_N}U_{\beta a b_N a'_N}\rho_\alpha\Delta\Big[\omega_{\alpha\beta}+(a'_N-b_N)\omega_N\Big]A_{b'_N a'_N}A_{a_M b_M} - \{3\}+\{4\},$$

where $\{3\},\{4\}$ designate a repetition of the first and the second sum with $N$ and $M$ swapped. Inspection of this relation shows that the oscillator frequency differences $(b'_N-a'_N)\omega_N$ and so on vanish through the effect of the Boson operator products in $U$. This introduces into $Q_{NM}$ a term (we set $a_N=a_N+1$ or $a_N, b_N=a_N+1$ or $a_N-1$, $a_M=a_M$, and $b_M=a_M+1$)

$$-(\Gamma'_N+\Gamma'_M)Q_{NM},$$

where

$$\Gamma'_N = \sum_{\alpha\beta}|\langle\alpha|D_N^{(2)}|\beta\rangle|^2\rho_\beta\delta(\omega_{\alpha\beta}) \quad (3.163)$$

is the phase relaxation rate in the golden rule form. Collecting all terms (Eqs.

3.161$b$–3.163) gives[235i]

$$\langle \dot{Q}_{NM} \rangle = \left[ i(\omega_N - \omega_M) - (\Gamma'_N + \Gamma'_M) \right] \langle Q_{NM} \rangle$$

$$- iC_{NM} \sum_{a_M} \left\{ (a_M + 1) \left[ (a_N + 1) \langle A_{a_N+1 a_N+1} A_{a_M a_M} \rangle - a_N \langle A_{a_N a_N} A_{a_M a_M} \rangle \right] \right.$$

$$\left. + a_M \left[ a_N \langle A_{a_N-1 a_N-1} A_{a_M a_M} \rangle - (a_N + 1) \langle A_{a_N a_N} A_{a_M a_M} \rangle \right] \right\}. \tag{3.164}$$

Equation 3.164 is easily solved if we only require the "stationary" solution (upper integration limit $t \to \infty$). Since we neglect the relatively slow energy relaxation process in the equation of motion of $\langle Q_{NM} \rangle$, the stationary approximation is valid. This yields (Note 11, Chapter 2)

$$\langle Q_{NM} \rangle = - iC_{NM} \sum \left[ \Gamma'_N + \Gamma'_M + i(\omega_N - \omega_M) \right] \left[ (\Gamma'_N + \Gamma'_M)^2 + (\omega_N - \omega_M)^2 \right]^{-1},$$

where $\Sigma$ is an abbreviation for the sum $\Sigma_{a_M} \{ \cdots \}$ appearing in Eq. 3.164. Inserting this result into Eq. 3.160$a$ and keeping only the term $C_{NM}$ of $\Lambda_{NM}$ (see Eq. 3.160$b$), modifies Eq. 3.158 to (see Eq. F.13, Appendix F)

$$\langle \dot{A}_{a_N a_N} \rangle = \left[ \Gamma_N^{(1)} + \sum_M \left( V_{NM}^{(1)} + W_{NM} \right) \right] \left[ (a_N + 1) \langle A_{a_N+1 a_N+1} \rangle - a_N \langle A_{a_N a_N} \rangle \right]$$

$$+ \Gamma_N^{(2)} \left[ a_N \langle A_{a_N-1 a_N-1} \rangle - (a_N + 1) \langle A_{a_N a_N} \rangle \right]$$

$$+ \sum_{a_M M} a_M \left\{ \left( V_{NM}^{(1)} + W_{NM} \right) \left[ (a_N + 1) \langle A_{a_M a_M} A_{a_N+1 a_N+1} \rangle \right. \right.$$

$$\left. - a_N \langle A_{a_M a_M} A_{a_N a_N} \rangle \right] + \left( V_{NM}^{(2)} + W_{NM} \right) \left[ a_N \langle A_{a_M a_M} A_{a_N-1 a_N-1} \rangle \right.$$

$$\left. \left. - (a_N + 1) \langle A_{a_M a_M} A_{a_N a_N} \rangle \right] \right\}, \tag{3.165}$$

with

$$W_{NM} = 2C_{NM}^2 (\Gamma'_N + \Gamma'_M) \left[ (\Gamma'_N + \Gamma'_M)^2 + (\omega_N - \omega_M)^2 \right]^{-1}.$$

(As before, the time dependence of the system is understood. Summation is over all $M \neq N$.)

As result, we see that operator $\langle A'_{a_N a_N} \rangle$ introduces into Eq. 3.158 an additional set of quantities $\langle A_{a_M a_M} A_{a_N \pm 1 a_N \pm 1} \rangle$, weighted by rate constant $W_{NM}$, which contribute direct vibrational energy transfer (quasi-resonant) and pure dephasing. Notice also that the smaller $|\omega_N - \omega_M|$, the larger the coefficient of direct oscillator coupling of strength $C_{NM}$. On the other hand, for $\omega_N \neq \omega_M$,

$W_{NM}$ contributes predominantly to dephasing (through $\Gamma'$). From Eqs. 3.163, 3.164, and 3.156c we see that dephasing can be understood as a *scattering of bath excitations by the dynamic system oscillators without exchange of energy between bath and oscillators*. In principle this is the very meaning we had attached to dephasing in Section 2 of this chapter. But here we assume the harmonic approximation (no higher than quadratic terms in the normal coordinates) and a long-time decay regime (no quasi-static effects of dephasing).

If we now take in Eq. 3.165 the trace with the canonical oscillator density matrix $\rho$ (Eq. K.1, Appendix K) over all system operators, for instance,

$$\mathrm{Tr}\rho\langle A_{a_N a_N}(t)\rangle = \mathrm{Tr}\rho(t)\langle A_{a_N a_N}(0)\rangle$$

$$= \mathrm{Tr}\rho(t)\langle|a_N\rangle\langle a_N|\rangle \equiv \sigma_{a_N a_N}(t) = \langle\langle A_{a_N a_N}(t)\rangle\rangle, \quad (3.166)$$

we obtain a rate-balancing relation for the probability of finding the subsystem oscillator $N$ in level $a_N$ while oscillator $N$ transfers (receives) energy directly or with participation of the bath to (from) any other oscillator level $a_M$ as well as transfers (obtains) vibrational energy to (from) the bath.

This partial result already shows the versatility of the model. First, once we know $\langle A_{a_N b_N}(t)\rangle$, we can express other interesting quantities, for instance, the expectation value of the transition moment operator **m** of oscillator $N$ between levels $a_N, b_N$,

$$\langle\langle m(t)\rangle\rangle = \langle a_N|m|b_N\rangle\sigma_{a_N b_N}(t), \quad (3.167a)$$

and, by Fourier transformation, obtain the corresponding susceptibility (see Section 1.3, Chapter 1). Of course, we could also construct correlation functions such as[235h]

$$\langle\langle A_{a_N b_N}(t)A_{b_N a_N}(t')\rangle\rangle. \quad (3.167b)$$

Second, we have seen that the general formalism lets us write equations of motion of two-oscillator operators (Eq. 3.161a). Their time evolution depends on three-oscillator operators. Hence, to follow the time evolution of oscillator subsystem $N$ which couples with $n$ other oscillators, we arrive at expressions that contain $n$-oscillator operators ("hierarchy of density matrices").[235b] Although such an $n$-oscillator process is difficult to evaluate numerically, two interesting situations can be readily treated.

1. Two oscillators interact by quasi-resonance in dilute solution. An example is the near-resonance coupling between the fundamental oscillator of liquid $N_2$ with the oscillator of a dopant, discussed in Section 6.5.3.

2. A large number of near-identical oscillators interact. Examples are vibrational relaxation processes in one-component liquids and in con-

centrated solutions (quasi-resonant vibrational energy transfer) or intramolecular vibrational energy exchange within a large molecule. In fact, the formalism permits us to dispense with the questionable assumption that the inactive intramolecular oscillators constitute a heat bath (Section 6.5.2).

In the following section we discuss direct and indirect energy exchange between two oscillators (process 1). Process 2 is discussed in Section 8.2.2.

**8.2.1. Direct and Indirect Energy Exchange between Two Oscillators.** The relations derived in the preceding section allow us to describe numerically the energy transfer between oscillator $N$ and any other oscillator $M$, as well as energy transfer between two coupled oscillators $N$, $M$ and any other oscillator $L$, if we succeed in "removing" the complicating multiple-oscillator densities. Let us try to accomplish this now.

To save writing, we set

$$\langle A_{j_N j_N} \rangle \equiv \rho_{j_N}; \qquad \langle A_{j_N j_N} A_{k_M k_M} \rangle \equiv \rho_{j_N, k_M}. \tag{3.168}$$

First, we examine under which conditions we can reproduce previously discussed, familiar results. For instance, let us consider Eq. 3.165 but disregard oscillator energy dissipation to the bath, $\Gamma_N^{(1)}, \Gamma_N^{(2)} = 0$. With the relation

$$\sum_{k_M} \rho_{j_N, k_M} = \sum_{k_M} \langle | j_N \rangle \langle j_N \| k_M \rangle \langle k_M | \rangle = \langle | j_N \rangle \langle j_N | \rangle = \rho_{j_N}, \tag{3.169}$$

(see Eq. F.13, Appendix F) we can expand in Eq. 3.165 the term

$$\sum_M \left( V_{NM}^{(1)} + W_{NM} \right) \left[ (j_N + 1) \rho_{j_N + 1} - j_N \rho_{j_N} \right]$$

to

$$\sum_{k_M} \sum_M \left( V_{NM}^{(1)} + W_{NM} \right) \left[ (j_N + 1) \rho_{j_N + 1, k_M} - j_N \rho_{j_N, k_M} \right]$$

and add it to the summation term $\Sigma_{k_M} \Sigma_M$ of Eq. 3.165. Equation 3.165 now reads

$$\dot{\rho}_{j_N} = \sum_{k_M} \sum_M W'_{NM} \left\{ (k_M + 1) \left[ (j_N + 1) \rho_{j_N + 1, k_M} - j_N \rho_{j_N, k_M} \right] \right.$$

$$\left. + k_M \left[ j_N \rho_{j_N - 1, k_M} - (j_N + 1) \rho_{j_N, k_M} \right] \right\}, \tag{3.170}$$

with

$$W'_{NM} = V_{NM}^{(1)} + W_{NM} = V_{MN}^{(2)} + W_{NM}.$$

The coefficient $W'_{NM}$ represents the transition probability per unit time for interaction of normal mode $N$ with normal mode $M$ for bath-modulated (Eq. 3.159) as well as direct energy exchange (Eq. 3.164). Summing over $k_M$, we get

$$\dot{\rho}_{j_N} = \sum_M W'_{NM} \left\{ \left( \sum_{k_M} k_M \rho_{k_M} + \sum_{k_M} \rho_{k_M} \right) \left[ (j_N + 1)\rho_{j_N+1} - j_N \rho_{j_N} \right] \right.$$

$$\left. + \sum_{k_M} k_M \rho_{k_M} \left[ j_N \rho_{j_N-1} - (j_N + 1)\rho_{j_N} \right] \right\}. \tag{3.171}$$

Assuming now that all $k_M$ are distributed by a Boltzmann distribution $\rho_{k_M} = \exp(-\hbar k_M \omega_M / k_B T) Z_M^{-1}$, the elementary sums in Eq. 3.171 are readily computed (see Eq. K.3, Appendix K). The result (we do not write it down here) corresponds to Eq. 3.110 with $K_n$ replaced by $W'_{NM}$. Notice that we have not used here a collision concept or explicitly defined joint transition probabilities (Eq. 3.109). Evidently, we have established a previous result in a more satisfying manner.

Let us now consider the rate-balancing equation of a two-oscillator probability density; it is more complicated but also more interesting than the one-oscillator probability.

We recall that $\rho_{j_N, k_M}(t)$ (Eq. 3.168) describes the time evolution of the joint probability of finding oscillator $N$ in state $j_N$ and oscillator $M$ in state $k_M$. We allow $N$ and $M$ to interact with each other and with any other oscillator $L$ by energy exchange, as well as to dissipate (receive) independently their energy to (from) the bath. The master equation for $\dot{\rho}_{j_N, k_M}(t)$ is obtained from the equation of motion, Eq. 3.161a, eliminating operator products as we discussed before (Section 8.2). The result is[235c]

$$\dot{\rho}_{j_N, k_M} = \Gamma_N^{(1)} \left[ (j_N + 1)\rho_{j_N+1, k_M} - j_N \rho_{j_N, k_M} \right] + \Gamma_N^{(2)} \left[ j_N \rho_{j_N-1, k_M} - (j_N + 1)\rho_{j_N, k_M} \right]$$

$$+ \Gamma_M^{(1)} \left[ (k_M + 1)\rho_{j_N, k_M+1} - k_M \rho_{j_N, k_M} \right]$$

$$+ \Gamma_M^{(2)} \left[ k_M \rho_{j_N, k_M-1} - (k_M + 1)\rho_{j_N, k_M} \right]$$

$$+ \sum_{L \neq M, N} \sum_{k_L} W'_{NL} \left\{ (k_L + 1) \left[ (j_N + 1)\rho_{j_N+1, k_M, k_L} - j_N \rho_{j_N, k_M, k_L} \right] \right.$$

$$\left. + k_L \left[ j_N \rho_{j_N-1, k_M, k_L} - (j_N + 1)\rho_{j_N, k_M, k_L} \right] \right\}$$

$$+ \sum_{L \neq M, N} \sum_{k_L} W'_{ML} \left\{ (k_L + 1) \left[ (k_M + 1)\rho_{j_N, k_M+1, k_L} - k_M \rho_{j_N, k_M, k_L} \right] \right.$$

$$\left. + k_L \left[ k_M \rho_{j_N, k_M-1, k_L} - (k_M + 1)\rho_{j_N, k_M, k_L} \right] \right\} + Y_{NM}^{(i)}, \tag{3.172}$$

where $i = 1$ or 2, with

$$Y_{NM}^{(1)} = W'_{NM}\{j_N(k_M+1)\rho_{j_N-1,k_M+1} - (j_N+1)k_M\rho_{j_N,k_M}$$

$$+ (j_N+1)k_M\rho_{j_N+1,k_M-1} - j_N(k_M+1)\rho_{j_N,k_M}\} \qquad (3.173a)$$

and

$$Y_{NM}^{(2)} = W'_{NM}\{(k_M+1)[(j_N+1)\rho_{j_N+1,k_M} - j_N\rho_{j_N,k_M}] + k_M[j_N\rho_{j_N-1,k_M}$$

$$- (j_N+1)\rho_{j_N,k_M}] + (j_N+1)[(k_M+1)\rho_{j_N,k_M+1} - k_M\rho_{j_N,k_M}]$$

$$+ j_N[k_M\rho_{j_N,k_M-1} - (k_M+1)\rho_{j_N,k_M}]\}. \qquad (3.173b)$$

Process $Y_{NM}^{(1)}$ relates to indirect (bath-modulated) oscillator-oscillator energy exchange ($N + M = $ constant), whereas $Y_{NM}^{(2)}$ relates to direct oscillator-oscillator energy exchange (without participation of the bath). Notice the appearance of three-oscillator probability densities in Eq. 3.172.

Comparison of Eq. 3.172 with Eq. 3.127 shows (apart from the equilibrium boundary conditions introduced into Eq. 3.127) that terms for (1) a direct exchange mechanism [term $Y^{(2)}$] and (2) coupling of $N, M$ to oscillators $L$ (terms $\sum W'_{NL}, \sum W'_{ML}$) are missing in Eq. 3.127. Also, process 2 describes coupling of the $j_N, k_M$ system to other oscillators which are not necessarily of a sufficiently large number to constitute an equilibrium bath or lattice (Section 6.5.2). Only if

$$\sum_{k_L} k_L\rho_{j_N,k_M,k_L} \approx \left(\sum_{k_L} k_L\rho_{k_L}\right)\rho_{j_N,k_M} \qquad (3.174)$$

may we take the first moment[239]

$$\sum_{k_L} k_L\rho_{k_L} \equiv \langle E_L\rangle \qquad (3.175)$$

(the time development of all occupation numbers of oscillators $k_L$ or its internal energy) into the lattice (bath) and obtain the corresponding terms for process 2 in Eq. 3.127, with its perturbation Hamiltonian given by Eq. 3.124. (The correspondence is $j_N, v_n; k_M, v_m; L, l$.)

The procedure of "disentangling" the indices of multioscillator operators by Eq. 3.174 requires an investigation into the validity of the factorization

$$\frac{d}{dt}\langle A_{a_N a_N} A_{a_M a_M}\rangle = \frac{d}{dt}\langle A_{a_N a_N}\rangle\langle A_{a_M a_M}\rangle$$

$$= \langle \dot{A}_{a_N a_N}\rangle\langle A_{a_M a_M}\rangle + \langle A_{a_N a_N}\rangle\dot{A}_{\langle a_M a_M\rangle}. \qquad (3.176)$$

The first equality in Eq. 3.176 is not trivial owing to the presence of term $\langle R'_{a_N a_N a_M a_M} \rangle$ (Eq. 3.161$a$). Assuming, however, that $\langle R'_{a_N a_N a_M a_M} \rangle$ is small, Eq. 3.161$a$ simplifies to

$$\frac{d}{dt} \langle A_{a_N a_N} A_{a_M a_M} \rangle = \langle R_{a_N a_N} A_{a_M a_M} \rangle + \langle A_{a_N a_N} R_{a_M a_M} \rangle.$$

On the other hand, note that

$$\langle R_{a_N a_N} A_{a_M a_M} \rangle = \langle \dot{A}_{a_N a_N} A_{a_M a_M} \rangle, \qquad \langle \dot{A}_{a_N a_N} \rangle = \langle R_{a_N a_N} \rangle.$$

This can be verified by comparing the terms of Eqs. 3.165 and 3.138 with the terms of $\langle R_{a_N a_N} A_{a_M a_M} \rangle$. (See Note 10.) Consequently

$$\langle A_{a_N a_N} A_{a_M a_M} \rangle = \langle A_{a_N a_N} \rangle \langle A_{a_M a_M} \rangle \tag{3.177}$$

is indeed equivalent to Eq. 3.176.

The physical conditions that make $\langle R'_{a_N a_N a_M a_M} \rangle$ small are the very conditions of main interest to us here, namely, the predominance of direct interoscillator energy exchange between two oscillators or indirect coupling between many oscillators. In other words, Eq. 3.176 is satisfied if relaxation takes place by low-order resonance processes of strong oscillator-oscillator coupling rather than by high-order resonance processes of direct energy dissipation from individual oscillators to the bath (weak coupling).[235b]

We describe first direct two-oscillator energy exchange. Therefore we must solve Eq. 3.173$b$.

To remove oscillator $M$ from Eq. 3.173$b$, we apply the analogue of Eq. 3.174 ("linearization"),

$$\sum_{k_M} k_M \rho_{j_N, k_M} \approx \left( \sum_{k_M} k_M \rho_{k_M} \right) \rho_{j_N}.$$

Multiplying Eq. 3.173$b$ out gives

$$\dot{\rho}_{j_N, k_M} = W'_{NM} \{ (j_N + 1) \rho_{j_N + 1} (k_M + 1) \rho_{k_M} + (j_N + 1) \rho_{j_N} (k_M + 1) \rho_{k_M + 1}$$

$$+ j_N \rho_{j_N - 1} k_M \rho_{k_M} + j_N \rho_{j_N} k_M \rho_{k_M - 1} - 2 \big[ j_N \rho_{j_N} (k_M + 1) \rho_{k_M}$$

$$+ (j_N + 1) \rho_{j_N} k_M \rho_{k_M} \big] \}. \tag{3.178}$$

Summation over $k_M$ transforms this into

$$\dot{\rho}_{j_N} = W_{NM} \{ \big[ (j_N + 1) \rho_{j_N + 1} - j_N \rho_{j_N} \big] (\langle E_M \rangle + 1)$$

$$+ \big[ j_N \rho_{j_N - 1} - (j_N + 1) \rho_{j_N} \big] \langle E_M \rangle \}, \tag{3.179}$$

where Eqs. 3.175 and its obvious extensions, such as

$$\sum_{k_M} (k_M + 1)\rho_{k_M} = \langle E_M \rangle + 1,$$

$$\sum_{k_M} k_M \rho_{k_M - 1} = \sum_{k_M} \{(k_M - 1)\rho_{k_M - 1} + \rho_{k_M}\}$$

$$= \langle E_M \rangle + 1, \qquad (3.180)$$

have been used. Furthermore, we have set $W'_{NM} = W_{NM}$ (Eq. 3.170) since we are concerned only with direct energy exchange. (See Eqs. 3.159 and 3.165.)

Notice that Eq. 3.179 is linear in $\rho$, in contrast to Eq. 3.178, and that it contains probabilities that pertain solely to oscillator $N$. All parameters belonging to oscillator $M$ are expressed by coupling coefficient $W_{NM}$ and average (time-dependent) occupation $\langle E_M \rangle$. (A similar relation can be written for $\dot{\rho}_{k_M}$.)

Incidentally, recasting Eq. 3.179 into

$$\dot{\rho}_{j_N}(t) = W_{NM}\langle E_M(t) \rangle \{(j_N + 1)\rho_{j_N + 1}(t) - (2j_N + 1)\rho_{j_N}(t)$$

$$+ j_N \rho_{j_N - 1}(t)\} + W_{NM}\{(j_N + 1)\rho_{j_N + 1}(t) - j_N \rho_{j_N}(t)\}$$

shows its formal equivalence to an early formulation[251] derived in terms of specific joint transition probabilities induced by binary collisions between $N$ and $M$.

Of course, to solve Eq. 3.179, we must know $\langle E_M(t) \rangle$. We obtain it by multiplying each side of Eq. 3.179 by $j_N$ and summing over $j_N$. This gives

$$\sum_{j_N} j_N \dot{\rho}_{j_N} = W_{NM} \sum_{j_N} \left\{ \left[ j_N(j_N + 1)\rho_{j_N + 1} - j_N^2 \rho_{j_N} \right] (\langle E_M \rangle + 1) \right.$$

$$\left. + \left[ j_N^2 \rho_{j_N - 1} - j_N(j_N + 1)\rho_{j_N} \right] \langle E_M \rangle \right\}$$

$$= W_{NM}\{-\langle E_N \rangle(\langle E_M \rangle + 1) + (\langle E_N \rangle + 1)\langle E_M \rangle\}, \qquad (3.181)$$

where

$$\sum_{j_N} \{j_N(j_N + 1)\rho_{j_N + 1} - j_N^2 \rho_{j_N}\}$$

has been expanded to

$$\sum_{j_N} \{(j_N + 1)^2 \rho_{j_N + 1} - (j_N + 1)\rho_{j_N + 1} - j_N^2 \rho_{j_N}\} = -\langle E_N \rangle$$

to apply Eq. 3.175. The other term of Eq. 3.181 is expanded accordingly.

Equation 3.181 then reads

$$\langle \dot{E}_N(t) \rangle = -W_{NM}\{\langle E_N(t) \rangle - \langle E_M(t) \rangle\}. \qquad (3.182)$$

We see that the time development of the average vibrational population of oscillator $N$ in the $N$-$M$ coupling scheme depends only on the mean energy-gradient between systems $N$ and $M$.[251] [The analogous relation is obtained for $\langle E_M(t)\rangle$; notice that Eq. 3.174 is symmetric in $j_N$, $k_M$.] Since the mean energy $\langle E\rangle$ of the two-oscillator system is constant,

$$(\tfrac{1}{2})\{\langle E_M(t)\rangle + \langle E_N(t)\rangle\} \equiv \langle E\rangle = \langle E_N(\infty)\rangle = \langle E_M(\infty)\rangle, \quad (3.183)$$

Eq. 3.182 yields

$$\langle E_N(t)\rangle = \langle E_N(\infty)\rangle + [\langle E_N(0)\rangle - \langle E_N(\infty)\rangle]\exp(-2W_{NM}t). \quad (3.184)$$

Now we can integrate Eq. 3.179. We do it with the help of a generating function,

$$G(x,t) = \sum_{j_N} x^{j_N}\langle A_{j_N j_N}(t)\rangle,$$

which leads to a hypergeometric function or its expansion of orthogonal polynomials. The derivation is of no interest here; the result is[235b]

$$\rho_{j_N}(t) = \sum_{k_N=0}^{\infty} \left(Z_N^{(1)}\right)^{j_N}\left(Z_N^{(2)}\right)^{k_N}\left(1+Z_N^{(1)}\right)^{-j_N-k_N-1}$$

$$\times F(-j_N, -k_N, 1; -S_N)\rho_{k_N}(t_0)$$

$$= \sum_{k_N=0}^{\infty}\sum_{q=0}^{j_N,k_N}\left(Z_N^{(1)}\right)^{j_N-q}\left(Z_N^{(2)}\right)^{k_N-q}\left(1+Z_N^{(1)}\right)^{-j_N-k_N-1}$$

$$\times \binom{j_N}{q}\binom{k_N}{q}\exp[-q\gamma_N(t)]\rho_{k_N}(t_0), \quad (3.185)$$

where $j_N$, $k_N$ are the momentary and initial excitations of oscillator $N$ with their respective contributions $\rho_{j_N}(t)$ and $\rho_{k_N}(t_0)$. Hence $\rho_{j_N}(t)$ is the conditional probability of finding oscillator $N$, in its $N$-$M$ coupling scheme, with excitation number $j_N$ at $t$ if it had excitation number $k_N$ at $t_0$.[236]

The quantities $Z_N^{(1)}$, $Z_N^{(2)}$, $\gamma_N$, and $S_N$ are time-dependent operator functions defined by

$$Z_N^{(1)}(t) = \int_{t_0}^{t} dt' \sum_M W_{NM}\langle E_M(t')\rangle\exp\{-[\gamma_N(t)-\gamma_N(t')]\},$$

$$Z_N^{(2)}(t) = 1 + Z_N^{(1)}(t) - \exp[-\gamma_N(t)],$$

$$\gamma_N(t) = \int_{t_0}^{t} dt \sum_M W_{NM}; \qquad S_N = \exp[-\gamma_N(t)]\left[Z_N^{(1)}Z_N^{(2)}\right]^{-1}. \quad (3.186)$$

Index $q$ runs from zero to the smaller of $j_N$, $k_N$. Recall that we assume $\langle E_M(t) \rangle$ is known (see Eq. 3.184).

Since $\langle E_M(t) \rangle$ depends on the initial value $\langle E_M(t_0) \rangle$ of all oscillators, we put[235b]

$$Z_N^{(1)} = Z_N^{(1)}(\{\langle E_M(t_0) \rangle\}, t) = Z_N^{(1)}(\{k_M\}, t),$$

where $\{\langle E_M(t_0) \rangle\}$ or $\{k_M\}$ designates the set of all $\langle E_M(t_0) \rangle$ or respective quantum numbers $k_M$. [The analogous expressions are written for $Z_N^{(2)}$, $\gamma_N$, and $\rho(t_0)$.] We therefore sum in Eq. 3.185 over all $\{k_M\}$. This step is equivalent to a trace operation (Eq. K.1, Appendix K) that leads from $\rho_{j_N}(t)$ to $\sigma_{j_N j_N}(t)$ [Eq. 3.166].

On the other hand, since we discuss here a two-oscillator system, there is only one other oscillator $M$. Furthermore, the direct $N$-$M$ process is most effective if initially oscillator $N$ is in excited level $n$ and oscillator $M$ is in its ground state. Consequently, we need not sum over $k_N$ and set in Eq. 3.185 $W_{NM} \equiv W$, $Z_N^{(1)} \equiv Z_1$, $Z_N^{(2)} \equiv Z_2$. Writing a few terms for $\rho_{j_N}(t)$, we get, accordingly, with $t_0 = 0$, $\rho_{k_N}(0) = \rho_{1_N}(0) = 1$,

$$\rho_{1_N}(t) = (1 + Z_1)^{-3} \{Z_1 Z_2 + \exp(-Wt)\},$$

$$\rho_{2_N}(t) = (1 + Z_1)^{-4} \{Z_1^2 Z_2 + 2Z_1 \exp(-Wt)\},$$

$$\rho_{0_N}(t) = (1 + Z_1)^{-2} Z_2, \text{ and so on.} \tag{3.187}$$

For $\rho_{k_M}(t)$ of oscillator $M$ we obtain (abbreviating $Z_M^{(1)} \equiv Z_1'$, etc.), with $\rho_{0_M}(0) = 1$,

$$\rho_{0_M}(t) = (1 + Z_1')^{-1},$$

$$\rho_{1_M}(t) = (1 + Z_1')^{-2} Z_1',$$

$$\rho_{2_M}(t) = (1 + Z_1')^{-3} Z_1'^2, \text{ and so on.} \tag{3.188}$$

Since $Z_N^{(1)}(0) = Z_N^{(2)}(0) = 0$ (Eq. 3.186), we find from Eqs. 3.187 and 3.188 that

$$\rho_{j_N}(0) = \rho_{k_M}(0) \begin{cases} = 1 & \text{for } j_N = 1, k_M = 0 \\ = 0 & \text{otherwise,} \end{cases}$$

in agreement with the initial boundary conditions. Notice that the oscillator excitation can rise *above* its initial value.

For $t \gg W^{-1}$, the level occupation probability $\rho_{j_N}(t)$ drops with excitation $j_N$ as

$$\rho_{j_N}(t \gg W^{-1}) = \left[ Z_1(t \gg W^{-1}) \right]^{j_N} \left[ 1 + Z_1(t \gg W^{-1}) \right]^{-j_N - 1}. \tag{3.189}$$

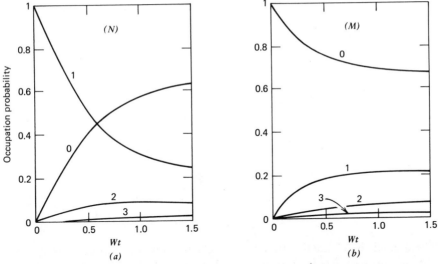

**Figure 3.25.** (*a*) Evolution of the time-dependent occupation probability of the energy levels of oscillator $N$, $j_N = 0$, 1, 2, and 3. The oscillator is engaged in direct energy exchange (no participation of the bath) with oscillator $M$. Initial state: $j_N = 1$, $k_M = 0$. (*b*) Concomitant evolution of the time-dependent probability of the energy levels $k_M = 0$, 1, 2, and 3 for oscillator $M$. Levels higher than $j_N$, $k_M > 3$ are too weakly populated to be shown on the scale of the figure. [H. Paerschke, K.-E. Süsse, and D.-G. Welsch, *Exp. Tech. Phys.* **25**, 481 (1977).]

(See Eq. 3.186 and 3.187.) The time regime $t \gg W^{-1}$ does not signify the *thermal* equilibrium state: This state is only attained for $t$ obeying $\Gamma^{(1)}t \gg 1$.[235c]

Figure 3.25 shows a numerical evaluation of Eqs. 3.187 and 3.188 on the basis of the definitions

$$Z_1 = \frac{n}{2}\{1 + \exp(-2Wt) - 2\exp(-Wt)\}$$

$$Z_1' = \frac{n'}{2}\{1 - \exp(-2Wt)\}$$

$$Z_2 = Z_1 + 1 - \exp(-Wt), \text{ and so on,}$$

(refer also to Eqs. 3.186 and 3.184). The left-hand side of the figure shows the time dependence of the occupation probabilities of oscillator $N$, the right-hand side that of oscillator $M$. We see that levels above the initial excitation ($n$, $n'$) become populated and that an "equilibrium" population (Eq. 3.189),

$$\rho_{j_N}(t \gg W^{-1}) = \left(\frac{n}{2}\right)^{j_N}\left(1 + \frac{n}{2}\right)^{-j_N - 1},$$

is reached for $t \gg W^{-1}$.

This concludes our discussion of direct $N$-$M$ oscillator coupling; we turn briefly to the indirect $N$-$M$ coupling process (lattice assisted).

We must solve Eq. 3.173a. Obviously, this relation can be decomposed into $n$ subrelations, each containing all matrix elements $\rho_{j_N, k_M}$ with $j_N + k_M = n$ ($n = 0, 1, 2, \ldots$). Assuming again that oscillator $N$ is initially in level $n$, oscillator $M$ in its ground state, we see that only levels with $j_N + k_M = n$ participate in the indirect coupling mechanism. All other levels remain unchanged. In contrast to direct energy transfer of two coupled oscillators, the indirect energy transfer does not populate $j_N$ above its initial excitation value.

The general solution of Eq. 3.173a is

$$\rho_{n-k_M, k_M}(t) = \sum_{q=0}^{n} C_q a_{k_M q} \exp[-q(q+1)t],$$

$$n = j_N + k_M. \tag{3.190}$$

The coefficients $C_q$ are determined from the boundary conditions, the $a_{k_M q}$ are obtained successively from recursion relations, with $a_{n q} = 1$.[235c, g]

With the help of Eq. 3.169 we obtain, setting $\rho_{j_N, k_M}(t) \equiv \rho_{j, k}^{(N, M)}(t)$,

$$\rho_{n-k, k}^{(N, M)}(t) = \rho_{n-k}^{(N)}(t) = \rho_k^{(M)}(t).$$

We shall not go into further detail about the indirect coupling mechanism. We discussed a simplified version of Eq. 3.190 (including independent energy dissipation of oscillators $N$ and $M$ to the lattice) in Section 6.5.3.

### 8.2.2.    Energy Exchange between Many Coupled Vibrations.

In the previous section we demonstrated a numerical evaluation of energy exchange between two oscillators. We now outline that the global vibrational relaxation model describes equally well pure vibrational dephasing and multiple-oscillator vibrational energy transfer (which can be quasi-resonant) in neat media or concentrated solutions.[235d, e] In these situations the formalism considers interaction of many closely spaced oscillator energy levels.

Discussing quasi-resonant energy transfer first, we start off with Eq. 3.179 but keep, for sake of generality, the energy dissipation terms with their coefficients $\Gamma_N^{(1)}$ and $\Gamma_N^{(2)}$ (Eq. 3.172). In other words, Eq. 3.179 is of the form

$$\dot{\rho}_{j_N} = \Upsilon_N^{(1)}(t)\left[(j_N + 1)\rho_{j_N+1} - j_N\rho_{j_N}\right]$$

$$+ \Upsilon_N^{(2)}(t)\left[j_N\rho_{j_N-1} - (j_N + 1)\rho_{j_N}\right], \tag{3.191}$$

with the generalized damping coefficients

$$\Upsilon_N^{(1)}(t) = \Gamma_N^{(1)} + \sum_M W'_{NM}(\langle E_M(t)\rangle + 1),$$

$$\Upsilon_N^{(2)}(t) = \Gamma_N^{(2)} + \sum_M W'_{MN}\langle E_M(t)\rangle, \qquad (M \neq N). \tag{3.192}$$

(Recall that $W'$ includes direct and indirect oscillator-oscillator coupling; see Eqs. 3.170, 3.165, and 3.159.)

To solve Eq. 3.191 for $\rho_{j_N}$, we must know the $\Upsilon_N^{(i)}(t)$; in other words, the $\langle E_N(t) \rangle$. The equation of motion for $\langle E_N(t) \rangle$ is obtained from Eqs. 3.191 and 3.192 by the same method we employed to calculate Eqs. 3.179 to 3.181. We find $(M \neq N)$

$$\langle \dot{E}_N \rangle = -\left( \Gamma_N^{(1)} - \Gamma_N^{(2)} \right) \langle E_N \rangle + \Gamma_N^{(2)}$$

$$+ \sum_M \left\{ W'_{MN} \langle E_M \rangle - W'_{NM} \langle E_N \rangle + \left( W'_{MN} - W'_{NM} \right) \langle E_N \rangle \langle E_M \rangle \right\}.$$

$$(3.193)$$

(Note that for $V_{NM} = 0$, $\Gamma_N = 0$, Eq. 3.193 gives Eq. 3.182—as it should.) Although Eq. 3.193 looks complicated, it can be readily solved for systems of many interacting oscillators. Since the strongest effects take place between closest-spaced levels, the boundary conditions

$$\Gamma_N^{(i)} - \Gamma_M^{(i)} \ll \Gamma_N^{(i)}, \Gamma_M^{(i)}$$

$$W'_{NM} - W'_{MN} \ll W'_{NM}, W'_{MN} \qquad (3.194)$$

must hold between adjacent oscillators $N$ and $M$. Using the trial function

$$\langle E_N \rangle = \langle \bar{E}_N \rangle \exp\left[ -\left( \Gamma_N^{(1)} - \Gamma_N^{(2)} \right)(t - t_0) \right]$$

$$+ \left[ \frac{\Gamma_N^{(2)}}{\Gamma_N^{(1)} - \Gamma_N^{(2)}} \right] \left\{ 1 - \exp\left[ -\left( \Gamma_N^{(1)} - \Gamma_N^{(2)} \right)(t - t_0) \right] \right\} \qquad (3.195)$$

yields an equation of motion for $\langle \bar{E}_N(t) \rangle$ [Note 11] that reads

$$\langle \dot{\bar{E}}_N(t) \rangle = \sum_M \left\{ W'_{MN} \langle \bar{E}_M(t) \rangle - W'_{NM} \langle \bar{E}_N(t) \rangle \right\}. \qquad (3.196)$$

We next solve this relation for $\langle \bar{E}_N(t) \rangle$.

We consider the quantities $\langle \bar{E} \rangle$ of the many, near-identical oscillators to represent a quasi-continuous density distribution of internal oscillator energies. Hence subscript $N$ is symbolic for the set $\{N\}$ of normal frequencies and positional coordinates of adjacent oscillators $N$. Since we assume that the oscillator energies between adjacent oscillators are near identical, the oscillators differ mainly by their spacial distribution (different environment). We then develop the density function $\langle \bar{E}_M(t) \rangle$ into a series *at* $\{N\}$, with the coefficients

$$\kappa_N^{(q)} = (q!)^{-1} \sum_M W'_{MN} (M - N)^q. \qquad (3.197a)$$

Breaking the series after the quadratic term (pairwise interactions) gives

$$\langle \dot{\bar{E}}(\{N\},t)\rangle = \kappa_N^{(1)}\frac{d}{dN}\langle \bar{E}(\{N\},t)\rangle + \kappa_N^{(2)}\frac{d^2}{dN^2}\langle \bar{E}(\{N\},t)\rangle. \quad (3.197b)$$

Assuming spacial isotropy of the medium, we can drop the term $\kappa_N^{(1)}$. The solution of Eq. 3.197 is a diffusion-type equation (see also Eq. 3.23a),[58]

$$\langle \bar{E}(\{N\},t)\rangle = \left[4\kappa_N^{(2)}\pi(t-t_0)\right]^{-1/2}\int_{-\infty}^{\infty}dN'\langle E(N',t_0)\rangle\exp\left[-\frac{(N-N')^2}{4\kappa_N^{(2)}(t-t_0)}\right]$$

$$(3.198)$$

with (see Eq. 3.195)

$$\langle \bar{E}(N,t_0)\rangle = \langle E(N,t_0)\rangle.$$

Using Eq. 3.198, we can now compute the time-dependent damping parameters in Eq. 3.191. We first insert Eq. 3.195 into Eq. 3.192 and obtain

$$\Upsilon_N^{(1)}(t) = \Gamma_N^{(1)} + \sum_M W'_{NM}$$

$$+ \sum_M W'_{NM}\Gamma_M^{(2)}\left(\Gamma_M^{(1)} - \Gamma_M^{(2)}\right)^{-1}\left\{1 - \exp\left[-\left(\Gamma_M^{(1)} - \Gamma_M^{(2)}\right)(t-t_0)\right]\right\}$$

$$+ \int dM\exp\left[-\left(\Gamma_M^{(1)} - \Gamma_M^{(2)}\right)(t-t_0)\right]\langle \bar{E}(M,t)\rangle W'_{NM}, \quad (N \neq M),$$

$$(3.199)$$

where integration over $M$ excludes the range occupied by oscillator $N$.

This result has an interesting corollary. At long times,

$$\langle \bar{E}(\{N\},t\to\infty)\rangle = 0.$$

Therefore,

$$\Upsilon_N^{(1)}(t\to\infty) = \Gamma_N^{(1)} + \sum_M W'_{NM}\left\{1 + \Gamma_M^{(2)}\left(\Gamma_M^{(1)} - \Gamma_M^{(2)}\right)^{-1}\right\}. \quad (3.200)$$

This represents the overall damping constant for relaxation of oscillator $N$ if all other oscillators $M$ in the $N$-$M$ pairwise interaction scheme are assigned to the dissipative system (bath); the $\langle \bar{E}_M(t)\rangle$ in Eqs. 3.191 are replaced by their thermal averages (Note 12). Consequently, arbitrary assignment of system oscillators $M$ to a Boltzmann heat bath leads to time-*independent* damping constants and purely exponential short-time behavior.

We now develop[235e] the integral of Eq. 3.199,

$$\int dM \exp\left[-\left(\Gamma_M^{(1)}-\Gamma_M^{(2)}\right)(t-t_0)\right]\langle\bar{E}(M,t)\rangle W'_{NM}\equiv\Gamma''_N(t).\quad(3.201)$$

This term is mainly responsible for nonexponential behavior. In particular, let us discuss interaction between identical normal modes of different molecules (resonance energy transfer). This means that $W'_{NM}$ depends only on the separation between molecules carrying the oscillators; variables $N, N'$ in Eq. 3.198 are to be replaced by position vectors $\mathbf{R}, \mathbf{R}'$,

$$\langle\bar{E}(\mathbf{R},t)\rangle=\left[4\pi\kappa_N^{(2)}(t-t_0)\right]^{-3/2}\int d^3\mathbf{R}'\langle\bar{E}(\mathbf{R}',t_0)\rangle\exp\left[-\frac{(\mathbf{R}-\mathbf{R}')^2}{4\kappa_N^{(2)}(t-t_0)}\right].$$

$$(3.202)$$

Defining $\langle E_{\mathbf{R}_N}(t_0)\rangle$ as operator of the total occupation of an oscillator positioned at $\mathbf{R}_N$ at instant $t_0$, we express by

$$\langle E(\mathbf{R},t_0)\rangle=\sum_M\langle E_{\mathbf{R}_M}(t_0)\rangle\,\delta(\mathbf{R}-\mathbf{R}_M)\qquad(3.203)$$

the boundary condition that oscillators of eigenfrequency $\omega_M$ and position $\mathbf{R}_M$ are initially excited. Inserting Eqs. 3.202 and 3.203 into Eq. 3.201 yields the time-dependent damping parameter

$$\Gamma''_N(t)=\exp\left[-\left(\Gamma_{\mathbf{R}_N}^{(1)}-\Gamma_{\mathbf{R}_N}^{(2)}\right)(t-t_0)\right]\sum_M\langle E_{\mathbf{R}_M}(t_0)\rangle W''_{\mathbf{R}_N\mathbf{R}_M}(t),\qquad(3.204)$$

where

$$W''_{\mathbf{R}_N\mathbf{R}_M}(t)=\left[4\pi\kappa_{\mathbf{R}_N}(t-t_0)\right]^{-1/2}R_M^{-1}\int_r^\infty dR\,RV(R)\left\{\exp\left[-\frac{(\mathbf{R}-\mathbf{R}_M)^2}{4\kappa_{\mathbf{R}_N}(t-t_0)}\right]\right.$$

$$\left.-\exp\left[-\frac{(\mathbf{R}+\mathbf{R}_M)^2}{4\kappa_{\mathbf{R}_N}(t-t_0)}\right]\right\}\qquad(3.205)$$

and (Eq. 3.197a)

$$\kappa_{\mathbf{R}_N}=\sum_M\frac{1}{6}|\mathbf{R}_M-\mathbf{R}_N|^2 W'(|\mathbf{R}_M-\mathbf{R}_N|).$$

$V(R)$ is the radial intermolecular coupling Hamiltonian. $W'_{NM}$ is defined in Eq. 3.170. To see how Eq. 3.205 comes about, note that taking the limit (we now abbreviate $\kappa_{R_N} \equiv \kappa$)

$$\lim_{R_M \to 0} \frac{1}{R_M} \left\{ \exp\left[ -\frac{(R-R_M)^2}{4\kappa t} \right] - \exp\left[ -\frac{(R+R_M)^2}{4\kappa t} \right] \right\} = \frac{R}{\kappa t} \exp\left( -\frac{R^2}{4\kappa t} \right)$$

(3.206)

leads to the usual volume element $4\pi R^2 \, dR$ and normalization factor $(4\pi\kappa t)^{-3/2}$ (see Section 2.2.1). Hence if we designate the intermolecular average distance by symbol $\sigma$ (we get its value from the density), we can define distance $r$ as the radius of the critical (spherical) volume

$$\frac{\sigma}{2} \leq r \leq \sigma$$

which is to be excluded for molecules positioned at $\mathbf{R}_M \neq \mathbf{R}_N$ (Fig. 3.2).

Equation 3.205 can be numerically evaluated once we assign the form of the radial intermolecular coupling Hamiltonian $V(R)$. We consider two representative situations, namely, an exponential, short-range repulsive Hamiltonian,

$$V_e(R) = \overline{W} \exp(-2\alpha R),$$

(Eqs. 3.62, 3.92, and 3.94), and a long-range dipole-dipole perturbation potential,

$$V_d(R) = \overline{\overline{W}} R^{-6}$$

(Eq. 3.21). Solution of Eq. 3.205 with Eq. 3.21 is analytical only for $W''_{\mathbf{R}_N \mathbf{R}_N}(t)$,[252a]

$$W''_{\mathbf{R}_N \mathbf{R}_N}(t) = \left\{ \frac{4\pi\overline{\overline{W}}}{[4\pi\kappa(t-t_0)]^{3/2}} \right\} \int_r^\infty dR \, R^{-4} \exp\left[ -\frac{R^2}{4\kappa(t-t_0)} \right]$$

$$= \left(\tfrac{4}{3}\right) V_d(r) \left(r^2 / [4\kappa(t-t_0)]\right)^3 \left\{ 2\left[ 1 - \Phi\left(r / [4\kappa(t-t_0)]^{1/2}\right) \right] \right.$$

$$+ \pi^{-1/2} \left[ 4\kappa(t-t_0)/r^2 \right]^{1/2} \left[ (4\kappa(t-t_0)/r^2) - 2 \right]$$

$$\left. \times \exp\left[ -r^2 / 4\kappa(t-t_0) \right] \right\},$$

(3.207)

where $\Phi(u)$ is the error function[252b] and Eq. 3.206 has been used.

Solution of Eq. 3.205 with the repulsive interaction Hamiltonian is analytical for both $W''_{\mathbf{R}_N \mathbf{R}_M}(t)$, $\mathbf{R}_M \neq 0$, and $W''_{\mathbf{R}_N \mathbf{R}_N}(t)$, $\mathbf{R}_M = 0$. For instance, expanding the integral into the form

$$c_1 \int dx \exp[f(x)](d/dt)f(x) + c_2 \int dx \exp[f(x)],$$

each term is readily integrated, giving (we abbreviate $4\kappa(t - t_0)$ by $\beta_N$)

$$W''_{\mathbf{R}_N \mathbf{R}_M}(t) = V_e(r)(1/2R_M)\left\{ (\beta_N/\pi)^{1/2} \right.$$

$$\times \left( \exp\left[ -(r - R_M)^2/\beta_N \right] - \exp\left[ -(r + R_M)^2/\beta_N \right] \right)$$

$$+ \exp\left[ \beta_N \alpha^2 + 2\alpha(r \mp R_M) \right] \left[ R_M \mp \beta_N \alpha \right]$$

$$\left. \times \left[ 1 - \Phi\left( [r \mp R_M + \beta_N \alpha]/\beta_N^{1/2} \right) \right] \right\}. \tag{3.208}$$

The $\pm$ signs signify that the second term is repeated with the indicated signs.

$W''_{\mathbf{R}_N \mathbf{R}_M}(t)$ characterizes a "back" or "retro" effect during vibrational resonance energy exchange between many oscillators. In particular, $W''_{\mathbf{R}_N \mathbf{R}_N}(t)$ describes the retro-effect for a situation where only the observed oscillator $N$ (at $\mathbf{R}_N$) is initially excited. On the other hand, $W''_{\mathbf{R}_N \mathbf{R}_M}(t)$ includes resonance energy transfer to the observed oscillator $N$ from those surrounding oscillators $(\mathbf{R}_M \neq \mathbf{R}_N)$ that are *also* excited at $t = 0$.[235e]

Figure 3.26 shows this in terms of $W''_{\mathbf{R}_N \mathbf{R}_M}(t)/V(\sigma)$, where $V(\sigma)$ is the value of either $V_e$ or $V_d$ at $R = \sigma$. The solid curves represent the repulsive, the dash-dot curves the long-range interactions. Curves 1 depict the situation where $\mathbf{R}_M = 0$. Only the oscillator $N$, at $\mathbf{R}_N = 0$, is excited at $t = 0$; initially, there is no retro-effect on oscillator $N$ from adjacent oscillators since $W''_{\mathbf{R}_N \mathbf{R}_N}(0) = 0$. Curves 2 simulate the case where the oscillator at $N$ is excited at $t = 0$ in an environment of already excited oscillators at positions $\mathbf{R}_M = \sigma$. In this case, initial transfer from $M$ to $N$ takes place, and $W''_{\mathbf{R}_N \mathbf{R}_M}(0)$ does not vanish.

We notice that the decay of $W''_{\mathbf{R}_N \mathbf{R}_N}(t)/V(\sigma)$ from its peak values (where $N \rightarrow M$ and $M \rightarrow N$ transfer processes balance) is slower for the $\exp(-R)$ than for the $R^{-6}$ potential; the back-effect on the observed oscillator at $N$ from adjacent oscillators at $M$ is more effective for short- than for long-range perturbation potentials. In other words, memory of the peak value is conserved longer for a short-range interaction.

The reason for this is easily understood: The oscillator interaction energy at $N$ stretches over a wider domain for a long-range than for the short-range potential. Therefore, the time interval for retro $M \rightarrow N$ transfer is prolonged in the environments of long-range potentials. Notice that for $t - t_0 \rightarrow \infty$, the excitation effects $W''_{\mathbf{R}_N \mathbf{R}_N}$ and $W''_{\mathbf{R}_N \mathbf{R}_M}$ have diffused away (see Eqs. 3.199, 3.200); $W''_{\mathbf{R}_N \mathbf{R}_M}(\infty) = 0$ for any $\mathbf{R}_M$.

If we should wish to bring in the concept of collisions as the agent that induces the (short-range) oscillator-oscillator interaction mechanism, it is clear

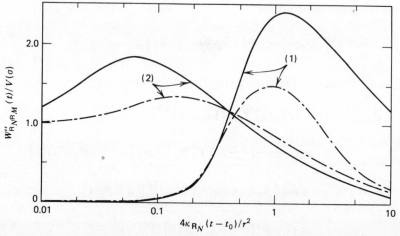

**Figure 3.26.** Time evolution of the general damping constant of oscillator $N$ for quasi-resonant energy exchange with many oscillators $M$ for an $\exp(-R)$ repulsive (solid curves) and a $R^{-6}$ attractive potential (dash-dot), respectively. Curves 1: Only oscillator $N$ is initially excited. Curves 2: Oscillators $N$, $M$ are initially excited, with $R_M = \sigma = 3.2$ Å. Number density: $3 \times 10^{22}$ cm$^{-3}$. [H. Paerschke, K.-E. Süsse, and D.-G. Welsch, *Ann. Phys.* **34**, 405 (1977).]

that the retro-effect displayed by $W''_{\mathbf{R}_N \mathbf{R}_N}(t)$ requires consideration of collective collision phenomena (Sections 2.2.2 and 3.3).

The global model of the time evolution of the vibrational population density for combined resonance energy transfer and energy dissipation is considerably more sophisticated than a simple combination of models discussed previously (Sections 6.1 and 3.3, respectively). The global model leads to realistic schemes of multilevel interactions with time-dependent damping parameters. The scheme naturally includes the process of back-diffusion of vibrational excitation to the initially excited oscillator.

**8.2.3. Phase Relaxation in the Harmonic Approximation.** Finally, we discuss the phase relaxation of an oscillator in a multilevel system. By a dephasing process we understand here the time evolution of the expectation value of the system oscillator $N$ to undergo transitions between any two adjacent levels (Note 13),

$$\langle\langle B_N(t) \rangle\rangle = \sum_{a_N} (a_N + 1)^{1/2} \mathrm{Tr}\,\rho \langle A_{a_N a_N + 1}(t) \rangle. \tag{3.209}$$

$B_N$ is the Boson operator and $\rho$ is the canonical density matrix of the oscillator system.[235a,d] We shall include *all* vibrational relaxation processes, pure dephasing (Section 2), quasi-resonant energy transfer (Section 3), and energy dissipation to the lattice (bath).

First, we give without derivation (but see Eq. 3.138) the equation of motion of $\langle\langle B_N(t)\rangle\rangle$,

$$\left[\frac{d}{dt} + i\left(\omega_N + \varepsilon_N^{(1)}\right) + \varepsilon_N^{(2)}\right]\langle\langle B_N\rangle\rangle - \sum_M (\eta_{NM} - iC_{NM})\langle\langle B_M\rangle\rangle = 0,$$

where

$$\varepsilon_N^{(2)} = \tfrac{1}{2}\left(\Gamma_N^{(1)} - \Gamma_N^{(2)} + \sum_M V_{NM}^{(1)}\right) + \Gamma_N',$$

$$\varepsilon_N^{(1)} = \mathrm{Im}\left\{\sum_{\alpha\beta}\left[\left|\langle\alpha|D_N^{(1)}|\beta\rangle\right|^2(\rho_\beta - \rho_\alpha)\Delta(\omega_{\alpha\beta} - \omega_N)\right.\right.$$

$$\left.\left. + \sum_M |\langle\alpha|D_{NM}|\beta\rangle|^2\rho_\beta\Delta(\omega_{\alpha\beta} - \omega_N + \omega_M)\right]\right\},$$

$$\eta_{NM} = \sum_{\alpha\beta}\left\{\langle\alpha|D_N^{(1)}|\beta\rangle\langle\beta|D_M^{(1)}|\alpha\rangle(\rho_\beta - \rho_\alpha)\Delta(\omega_{\alpha\beta} + \omega_M)\right.$$

$$+ \rho_\beta\langle\alpha|D_{NM}|\beta\rangle\left[-\langle\beta|D_M^{(2)}|\alpha\rangle\Delta(\omega_{\alpha\beta})\right.$$

$$\left.\left. + \langle\beta|D_N^{(2)}|\alpha\rangle\Delta(\omega_{\alpha\beta} - \omega_N + \omega_M)\right]\right\} \qquad \text{with } M \neq N. \quad (3.210)$$

Function $\Delta(\Omega)$ and all other operators have been defined previously in Eqs. 3.153, 3.155, 3.159, and 3.163 (see also Eq. 3.160$b$). Notice that the rate expression indeed includes scattering of bath excitations on the dynamic oscillator system (pure dephasing, $\Gamma_N'$), quasi-resonant vibrational energy transfer between two system oscillators ($\eta_{NM}, C_{NM}$), and energy dissipation to the bath ($\Gamma_N^{(i)}$).

For vanishing $N$-$M$ interactions, the solution of Eq. 3.210 is simply

$$\langle\langle B_N(t)\rangle\rangle = \langle\langle B_N(t_0)\rangle\rangle\exp\left\{\left[-i\left(\omega_N + \varepsilon_N^{(1)}\right) - \varepsilon_N^{(2)}\right](t - t_0)\right\}, \quad (3.211)$$

where $\varepsilon_N^{(1)}$ is a frequency shift. We see that this leads to a pure exponential decay of the phase relaxation $\exp(-\varepsilon_N^{(2)}t)$, since we have employed time-independent relaxation rates $\Gamma_N^{(i)}, V_{NM}^{(i)}$, as in Section 8.2.2. [Note 14.]

To readily obtain a general solution of Eq. 3.210, we assume that adjacent oscillators obey essentially the same time evolution. This allows us to develop $\langle\langle B_M(t)\rangle\rangle$ in Eq. 3.210 into a power series at position $N$ (refer also to Section 8.2.2, where we used the same stratagem). Breaking the series after the quadratic term and neglecting again the linear term (isotropy) modifies Eq.

3.210 to

$$\left\{ \frac{d}{dt} + i\bar{\omega}_N + \bar{\varepsilon}_N \right\} \langle\langle B(N, t) \rangle\rangle = \kappa_N \frac{d^2}{dN^2} \langle\langle B(N, t) \rangle\rangle$$

where (compare with Eq. 3.197a)

$$\kappa_N = \frac{1}{2} \sum_M (\eta_{NM} - iC_{NM})(M - N)^2,$$

$$\bar{\omega}_N = \omega_N + \varepsilon_N^{(1)} + \sum_M \{C_{NM} - \mathrm{Im}(\eta_{NM})\},$$

$$\bar{\varepsilon}_N = \varepsilon_N^{(2)} - \sum_M \mathrm{Re}(\eta_{NM}), \qquad (M \neq N). \tag{3.212}$$

Symbol $\langle\langle B(N, t) \rangle\rangle$ represents the continuous density function of the many, closely spaced system oscillators.

Parameters $\bar{\omega}_N$, $\bar{\varepsilon}_N$, and $\kappa_N$ depend only weakly on $N$. The solution of Eq. 3.212 is then[58]

$$\langle\langle B(N, t) \rangle\rangle = \exp\{-(i\bar{\omega}_N + \bar{\varepsilon}_N)(t - t_0)\}[4\pi\kappa_N(t - t_0)]^{-1/2}$$

$$\times \int_{-\infty}^{\infty} dN' \langle\langle B(N', t_0) \rangle\rangle \exp\left[ -\frac{(N - N')^2}{4\kappa_N(t - t_0)} \right]. \tag{3.213}$$

(Notice the resemblance of Eq. 3.213 with Eq. 3.198; they are formally identical but for the complex phase factor.)

Let us assume that at $t_0$ only the phase of a single oscillator, located at $\mathbf{R}_0$, is excited (see also Eq. 3.203),

$$\langle\langle B(\mathbf{R}, t_0) \rangle\rangle = \langle\langle B(\mathbf{R}_0, t_0) \rangle\rangle \delta(\mathbf{R} - \mathbf{R}_0).$$

This modifies Eq. 3.213 to

$$\langle\langle B_{\mathbf{R}_0}(t) \rangle\rangle = 4\pi \langle\langle B_{\mathbf{R}_0}(t_0) \rangle\rangle [4\pi\kappa_{\mathbf{R}_0}(t - t_0)]^{-3/2}$$

$$\times \exp\left[ -(\bar{\varepsilon}_{\mathbf{R}_0} + i\bar{\omega}_{\mathbf{R}_0})(t - t_0) \right]$$

$$\times \int_0^{R'} dR\, R^2 \exp\left[ -\frac{R^2}{4\kappa_{\mathbf{R}_0}(t - t_0)} \right], \tag{3.214}$$

following the steps from Eq. 3.202 to 3.206. Integration leads to[252a] (integrate $\int dx \{x \exp(-x^2)\} x$ by parts)

$$\langle\langle B_{\mathbf{R}_0}(t) \rangle\rangle = \langle\langle B_{\mathbf{R}_0}(t_0) \rangle\rangle \exp\left[ -(\bar{\varepsilon}_{\mathbf{R}_0} + i\bar{\omega}_{\mathbf{R}_0})(t - t_0) \right] G_{\mathbf{R}_0}(t - t_0),$$

$$G_{\mathbf{R}_0}(t - t_0) = \Phi\{ [\bar{\gamma}_{\mathbf{R}_0}(t - t_0)]^{-1/2} \} - 2[\pi\bar{\gamma}_{\mathbf{R}_0}(t - t_0)]^{-1/2}$$

$$\times \exp\{ -[\bar{\gamma}_{\mathbf{R}_0}(t - t_0)]^{-1} \}, \tag{3.215}$$

where

$$\bar{\gamma}_{\mathbf{R}_0} = \frac{4\kappa_{\mathbf{R}_0}}{R'^2}, \qquad R' = \left(\frac{3}{4\pi}\right)^{1/3}\sigma = 0.62\sigma,$$

is the relaxation constant for interoscillator energy exchange of oscillators separated by average distance $\sigma$. $\Phi(u)$ is the error function of argument $u$.

We can easily establish two limiting situations in Eq. 3.215. If $\bar{\varepsilon} \gg |\bar{\gamma}|$, interaction of the bath with each oscillator, independent of the remaining oscillators, mainly affects the global phase relaxation. [See the definitions of $\varepsilon_N^{(2)}$ (Eq. 3.210), $\bar{\varepsilon}_N$ and $\kappa_N$ (Eq. 3.212), and $\bar{\gamma}$ (Eq. 3.215).] Time evolution of $\langle\langle B \rangle\rangle$ is exponential [$G(t) \sim 1$] and occurs by depopulation-population of oscillator $N$ to the bath ($\Gamma_N^{(1)}$, $\Gamma_N^{(2)}$) as well as by pure dephasing ($\Gamma_N'$). In essence, this is the same physical situation as pictured by Eq. 3.211 (negligible $N$-$M$ interaction).[253]

For the opposite boundary condition, $\bar{\varepsilon} \ll |\bar{\gamma}|$, bath-assisted or direct oscillator-oscillator energy exchange predominates. Now $G(t)$ determines the (diffusional) time evolution of $\langle\langle B(t) \rangle\rangle$. We get ($t_0 = 0$) [Note 15]

$$\langle\langle B(t) \rangle\rangle\langle\langle B(0) \rangle\rangle^{-1} = 1 - (\bar{\gamma}t + 2)(\pi\bar{\gamma}t)^{-1/2}\exp\left[-(\bar{\gamma}t)^{-1}\right]$$

for

$$|\bar{\gamma}|t \ll 1$$

and

$$\langle\langle B(t) \rangle\rangle\langle\langle B(0) \rangle\rangle^{-1} = \left(\tfrac{4}{3}\right)\pi^{-1/2}(\bar{\gamma}t)^{-3/2}$$

for

$$|\bar{\gamma}|t \gg 1. \tag{3.216}$$

This concludes our discussion of the global vibrational relaxation model. Although its formalisms are complicated and difficult to derive, they allow us to apply our physical insight and to introduce realistic assumptions at key points in the calculations. Obviously, the potentiality of the model is large and encompassing.

## 9. CONCLUSIONS

It is appropriate to conclude this chapter with another critical look at the present state of vibrational relaxation, in particular, of relaxation of multilevel systems, since this is the basis for future developments.

To repeat what was said at the beginning of this chapter: There is no dearth of theories of vibrational relaxation for condensed-phase systems. In fact, the discussion of quite a few had to be left out for want of space and time.[254] (However, with the knowledge the reader has gained, he should be able to work his way through the various formalisms published in the literature.) In contrast to this, there is a lack of judicious and pertinent experimentation for testing, comparing, and weighing the various theoretical concepts. Cries of alarm that this and that interfering, perturbing, or neglected effect mars a theoretical approach should not be left on paper, but ought to be verified or rejected by the experiment.

For instance, there is considerable room for more work to establish how significant rotation-vibration interaction really is. Indeed, it is important to define, first, the time scales involved, second, what stage of interaction is to be understood to constitute a significant coupling effect. Coriolis interaction? Strong modulation of a vibrational intermolecular perturbation Hamiltonian by its coupling to orientational-translational lattice phonons? These two examples are coupling schemes below the maximal effect—the complete breakdown of statistical independence between rotational and vibrational system state vectors over the whole time interval of motion.

A largely neglected area is spectroscopic investigation of the dynamics of polymers and glassy systems. Although some papers in this area have been published,[235f, 255] much needs be done to obtain, for instance, information on the time scales of fluctuations of polymer skeletal motion, on the nature and extent of side group mobility, on effects of regions of crystallinity, and on manifestations of coupling of internal vibrational modes within few-atom groups with the lattice motions of the whole polymer skeleton.

Other generally interesting phenomena that deserve more attention are interactions of molecules with "external" and "internal" surfaces. By evaluating the band profiles of vibrational modes of molecules adsorbed to varying degrees of strength (physical adsorption, chemisorption), or of molecules trapped within rigid cavities (zeolites), valuable information about the dynamics of surface structure and molecular mobility is within reach.[256] Of course, in such systems the adsorbate-adsorbent interactions may be strongly anisotropic, requiring modification of existing theories. Furthermore, for species adsorbed on well-defined metal surfaces or on larger metal crystallites (possibly dispersed on a dielectric support), the inherent properties of the metallic state must be considered. Experimentally, this may require reflection or scattering—rather than absorption—measurements. Theoretically, this demands probing of any significant interaction of the oscillator coordinate with the electrons of the Fermi sea.[257]

For obvious reasons, namely, the possibility to separate a vibrational from an orientational relaxation process, Raman spectroscopic fluctuation studies of vibrational relaxation have been greatly preferred over infrared work. In fact, at the present, infrared studies are rarely found in the literature.

There is no compelling reason why this situation should remain. On one hand, systems can be chosen judiciously. On the other hand, methods of averaging over relatively slowly relaxing orientational coordinates (usually the "tumbling motion") are available. Any rapid orientational motion ("spinning motion") can be readily expressed by an exponential factor (damping coefficient) as long as it does not possess an appreciable potential barrier. Therefore a separation of vibrational and orientational dynamic variables is no longer necessary.[258]

All these worthwhile subjects and topics, together with the previously mentioned significance of vibrational relaxation in regard to reaction dynamics and transition state theory (in our opinion its most important application to chemistry), show that the future of vibrational relaxation is full of promise and excitement. And—as is well known—predictions about the future of a scientific discipline more often than not underestimate its potential and realization.

# NOTES

1. Draw a square coordinate system with coordinate axes $t'$ and $t''$ and enter the diagonal $t' - t'' = \bar{t} = 0$. Note then that integration over the square with elements $dt'\,dt''$ from $t', t'' = 0,0$ to $t,t$ is equivalent to twice integrating over the upper triangle with element $(t - \bar{t})\,d\bar{t}$ from $\bar{t} = 0$ to $\bar{t} = t$.

2. A priori, the dispersion, $\langle |\omega_1(0)|^2 \rangle^c$ (Eq. H.2$b$, Appendix H), and not the mean squared value $\langle |\omega_1(0)|^2 \rangle$ is the primary quantity. However, for ready application and simple meaning, we always set $\langle |\omega_1(0)| \rangle^c = 0$. This is easily accomplished by including $\langle |\omega_1(0)| \rangle$ into the zero-order Hamiltonian ("renormalization"). (See also Section 5.4, Chapter 2.)

3. Multiply the integral of Eq. 3.3 out, recall that

$$\frac{d}{dt} \int_0^t dt\, f(t) = f(t),$$

and note that

$$\hat{\psi}(t)\langle |\omega_1(0)|^2 \rangle = \langle \omega_1(0)\omega_1(t) \rangle.$$

4. For instance, thermal neutrons have a de Broglie wavelength of $\lambda$ approximately equal to $\hbar/2Mk_BT$ approximately 1 Å or $|\mathbf{k}|$ approximately $10^8$ cm$^{-1}$, whereas laser photons have $k$ approximately $10^5$ cm$^{-1}$ ($M$ = neutron mass).

5. To keep a sphere stable against displacements in one direction, three spheres in that direction are needed to block it. With three more spheres in the opposite direction, the average coordination number in the packing is 6.0. (See "Statistical geometrical approach to random packing density of equal spheres," K. Gotoh and J. L. Finney, *Nature* **252**, 202 (1974).)

6. We have not considered here that the 923-cm$^{-1}$ band for the solid-state phases of the neopentane is inhomogeneously broadened by factor and site group splittings (Section 3.3, Chapter 2). Decomposition of the

multiple-oscillator profile must be performed if quantitative information on the orientational and vibrational dynamics is desired (Section 2.4, this chapter).

7. The coefficient $\gamma^{1/2}$ corresponds to the value of the average (over $H_0$) of the normal coordinate $Q$ (cm $g^{1/2}$) between the ground and first excited vibrational levels. If $Q$ is a displacement coordinate [cm], $\gamma = \hbar/2\alpha\mu_M$ (see Ref. 175).

8. The root-mean-squared average frequency shift $\langle|\omega_1(0)|^2\rangle^{1/2}$ of Eq. 3.11 arises from any intermolecular potential $\langle|\omega_1(0)|^2\rangle^{1/2} = \hbar^{-1}\langle|H_1(0)|^2\rangle^{1/2}$. Modulation time $\tau_c$ is the correlation time of $\psi(t)$ (Eq. 3.10), whereas $\tau_R^{(j)}$ is, specifically, an orientational correlation time of a tensor of rank $j$.

9. We use the concept of "bath" (heat bath) and "lattice" (an expression introduced by nuclear magnetic relaxation) interchangeably. In a condensed phase, little is gained in assigning, for instance, "bath" to the average rotatory-translatory motions of the individual molecules and "lattice" to the collective translatory-rotatory motions in the equilibrium ensemble (phonons); any difference is immaterial since we do not generally care about such *detailed* dynamics of the equilibrium system. On the other hand, in situations where the bath is represented by molecular (internal) oscillators or, indeed, by condensed-phase low-frequency phonons, the corresponding wording in the text will bring out this difference.

10. Since we do not make explicit use of $\langle R_{a_N a_N} A_{a_M a_M}\rangle$ and $\langle R'_{a_N a_N a_M a_M}\rangle$ elsewhere, we have not written out their (lengthy) expressions.

11. Differentiating Eq. 3.195 with respect to $t$ ($t_0 = 0$) gives

$$\langle\dot{E}_N\rangle = \left\{-\left(\Gamma_N^{(1)} - \Gamma_N^{(2)}\right)\langle\bar{E}_N\rangle + \Gamma_N^{(2)} + \langle\dot{\bar{E}}_N\rangle\right\}\exp\left[-\left(\Gamma_N^{(1)} - \Gamma_N^{(2)}\right)t\right].$$

In Eq. 3.193, replace $\langle\dot{E}_N\rangle$ by this result and substitute $\langle E_N\rangle, \langle E_M\rangle$ by Eq. 3.195. Solving for $\langle\bar{E}_N\rangle$, keeping in mind Eqs. 3.194, leads to Eq. 3.196.

12. Write Eq. 3.193 for $M$, set $\langle\dot{E}_M\rangle = 0$ (equilibrium), and find

$$\langle E_M\rangle_{\text{equil}} = \Gamma_N^{(2)}\left\{\Gamma_N^{(1)} - \Gamma_N^{(2)}\right\}^{-1}.$$

Inserting this result into Eq. 3.199 gives Eq. 3.200.

13. Let the Boson operator $B_N$ operate on state $|a_N\rangle$,

$$B_N|a_N\rangle = a_N^{1/2}|a_N - 1\rangle.$$

Multiplying from the right by $\langle a_N|$, summing over all states, and taking the ensemble average gives

$$\langle\langle B_N\rangle\rangle = \sum_{a_N} a_N^{1/2}\text{Tr}\,\rho|a_N - 1\rangle\langle a_N|$$

because of the closure condition, $\sum_{a_N} |a_N\rangle\langle a_N| = 1$. Therefore

$$\langle\langle B_N(t)\rangle\rangle = \sum_{a_N} a_N^{1/2} \mathrm{Tr}\rho\langle A_{a_N-1a_N}\rangle$$

$$= \sum_{a_N} (a_N + 1)^{1/2} \mathrm{Tr}\rho\langle A_{a_N a_N+1}\rangle,$$

where $\rho$ is the density matrix over the oscillator system, $\langle A_{a_N a_N+1}\rangle$ is the noise-averaged system operator (see also Eq. 3.166).

14. Under conditions of vanishing $N$-$M$ interactions,

$$V_{NM}^{(i)} \to 0,$$

$$|\langle\alpha|D_{NM}|\beta\rangle|^2 \ll |\langle\alpha|D_N|\beta\rangle|^2,$$

$$\eta_{NM} \to 0, C_{NM} \to 0.$$

15. From Formula 8.254 of Ref. 252 and the definition of the $\Gamma$ function of argument $x$,

$$\Gamma(x) = \int_0^\infty dy \exp(-y) y^{x-1},$$

we get, using Formulas 3.361(2) and 3.371 of Ref. 252,

$$\Phi(\sqrt{x}) = 1 - \left(\frac{1}{\pi}\right)^{1/2} \exp(-x)\left\{x^{-1/2} - \frac{1}{2}x^{-3/2} + -\right\}.$$

Inserting this into $G(t)$ of Eq. 3.215 and setting $x = 1/\bar{\gamma}t$ with $|\bar{\gamma}|t \ll 1$ gives the upper relation for $\langle\langle B\rangle\rangle$ in Eq. 3.216.

For the condition $|\bar{\gamma}|t \gg 1$, we need the error function for small arguments, Formula 8.253(1) of Ref. 252,

$$\Phi([\bar{\gamma}t]^{-1/2}) = \frac{2}{\sqrt{\pi}}\left\{(\bar{\gamma}t)^{-1/2} - \frac{1}{3}(\bar{\gamma}t)^{-3/2} + -\right\}.$$

Inserting this value into $G(t)$ of Eq. 3.215 and approximating its exponential by $1 - (\bar{\gamma}t)^{-1}$ yields the lower relation for $\langle\langle B\rangle\rangle$ in Eq. 3.216.

# APPENDIX A

# SYMMETRIZED CORRELATION FUNCTION, PRINCIPLE OF DETAILED BALANCE, TIME REVERSAL PROPERTIES

We give here frequently encountered forms of autocorrelation functions and some important relations among them.

We first ask: How are the Fourier transforms of $\langle A(t)A(0)\rangle$ and $\langle A(0)A(t)\rangle$ related? To see this, we rewrite Eq. 1.27 of Chapter 1 ($\hbar = 1$, $\beta = \hbar/kT$)

$$\mathcal{I}(\omega) = \int_{-\infty}^{\infty} dt \exp(i\omega t) \sum_{nm} (m|A|n)$$
$$\times (n|\exp(iHt)A\exp(-iHt)|m)\rho_m\exp(\beta\omega_{mn})$$

by commuting the matrix elements and using $\rho_m = \rho_n\exp(-\beta\omega_{mn})$. Again, the delta function permits us to take the exponential out of the integrand (Eq. 1.24):

$$\mathcal{I}(\omega) = \exp(\beta\omega) \int_{-\infty}^{\infty} dt \exp(i\omega t)\langle A(0)A(t)\rangle. \qquad (A.1)$$

Now we replace $t$ by $-t$ and use the time-shift invariance (Eq. 1.18a). Hence

$$\mathcal{I}(\omega) = \exp(\beta\omega) \int_{-\infty}^{\infty} dt \exp(-i\omega t)\langle A(0)A(-t)\rangle$$
$$= \exp(\beta\omega) \int_{-\infty}^{\infty} dt \exp(-i\omega t)\langle A(t)A(0)\rangle \qquad (A.2)$$

353

(shift by $+t$). Comparison of the last equality with Eq. 1.27 shows that

$$\exp(\beta\omega)\int_{-\infty}^{\infty} dt\exp(-i\omega t)\langle A(t)A(0)\rangle = \int_{-\infty}^{\infty} dt\exp(i\omega t)\langle A(t)A(0)\rangle,$$

(A.3)

or

$$\mathscr{I}(-\omega) = \exp\left(-\frac{\hbar\omega}{kT}\right)\mathscr{I}(\omega).$$

Equation A.3 describes a quantal effect: For $\hbar \to 0$, $\mathscr{I}(-\omega) = \mathscr{I}(\omega)$: The band contour is symmetric.

The general idea behind Eq. A.3 is the "principle of detailed balance": We discuss it in more detail further into our discussions.

Next, we describe relations between the so-called *symmetrized correlation function*

$$\tfrac{1}{2}\langle A(0)A(t) + A(t)A(0)\rangle \equiv \langle [A(0), A(t)]_+\rangle \tag{A.4}$$

and the commutator correlation function

$$\langle [A(0), A(t)]\rangle = \langle A(0)A(t) - A(t)A(0)\rangle.$$

The development of $\langle [A(t), A(0)]_+\rangle$ in terms of the matrix elements of eigenstates of $H$ is readily written down. We merely replace in the corresponding development of $\langle [A(t), A(0)]\rangle$ (Section 1.4 of Chapter 1) the factor $1 - \exp(-\beta\omega)$ by $1 + \exp(-\beta\omega)$:

$$\langle [A(t), A(0)]\rangle$$

$$= \sum_{nm} \rho_n [1 - \exp(-\beta\omega_{mn})](n|A(0)|m)(m|A(0)|n)\exp(i\omega_{nm}t),$$

$$\langle [A(t), A(0)]_+\rangle$$

$$= \frac{1}{2}\sum_{nm} \rho_n [1 + \exp(-\beta\omega_{mn})](n|A(0)|m)(m|A(0)|n)\exp(i\omega_{nm}t).$$

(A.5)

Referring to Eq. 1.22 (Chapter 1), we see that

$$\chi''(\omega) = \left\{\frac{1 - \exp(-\beta\omega)}{1 + \exp(-\beta\omega)}\right\}\int_{-\infty}^{\infty} dt\exp(i\omega t)\langle [A(t), A(0)]_+\rangle$$

$$= \tanh\left(-\frac{\hbar\omega}{2kT}\right)\int_{-\infty}^{\infty} dt\exp(i\omega t)\langle [A(t), A(0)]_+\rangle. \tag{A.6}$$

Now we give some useful theorems. First, we require that

$$\mathscr{I}(\omega) = \mathscr{I}^*(\omega)$$

since the spectrum is real, but $\langle A(t)A(0)\rangle$ is, a priori, complex.

$$\int_{-\infty}^{\infty} dt \exp(i\omega t)\langle A(t)A(0)\rangle = \int_{-\infty}^{\infty} dt \exp(-i\omega t)\langle A(t)A(0)\rangle^*$$

(replace $t$ by $-t$ on the right-hand side)

$$= \int_{-\infty}^{\infty} dt \exp(i\omega t)\langle A(-t)A(0)\rangle^*$$

(shift time axis by $+t$ under time-shift invariance)

$$= \int_{-\infty}^{\infty} dt \exp(i\omega t)\langle A(0)A(t)\rangle^*. \tag{A.7}$$

Consequently,

$$\langle A(t)A(0)\rangle = \langle A(-t)A(0)\rangle^*$$
$$= \langle A(0)A(t)\rangle^*. \tag{A.8}$$

This means $\langle A(0)A(t)\rangle$ is Hermitian.

Second, let us write the complex correlation function as a sum of its real and imaginary parts,

$$\langle A(t)A(0)\rangle = X(t) + iY(t).$$

Inserting this and the corresponding expression $\exp(\pm i\omega t) = \cos(\omega t) \pm i\sin(\omega t)$ into the first and third relation of Eqs. A.7, we obtain (Re = real, Im = imaginary)

$$\mathrm{Re}\langle A(-t)A(0)\rangle = \mathrm{Re}\langle A(t)A(0)\rangle$$
$$\mathrm{Im}\langle A(-t)A(0)\rangle = -\mathrm{Im}\langle A(t)A(0)\rangle. \tag{A.9a}$$

In other words, the real (imaginary) part of $\langle A(t)A(0)\rangle$ is an even (odd) function of time.

Next we establish the time reversal property of the symmetrized product. We find

$$\langle [A(-t), A(0)]_+\rangle = \tfrac{1}{2}\{\langle A(-t)A(0)\rangle + \langle A(0)A(-t)\rangle\}$$
$$= \tfrac{1}{2}\{\langle A(0)A(t)\rangle + \langle A(t)A(0)\rangle\}$$
$$= \langle [A(t), A(0)]_+\rangle$$
$$= \tfrac{1}{2}\{\langle A(t)A(0)\rangle^* + \langle A(0)A(t)\rangle^*\}$$
$$= \langle [A(t), A(0)]_+\rangle^*. \tag{A.9b}$$

The symmetrized correlation function is an even and real function of time. Hence

$$\text{Re}\langle A(0)A(t)\rangle = \langle [A(0), A(t)]_{+}\rangle. \tag{A.10}$$

Similarly,

$$\text{Im}\langle A(t)A(0)\rangle = \frac{1}{2i}\langle [A(t), A(0)]\rangle. \tag{A.11}$$

We shall deal with many systems in which the energy difference between levels involved in the transitions is considerably smaller than $kT$. We consider such systems to behave "classically." For instance, nearly all pure rotational spectra at room temperature will fall into this category. Now, if $\hbar\omega_{mn} \ll kT$, the factor $\exp(-\beta\omega_{mn})$ in Eqs. A.6 can be approximated by the linear term of its infinite series expansion. Then we readily obtain the pair

$$\chi''(\omega) \approx \frac{\hbar\omega}{2kT}\int_{-\infty}^{\infty} dt\exp(i\omega t)\langle [A(t), A(0)]_{+}\rangle \tag{A.12}$$

and

$$\chi''(\omega) \approx \frac{\hbar\omega}{2kT}\int_{-\infty}^{\infty} dt\exp(i\omega t)\langle A(t)A(0)\rangle$$

$$= \frac{\hbar\omega}{2kT}\int_{-\infty}^{\infty} dt\exp(-i\omega t)\langle A(0)A(t)\rangle \tag{A.13}$$

instead of Eq. 1.22, Chapter 1. Note that under the condition $\hbar\omega \ll kT$, Eqs. A.12 and A.13 are equivalent,

$$\langle [A(t), A(0)]_{+}\rangle \approx \langle A(0)A(t)\rangle^{*} \approx \langle A(0)A(t)\rangle.$$

As a consequence, the spectral density $\mathcal{I}(\omega)$ or the susceptibility $\chi''(\omega)$ should be near symmetric (see Eq. A.3). If not, $\langle A(t)A(0)\rangle$ is complex (Eq. A.11).

It is instructive to show that the Fourier transform of a complex correlation function leads to a real spectral density—as it should (see above). Write

$$\mathcal{I}(\omega) = \int_{-\infty}^{\infty} dt\,[\cos(\omega t) + i\sin(\omega t)][X(t) + iY(t)], \tag{A.14}$$

by splitting the exponential and the correlation function into their real and purely imaginary parts. Multiplying out gives

$$\mathcal{I}(\omega) = \int_{-\infty}^{\infty} dt\,\{\cos(\omega t)X(t) - \sin(\omega t)Y(t)$$

$$+ i[\cos(\omega t)Y(t) + \sin(\omega t)X(t)]\}. \tag{A.15}$$

Since the purely imaginary terms under the integral are odd functions of the time (see the foregoing time-reversal relation, Eqs. A.9a), they drop out. Hence we are left with a result that is a real quantity.

This shows, for an asymmetric band shape, that

$$\text{Re}\langle A(0)A(t)\rangle \neq \int_{-\infty}^{\infty} d\omega \cos(\omega t) \mathscr{I}(\omega)$$

$$\text{Im}\langle A(0)A(t)\rangle \neq \int_{-\infty}^{\infty} d\omega \sin(\omega t) \mathscr{I}(\omega). \tag{A.16}$$

Since spectral densities of isolated oscillators are usually symmetric, we shall have no opportunity to be greatly concerned about the imaginary part of the correlation function. We shall, almost always, find situations where

$$\text{Re}\langle A(0)A(t)\rangle \gg \text{Im}\langle A(0)A(t)\rangle. \tag{A.17}$$

We compute the real part of the correlation functions by inverting Eq. A.6:

$$\text{Re}\langle A(t)A(0)\rangle = \langle [A(t), A(0)]_+ \rangle$$

$$= \int_{-\infty}^{\infty} d\omega' \chi''(\omega_{mn}^0 + \omega') \left\{ \frac{1 + \exp(-\beta\omega)}{1 - \exp(-\beta\omega)} \right\} \exp(-i\omega' t)$$

$$= \int_{-\infty}^{\infty} d\omega' \mathscr{I}(\omega_{mn}^0 + \omega')[1 + \exp(-\beta\omega)]\exp(-i\omega' t),$$

$$\tag{A.18}$$

where we have set $\omega' = \omega - \omega_{mn}^0$ as the frequency displacement or "shift" with respect to the band center $\omega_{mn}^0$. (Note that the last equality in Eq. A.18 follows from Eq. 1.3 and the definition of the spectral density of Eq. 1.25, Chapter 1.)

Equation A.18 expresses that we have, effectively, taken a symmetrized spectral density $\mathscr{I}_s(\omega)$:

$$\mathscr{I}_s(\omega) = \tfrac{1}{2}\left\{ \mathscr{I}(\omega_{mn}^0 + \omega') + \mathscr{I}(\omega_{mn}^0 - \omega') \right\}$$

$$= \tfrac{1}{2}\left\{ \mathscr{I}(\omega_{mn}^0 + \omega') + \exp(-\beta\omega)\mathscr{I}(\omega_{mn}^0 + \omega') \right\}$$

$$= (\tfrac{1}{2})[1 + \exp(-\beta\omega)]\mathscr{I}(\omega_{mn}^0 + \omega'). \tag{A.19}$$

Hence

$$\text{Re}\langle A(t)A(0)\rangle = \text{Re}\langle A(0)A(t)\rangle$$

$$= 2\int_0^{\infty} d\omega' \cos(\omega' t) \mathscr{I}_s(\omega_{mn}^0 + \omega'). \tag{A.20}$$

Of course, if the spectral density is symmetric, $\mathscr{I}_s(\omega)$ equals $\mathscr{I}(\omega)$.

In summary, we shall take as the quantum-mechanical autocorrelation function the symmetrized product $\langle [A(0), A(t)]_+ \rangle$, which, as we discussed, is real and an even function of $t$. As we have shown, it cannot be identified with the complex quantum-mechanical correlation function $\langle A(0)A(t) \rangle$. Furthermore, we have encountered the classical correlation function, which is also a real and even function of $t$. The classical correlation function also cannot be identified with $\langle A(0)A(t) \rangle$ since it does not obey the condition of detailed balance (see Eq. A.3). The approach of the classical correlation function to the quantum-mechanical correlation function has been of great interest, but this is not within the range of our experimental situations. It is sufficient to say that replacing the time $t$ in the classical correlation function by Egelstaff's complex "$y$-time"

$$ y = \left( t^2 - \frac{i\hbar t}{kT} \right)^{1/2}, $$

leads to a classical correlation function $\langle A(0)A(t) \rangle_{cl}$ which agrees with the quantum-mechanical correlation function $\langle A(0)A(t) \rangle$ up to $t^4$. [B. J. Berne, J. Jortner, and R. G. Gordon, *J. Chem. Phys.* **47**, 1600 (1967); Ref. 14.]

The reader should be aware of but not unduly worried about these differences. We repeat: most observed band profiles are near symmetric. The difference between $\langle A(0)A(t) \rangle$, $\langle A(t)A(0) \rangle$, $\langle A(0)A(t) \rangle_{cl}$, and $\langle [A(0), A(t)]_+ \rangle$ thus remains somewhat academic.

If indeed strongly asymmetric band profiles are observed, we must analyze whether the asymmetry is caused by overlap of several symmetric components of different intensity (hot bands, isotopic species, frequency "splittings" in solution spectra due to a mixture of "complexed" and "free" molecules) or whether the asymmetry is inherent. In the latter case, a modification of the theory is required (see Section 4, Chapter 3).

# APPENDIX B

# TIME DERIVATIVES OF CORRELATION FUNCTIONS AND SHORT-TIME BEHAVIOR OF THE ORIENTATIONAL CORRELATION FUNCTION OF THE *J*-DIFFUSION MODEL

We first show that the correlation function $\langle A(0)A(t)\rangle$ obeys the relations

$$\frac{d}{dt}\langle A(0)A(t)\rangle = -\langle \dot{A}(0)A(t)\rangle \tag{B.1}$$

and

$$\frac{d^2}{dt^2}\langle A(0)A(t)\rangle = -\langle \dot{A}(0)\dot{A}(t)\rangle. \tag{B.2}$$

Employing the Liouville formalism $(d/dt)A = iLA$ (Eq. 2.139, Chapter 2), we write $[A(0) \equiv A]$

$$\frac{d}{dt}\langle A|A(t)\rangle = \frac{d}{dt}\langle A|\exp(iLt)|A\rangle$$

$$= \langle \dot{A}|\exp(iLt)|A\rangle + \langle A|iL\exp(iLt)|A\rangle + \langle A|\exp(iLt)|\dot{A}\rangle$$

$$= \langle iLA|\exp(iLt)|A\rangle - \langle iLA|\exp(iLt)|A\rangle + \langle A|\exp(iLt)|\dot{A}\rangle$$

$$= \langle A(0)\dot{A}(t)\rangle = \langle \dot{A}(t)A(0)\rangle^* = \langle \dot{A}(0)A(-t)\rangle^*$$

**359**

by time-shift invariance (see Appendix A). Hence

$$\langle \dot{A}(0)A(-t)\rangle^* = \langle iLA|\exp(-iLt)|A\rangle^*$$

$$= -\langle iLA|\exp(iLt)|A\rangle = -\langle \dot{A}(0)A(t)\rangle.$$

We note that, as a consequence, a classical correlation function obeys

$$\langle \dot{A}(0)A(t)\rangle|_{t=0} = 0. \tag{B.3}$$

Equation B.2 is obtained in the same manner by differentiating $-\langle \dot{A}(0)A(t)\rangle$.

It is easy to verify that the $J$- and $M$-model correlation functions approach the corresponding free rotor correlation functions for $t \to 0$. We show this for a totally symmetric species ($k = 0$) of an infrared band ($j = 1$) of a linear or symmetric top in the $J$-diffusion limit. From Eq. 2.37 (Chapter 2) we write, up to order $\tau_J^{-2}$,

$$G_J^{(1,0)}(t) = \exp\left(-\frac{t}{\tau_J}\right)\left\{F^{(1,0)}(t) + \tau_J^{-1}\int_0^t dt_1\, F^{(1,0)}(t-t_1)F^{(1,0)}(t_1)\right.$$

$$\left. + \tau_J^{-2}\int_0^t dt_2\, F^{(1,0)}(t-t_2)\int_0^{t_2} dt_1\, F^{(1,0)}(t_2-t_1)F^{(1,0)}(t_1)\right\}.$$

With Eq. 2.4, Chapter 2,

$$F^{(1,0)}(t) = 1 - \left(\frac{k_BT}{I_x}\right)t^2 + O(t^n),$$

we obtain, after some algebra,

$$G(t) = \exp\left(-\frac{t}{\tau_J}\right)\left\{1 - (k_BT/I_x)t^2 + \tau_J^{-1}\left[t - \frac{(2k_BT/I_x)t^3}{3}\right]\right.$$

$$\left. + \tau_J^{-2}\left[\frac{t^2}{2} - \frac{(k_BT/I_x)t^4}{4}\right]\right\}.$$

Differentiation leads to (see also Eq. B.3)

$$\frac{d}{dt}G(t)|_{t=0} = 0, \tag{B.4}$$

$$\frac{d^2}{dt^2}G(t)|_{t=0} = -2\left(\frac{k_BT}{I_x}\right). \tag{B.5}$$

Therefore the memory function formalism leads to the correct short-time behavior of $G(t)$.

Incidentally, if the series development of $G(t)$ is broken off with the term $\tau_J^{-1}$, Eq. B.5 reads[64]

$$\frac{d^2}{dt^2}G(t)|_{t=0} = -2\left(\frac{k_BT}{I_x}\right) - \tau_J^{-2}.$$

Usually the first term exceeds the second by orders of magnitude.

# APPENDIX C

# SYMMETRY-BASED DEVELOPMENT OF THE TRANSITION KERNEL FOR NONINERTIAL ORIENTATIONAL MOTION

In this appendix we outline the symmetry-based development of the transition kernel $A(\Omega, \Omega')$ for noninertial motion introduced with Eq. 2.55 of Chapter 2.[38, 71a] Symmetry enters in two ways: first, if we assume isotropy in the medium. This means that the *absolute* orientations of the space- and molecule-fixed coordinate systems are immaterial. Second, symmetry enters if we have molecular symmetry. This means we consider here—in particular—the simplifications arising from symmetric-top point group symmetry (at least a threefold rotation axis). This generates certain equivalent orientations of the body—hence the transition rate kernel is invariant to the corresponding symmetry operations (rotations).

We assume isotropy and develop the transition kernel $A(\Omega - \Omega')$ (see Eq. 2.56) as a series in terms of Wigner rotation matrices $\mathbf{D}(\Omega)$. This is convenient since the $\mathbf{D}(\Omega - \Omega')$ determine the relative orientation of the body with respect to the space-fixed coordinate system:

$$A(\Omega - \Omega') = \sum_{j=0}^{\infty} \sum_{m=-j}^{j} \sum_{n=-j}^{j} c_{nm}^{(j)} D_{mn}^{(j)}(\Omega - \Omega').$$

This can be written in terms of the individual angles (see Eq. 1.90, Chapter 1)

$$D_{mn}^{(j)}(\Omega - \Omega') = \sum_{m'} D_{mm'}^{(j)}(-\Omega') D_{m'n}^{(j)}(\Omega),$$

$$A(\Omega - \Omega') = \sum_{jmnm'} c_{nm}^{(j)} D_{mm'}^{(j)}(-\Omega') D_{m'n}^{(j)}(\Omega)$$

$$= \sum_{jmnm'} c_{nm}^{(j)} D_{m'm}^{(j)*}(\Omega') D_{m'n}^{(j)}(\Omega)$$

$$= \sum_{j=0}^{\infty} \text{Tr}\big[\mathbf{D}^{(j)}(\Omega)\mathbf{c}^{(j)}\mathbf{D}^{(j)}(-\Omega')\big],$$

where the invariance of the trace operation to cyclic permutation has been used. Therefore

$$\int d\Omega\, A(\Omega - \Omega') D_{pq}^{(j)*}(\Omega) = \left[\frac{8\pi^2}{2j+1}\right] \sum_{k} c_{qk}^{(j)} D_{pk}^{(j)*}(\Omega'), \tag{C.1}$$

using the orthogonality of the $D$ (see Eq. 4.60 of Ref. 37). We see that the right-hand side of Eq. C.1 involves only one index of the $D$. Hence, for fixed $j$ and $m$, the $i$th eigenfunction ($i = -j, -j+1,\ldots,j$) of $A(\Omega - \Omega')$ can be written

$$^{(i)}\psi_m^{(j)}(\Omega) = \sum_n {}^{(i)}b_{nm}^{(j)} D_{mn}^{(j)*}(\Omega). \tag{C.2}$$

Then

$$\int d\Omega\, A(\Omega - \Omega')^{(i)}\psi_m^{(j)}(\Omega) = {}^{(i)}\lambda_m^{(j)}\,{}^{(i)}\psi_m^{(j)}(\Omega'), \tag{C.3}$$

with eigenvalue $^{(i)}\lambda_m^{(j)}$. Substituting Eq. C.2 into Eq. C.3 and comparing the result with Eq. C.1 gives a relation between the expansion coefficients $c$, the eigenfunction expansion coefficients $b$, and the eigenvalues $\lambda$,

$$^{(i)}\lambda_m^{(j)} \sum_n {}^{(i)}b_{nm}^{(j)} D_{mn}^{(j)*}(\Omega') = {}^{(i)}\lambda_m^{(j)} \sum_k {}^{(i)}b_{km}^{(j)} D_{mk}^{(j)*}(\Omega')$$

$$= \frac{8\pi^2}{2j+1} \sum_{kn} {}^{(i)}b_{nm}^{(j)} c_{nk}^{(j)} D_{mk}^{(j)*}(\Omega'). \tag{C.4}$$

The Wigner rotation matrices are linearly independent. Therefore Eq. C.4 holds for each $k$. It follows that

$$\frac{8\pi^2}{2j+1} \sum_n {}^{(i)}b_{nm}^{(j)} c_{nk}^{(j)} = {}^{(i)}\lambda_m^{(j)}\,{}^{(i)}b_{km}^{(j)}. \tag{C.5}$$

For $j$ fixed, notice that $^{(i)}b_{nm}^{(j)}$ is the $n$th component of the $i$th left eigenvector of matrix $\mathbf{c}^{(j)}$ with eigenvalue $[(2j+1)/8\pi^2]^{(i)}\lambda_m^{(j)}$ — independent of index $m$. For each $j$ there are then, at most, $2j+1$ distinct eigenvalues of $A(\Omega - \Omega')$. Hence Eq. C.3 is

$$\int d\Omega\, A(\Omega - \Omega')^{(i)}\psi_m^{(j)}(\Omega) = {}^{(i)}\lambda^{(j)\,(i)}\psi_m^{(j)}(\Omega'). \qquad (C.6)$$

Further simplification of the $\psi$ depends on the symmetry of the rotating body. For a symmetric top, each matrix $\mathbf{c}^{(j)}$ is diagonal.[38] Consequently each normalized $^{(i)}\psi_m^{(j)}(\Omega)$ consists of only one rotation matrix (see Eq. C.2),

$$^{(i)}\psi_m^{(j)}(\Omega) = {}^{(i)}b_m^{(j)}D_{mi}^{(j)*}(\Omega) = \left[\frac{2j+1}{8\pi^2}\right]^{1/2} D_{mi}^{(j)*}(\Omega), \qquad (C.7)$$

with the $i$th eigenvalue and expansion coefficient (Eq. C.5)

$$^{(i)}\lambda^{(j)} = \left(\frac{8\pi^2}{2j+1}\right)c_{ii}^{(j)},$$

$$^{(i)}b_m^{(j)} = \left(\frac{2j+1}{8\pi^2}\right)^{1/2}\delta_{mi}. \qquad (C.8)$$

Since

$$c_{nm}^{(j)} = (-1)^{m-n}c_{-m-n}^{(j)},$$

there are only $j+1$ eigenvalues for the symmetric top.

For spherical point group symmetry, $\mathbf{c}^{(j)}$ is independent of $m$; there is only one eigenvalue $\lambda^{(j)}$.

Using the eigenfunctions and eigenvalues of $A(\Omega - \Omega')$, we write (see Eq. 2.57, Chapter 2)

$$P(\Omega, t; \Omega_0) = \sum_{jim} {}^{(i)}\psi_m^{(j)*}(\Omega_0)\,{}^{(i)}\psi_m^{(j)}(\Omega)\exp\left({}^{(i)}\lambda_m^{(j)}t\right), \qquad (C.9)$$

with $P(\Omega, 0; \Omega_0) = \delta(\Omega - \Omega_0)$ as the initial condition. For the symmetric top, substitution of $^{(i)}\psi_m^{(j)}(\Omega)$ by Eq. C.7 and of $^{(i)}\lambda^{(j)}$ by Eq. C.8 gives

$$P(\Omega, t; \Omega_0) = \sum_j^\infty \frac{2j+1}{8\pi^2}\sum_{im}D_{mi}^{(j)*}(\Omega)D_{mi}^{(j)}(\Omega_0)\exp\left[\frac{8\pi^2 c_{ii}^{(j)}t}{2j+1}\right]. \qquad (C.10)$$

The resulting orientational correlation function is (see Eq. 2.58$a$, Chapter 2)

$$\mathscr{D}_{nm,n'm'}^{(jj')}(t) = \langle D_{nm}^{(j)*}(\Omega(0))D_{n'm'}^{(j')}(\Omega(t))\rangle$$

$$= \int d\Omega_0\, P(\Omega_0)D_{nm}^{(j)*}(\Omega_0)\int d\Omega\, P(\Omega,t;\Omega_0)D_{n'm'}^{(j')}(\Omega)$$

$$= (2j+1)^{-1}\delta_{jj'}\delta_{mm'}\delta_{nn'}\exp(^{(m)}\lambda^{(j)}t). \qquad (C.11)$$

The last equality in Eq. C.11 follows from the orthogonality relation of the **D** with $\int d\Omega_0 = 8\pi^2$.

Our choice of $A(\Omega)$, hence of matrix **c**, determines the particular model of noninertial orientational motion in the (macroscopically) isotropic medium. For instance, the eigenvalues for the Debye model are given by Eq. 2.47 (Chapter 2) for the spherical and by Eq. 2.48 for the symmetric top.

# FREE ROTOR CORRELATION FUNCTION WITH CORIOLIS COUPLING

We derive here the free rotor infrared correlation function of the spherical top with Coriolis coupling.[79]

The expressions for the rotational Hamiltonian and energy of a spherical molecule in the presence of Coriolis coupling with coupling constant $\zeta$ are

$$H = \frac{(\mathbf{J} - \boldsymbol{\kappa})^2}{2I} = \frac{\mathbf{J}^2}{2I} - \frac{\mathbf{J} \cdot \boldsymbol{\kappa}}{I} + \frac{\boldsymbol{\kappa}^2}{2I}$$

$$E = \frac{\hbar^2}{2I}\left[ J(J+1) + \kappa(\kappa+1) \right] - \frac{\zeta}{I}(\mathbf{J} \cdot \boldsymbol{\kappa}). \tag{D.1}$$

The second term of $H$ represents the Coriolis coupling of the vibrational angular momentum $\boldsymbol{\kappa}$ with the total angular momentum $\mathbf{J}$. The third term is combined with the pure vibrational energy term $\hbar\omega(v + \frac{1}{2})$ and is of no further interest. Vector $\boldsymbol{\kappa}$, which has the eigenvalues $\hbar l \zeta$ with $l = v, v - 2, \ldots, -v$, is defined in the molecule-fixed coordinate system; Eq. D.1 shows that $\boldsymbol{\kappa}$ rotates around the total angular momentum $\mathbf{J}$. This implies that $\boldsymbol{\kappa}$ rotates with the same angular frequency as the rotational transition moment $\hat{\mathbf{m}}(t)$. Hence we can write the *orientational* correlation function

$$\langle \boldsymbol{\kappa}(0) \cdot \boldsymbol{\kappa}(t) \rangle \langle \boldsymbol{\kappa}(0) \cdot \boldsymbol{\kappa}(0) \rangle^{-1} = \langle \hat{\mathbf{m}}(0) \cdot \hat{\mathbf{m}}(t) \rangle. \tag{D.2}$$

The classical equation of motion in the molecule-fixed frame is

$$\frac{d\boldsymbol{\kappa}(t)}{dt} = -\frac{\zeta}{I}[\mathbf{J}(t) \times \boldsymbol{\kappa}(t)], \tag{D.3}$$

where the right-hand side of Eq. D.3 represents the torque exerted on the molecule by the Coriolis force. Changing to a laboratory-fixed frame of reference adds the term (see Eq. 4-102 in H. Goldstein, *Classical Mechanics*, Addison-Wesley, Mass., 1953)

$$\frac{d\kappa(t)}{dt} = \omega \times \kappa(t) = \frac{\mathbf{J}}{I} \times \kappa(t).$$

Hence

$$\frac{d\kappa(t)}{dt} = \frac{1-\zeta}{I}\left[\mathbf{J}(t) \times \kappa(t)\right]. \tag{D.4}$$

For the free molecule, $\mathbf{J}(t) = $ constant. We let $\mathbf{J}$ fall along the $z$ axis of the molecule frame: $J_z = J$. Solving the axial vector product of Eq. D.4 for all three components of $\kappa$, remembering that $J_x = J_y = 0$, gives for the $z$ component

$$\frac{d\kappa_z(t)}{dt} = 0$$

since there is no torque reorienting direction $z$. For the $x$, $y$ components, we use complex notation, obtaining

$$\frac{d\kappa_\pm(t)}{dt} = -i\omega_0\kappa_\pm(t)$$

with $(1-\zeta)J/I \equiv \omega_0$ and $\kappa_\pm = \kappa_x \pm i\kappa_y$. Integration gives

$$\kappa_z(t) = \kappa_z(0); \qquad \kappa_\pm(t) = \kappa_\pm(0)\exp(\mp i\omega_0 t)$$

and the correlation functions

$$\langle \kappa_+(0)\kappa_+(t)\rangle \langle |\kappa_+(0)|^2\rangle^{-1} = \exp(-i\omega_0 t),$$

$$\langle \kappa_z(0)\kappa_z(t)\rangle \langle |\kappa_z(0)|^2\rangle^{-1} = 1. \tag{D.5}$$

Equation D.2 is therefore

$$\langle \kappa(0) \cdot \kappa(t)\rangle = \langle \kappa_x(0)\kappa_x(t) + \kappa_y(0)\kappa_y(t) + \kappa_z(0)\kappa_z(t)\rangle$$

$$= \langle \tfrac{1}{2}\kappa_-(0)\kappa_+(t) + \tfrac{1}{2}\kappa_+(0)\kappa_-(t) + \kappa_z(0)\kappa_z(t)\rangle_{J,\theta,\varphi} \tag{D.6}$$

using $\kappa_x = (\tfrac{1}{2})(\kappa_+ + \kappa_-)$ and $\kappa_y = (\tfrac{1}{2})(\kappa_+ - \kappa_-)/i$.

Now we average over all original orientations $\theta, \varphi$. Notice that the linear combinations $\kappa_\pm$ and $\kappa_z$ are proportional to the spherical harmonics $Y^1_{\pm 1}(\theta, \varphi)$ and $Y^1_0(\theta, \varphi)$, respectively (see Ref. 39, p. 124). Hence

$$\kappa_z(0) \propto \cos\theta; \qquad \kappa_\pm(0) \propto \sin\theta\exp(\pm i\varphi).$$

Insertion into Eq. D.6 and remembering Eq. D.5 gives

$$\langle \boldsymbol{\kappa}(0)\cdot\boldsymbol{\kappa}(t)\rangle\langle|\boldsymbol{\kappa}(0)|^2\rangle^{-1} = \langle \hat{\boldsymbol{\kappa}}(0)\cdot\hat{\boldsymbol{\kappa}}(t)\rangle$$

$$= \langle \tfrac{1}{2}\sin^2\theta\exp(-i\omega_0 t) + \tfrac{1}{2}\sin^2\theta\exp(i\omega_0 t) + \cos^2\theta\rangle_{J,\theta,\varphi}. \quad (D.7)$$

Integrating with the isotropic distribution function

$$(4\pi)^{-1}\int_0^\pi\int_0^{2\pi}\sin\theta\,d\theta\,d\varphi$$

gives

$$\langle \hat{\boldsymbol{\kappa}}(0)\cdot\hat{\boldsymbol{\kappa}}(t)\rangle_J = \tfrac{1}{3}\exp(-i\omega_0 t) + \tfrac{1}{3}\exp(i\omega_0 t) + \tfrac{1}{3} = \tfrac{2}{3}\cos(\omega_0 t) + \tfrac{1}{3}.$$

Finally we average over all $J$ with the normalized distribution function $(2/\pi)^{1/2}J^2(k_BIT)^{-3/2}\exp(-J^2/2k_BIT)$, neglecting the contributions from $\kappa$ (see Eq. 2.39, Chapter 2, for $\xi \to 0$, or see Ref. 78):

$$\langle \hat{\boldsymbol{\kappa}}(0)\cdot\hat{\boldsymbol{\kappa}}(t)\rangle = \frac{2}{3}\left(1 - \tilde{\tilde{t}}^2\right)\exp\left(-\frac{\tilde{\tilde{t}}^2}{2}\right) + \frac{1}{3},$$

$$\tilde{\tilde{t}} \equiv \left(\frac{k_BT}{I}\right)^{1/2}(1-\zeta)t \equiv (1-\zeta)t^*. \quad (D.8)$$

The negative value of the second derivative of Eq. D.8, taken at $t = 0$, gives the classical theoretical value of the rotational infrared second spectral moment of a spherical top molecule in the presence of Coriolis interaction (see Sections 4, Chapter 2):

$$M(2) = \frac{2k_BT}{I}(1-\zeta)^2. \quad (D.9)$$

(This discussion has been taken from the doctoral dissertation of K. Müller, Eidgenössische Hochschule Zürich, No. 5378 (1974), in particular from his private communication of July 9, 1973.)

# APPENDIX E

# CLASSICAL ROTATIONAL SECOND SPECTRAL MOMENTS FOR FIRST- AND SECOND-ORDER TENSORS

## 1. SYMMETRIC TOP MOLECULES

### 1.1. Vanishing Coriolis Interaction

We derive here the general formulation for the second spectral rotational moment $M(2)$ for first-order (infrared and far-infrared) and second-order (Raman, Rayleigh) rotational transition moments. We use the relation

$$M(2)^{(j,k)} = -\frac{d^2}{dt^2}F^{(j,k)}(t)\bigg]_{t=0},$$

where $F^{(j,k)}$ is the classical free rotor correlation function of a tensor of order (or rank) $j$ and component $k$ (see Eq. 2.72 of Chapter 2). The derivations are restricted to symmetric top molecules.

$F^{(j,k)}(t)$ is given by Eq. 2.41 or 2.42, noting Eq. 2.37,

$$F^{(j,k)}(t^*) = \int_0^\infty dL \int_{-1}^1 dx \left(\frac{1+\xi}{2\pi}\right)^{1/2} L^2 \exp\left[-\frac{1}{2}L^2(1+\xi x^2)\right]$$

$$\times \sum_{a=-j}^{j} \left[d_{ka}^{(j)}(x)\right]^2 \exp\left[-iL(a+k\xi x)t^*\right]. \tag{E.1}$$

Parameters are in reduced units: $L = J/(I_x k_B T)^{1/2}$, $x = \cos\theta$, $t^* = t/(I_x/k_B T)^{1/2}$, $\omega^* = \omega/(k_B T/I_x)^{1/2}$, and $\xi = (I_x/I_z)-1$. The coordinates $x$, $y$, and $z$ denote the inertial axes of the molecule; direction $z$ is taken along the $C_3$ symmetry axis. Differentiating twice with respect to $t^*$ and setting, in

**371**

the result, $t^* = 0$, we get

$$-\frac{d^2}{dt^2} F^{(j,k)}(t^*)\Big|_{t=0} = \left(\frac{1+\xi}{2\pi}\right)^{1/2} \int_0^\infty dL \int_{-1}^1 dx\, L^4 \exp\left\{-\frac{1}{2} L^2(1+\xi x^2)\right\}$$

$$\times \sum_{a=-j}^{j} \left[d_{ka}^{(j)}(x)\right]^2 (a+k\xi x)^2. \qquad (E.2)$$

The integral over $L$ is of the well-known type

$$\int_0^\infty y^{2n} e^{-by^2}\, dy = \frac{1\cdot 3\cdot 5\cdots(2n-1)}{2^{n+1} b^n}\left(\frac{\pi}{b}\right)^{1/2}.$$

Hence in units of $k_B T/I_x$

$$M(2)^{(j,k)} = (3/2)(1+\xi)^{1/2} \int_{-1}^1 dx\, (1+\xi x^2)^{-5/2} \sum_{a=-j}^{j} \left[d_{ka}^{(j)}(x)\right]^2 (a+k\xi x)^2,$$

$$(E.3)$$

where the $d_{ka}^{(j)}$ are defined by

$$d_{m'm}^{(j)}(\beta) = \left( jm' |\exp(-i\beta J_y)| jm \right) \qquad (E.4a)$$

in terms of the rotation matrix elements in the three Euler angles $\alpha, \beta, \gamma$:

$$D_{m'm}^{(j)}(\alpha, \beta, \gamma) = \exp(-im'\alpha)\left( jm' |\exp(-i\beta J_y)| jm \right)\exp(-im\gamma). \qquad (E.4b)$$

The matrix elements follow. (Note that the phase relations in Eqs. E.4a and E.4b are opposite in Refs. 37 and 39. This has no bearing on the results here.)

With the help of these elements, we can compute in Eq. E.3 the sum over index $a$ from $-j$ to $j$ with the desired index $k$ (0, 1 or 2). For instance, the sum amounts, for $j = 2$, $k = 2$, to

$$\sum_{-a}^{a} \left[d_{2,a}^{(2)}(x)\right]^2 (a+2\xi x)^2$$

$$= \frac{1}{16}(-2+2\xi x)^2(1-x)^4 + \frac{1}{4}(-1+2\xi x)^2(1-x^2)(1-x)^2 + \frac{3}{8}(2\xi x)^2$$

$$\times (1-x^2)^2 + \frac{1}{4}(1+2\xi x)^2(1-x^2)(1+x)^2 + \frac{1}{16}(2+2\xi x)^2(1+x)^4,$$

$$(E.5)$$

**Rotation Matrix $D_{m'm}^{(j)}(0,\cos\theta,0)=d_{m'm}^{(j)}(x)$ for $j=1,2$**

$j=1$

|  | 1 | 0 | $-1$ |
|---|---|---|---|
| 1 | $\frac{1}{2}(1+x)$ | $\frac{1}{\sqrt{2}}(1-x^2)^{1/2}$ | $\frac{1}{2}(1-x)$ |
| 0 | $-\frac{1}{\sqrt{2}}(1-x^2)^{1/2}$ | $x$ | $\frac{1}{\sqrt{2}}(1-x^2)^{1/2}$ |
| $-1$ | $\frac{1}{2}(1-x)$ | $-\frac{1}{\sqrt{2}}(1-x^2)^{1/2}$ | $\frac{1}{2}(1+x)$ |

$j=2$

|  | 2 | 1 | 0 | $-1$ | $-2$ |
|---|---|---|---|---|---|
| 2 | $\frac{1}{4}(1+x)^2$ | $\frac{1}{2}(1+x)(1-x^2)^{1/2}$ | $\frac{\sqrt{6}}{4}(1-x^2)$ | $\frac{1}{2}(1-x)(1-x^2)^{1/2}$ | $\frac{1}{4}(1-x)^2$ |
| 1 | $-\frac{1}{2}(1+x)(1-x^2)^{1/2}$ | $\frac{1}{2}(2x^2+x-1)$ | $\frac{\sqrt{6}}{2}x(1-x^2)^{1/2}$ | $-\frac{1}{2}(2x^2-x-1)$ | $\frac{1}{2}(1-x)(1-x^2)^{1/2}$ |
| 0 | $\frac{\sqrt{6}}{4}(1-x^2)$ | $-\frac{\sqrt{6}}{2}x(1-x^2)^{1/2}$ | $\frac{1}{2}(3x^2-1)$ | $\frac{\sqrt{6}}{2}x(1-x^2)^{1/2}$ | $\frac{\sqrt{6}}{4}(1-x^2)$ |
| $-1$ | $-\frac{1}{2}(1-x)(1-x^2)^{1/2}$ | $-\frac{1}{2}(2x^2-x-1)$ | $-\frac{\sqrt{6}}{2}x(1-x^2)^{1/2}$ | $\frac{1}{2}(2x^2+x-1)$ | $\frac{1}{2}(1+x)(1-x^2)^{1/2}$ |
| $-2$ | $\frac{1}{4}(1-x)^2$ | $-\frac{1}{2}(1-x)(1-x^2)^{1/2}$ | $\frac{\sqrt{6}}{4}(1-x^2)$ | $-\frac{1}{2}(1+x)(1-x^2)^{1/2}$ | $\frac{1}{4}(1+x)^2$ |

which simplifies to

$$1+(3+8\xi+4\xi^2)x^2.$$

(Odd terms and terms of order higher than two in $x$ cancel. We compute, essentially, an average rotational kinetic energy—see Sections 4 of Chapter 2.)

We rewrite this result in a more useful form,

$$\tfrac{1}{2}\{[2(2+1)-2^2](1-x^2)+2(2^2)(1+\xi)^2x^2\},$$

an expression that originates from the general relation

$$\sum_{a=-j}^{j}[d_{ka}^{(j)}(x)]^2(a+k\xi x)^2=\frac{1}{2}\{[j(j+1)-k^2](1-x^2)+2k^2(1+\xi)^2x^2\}.$$

$$(E.6)$$

This formulation can be readily understood by noting that the symmetry of a symmetric top molecule permits us to decompose $M(2)$ into a component along the $C_3$ symmetry axis ($=z$) and a component in the degenerate $x-y$ plane of the inertial axis system of the molecule.

Integration of Eq. E.3 over $x$ can now be performed analytically, employing the basic integrals

$$\int_{-1}^{1}\frac{1-x^2}{(1+\xi x^2)^{5/2}}dx=\frac{4}{3}(1+\xi)^{-1/2};\ \int_{-1}^{1}\frac{x^2}{(1+\xi x^2)^{5/2}}dx=\frac{2}{3}(1+\xi)^{-3/2}.$$

Hence from Eqs. E.3 and E.6, we obtain

$$M(2)^{(j,k)}=\frac{3}{2}(1+\xi)^{1/2}\left\{\int_{-1}^{1}dx\frac{1}{2}[j(j+1)-k^2](1+\xi x^2)^{-5/2}(1-x^2)\right.$$

$$\left.+k^2(1+\xi)^2\int_{-1}^{1}dx(1+\xi x^2)^{-5/2}x^2\right\}$$

$$=j(j+1)-k^2+k^2(1+\xi) \qquad (E.7)$$

or

$$=k_BT\{[j(j+1)-k^2]I_x^{-1}+k^2I_z^{-1}\}$$

in dimensioned units. (As to be expected, the appearance of Eq. E.7 is related to that of the rotational energy of a symmetric top; see Eq. 4.23 in Ref. 37.)

The $j+1$ linearly independent (see Appendix C) classical rotational second spectral moments of a tensor of rank $j$ of the symmetric top, without Coriolis interaction, are

$$M(2)^{(1,0)} = k_B T \left( I_x^{-1} + I_x^{-1} \right) = 2 k_B T \left( B_x + B_x \right)$$

$$M(2)^{(1,1)} = k_B T \left( I_x^{-1} + I_z^{-1} \right) = 2 k_B T \left( B_x + B_z \right)$$

$$M(2)^{(2,0)} = k_B T \left( 3 I_x^{-1} + 3 I_x^{-1} \right) = 2 k_B T \left( 3 B_x + 3 B_x \right) \qquad \text{(E.8)}$$

$$M(2)^{(2,1)} = k_B T \left( 5 I_x^{-1} + I_z^{-1} \right) = 2 k_B T \left( 5 B_x + B_z \right)$$

$$M(2)^{(2,2)} = k_B T \left( 2 I_x^{-1} + 4 I_z^{-1} \right) = 2 k_B T \left( 2 B_x + 4 B_z \right).$$

Moments of inertia are in units of g cm$^2$, rotational constants in cm$^{-1}$. Hence $M(2)$ in terms of the $I$ is in units of radian$^2$ sec$^{-2}$ with $k_B = 1.38 \times 10^{-16}$ erg degree$^{-1}$, $M(2)$ in terms of the $B$ is in units of cm$^{-2}$ with $k_B = 0.695$ cm$^{-1}$ degree$^{-1}$.

### 1.2. Presence of Coriolis Interaction (Coupling Parameter $\zeta$).

The band contours of (doubly) degenerate infrared and Raman vibrational transitions of symmetric top molecules may be broadened ($\zeta < 0$) or narrowed ($\zeta > 0$) by Coriolis coupling. The $M(2)$ in the presence of this interaction are computed from the selection rules $\Delta K_z$ of the component $K_z$ of the rotational angular momentum $\mathbf{K} = \mathbf{J} - \boldsymbol{\kappa}$ along the $C_3$ axis ($= z$), and of the subband frequencies $\Delta_{\text{sub}}$. (A subband is a set of vibrational-rotational transitions with fixed $K_z$.)

The following quantities $\Delta K_z$ and $\Delta_{\text{sub}}$ have been computed [I. M. Mills, *Mol. Phys.* **8**, 363 (1964)]:

---

Point Groups $C_3, C_{3v}, C_{3h}, D_3$
Vibrational Transition: $A \rightarrow E$

---

First-order tensor:
$\qquad \Delta K_z = \pm 1 \qquad \Delta_{\text{sub}} = 2\{B_x - B_z(1 - \zeta)\}.$
Second-order tensor:
$\qquad \Delta K_z = \pm 1 \qquad \Delta_{\text{sub}} = 2\{B_x - B_z(1 - \zeta)\}$
$\qquad \Delta K_z = \pm 2 \qquad \Delta_{\text{sub}} = 4\{B_x - B_z(1 + \zeta/2)\}.$

---

Hence to correct for Coriolis coupling we just substitute $I_z^{-1}$ by $I_z^{-1}(1 - \zeta)$ for $k = \pm 1$ and by $I_z^{-1}(1 + \zeta/2)$ for $k = \pm 2$. This replaces $\xi = (I_x/I_z) - 1$ in the time exponential of Eq. E.1 by $\xi - \zeta(1 + \xi)$ for $k = \pm 1$ and by $\xi + (\zeta/2)(1 + \xi)$ for $k = \pm 2$. The probability distribution function over $L$ is *not* affected.

It is not difficult to see that this modification turns up in the factor $(1 + \xi)^2$ before the second integral of Eq. E.7 (see also Eq. E.6). The resulting second moments are

$$M(2)^{(1,1)} = k_B T \left\{ I_x^{-1} + (1 - \zeta)^2 I_z^{-1} \right\}$$

$$M(2)^{(2,1)} = k_B T \left\{ 5 I_x^{-1} + (1 - \zeta)^2 I_z^{-1} \right\} \tag{E.9}$$

$$M(2)^{(2,2)} = k_B T \left\{ 2 I_x^{-1} + 4 \left( 1 + \frac{\zeta}{2} \right)^2 I_z^{-1} \right\}.$$

The same results, by a different route, were obtained by M. Gilbert, P. Nectoux, and M. Drifford [*J. Chem. Phys.* **68**, 679 (1978)].

## 2. SPHERICAL TOP MOLECULES

In the absence of Coriolis interaction, extrapolation of Eqs. E.8 to spherical tops is trivial,

$$M(2)^{(j,2)} = M(2)^{(j,1)} = M(2)^{(j)}. \tag{E.10}$$

For infrared bands (only triply degenerate modes are allowed) with Coriolis interaction, we must use Eq. D.9 of Appendix D. Triply degenerate Raman contours, with or without Coriolis interaction,[84a] have been described in the literature. Doubly degenerate Raman modes of spherical tops have zero Coriolis interaction (see Ref. 33, p. 447).

## APPENDIX F

# THE KUBO OPERATOR AND OTHER SUPEROPERATORS

We give here an important relation using the Kubo operator and touch upon the related subject of superoperators. First, we repeat the definition

$$R^\times A \equiv [R, A] = RA - AR. \tag{F.1}$$

Now, operating with $\exp(R^\times)$ on vector $A$ gives

$$\{\exp(R^\times)\}A = \{1 + R^\times + (\tfrac{1}{2})(R^\times)^2\}A + \cdots$$

$$= A + R^\times A + (\tfrac{1}{2})R^\times R^\times A + \cdots$$

$$= A + RA - AR + (\tfrac{1}{2})(R^2 A - 2RAR + AR^2) + \cdots.$$

On the other hand (also up to second order),

$$\exp(R)A\exp(-R) = \{1 + R + (\tfrac{1}{2})R^2\}A\{1 - R + (\tfrac{1}{2})R^2\}$$

$$= A + RA + (\tfrac{1}{2})R^2 A - AR - RAR + (\tfrac{1}{2})AR^2.$$

Hence

$$\{\exp(R^\times)\}A = \exp(R)A\exp(-R). \tag{F.2}$$

It is not difficult to see that the matrix elements of such operators need four instead of two labels (tetradic operators): Put

$$(mn|R^\times|m'n') = R_{mm'}\delta_{nn'} - \delta_{mm'}R_{n'n} \tag{F.3}$$

where $\delta$ is a delta function. Then we obtain indeed Eq. F.1:

$$(m|R^{\times}A|n) = \sum_{m'n'} (mn|R^{\times}|m'n')(m'|A|n')$$

$$= \sum_{m'n'} \{R_{mm'}\delta_{nn'} - \delta_{mm'}R_{n'n}\}A_{m'n'}$$

$$= \sum_{m'n'} \{R_{mm'}A_{m'n} - A_{mn'}R_{n'n}\}$$

$$= (m|RA - AR|n). \tag{F.4}$$

If we replace $R_{n'n}$ by its complex conjugate $R_{nn'}^{*}$, we can abbreviate the relation by writing

$$(R^{\times}A)_{mn} = (RA - R^{*}A)_{mn}, \tag{F.5}$$

stressing thereby that $R$ and $R^{*}$ operate on *different* variables: The nonasterisk operator effects the rows, the asterisk operator the columns of matrix $A$. (Refer, for instance, to Eq. 1.74 in Chapter 1.)

The space of these "superoperators" is the so-called double or Liouville $L$ space, defined as a Hilbert space of quantum-mechanical operators—in contrast to ordinary Hilbert space (on vectors). In other words, in $L$ space ordinary operators become vectors. They are expanded in a basis written $|m\rangle\langle n|$ in Dirac notation. An operator in $L$ space is then written

$$A = \sum_{mn} A_{mn}|m\rangle\langle n|. \tag{F.6}$$

Double-bracket notation is also common:

$$|A\rangle\rangle = \sum_{mn} A_{mn}|mn\rangle\rangle. \tag{F.7}$$

The scalar product of two vectors is defined as

$$\langle\langle B|A\rangle\rangle = \mathrm{Tr}\{wB^{\dagger}A\} \qquad \left(B_{ij}^{\dagger} = B_{ji}^{*}\right), \tag{F.8}$$

where $w$ is a metric (weighting) operator. We recognize that Eq. F.8 represents a correlation function if $w$ is a statistical density matrix. (See, for instance, Eq. 1.6 in Chapter 1.)

Matrix elements of superoperators operating on $L$-space vectors (written frequently in script capital letters) are correspondingly defined:

$$\langle\langle mn|\mathcal{O}|m'n'\rangle\rangle = \mathrm{Tr}\{w|n\rangle\langle m|\mathcal{O}(|m'\rangle\langle n'|)\}. \tag{F.9}$$

$L$-space operators can also be expressed in terms of bilinear forms of ordinary Hilbert-space operators: Formally,

$$\mathcal{O} = AB^*$$

and we define

$$(\mathcal{O}X)_{mn} \equiv \sum_{m'n'} \mathcal{O}_{mn;\,m'n'} X_{m'n'}$$

$$= \sum_{m'n'} (AB^*)_{mn;\,m'n'} X_{m'n'}$$

$$= \sum_{m'n'} A_{mm'} B^*_{nn'} X_{m'n'}$$

$$= \sum_{m'n'} A_{mm'} X_{m'n'} B^\dagger_{n'n}$$

$$= (AXB^\dagger)_{mn}. \tag{F.10}$$

For instance, if $A = \exp(iHt/\hbar)$, $B^\dagger = \exp(-iHt/\hbar)$, we obtain

$$\exp\left(\frac{iHt}{\hbar}\right) X \exp\left(-\frac{iHt}{\hbar}\right) = \left\{\exp\left(\frac{iHt}{\hbar}\right)\left[\exp\left(-\frac{iHt}{\hbar}\right)\right]^*\right\} X$$

$$\equiv \exp\left(\frac{i\mathcal{H}t}{\hbar}\right) X, \tag{F.11}$$

where

$$\mathcal{H} = H - H^* = H^\times.$$

(Compare with Eq. F.2 and F.5.)

Let us now see whether the definitions of Eq. F.9 and Eq. F.10 are equivalent. We write first

$$\langle\langle mn|\mathcal{O}|m'n'\rangle\rangle = \sum_{kl} \langle\langle mn|kl\rangle\rangle \mathcal{O}_{kl;\,m'n'}$$

$$= \sum_{kl} \mathrm{Tr}\{w|n\rangle\langle m\|k\rangle\langle l|\} \mathcal{O}_{kl;\,m'n'}, \tag{F.12}$$

where the last equality follows from Eq. F.9. Inspection of Eq. F.12 shows that it corresponds to Eq. F.10 if

$$\sum_{kl} \mathrm{Tr}\{w|n\rangle\langle m\|k\rangle\langle l|\} = \delta_{mk}\,\delta_{nl} \tag{F.13}$$

—in other words, if an orthonormal metric is used.

The usefulness for us of employing such operators is that they make a "two-state" case very similar to a "one-state" case. The two-state case may be regarded as a new function space which is associated with the initial state and the *complex conjugate* of the final state. The Hamiltonian of this double space is such that its (double space) eigenvalue is the difference of the initial and final single states from which it was formed—see, for instance, Eq. F.11 and Eq. 2.120 of Chapter 2: Hence each eigenvalue of the double space corresponds to a *line* of the molecular transitions. [For more detail, see M. Baranger, *Phys. Rev.* **111**, 494 (1958); U. Fano, *Phys. Rev.* **131**, 259 (1963); U. Fano, *Rev. Mod. Phys.* **29**, 74 (1957); A. Ben-Reuven, *Adv. Chem. Phys.* **33**, 235 (1975), from which our brief description was taken.]

# THE INTERACTION REPRESENTATION

We describe here the interaction representation on the example of a perturbation Hamiltonian. Setting the total Hamiltonian

$$H = H_0 + H_1,$$

where $H_0$ is the zero-order and $H_1$ the (time dependent) perturbation Hamiltonian, we construct the two time evolution operators

$$T(t) = \exp\left(-\frac{iHt}{\hbar}\right), \qquad T_0(t) = \exp\left(-\frac{iH_0 t}{\hbar}\right).$$

Assuming that the evolution of the system in zero order (under $T_0$) is at least approximately known, we set

$$T = T_0 T_1$$

and search for the meaning of $T_1$. The Schrödinger equation for $T$, $i\hbar(d/dt)T \equiv i\hbar\dot{T} = HT$, gives

$$i\hbar\left(\dot{T}_0 T_1 + T_0\dot{T}_1\right) = HT_0 T_1.$$

Multiplying from the left by $T_0^\dagger$ (where $T_{ij}^\dagger = T_{ji}^*$), we get

$$i\hbar\left(T_0^\dagger\dot{T}_0 T_1 + \dot{T}_1\right) = T_0^\dagger HT_0 T_1,$$

which leads to

$$i\hbar\dot{T}_1 = T_0^\dagger\left(HT_0 - i\hbar\dot{T}_0\right)T_1$$

$$= T_0^\dagger\left(HT_0 - H_0 T_0\right)T_1$$

$$= \left(T_0^\dagger H_1 T_0\right)T_1$$

$$= H_1^I T_1. \tag{G.1}$$

Hence

$$H_1^I = T_0^\dagger H_1 T_0 = \left(T_0^\dagger\right)^\times H_1, \tag{G.2}$$

where the last equality follows from Eq. F.2 of Appendix F.

We then call

$$T_1(t) = \exp\left(-\frac{iH_1^I t}{\hbar}\right) \tag{G.3}$$

the time evolution operator of the perturbation Hamiltonian $H_1$ in an interaction representation which moves under the zero-order Hamiltonian $H_0$.

Note that the time variation of $T_1(t)$ is slow since $HT_0 \approx H_0 T_0$; this is the reason why a perturbation development of Eq. G.1 is convenient and useful.

To get a better overview, we write out the time evolution operator $T(t) = T_0(t)T_1(t)$, assuming the Hamiltonian $H = H_0 + H_1(t)$, with $H_0 \equiv H_0(0)$ and $\hbar = 1$. From Eqs. G.2 and G.3 we get

$$T(t) = \exp(-iH_0 t)\exp\left\{-i\int_0^t dt' \exp(iH_0 t')H_1(t')\exp(-iH_0 t')\right\}. \tag{G.4}$$

The formal solution of the equation of motion of an observable $X(t)$ can then be written in the form (see Eqs. F.1 and F.2)

$$X(t) = T^\dagger X T$$

$$= \exp\left\{i\int_0^t dt'\, H_1^I(t')\right\}\exp(iH_0 t)X\exp(-iH_0 t)\exp\left\{-i\int_0^t dt'\, H_1^I(t')\right\}$$

$$\equiv \exp\left\{i\int_0^t dt'\, H_1^{I\times}(t')\right\}\exp(iH_0^\times t)X, \text{ with } X \equiv X(0). \tag{G.5}$$

[See also R. P. Feynman, *Phys. Rev.* **84**, 108 (1951), in particular Section 4, which gives a method for disentangling exponentials of noncommuting operators.]

The Schrödinger equation for state vector $x(t)$,

$$i\hbar\dot{x}(t) = \{H_0 + H_1(t)\}x(t),$$

is readily transformed into

$$i\hbar\dot{x}^I(t) = \exp\left(\frac{iH_0 t}{\hbar}\right)H_1(t)\exp\left(-\frac{iH_0 t}{\hbar}\right)x^I(t)$$

$$= H_1^I(t)x^I(t)$$

by defining

$$x(t) = \exp\left(-\frac{iH_0 t}{\hbar}\right) x^I(t),$$

substituting back, differentiating, and multiplying from the left by $\exp(iH_0 t/\hbar)$. [See Eq. G.1.] A state vector in the interaction representation moves only under the perturbation Hamiltonian, not under the total Hamiltonian. To summarize: for a state vector $x(t)$,

$$x^I(t) = \exp\left(\frac{iH_0 t}{\hbar}\right) x(t), \qquad (G.6)$$

for observable $X(t)$,

$$X^I(t) = \exp\left(\frac{iH_0^\times t}{\hbar}\right) X(t), \qquad (G.7)$$

$$\dot{X}^I(t) = \frac{i}{\hbar}\left[H_0, X^I(t)\right]. \qquad (G.8)$$

# APPENDIX H

# ON CUMULANT AVERAGES AND THE CENTRAL LIMIT THEOREM

## 1. CUMULANTS

The concept of "cumulant" or "connected" averages can get very complicated, particularly in the case of multivariate quantities and under the a priori restriction to noncommuting operators in the average. We shall thus keep our discussion on the lowest level compatible with the complexity of the material.

We define several equalities of averages of exponential series without questioning the existence of the "moments" in the development or assurance of convergence and see where it leads.

$$\langle \exp(ia_n t) \rangle = \sum_0^\infty \frac{\langle a_n \rangle}{n!}(it)^n = \exp\left\{ \sum_1^\infty \frac{\langle \kappa_n \rangle}{n!}(it)^n \right\}$$

$$= \exp\{\langle \exp(i\kappa_n t) - 1 \rangle\}. \tag{H.1}$$

The first and the third equalities are obvious. To get the second, we (1) call

$$1 + \sum_1^\infty \frac{a_n}{n!}(it)^n = \exp(ia_n t) \equiv \varphi(t);$$

(2) we remember the series development of the natural logarithm of $1 + z$,

$$\ln(1+z) = z - \frac{1}{2}z^2 \pm \cdots \pm \frac{z^k}{k} \pm \cdots;$$

(3) we set

$$1 + z = \varphi(t)$$

and obtain

$$\ln \varphi(t) = \ln\left\{ 1 + \sum_1^\infty \frac{\langle a_n \rangle}{n!}(it)^n \right\};$$

(4) set this *formally* equal to

$$\sum_1^\infty \frac{\langle \kappa_n \rangle}{n!} (it)^n;$$

and (5) get $\varphi(t)$ back by exponentiation.

Developing $\langle \exp(ia_n t) \rangle$ and $\exp\langle\langle \exp(i\kappa_n t)-1\rangle\rangle$ into their power series, multiplying out, and comparing coefficients of equal powers of $t$, yields the cumulant average $\langle \kappa_n \rangle$ of order $n$

$(a)$ $\langle \kappa_1 \rangle = \langle a_1 \rangle$

$(b)$ $\langle \kappa_2 \rangle = \langle a_2 \rangle - \langle a_1 \rangle^2$

$(c)$ $\langle \kappa_3 \rangle = \langle a_3 \rangle - 3\langle a_2 \rangle\langle a_1 \rangle + 2\langle a_1 \rangle^3$  \hfill (H.2)

$(d)$ $\langle \kappa_4 \rangle = \langle a_4 \rangle - 3\langle a_2 \rangle^2 - 4\langle a_1 \rangle\langle a_3 \rangle + 12\langle a_1 \rangle^2\langle a_2 \rangle - 6\langle a_1 \rangle^4$,

and so forth.

We recognize from Eq. H.2$b$ that the second cumulant $\langle \kappa_2 \rangle$ is equal to the second moment $\langle a_2 \rangle$ taken about the first moment $\langle a_1 \rangle$. Only if $\langle a_1 \rangle = 0$ is $\langle a_2 \rangle = \langle \kappa_2 \rangle$. The quantity $\langle \kappa_2 \rangle$ is the well-known mean square fluctuation of a quantity $X$ about its mean $\langle X \rangle$, the so-called dispersion:

$$\langle \kappa_2 \rangle = \langle (X - \langle X \rangle)^2 \rangle = \langle X^2 - 2X\langle X \rangle + \langle X \rangle^2 \rangle$$

$$= \langle X^2 \rangle - 2\langle X \rangle\langle X \rangle + \langle X \rangle^2$$

$$= \langle X^2 \rangle - \langle X \rangle^2. \hfill (H.3)$$

Notice that $\langle \cdots \rangle$ denotes here any prescription of averaging. For instance, in Eq. 2.73, Chapter 2, we used the spectral density as probability distribution.

Equations H.2 are readily extended to multivariate quantities (from now on we also denote a cumulant average by superscript $c$),

$$\langle X_i \rangle^c = \langle X_i \rangle,$$

$$\langle X_i X_j \rangle^c = \langle X_i X_j \rangle - \langle X_i \rangle\langle X_j \rangle, \hfill (H.4)$$

$$\langle X_i X_j X_k \rangle^c = \langle X_i X_j X_k \rangle - \{\langle X_i \rangle\langle X_j X_k \rangle + \langle X_j \rangle\langle X_k X_i \rangle + \langle X_k \rangle\langle X_i X_j \rangle\}$$

$$+ 2\langle X_i \rangle\langle X_j \rangle\langle X_k \rangle,$$

and so forth. Notice now how statistical independence of any two $X$ causes the *cumulant* average to *vanish*. For instance,

$$\langle X_i X_j \rangle^c = \langle X_i \rangle\langle X_j \rangle - \langle X_i \rangle\langle X_j \rangle.$$

Second, if $Z$ is a sum of *independent* variables $X, Y, \ldots$, we can write (see Eq. H.1)

$$\ln\langle \exp(iZt)\rangle = \ln\langle \exp(i[X + Y + \cdots]t)\rangle$$

$$= \ln\{\langle \exp(iXt)\rangle\langle \exp(iYt)\rangle \cdots\}$$

$$= \ln\langle \exp(iXt)\rangle + \ln\langle \exp(iYt)\rangle + \cdots.$$

In other words, the *cumulant* averages of any order simply *add*:

$$\langle \kappa_n(Z)\rangle = \langle \kappa_n(X)\rangle + \langle \kappa_n(Y)\rangle + \cdots. \tag{H.5}$$

Inspection of Eq. H.2 shows that the individual second (and third) moments, if taken about the mean $\langle \kappa_1\rangle = 0$, also add—a relation we made use of in Sections 4.2.1 and 4.2.2 in Chapter 2.

The cumulant expansion of an exponential $\langle \exp_0(i\int dt' F(t'))\rangle$ is readily performed. If $F(t)$ is small, we obtain (see Eq. 2.125 to 2.128, Chapter 2)

$$\left\langle \exp_0\left\{ i\int_0^t dt' F(t')\right\}\right\rangle$$

$$= \exp_0\left\{ \sum_1^\infty \frac{i^n}{n!} \int_0^t dt_1 \int_0^t dt_2 \cdots \int_0^t dt_n \langle F(t_1)F(t_2)\cdots F(t_n)\rangle^c\right\}$$

$$= \exp_0\left\{ i\int_0^t dt_1 \langle F(t_1)\rangle^c - \frac{1}{2}\int_0^t\int_0^t dt_1\, dt_2 \langle F(t_1)F(t_2)\rangle^c\right.$$

$$\left. - \frac{i}{6}\int_0^t\int_0^t\int_0^t dt_1\, dt_2\, dt_3 \langle F(t_1)F(t_2)F(t_3)\rangle^c + \cdots\right\}. \tag{H.6}$$

We need not go beyond the second-order term since we assume that the probability distribution of $F(t)$ is Gaussian[127] (see also Section 2). Furthermore, we will be usually in a situation where the frequency shift vanishes,

$$\int_0^t dt' \langle F(t')\rangle^c = \int_0^t dt' \langle F(t')\rangle = 0 \tag{H.7}$$

—either a priori, or by construction; for example, see Eq. 2.110 in Chapter 2. Consequently, the cumulant average $\langle F(t_1)F(t_2)\rangle^c$ becomes an ordinary average (see Eq. H.2b). Therefore

$$\left\langle \exp_0\left\{ i\int_0^t dt' F(t')\right\}\right\rangle = \exp_0\left\{ -\frac{1}{2}\int_0^t\int_0^t dt\, dt' \langle F(t)F(t')\rangle\right\}$$

$$= \exp_0\left\{ -\int_0^t\int_0^{t'} dt\, dt' \langle F(t)F(t')\rangle\right\}. \tag{H.8}$$

## 2. CENTRAL LIMIT THEOREM

We go into the fundamentals of this subject only in the briefest detail, keeping a little more extended discussion for a numerical "proof" of the validity of the central limit theorem in instances of interest to us.

The central limit theorem says that a random variable $X$ that is a sum of $n$ independent contributions approaches the normal distribution (Gaussian) for $n \to \infty$:

$$P\left(\frac{X-a}{a}\right) = \frac{1}{\sqrt{2\pi}\,\sigma} \int_{-\infty}^{X} dX \exp\left\{-\frac{(X-a)^2}{2\sigma^2}\right\} \tag{H.9}$$

where $\sigma^2 = \langle (X - \langle X \rangle)^2 \rangle$ is the dispersion and $a = \langle X \rangle$ the expectation value of $X$ (see Eq. H.3). The question is: Which is the smallest value of $n$ to satisfy Eq. H.9? The answer is that $n = 10–12$ appears to be sufficient.[127a]

Yet it can be shown "operationally" that even smaller values suffice.[127b] If a repeated convolution of only four rectangular step functions of unit height and bases of 1, 2, 3, and 4 units, respectively, is performed, the resulting convolution contour is nearly a Gaussian. In other words, the combined effect of four *independent*, different rectangular pulses gives a near-normal distribution.

Even in situations where (reasonable) functions with *no symmetry* are chosen, four to five convolution steps lead to a rather "decent" looking Gaussian profile. Self-convolution of $\exp(-X)\text{step}(X)$, where

$$\text{step}(X) = \begin{cases} 0, & X < 0 \\ \frac{1}{2}, & X = 0, \\ 1, & X > 0 \end{cases} \tag{H.10}$$

furnishes an instructive example.[127b]

Henceforth we assume that the central limit theorem is essentially obeyed in all systems where the probe molecule sees the effect of, at least, four independent nearest neighbor interactions. Some opposite opinions expressed in the literature[18a] appear overly concerned.

Note that for a molecule to be stable against displacements in a given direction, three molecules in that direction are needed to block the displacement (see Note 5, Chapter 3).

# APPENDIX J

# RELATION BETWEEN BANDWIDTH, CORRELATION TIME, AND RELAXATION TIME

Two questions are answered here:

1. How is a true vibrational energy relaxation time, obtained from an induced process or theoretically computed, related to the corresponding spontaneous bandwidth increment (or the correlation time of the associated correlation function)?

2. How is the vibrational dephasing time, observed from an induced process or theoretically computed, related to the corresponding spontaneous bandwidth increment?

For a spontaneous vibrational energy relaxation process, we assume a Lorentzian band profile. This approximation is in tune with the usually long decay times observed. Furthermore, a Lorentzian corresponds to an exponential vibrational population decay observed in picosecond pulsing experiments (induced scattering).

The Lorentzian spectral density and the exponential correlation function (mutual Fourier transforms) are

$$\mathscr{I}(\omega) = 2\gamma\{\gamma^2 + (\omega - \omega_0)^2\}^{-1},$$

$$C(t) = \exp(-\gamma|t|) = \exp\left(-\frac{|t|}{\tau}\right), \tag{J.1}$$

where

$$\gamma \equiv \tau^{-1}; \qquad \tau = \int_0^\infty dt\, C(t).$$

Notice that $\tau$ or $\gamma$ have, as yet, no particular physical meaning in terms of a mechanism or a relaxation process. In other words, $\tau$ or $\gamma$ are merely band-

**389**

broadening parameters: Parameter $\gamma$ is the "damping constant," $\tau$ the "correlation time," and $\omega_0$ the center frequency. Obviously, if the frequency is shifted off center by $\omega - \omega_0 = \gamma$, $\mathcal{I}(\omega) \rightarrow (\frac{1}{2})\mathcal{I}(\omega)$. Hence $2\gamma$ is the *full* width at *half*-peak height of the profile ("halfwidth") $\Delta\nu_{1/2}$. It follows

$$\gamma = \tfrac{1}{2}\Delta\nu_{1/2} = \tau^{-1}, \tag{J.2}$$

or

$$\tau = \frac{2}{\Delta\nu_{1/2}} = \frac{1}{\pi c \Delta_{1/2}}$$

with $\Delta_{1/2}$ in units of $cm^{-1}$ and $c$ the velocity of light.

To introduce $\gamma$ as a physical parameter, we write Newton's equation of motion of oscillator amplitude $Q$ (Ref. 29, p. 32)

$$\ddot{Q} = -\omega_0^2 Q - \gamma\dot{Q}, \qquad \gamma \ll Q_0$$

where $\gamma$ is the amplitude damping constant. An approximate solution (within $\frac{1}{4}\gamma^2 Q^2$) is

$$Q(t) = Q_0 \exp\left(-\frac{\gamma t}{2}\right)\exp(-i\omega_0 t), \tag{J.3}$$

as is verified by differentiation. The oscillator energy (averaged over one period $2\pi/\omega_0$) is

$$E = \tfrac{1}{2}\left(\dot{Q}^2 + \omega_0^2 Q^2\right),$$

which gives, with Eq. J.3,

$$E = E_0 \exp(-\gamma t), \tag{J.4}$$

where $E_0$ is a constant. The reciprocal of $\gamma$ is the energy dissipation time, namely, the time the oscillator energy needs to drop to $E_0/e$,

$$\frac{1}{\gamma} \equiv \tau_{ED}. \tag{J.5}$$

Writing the Fourier transform of the amplitude correlation of Eq. J.3 (see also Eq. A.14, Appendix A),

$$\mathrm{Re}\int_{-\infty}^{\infty} dt\,\exp(i\omega t)Q(t)Q_0 = \mathcal{I}(\omega),$$

gives (leaving out uninteresting factors)

$$\mathrm{Re}\left[i(\omega - \omega_0) + \frac{\gamma}{2}\right]^{-1} = \frac{\gamma}{2}\left[(\omega - \omega_0)^2 + \frac{\gamma^2}{4}\right]^{-1} \tag{J.6}$$

where (see Eq. J.1)

$$\tfrac{1}{2}\Delta\nu_{1/2} = \frac{\gamma}{2} = \tau^{-1}$$

or

$$\tau = \frac{2}{\Delta\nu_{1/2}}. \qquad (J.7)$$

Hence if we evaluate a band profile in terms of a correlation time $\tau$ by way of Eq. J.1 or J.7, the true oscillator energy relaxation time $\tau_{ED} = \gamma^{-1}$ is only one-half the value of the correlation time $\tau$.

Indeed, if we use the energy uncertainty relation on a bandwidth,

$$\Delta E\tau' \sim \hbar; \qquad \tau' \sim \frac{1}{\Delta\nu_{1/2}}, \qquad (J.8)$$

we obtain at once the "correct" result,

$$\tau_{ED} = \tau'. \qquad (J.9)$$

In an ultrashort pulsing experiment, we observe the decay of the anti-Stokes signal and thus directly probe the time development of $|Q(t)|^2$. Hence its decay time equals $\gamma^{-1} = \tau_{ED}$ (see Eqs. J.3 and J.4).

The same relations hold for a dephasing process; we do not derive them here.[210]

In summary: The vibrational amplitude *correlation* time $\tau_{sp}$ of the correlation function $C(t)$ that describes a *spontaneous* energy dissipation process,

$$\tau_{sp} = \int_0^\infty dt\, C(t),$$

is related to the *relaxation* time $\tau_{ED}$ of vibrational energy $E(t)$,

$$\tau_{ED} = \int_0^\infty dt\, E(t),$$

by $\tau_{sp} = 2\tau_{ED}$. In an *induced* dissipation process (picosecond pulsing) $\tau_{ED}$ is obtained *directly* from the data (see Fig. 3.17).

# APPENDIX K

# AVERAGE VIBRATIONAL ENERGY OF A QUANTAL OSCILLATOR

The density matrix under the vibrational coordinate of the vibrational Hamiltonian $H_0$ in a diagonal representation is ($\omega_0 \equiv \alpha$)

$$( \, |\rho_{H_0}| \, ) = \frac{\exp(-H_0/k_BT)}{\mathrm{Tr}\exp(-H_0/k_BT)}$$

$$= \frac{\exp[-(v+\tfrac{1}{2})(\hbar\alpha/k_BT)]}{\displaystyle\sum_{v=0}^{\infty} \exp[-(v+\tfrac{1}{2})(\hbar\alpha/k_BT)]}$$

$$= \exp\left(-\frac{v\hbar\alpha}{k_BT}\right)\left[1-\exp\left(-\frac{\hbar\alpha}{k_BT}\right)\right], \qquad (\mathrm{K}.1)$$

noting that the sum is equivalent to a simple geometric series. Therefore

$$\mathrm{Tr}\rho_{H_0}\left(\frac{H_0}{\hbar\alpha}\right) = \left[1-\exp\left(-\frac{\hbar\alpha}{k_BT}\right)\right]\sum_{v=0}^{\infty}\exp\left(-\frac{v\hbar\alpha}{k_BT}\right)\left(v+\frac{1}{2}\right)$$

$$= \left[1-\exp\left(-\frac{\hbar\alpha}{k_BT}\right)\right]\left\{\frac{1}{2}\left[1-\exp\left(-\frac{\hbar\alpha}{k_BT}\right)\right]^{-1} + \sum_{v=0}^{\infty} v\left[\exp\left(-\frac{\hbar\alpha}{k_BT}\right)\right]^{v}\right\}$$

$$= \frac{1}{2}+\left[1-\exp\left(-\frac{\hbar\alpha}{k_BT}\right)\right]\exp\left(-\frac{\hbar\alpha}{k_BT}\right)\left[1-\exp\left(-\frac{\hbar\alpha}{k_BT}\right)\right]^{-2},$$

$$(\mathrm{K}.2)$$

noting that the sum is a series of the form

$$\sum_{k=0}^{\infty} k x^k = \frac{x}{(1-x)^2}.$$

It follows:

$$\text{Tr}\rho_{H_0}\left(\frac{H_0}{\hbar\alpha}\right) = \frac{1}{2} + \left[\exp\left(\frac{\hbar\alpha}{k_B T}\right) - 1\right]^{-1} \equiv \frac{1}{2}\coth\left(\frac{\hbar\alpha}{2k_B T}\right). \qquad (K.3)$$

For high temperatures ($\hbar\alpha \ll k_B T$), we obtain the classical result

$$\text{Tr}\rho_{H_0}(H_0) = k_B T.$$

Other useful relations are readily obtained by employing the population representation of a harmonic oscillator in terms of the creation and destruction operators $a^\dagger$, $a$. Since[216]

$$a^\dagger|v) = (v+1)^{1/2}|v+1); \; a|v) = v^{1/2}|v-1), \qquad (K.4)$$

where $|v)$ is an eigenvector of $a^\dagger a$ and $v$ the corresponding eigenvalue, we obtain

$$(v|aa^\dagger|v) = (v+1); \qquad (v|a^\dagger a|v) = v.$$

Writing a correlation function,

$$\langle a(t)a^\dagger(0)\rangle \equiv \langle a(0)a^\dagger(0)\rangle\exp(-i\omega_a t), \qquad (K.5)$$

we get (see Eqs. K.1, K.2)

$$\langle a(0)a^\dagger(0)\rangle = \text{Tr}\rho_{H_a}(a(0)a^\dagger(0))$$

$$= \left[1 - \exp\left(-\frac{\hbar\omega_a}{k_B T}\right)\right]\sum_v \exp\left(-\frac{v\hbar\omega_a}{k_B T}\right)(v+1)$$

$$= 1 + \left[\exp\left(\frac{\hbar\omega_a}{k_B T}\right) - 1\right]^{-1}$$

$$= 1 + \bar{N}_a, \qquad (K.6)$$

where

$$\bar{N}_a = \exp\left(-\frac{\hbar\omega_a}{k_B T}\right)\left[1 - \exp\left(-\frac{\hbar\omega_a}{k_B T}\right)\right]^{-1} \qquad (K.7)$$

is the (Bose) equilibrium distribution of the oscillators, and

$$Z_a = \left[1 - \exp\left(-\frac{\hbar\omega_a}{k_B T}\right)\right]^{-1} \tag{K.8}$$

the "sum of states." Analogously, from $\langle a^\dagger(t)a(0)\rangle = \langle a^\dagger(0)a(0)\rangle\exp(i\omega_a t)$,

$$\langle a^\dagger(0)a(0)\rangle = \bar{N}_a. \tag{K.9}$$

It follows that the symmetrized correlation function (see Appendix A),

$$\langle[a^\dagger, a]_+\rangle = \tfrac{1}{2}\langle a^\dagger a + aa^\dagger\rangle,$$

yields Eq. K.3—as it must.

# APPENDIX L

# SPECTRAL DENSITY OF TRANSLATIONAL DIFFUSION

We outline here the derivation of Eq. 3.86$b$ of Chapter 3. The Fourier transform $\mathscr{I}(\omega)$ of Eq. 3.24 is readily obtained with the help of some transformations. Setting

$$\tau_T^{-1} = \frac{\sigma^2}{2D}; \qquad -i\omega = s,$$

we get for $\mathscr{I}(\omega)$ [in units of $\sigma^3/\rho$]

$$\mathscr{I}(s) = \operatorname{Re} \int_0^\infty dt \int_0^\infty du\, u^{-1} J_{3/2}^2(u) \exp\left(-\frac{u^2 t}{\tau_T}\right) \exp(-st). \qquad (L.1)$$

Differentiating the integral in $u$ with respect to $t$ and then integrating over $u$ gives

$$-\tau_T^{-1} \int_0^\infty du\, J_{3/2}^2(u) \exp\left(-\frac{u^2 t}{\tau_T}\right) u = -\frac{1}{2t} \exp\left(-\frac{\tau_T}{2t}\right) I_{3/2}\left(\frac{\tau_T}{2t}\right)$$

$$\equiv \frac{d}{dt} \varphi(t) \equiv F(t), \qquad (L.2)$$

where $I_{3/2}$ is a modified Bessel function.[252c]
Writing the Laplace transform of $F(t)$,

$$\tilde{F}(s) = \int_0^\infty dt\, F(t) \exp(-st), \qquad (L.3)$$

we get[143a]

$$\tilde{F}(s) = -\varphi(0) + s \int_0^\infty dt\, \varphi(t) \exp(-st)$$

$$= -\varphi(0) + s\mathscr{I}(s),$$

where

$$\mathscr{I}(s) = \int_0^\infty dt\, \varphi(t) \exp(-st) = \left[\tilde{F}(s) + \tfrac{1}{3}\right] s^{-1} \tag{L.4}$$

with

$$\varphi(0) = \tfrac{1}{3} \tag{L.5}$$

is the desired result.[252d]

Insertion of Eq. L.2 into L.3 and then using Eq. L.4 gives

$$s\mathscr{I}(s) = \frac{1}{3} - \frac{1}{2} \int_0^\infty dt\, t^{-1} \exp\left(-\frac{\tau_T}{2t}\right) I_{3/2}\left(\frac{\tau_T}{2t}\right) \exp(-st). \tag{L.6}$$

The integral is known (see Ref. 51, Formula 4.17.4),

$$\tilde{F}(s) = -K_{3/2}\{(s\tau_T)^{1/2}\} I_{3/2}\{(s\tau_T)^{1/2}\}.$$

Hence

$$\mathscr{I}(s) = (3s)^{-1} - s^{-1} K_{3/2}\{(s\tau_T)^{1/2}\} I_{3/2}\{(s\tau_T)^{1/2}\}, \tag{L.7}$$

which gives Eq. 3.86$b$ in Chapter 3.[252e]

Neglecting in Eq. 3.86$b$ all terms beyond the second (see Eq. 3.87) readily leads to Eq. 3.88$b$.

# REFERENCES

1. See, for instance, R. Paul and G. G. Fuller, "Applications of field theoretical methods to the calculation of infrared band shapes of molecules in strongly interacting solvents," *J. Chem. Phys.* **64**, 3809 (1976). In this paper, a diatomic molecule is treated as a composite of elementary fermions. The solvent is a structureless entity. The solute–solvent interaction is described by a mathematically expedient potential, expressed as a parametrized series of Gaussian functions in appropriate center of mass separation and internal coordinates.

2. (a) R. Kubo, *J. Phys. Soc. Jap.* **12**, 570 (1957); (b) R. Kubo, *Rep. Prog. Phys.* **29**, Pt. I, 255 (1966); (c) B. J. Berne and G. D. Harp, *Adv. Chem. Phys.* **17**, 63 (1970).

3. (a) R. G. Gordon, *Adv. Magn. Resonance* **3**, 1 (1968) and references cited therein; (b) H. Shimizu, *J. Chem. Phys.* **43**, 2453 (1965).

4. A. Messiah, *Quantum Mechanics I*, Wiley, New York, 1958; (a) pp. 135–138; (b) 331–334.

5. S. Fujita, *Introduction to Non-Equilibrium Quantum Statistical Mechanics*, W. B. Saunders Co., Philadelphia, 1966, p. 123. For a brief description of the interaction "representation," see Section 1.2.

6. A. Abragam, *The Principles of Nuclear Magnetism*, Clarendon Press, Oxford, 1961, p. 100.

7. H. B. Callen and T. A. Welton, *Phys. Rev.* **83**, 34 (1951).

8. P. A. M. Dirac, *The Principles of Quantum Mechanics*, Clarendon Press, Oxford, 1958, p. 176.

9. R. L. Fulton, *J. Chem. Phys.* **55**, 1386 (1971).

10. (a) P. van Konynenburg and W. A. Steele, *J. Chem. Phys.* **56**, 4776 (1972); (b) W. G. Rothschild, G. J. Rosasco, and R. C. Livingston, *J. Chem. Phys.* **62**, 1253 (1975); (c) L. C. Rosenthal and H. L. Strauss, *J. Chem. Phys.* **64**, 282 (1976).

11. J. Vincent-Geisse, J. Soussen-Jacob, C. Breuillard, J. C. Briquet, and T. Nguyen-Tan, *Mol. Phys.* **43**, 145 (1977).

12. K. D. Möller and W. G. Rothschild, *Far-Infrared Spectroscopy*, Wiley-Interscience, New York, 1971, see Chapter 10.

13. R. G. Gordon, *J. Chem. Phys.* **43**, 1307 (1965).

14. B. Keller and F. Kneubühl, *Helv. Phys. Acta* **45**, 1127 (1972).

15. R. Loudon, *The Quantum Theory of Light*, Clarendon Press, Oxford, 1973, p. 267.

16. H. D. Dardy, Doctoral dissertation, The Catholic University of America, Washington, D. C., 1972 (University Microfilms, Ann Arbor, Mich., No 72-21, 565).

17. (a) L. A. Nafie, P. Stein, B. Fanconi, and W. L. Peticolas, *J. Chem. Phys.* **52**, 1584 (1970); (b) L. A. Nafie and W. L. Peticolas, *J. Chem. Phys.* **57**, 3145 (1972).

18. (a) S. Bratos and E. Marechal, *Phys. Rev.* **A4**, 1078 (1971); (b) F. J. Bartoli and T. A. Litovitz, *J. Chem. Phys.* **56**, 413 (1972). Both publications were received within two days in the respective journals and are to be considered the basic references on the VV-VH Raman experiment and its wider ramifications to molecular dynamics.

19. E. B. Wilson, Jr., J. C. Decius, and P. C. Cross, *Molecular Vibrations*, McGraw-Hill, New York, 1955, Appendix X.

20. J. Schroeder, V. H. Schiemann, P. T. Sharko, and J. Jonas, *J. Chem. Phys.* **66**, 3215 (1977).

21. See Ref. 10*b*, Section IVA1, Fig. 4.

22. W. C. Mundy, L. Gutierrez, and F. H. Spalding, *J. Chem. Phys.* **59**, 2173 (1973).

23. G. D. Enright, G. I. A. Stegeman, and B. P. Stoicheff, *J. Phys.* **33**, Colloque Cl, Suppl. to N° 2-3, Cl-C207 (1972); see Fig. 1.

24. S. J. Tsay and D. Kivelson, *Mol. Phys.* **29**, 1 (1975); N. D. Gershon and I. Oppenheim, *J. Chem. Phys.* **59**, 1337 (1973).

25. (a) L. I. Schiff, *Quantum Mechanics*, McGraw-Hill, New York 1955, pp. 249–252; (b) L. D. Landau and E. M. Lifshitz, *Quantum Mechanics*, Pergamon Press, London 1958, pp. 140–147; (c) Ref. 6, pp. 272–274.

26. A detailed account is given in *Concepts in Quantum Mechanics*, F. A. Kaempffer, Academic Press, New York 1965, pp. 43–46, 53–54. See also Ref. 6, pp. 105–106, Ref. 8, p. 173, and Ref. 4, p. 321.

27. See Ref. 4, Eqs. A.15, p. 469.

28. R. G. Gordon, *J. Chem. Phys.* **40**, 1973 (1964).

29. W. Heitler, *The Quantum Theory of Radiation*, Clarendon Press, Oxford 1954. See. Eq. 7, V, § 19.

30. The derivation of this result can be found in the article by J. L. Birman, "Theory of Crystal Space Groups and Infrared and Raman Lattice Processes of Insulating Crystals," *Encyclopedia of Physics*, Vol. XXV/2*b*, S. Flügge, Ed., Springer, New York 1974, pp. 282–289. We have outlined here the major ideas.

31. R. G. Gordon, *J. Chem. Phys.* **42**, 3658 (1965).

32. R. Loudon, *Adv. Phys.* **13**, 423 (1964); see table on pp. 440–441.

33. G. Herzberg, *Infrared and Raman Spectra of Polyatomic Molecules*, Van Nostrand Reinhold, New York 1945; see pp. 254–256, 260–261.

34. W. G. Rothschild, *J. Chem. Phys.* **53**, 990 (1970). In this paper no account was taken of any vibrational relaxation.

35. See Eq. 2.27 of Ref. 31.

36. J. Koningstein, *Introduction to the Theory of the Raman Effect*, D. Reidel Publishing Co., Dordrecht, 1972. See pp. 70–75.

37. M. E. Rose, *Elementary Theory of Angular Momentum*, Wiley, New York, 1957, pp. 64–66.

38. R. I. Cukier and K. Lakatos-Lindenberg, *J. Chem. Phys.* **57**, 3427 (1972).

39. A. R. Edmonds, *Angular Momentum in Quantum Mechanics*, Princeton University Press, New Jersey, 1957; see p. 63 and recall that $(ab)^* = b^*a^*$.

40. See Ref. 6, pp. 270–271; 299.

41. See. Ref. 37, Eq. 4.60 on p. 75.

42. T. Gierke, *J. Chem. Phys.* **65**, 3873 (1976).

43. B. J. Berne and R. Pecora, *Dynamic Light Scattering: With Applications to Chemistry, Biology, and Physics*, Wiley-Interscience, New York, 1976.

44. See, for instance, V. N. Smirnov, *Opt. Spectrosc.* **7**, 302 (1959); N. I. Resaev and A. S. Andreev, *Opt. Spectrosc.* **7**, 72 (1959).

45. A. V. Rakov, *Opt. Spectrosc.* **7**, 128 (1959).

46. W. G. Rothschild, unpublished data; see also K. Müller and F. Kneubühl, *Helv. Phys. Acta* **49**, 702 (1976).

47. W. G. Rothschild, in *Molecular Motions in Liquids*, J. Lascombe, Ed., D. Reidel Publishing Co., Dordrecht, 1974, p. 247.

48. Dinesh and M. T. Rogers, *J. Magn. Resonance* **7**, 30 (1972).

49. A. Laubereau and W. Kaiser, *Chemical and Biological Applications of Lasers*, Vol. 2, Academic Press, New York, 1977, p. 87.

50. M. Schubert and B. Wilhelmi, *Einführung in die Nichtlineare Optik*, Vol. 1, B. G. Teubner Verlagges., Leipzig, 1971. Apparently this excellent book is not available in English. See also N. Bloembergen, *Am. J. Phys.* **35**, 989 (1967).

51. A. Erdelyi, Ed., *Tables of Integral Transforms*, McGraw-Hill, New York, 1954. See Formulas 4.1.5, 4.7.1 and set the Laplace variable to $i\omega$.

52. R. G. Gordon, *J. Chem. Phys.* **44**, 1830 (1966).

53. R. E. D. McClung, *J. Chem. Phys.* **57**, 5478 (1972). This paper includes many references to the pioneering work of Gordon and others.

54. See Ref. 33, Section I.2.

55. Ming Chen Wang and G. E. Uhlenbeck, *Rev. Mod. Phys.* **17**, 323 (1945). Reprinted in *Selected Papers on Noise and Stochastic Processes*, Dover, New York, 1954; see pp. 114–115.

56. P. Debye, *Polar Molecules*, Dover, New York, 1929.

57. See Ref. 37, Eq. 4.20.

58. S. Chandrasekhar, *Rev. Mod. Phys.* **15**, 1(1943); see p. 20. Also reprinted in the Dover publication of Ref. 55, p. 22.

59. H. Mori, *Prog. Theor. Phys.* **33**, 423 (1965); **34**, 399 (1965).

60. See Ref. 51, Formulas 4.1.8 and 4.1.20.

61. See Ref. 51, Formula 4.1.20.

62. R. Zwanzig, *J. Chem. Phys.* **33**, 1338 (1960).

63. See Ref. 4, p. 260.

64. F. Bliot, C. Abbar, and E. Constant, *Mol. Phys.* **24**, 241 (1972).

65. See Ref. 51, Formula 4.1.5.

66. A. G. St. Pierre and W. A. Steele, *J. Chem. Phys.* **57**, 4638 (1972).

67. Y. Guissani, J. C. Leicknam, and S. Bratos, *Phys. Rev.* **A16**, 2072 (1977).

68. F. Bliot and E. Constant, *Chém. Phys. Lett.* **18**, 253 (1973).

69. See Ref. 5, p. 34.

70. See, for instance, Ref. 6, p. 298.

71. (a) R. I. Cukier, *J. Chem. Phys.* **60**, 734 (1974); (b) L. D. Favro, *Phys. Rev.* **119**, 53 (1960).

72. R. G. Gordon, *J. Chem. Phys.* **41**, 1819 (1964).

73. See pp. 34–35 of Ref. 5.

74. K. T. Gillen, *J. Chem. Phys.* **56**, 1573 (1972).

75. P. S. Hubbard, *Phys. Rev.* **131**, 1155 (1963).

76. See Ref. 2c, Eqs. 310 and 314.

77. G. Lévi, M. Chalaye, F. Marsault-Hérail, and J. P. Marsault, *Mol. Phys.* **24**, 1217 (1972).

78. R. E. D. McClung, *J. Chem. Phys.* **55**, 3459 (1971).

79. K. Müller and F. Kneubühl, *Chem. Phys.* **8**, 468 (1975).

80. See Ref. 33, p. 374 (linear molecules).

81. See Ref. 33, p. 429 (symmetric top), p. 454 (spherical top).

## 402    REFERENCES

82. (a) R. E. Lechner, J. M. Rowe, K. Sköld, and J. J. Rush, *Chem. Phys. Lett.* **4**, 444 (1969); (b) T. Mansson, L. G. Olsson, and K. E. Larsson, *J. Chem. Phys.* **66**, 5817 (1977).

83. R. C. Livingston, W. G. Rothschild, and J. J. Rush, *J. Chem. Phys.* **59**, 2498 (1973).

84. (a) H. Ziebert, *Z. anorg. allg. Chem.* **268**, 177 (1952); (b) Ph. Depondt, M. Debeau, and R. M. Pick, *J. Chem. Phys.* **77**, 2779 (1982).

85. E. O. Stejskal, D. E. Woessner, T. C. Farrar, and H. S. Gutowsky, *J. Chem. Phys.* **31**, 55 (1959).

86. See Ref. 12, Chapter 4.

87. T. Springer, *Quasielastic Neutron Scattering for the Investigation of Diffusive Motions in Solids and Liquids*, Springer, New York, 1972. See particularly Sections 2 and 7.

88. Chr. Steenbergen and L. A. de Graaf, *Physica* **96B**, 1 (1979).

89. C. Brot and B. Lassier-Govers, *Ber. Bunsenges. Phys. Chem.* **80**, 31 (1976).

90. W. G. Rothschild, *J. Chem. Phys.* **42**, 694 (1965); see Figs. 1–3. With great care, it is also possible to observe the $CH^{35}Cl_3$ and $CH^{35}Cl_2^{37}Cl$ peaks in the liquid phase of the infrared.

91. See Ref. 12, Chapter 13.

92. W. G. Rothschild, "Vibrational Relaxation and Orientational Disorder in Crystalline Neopentane," Gordon Research Conference on Orientational Disorder in Crystals, January 9–13, 1978, Santa Barbara, California. (Unpublished).

93. R. G. Gordon, *J. Chem. Phys.* **39**, 2788 (1963).

94. R. G. Gordon, *J. Chem. Phys.* **40**, 1973 (1964).

95. See Eq. 14 of Ref. 93.

96. I. Laulicht and S. Meirman, *J. Chem. Phys.* **59**, 2521 (1973).

97. E. Brindeau, S. Bratos, and J. C. Leicknam, *Phys. Rev.* **A6**, 2007 (1972).

98. H. F. Shurell, *J. Chem. Phys.* **58**, 5807 (1973).

99. W. G. Rothschild, *J. Chem. Phys.* **49**, 2250 (1968).

100. C. J. Montrose, T. G. Copeland, T. A. Litovitz, and R. A. Stuckart, *Mol. Phys.* **34**, 573 (1977).

101. H. D. Dardy, V. Volterra, and T. A. Litovitz, *J. Chem. Phys.* **59**, 4491 (1973).

102. See Fig. 7 of Ref. 10b. The theoretical presentations in its Section IV C are wrong (except Eqs. 7 and 8) and superseded by the present discussion.

103. C. Dreyfus and J. Vincent-Geisse, *Chem. Phys. Lett.* **21**, 170 (1973).

104. W. G. Rothschild, *J. Chem. Phys.* **57**, 991 (1972); see Section III A1.

105. W. G. Rothschild, J. Devaure, R. Cavagnat, and J. Lascombe, unpublished result on the pressure dependence of the Raman $I_{VV}$ and $I_{VH}$ contours of the C–D and C–I stretching fundamentals of $CDCl_3$ and $CH_3I$, respectively.

106. G. Lévi, J. P. Marsault, F. Marsault-Hérail, and R. E. D. McClung, *J. Chem. Phys.* **63**, 3543 (1975). The second term in Eq. 30 of this paper should be additive. Compare Eq. 30, for $x = 0$, with the relation on top of p. 248 of Ref. 64.

107. R. B. Wright, M. Schwartz, and C. H. Wang, *J. Chem. Phys.* **58**, 5125 (1973). Note that Eq. A.3 in this paper should be raised to the third power.

108. D. R. Jones, H. C. Andersen, and R. Pecora, *Chem. Phys.* **9**, 339 (1975), and references cited therein.

109. M. Constant, *Thèse Docteur ès Sciences Physiques*, Université de Lille, No. 410, 24 April 1978; M. Constant, R. Fauquembergue, and P. Descheerder, *J. Chem. Phys.* **64**, 667 (1976).

110. (a) W. G. Rothschild, *Chem. Phys. Lett.* **9**, 149 (1971); (b) W. G. Rothschild, *J. Chem. Phys.* **55**, 1402 (1971) and references cited therein.

111. (a) L. W. Reeves and W. G. Schneider, *Can. J. Chem.* **35**, 251 (1957); (b) H. Nomura, S. Koda, and Y. Miyahara, *J. Chem. Phys.* **65**, 4339 (1976).

112. R. Arndt and R. E. D. McClung, *J. Chem. Phys.* **69**, 4280 (1978); see Table IV.

113. See, for instance, (a) G. D. Patterson and J. E. Griffiths, *J. Chem. Phys.* **63**, 2406 (1975); (b) J. Soussen-Jacob, E. Dervil, and J. Vincent-Geisse, *Mol. Phys.* **28**, 935 (1974); see Table 7.

114. E. N. Ivanov, *Sov. Phys. JETP* **18**, 1041 (1964).

115. See References 37, Eq. 4.9 and Ref. 19, p. 357, Eq. 28.

116. B. Keller and F. Kneubühl, *Chem. Phys. Lett.* **9**, 178 (1971).

117. W. G. Rothschild, *J. Chem. Phys.* **53**, 3265 (1970); *ibid.*, **51**, 5187 (1969) and Erratum to this, **52**, 6453 (1970).

118. (a) J. H. Campbell, J. F. Fisher, and J. Jonas, *J. Chem. Phys.* **61**, 346 (1974); (b) M. Constant and R. Fauquembergue, *J. Chem. Phys.* **58**, 4030 (1973).

119. J. E. Griffiths, *Chem. Phys. Lett.* **21**, 354 (1973).

120. D. Richon, D. Patterson, and G. Turrell, *Chem. Phys.* **16**, 61 (1976).

121. D. Robert and L. Galatry, *J. Chem. Phys.* **55**, 2347 (1971); D. Robert, Thèse Docteur ès Sciences Physiques, Université de Besançon, October 17, 1967, C.N.R.S. N° A.O. 1554.

122. R. Kubo, *Lect. Theor. Phys.* **1**, 120 (1959).

123. L. Bonamy, D. Robert, and L. Galatry, *J. Mol. Struct.* **1**, 91, 139 (1967).

124. C. H. Townes and A. L. Schawlow, *Microwave Spectroscopy*, McGraw-Hill, New York 1955; see Eq. 13 in Section 10.

125. Measuring the temperature dependence of a correlation time and plotting its logarithm versus the reciprocal absolute temperature almost always gives a straight line whose slope yields the "constant" parameter in the exponential. That is all—and what to do with it in terms of molecular dynamics of molecules?

126. R. Kubo, in *Fluctuations, Relaxation, and Resonance in Magnetic Systems*, D. Ter Haar, Ed., Plenum Press, New York, 1962, p. 23.

127. (a) R. Kubo, *J. Phys. Soc. Jap.* **17**, 1100 (1962); (b) J. D. Gaskill, *Linear Systems, Fourier Transforms, and Optics*, Wiley, New York, 1978; see pp. 163–165.

128. A. M. Goulay-Bize and J. Vincent-Geisse, *J. Chem. Phys.* **73**, 4203 (1980).

129. S. Claesson and D. R. Jones, *Chem. Scripta* **9**, 103 (1976).

130. G. R. Alms, D. R. Bauer, J. I. Brauman, and R. Pecora, *J. Chem. Phys.* **58**, 5570 (1973).

131. K. Brenkert, Jr., *Elementary Theoretical Fluid Dynamics*, Wiley, New York, 1960; see p. 205.

132. (a) J. T. Hynes, R. Kapral, and M. Weinberg, *J. Chem. Phys.* **69**, 2725 (1979); (b) Chih-Ming Hu and R. Zwanzig, *J. Chem. Phys.* **60**, 4354 (1974).

133. (a) W. G. Rothschild, *Macromolecules* **1**, 43 (1968); (b) A. M. Goulay-Bize, E. Dervil, and J. Vincent-Geisse, *Chem. Phys. Lett.* **69**, 319 (1980).

134. See Ref. 39, p. 66, Eq. 4.8.2.

135. A. H. Narten, *J. Chem. Phys.* **48**, 1630 (1968).

136. N. Metropolis, A. W. Rosenbluth, M. N. Rosenbluth, and A. H. Teller, *J. Chem. Phys.* **21**, 1087 (1953).

137. O. Steinhauser and M. Neumann, *Mol. Phys.* **37**, 1921 (1979).

138. J. R. Sweet and W. A. Steele, *J. Chem. Phys.* **47**, (a) 3022; (b) 3029 (1967).

139. D. L. VanderHart, *J. Chem. Phys.* **60**, 1858 (1974).

140. H. Bertagnolli, D. O. Leicht, M. D. Zeidler, and P. Chieux, *Mol. Phys.* **36**, 1769 (1978).

141. H. Michel and E. Lippert, "Acetonitrile—The structure of the liquid and solid phases and the nature of the liquid-solid phase transitions," in A. D. Buckingham, E. Lippert, and S. Bratos, Eds., *Organic Liquids: Structure, Dynamics, and Chemical Properties*, p. 293, Wiley, New York, 1978.

142. (a) T. Keyes and D. Kivelson, *J. Chem. Phys.* **56**, 1057 (1972); (b) D. Kivelson and T. Keyes, *J. Chem. Phys.* **57**, 4599 (1972).

**404**     REFERENCES

143. See Reference 51; (a) Formula 4.1.8; (b) 5.2.3; (c) 4.1.8, 4.1.20, 4.5.1; (d) 5.2.4; (e) 5.2.1; (f) 4.5.3; (g) 4.2.1, 4.1.9; (h) 4.5.1; (i) 4.1.5, 4.1.20.

144. J. E. Griffiths, "Molecular Reorientation of Symmetric Top Molecules in the Liquid State," in J. R. Durig, Ed., *Vibrational Spectra and Structure*, Vol. 6, Elsevier, New York 1977.

145. (a) The result is known from stochastic modeling, D. Frenkel, G. H. Wegdam, and J. van der Elsken, *J. Chem. Phys.* **57**, 2691 (1972); see also Ref. 52. (b) R. D. Mountain, *J. Res. Nat. Bur. Stand.* **78A**, 413 (1974).

146. B. C. Sanctuary and D. Richon, *J. Chem. Phys.* **69**, 3782 (1978).

147. (a) G. Turrell, *Chem. Phys. Lett.* **49**, 289 (1977); (b) J. S. Anderson, W. E. Vaughan, L. C. Rosenthal, and H. L. Strauss, *J. Chem. Phys.* **65**, 2481 (1976).

148. W. A. Steele, *J. Chem. Phys.* **38**, 2404 (1963); *ibid.*, 2411; *Adv. Chem. Phys.* **34**, 1 (1976).

149. K. Mishima, *J. Phys. Soc. Jap.* **31**, 1796 (1971).

150. P. S. Hubbard, *Phys. Rev.* **A6**, 2421 (1972).

151. See the Erratum, Reference 117.

152. (a) R. Fauquembergue, P. Descheerder, and M. Constant, *J. Phys.* **38**, 707 (1977); (b) M. Constant and R. Fauquembergue, *J. Chem. Phys.* **72**, 2459 (1980).

153. B. J. Alder, H. L. Strauss, and J. J. Weis, *J. Chem. Phys.* **59**, 1002 (1973).

154. F. Vesely, *Computerexperimente an Flüssigkeitsmodellen*, Physik Verlag, Weinheim, 1978.

155. R. M. van Aalst, J. van der Elsken, D. Frenkel, and G. H. Wegdam, *Faraday Symp. Chem. Soc.*, No. 6, 1972, p. 94.

156. H. Langer and H. Versmold, *Ber. Bunsenges. Phys. Chem.* **83**, 510 (1979).

157. This is the simplest, nontrivial, angle-dependent process in first-order spectra. See H. A. Posch, *Mol. Phys.* **37**, 1059 (1959), for an appreciation of the complexity of collision-induced scattering.

158. See Ref. 12, Chapter 12, p. 447.

159. K. Lindenberg and R. I. Cukier, *J. Chem. Phys.* **62**, 3271 (1975).

160. G. J. Davies and M. Evans, *J. Chem. Soc. Faraday Trans. II*, **72**, 1194 (1976). Note that their $K_{n-1}$ corresponds to our $K_n$.

161. M. W. Evans, G. J. Evans, J. Yarwood, P. L. James, and R. Arndt, *Mol. Phys.* **38**, 699 (1979).

162. W. Feller, *An Introduction to Probability Theory and Its Applications*, Vol. I, Wiley, New York, 1968; see pp. 372 and 419.

163. D. R. Cox, *Renewal Theory*, Wiley, New York, 1962; see Ch. 1, 2, and 5.

164. D. Frenkel and G. H. Wegdam, *J. Chem. Phys.* **61**, 4671 (1974).

165. J. Lascombe and M. Perrot, *Faraday Disc. Chem. Soc.* **66**, 216 (1978); J. Lascombe, unpublished calculations.

166. H. G. Hertz and M. D. Zeidler, in P. Schuster, G. Zundel, and C. Sandorfy, Eds., *The Hydrogen Bond, Recent Developments in Theory and Experiments*, Vol. 3, North-Holland, New York, 1976, (a) p. 1053; (b) pp. 1052–1053, (c) p. 1050, (d) p. 1055.

167. C. H. Chatzidimitriou-Dreismann, *J. Mol. Struct.* **84**, 213 (1982).

168. J. Manz, *J. Am. Chem. Soc.* **102**, 1801 (1980).

169. (a) V. Mazzacurati, M. Nardone, and G. Signorelli, *J. Chem. Phys.* **66**, 5380 (1977); (b) Ref. 12, Section H, Chapter 12.

170. J. H. Hildebrand, *Faraday Disc. Chem. Soc.* **66**, 151, 180 (1978).

171. S. Bratoz, J. Rios, and Y. Guissani, *J. Chem. Phys.* **52**, 439 (1970). See also H. Morawitz and K. B. Eisenthal, *J. Chem. Phys.* **55**, 887 (1971).

172. R. Kopelman, in E. C. Lim, Ed., *Excited States*, Academic, New York, 1975, Vol. 2, p. 33.

173. S. Bratos, Y. Guissani, and J. C. Leicknam, in *Intermolecular Spectroscopy and Dynamical Properties of Dense Systems*, p. 375, LXXV Corso, Soc. Ital. di Fisica, Bologna, 1980; D. W. Oxtoby, *Adv. Chem. Phys.* **40**, 1 (1979).

174. J. Vincent-Geisse, *Spectrochim. Acta* **24A**, 1 (1968).

175. See Ref. 19, App. III.

176. P. W. Anderson, *J. Phys. Soc. Jap.* **9**, 316 (1954).

177. W. G. Rothschild, *J. Chem. Phys.* **65**, 455 (1976).

178. M. Constant, M. Delhaye, and R. Fauquembergue, *C.R. Acad. Sci.* **271B**, 1177 (1970).

179. (a) W. Kauzmann, *Quantum Chemistry*, Academic, New York 1957, pp. 505–507; (b) C. G. Gray, *Can. J. Phys.* **54**, 505 (1976); (c) K. A. Valiev, *Opt. Spectrosc.* **11**, 253 (1961); (d) P. C. M. van Woerkom, J. de Bleyser, M. de Zwart, and J. C. Leyte, *Chem. Phys.* **4**, 236 (1974); we follow their derivation here.

180. (a) See Ref. 6, pp. 300–302; (b) B. J. Berne, in *Physical Chemistry. An Advanced Treatise*, Vol. 8B, Academic, New York 1971; see pp. 578–579 (*dt* should be replaced by *d*$\tau$ in Eq. 2.112).

181. M. D. Zeidler, *Ber. Bunsenges. Phys. Chem.* **69**, 659 (1965); see Table 1.

182. K. A. Valiev, *Sov. Phys. JETP* **11**, 883 (1960); see figure on p. 886 and Eqs. 18, 22.

183. (a) F. Kohler, *The Liquid State*, Verlag Chemie, Weinheim, 1972; (b) F. Kohler and E. Wilhelm, *Adv. Mol. Relaxation Processes* **8**, 195 (1976); (c) D. Chandler, *J. Chem. Phys.* **60**, 3500 (1974).

184. (a) See, for instance, W. J. Moore, *Physical Chemistry*, Prentice-Hall, New Jersey, 1955, pp. 172, 178; (b) S. Chapman and T. G. Cowling, *The Mathematical Theory of Nonuniform Gases*, Cambridge University Press, Cambridge 1970; Ch. 16.

185. The conclusions reached about the nature of the vibrational relaxation process in Ref. 118*a* have been modified. At that time, it was not recognized that the $\nu_3$ mode relaxed also by vibrational dephasing.

186. K. Lindenberg and R. I. Cukier, *J. Chem. Phys.* **67**, 568 (1977).

187. R. F. Snider, *J. Chem. Phys.* **68**, 5118 (1978).

188. For the detailed derivation, see Ref. 2*c*, p. 194, Ref. 180*b*, p. 627, and R. Zwanzig, *J. Chem. Phys.* **40**, 2527 (1964).

189. J. W. Lyklema, *Physica* **96A**, 573 (1979).

190. G. Döge, R. Arndt, and A. Khuen, *Chem. Phys.* **21**, 53 (1977).

191. N. Trisdale and M. Schwartz, *Chem. Phys. Lett.* **68**, 461 (1979). The total vibrational relaxation mechanism of $\nu_1$ is, a priori, ascribed to dephasing (see Table 1).

192. S. F. Fischer and A. Laubereau, *Chem. Phys. Lett.* **35**, 6 (1975).

193. W. G. Rothschild and P. V. Huong, unpublished results, University of Bordeaux, Laboratoire de Spectroscopie Infrarouge, (1976).

194. J. E. Griffiths, *J. Chem. Phys.* **59**, 751 (1973).

195. S. Miller and J. H. R. Clarke, *Chem. Phys. Lett.* **56**, 235 (1978); R. M. Lynden-Bell, *Mol. Phys.* **36**, 1529 (1978).

196. (a) J. Schroeder, V. H. Schiemann, and J. Jonas, *Mol. Phys.* **34**, 1501 (1977); (b) M. Moradi-Araghi and M. Schwartz, *J. Chem. Phys.* **71**, 166 (1979); *ibid.*, **68**, 5548 (1978).

197. W. G. Rothschild, *J. Chem. Phys.* **65**, 2958 (1976).

198. W. R. L. Clements and B. P. Stoicheff, *Appl. Phys. Lett.* **12**, 246 (1968).

199. D. W. Oxtoby, D. Levesque, and J. J. Weis, *J. Chem. Phys.* **68**, 5528 (1978).

200. A. Laubereau, *Chem. Phys. Lett.* **27**, 600 (1974).

201. (a) P. S. Y. Cheung and J. G. Powles, *Mol. Phys.* **30**, 921 (1975); (b) J. Barojas, D. Levesque, and B. Quentrec, *Phys. Rev.* **A7**, 1092 (1973).

## 406    REFERENCES

202. S. S. Cohen and R. E. Wilde, *J. Chem. Phys.* **68**, 1138 (1978). Parameter $N_2$ is defined in accordance with Kubo's "coefficient of excess" (Ref. 126).

203. W. Schindler and H. A. Posch, *Chem. Phys.* **43**, 9 (1979); P. Fröhlich and H. A. Posch, *Mol. Phys.* **36**, 1421 (1978); see Table 2.

204. G. N. Robertson, and J. Yarwood, *Chem. Phys.* **32**, 267 (1978); J. Yarwood, R. Ackroyd, and G. N. Robertson, *ibid.*, **32**, 283 (1978).

205. See Ref. 12, Chapter 6.

206. S. Bratos, *J. Chem. Phys.* **63**, 3499 (1975).

207. P. Reich, A. Reklat, G. Seifert, and Th. Steiger, *Acta Phys. Pol.* **A58**, 665 (1980).

208. (a) H. C. Yao and W. G. Rothschild, *J. Chem. Phys.* **68**, 4774 (1978); (b) See Ref. 33, p. 146.

209. J. W. Gadzuk, *Phys. Rev.* **24B**, 1651 (1981).

210. A. Laubereau, G. Wochner, and W. Kaiser, *Phys. Rev.* **13A**, 2212, (1975); W. Zinth, H.-J. Polland, A. Laubereau, and W. Kaiser, *Appl. Phys.* **B26**, 77 (1981).

211. (a) W. G. Rothschild, J. Soussen-Jacob, J. Bessière, and J. Vincent-Geisse, unpublished data, University of Paris VI, May 1982. (b) H. Takeuchi, J.-L. Bribes, I. Harada, and T. Shimanouchi, *J. Raman Spectrosc.* **4**, 235 (1976).

212. G. Döge, *Z. Naturforsch.* **28a**, 919 (1973). This is the first quantitative account of applying the principles of the motional narrowing phenomenon from nuclear magnetic relaxation to a vibrational relaxation process in liquids (resonance coupling).

213. R. Wertheimer, *Chem. Phys. Lett.* **52**, 224 (1977); see Eq. 12.

214. T. Tokuhiro and W. G. Rothschild, *J. Chem. Phys.* **62**, 2150 (1975).

215. P. S. Hubbard, *Rev. Mod. Phys.* **33**, 249 (1961).

216. See Ref. 4, Chapter 12.

217. See Ref. 37, (a) Eq. 4.61; (b) Eq. 4.34.

218. J. P. Hawranek and R. N. Jones, *Spectrochim. Acta* **32A**, 111 (1976).

219. W. G. Rothschild and T. Tokuhiro, unpublished results, 1975.

220. R. LeSar and R. Kopelman, *J. Chem. Phys.* **66**, 5035 (1977).

221. D. N. Terpilovskii, *Opt. Spectrosc.* **28**, 380 (1970).

222. (a) J. P. Perchard, W. F. Murphy, and H. J. Bernstein, *Mol. Phys.* **23**, 519 (1972); (b) Y. Guissani and J. C. Leicknam, *Can. J. Phys.* **51**, 938 (1973).

223. W. G. Rothschild and H. C. Yao, *J. Chem. Phys.* **74**, 4186 (1981). The infrared absorption coefficient in Eq. 13 should read $\ln[T_0(\omega)/T(\omega)]$, where $T$, $T_0$ are the transmittance of the catalyst with and without adsorbed CO, respectively.

224. G. Döge, R. Arndt, H. Buhl, and G. Bettermann, *Z. Naturforsch.* **35a**, 468 (1980).

225. (a) A. F. Bondarev and A. I. Mardaeva, *Opt. Spectrosc.* **35**, 167 (1973); (b) T. Fujiyama, M. Kakimoto, and T. Suzuki, *Bull. Chem. Soc. Jap.* **49**, 606 (1976); (c) L. D. Landau and E. M. Lifschitz, *Statistische Physik*, Akademie-Verlag, Berlin, 1971, Chapter 12 (see Eq. 117.2).

226. R. R. Alfano and S. L. Shapiro, *Phys. Rev. Lett.* **29**, 1655 (1972).

227. K. A. Valiev, *Sov. Phys. JETP* **13**, 1287 (1961). This is one of the first publications on the effects of vibrational relaxation on infrared and Raman bandwidths of liquids.

228. See Ref. 19, p. 72.

229. The published values of $\Delta \nu = 25$ cm$^{-1}$ (Ref. 227) and 0.15 cm$^{-1}$ (Ref. 214) are numerically in error.

230. J. A. Pople, W. G. Schneider, and H. J. Bernstein, *High-Resolution Nuclear Magnetic Resonance*, McGraw-Hill, New York, 1959, p. 205.

231. P. R. Monson, S. Patumtevapibal, K. J. Kaufmann, and G. W. Robinson, *Chem. Phys. Lett.* **28**, 312 (1974).

232. See Ref. 12, Ch. 8.

233. K. Spanner, A. Laubereau, and W. Kaiser, *Chem. Phys. Lett.* **44**, 88 (1976).

234. H. R. Telle and A. Laubereau, *Chem. Phys. Lett.* **94**, 467 (1983).

235. (a) H. Paerschke, K.-E. Süsse, and D.-G. Welsch, *Exp. Tech. Phys.* **25**, 185 (1977); (b) *ibid.*, **25**, 279 (1977); (c) *ibid.*, **25**, 481 (1977); (d) *ibid.*, **25**, 489 (1977); (e) H. Paerschke, K.-E. Süsse, and D.-G. Welsch, *Ann. Phys.* **34**, 405 (1977); (f) H. Paerschke, K.-E. Süsse, and D.-G. Welsch, *Phys. Stat. Sol.* (B) **98**, 253 (1980); (g) The right-hand side of the formula should be as follows: Change subscript of the first (second) coefficient to $n - l + 1$, $k(n - l + 2, k)$. (Private communication, Dr. D.-G. Welsch, August 14, 1981.); (h) K.-E. Süsse, W. Vogel, and D.-G. Welsch, *Phys. Stat. Sol.* (B) **99**, 91 (1980); (i) Eq. 3.164 supersedes Eq. 10 of Ref. 235*b*. (Private communication, Drs. K.-E. Süsse and D.-G. Welsch, December 23, 1982.)

236. I. Oppenheim, K. E. Shuler, and G. H. Weiss, *Adv. Mol. Relaxation Processes* **1**, 13 (1967–1968).

237. S. H. Lin, *J. Chem. Phys.* **61**, 3810 (1974); *ibid.*, **65**, 1053 (1976).

238. R. W. Zwanzig, "Statistical Mechanics of Irreversibility," in *Lectures of the Summer Institute of Theoretical Physics*, University of Colorado, 1960, p. 106. See also D. J. Diestler and R. S. Wilson, *J. Chem. Phys.* **62**, 1572 (1975).

239. E. W. Montroll and K. E. Shuler, *J. Chem. Phys.* **26**, 454 (1957).

240. G. R. Fleming, O. L. J. Gijzeman, and S. H. Lin, *J. Chem. Soc. Faraday Trans. II*, **70**, 37 (1974).

241. See also Ref. 158, Eq. 12.59.

242. A. Laubereau, G. Kehl, and W. Kaiser, *Opt. Comm.* **11**, 74 (1974).

243. (a) W. F. Calaway and G. E. Ewing, *Chem. Phys. Lett.* **30**, 485 (1975); (b) *J. Chem. Phys.* **63**, 2842 (1975).

244. (a) D. W. Chandler and G. E. Ewing, *J. Chem. Phys.* **73**, 4904 (1980); (b) S. R. J. Brueck and R. M. Osgood, *J. Chem Phys.* **68**, 4941 (1978).

245. K. F. Herzfeld and T. A. Litovitz, *Absorption and Dispersion of Ultrasonic Waves*, Academic, New York, 1959.

246. K. Takagi, P.-K. Choi, and K. Negishi, *J. Chem. Phys.* **74**, 1424 (1981).

247. A. Laubereau, S. F. Fischer, K. Spanner, and W. Kaiser, *Chem. Phys.* **31**, 335 (1978).

248. See Ref. 34, Fig. 4. The infrared correlation time of the $\nu_4$ mode is about $0.8 \times 10^{-12}$ sec.

249. W. M. Gelbart, S. A. Rice, and K. F. Freed, *J. Chem. Phys.* **57**, 4699 (1971); K. G. Kay, *ibid.*, **61**, 5205 (1974).

250. A. Nitzan and J. Jortner, *Mol. Phys.* **25**, 713 (1973). (See, in particular, Eqs. 2.6.)

251. K. E. Shuler and G. H. Weiss, *J. Chem. Phys.* **45**, 1105 (1966); see Eq. 10.

252. I. S. Gradshteyn and I. M. Ryzhik, *Table of Integrals, Series, and Products*, Academic Press, New York, 1965; (a) Integral 3.461.1; (b) 8.250.1; (c) 6.633.2; (d) 8.464.3; (e) 8.467, 8.468.

253. D. J. Diestler, *Chem. Phys. Lett.* **39**, 39 (1976). In this paper the two-level approximation for uncorrelated simultaneous vibrational dephasing and energy dissipation to the lattice is developed, using the Langevin equation–projection operator method.

254. B. P. Hills, *Mol. Phys.* **35**, 793 (1978); S. I. Temkin and A. I. Burshtein, *Chem. Phys. Lett.* **66**, 52 (1979); *ibid.*, **66**, 57 (1979); S. Mukamel, *Chem. Phys.* **37**, 33 (1979); R. K. Wertheimer, *Mol. Phys.* **38**, 797 (1979); *Chem. Phys.* **45**, 415 (1980); R. M. Jul'met'ev, "Application of N. N. Bogoliubov's Idea of the Relaxation Time Hierarchy to Vibrational Relaxation in Liquids," *Ukr. J. Phys.* **21**, 1761 (1976); C. H. Wang, *Mol. Phys.* **33**, 207 (1977); P. A. Madden and R. M. Lynden-Bell, *Chem. Phys. Lett.* **38**, 163 (1976). These references are by no means exhaustive but give a cross section of various theoretical approaches.

## 408    REFERENCES

255. V. Mazzacurati, M. Nardone, and G. Signorelli, *J. Chem. Phys.* **66**, 5380 (1977); F. L. Galeener and P. N. Sen, *Phys. Rev.* **B17**, 1928 (1978).

256. E. Cohen de Lara and Y. Delaval, *J. Chem. Soc. Faraday Trans. II*, **74**, 790 (1978); Y. Delaval and E. Cohen de Lara, *J. Chem. Soc. Faraday Trans. I*, **77**, 869, 879 (1981).

257. J. W. Gadzuk, *Chem. Phys. Lett.* **80**, 5 (1981); B. N. J. Persson and M. Persson, *Solid State Comm.* **36**, 175 (1980). These papers describe coupling of the vibrational coordinate of the adsorbed molecule to substrate electron-hole pairs. H. Metiu and W. E. Palke, *J. Chem. Phys.* **69**, 2574 (1978), discuss coupling to the electronic degrees of freedom.

258. W. G. Rothschild, J. Soussen-Jacob, J. Bessière, and J. Vincent-Geisse, *J. Chem. Phys.* **79**, 3002 (1983); see also *Chem. Phys. Lett.* **96**, 43 (1983).

# INDEX

RETURN CHEMISTRY 42 171
Tc... Library 542-3753